Databook on Mechanical and Thermophysical Properties of Particulate-Reinforced Ceramic Matrix Composites

Databook
on
Mechanical and Thermophysical Properties of Particulate-Reinforced Ceramic Matrix Composites

Christian X. Campbell
and
Said K. El-Rahaiby

Ceramics Information Analysis Center
Center for Information and Numerical Data Analysis and Synthesis
Purdue University
2595 Yeager Road
West Lafayette, IN 47906-1398

and

The American Ceramic Society
735 Ceramic Place
Westerville, OH 43081-8720

Campbell, Christian X., 1970-
Databook on mechanical and thermophysical properties of particulate-reinforced ceramic matrix composites / Christian X. Campbell and Said K. El-Rahaiby.
p. cm.
Includes bibliogrphical references and index.
ISBN 0-931682-57-6 (hard bound)
1. Ceramic-matrix composites--Mechanical properties. 2. Ceramic-matrix composites--Thermal properties. I. El-Rahaiby, Said K., 1950- II. Title.
TA418.9.C6C26 1995
620.1'4--dc20 95-48874
CIP

Contents

		Page
1.	Introduction	1
2.	Data Analysis	1
3.	Materials Overview	3
4.	Mechanical Property Measurement Methods	7
5.	Thermophysical Property Measurement Methods	15
6.	Unit Conversion Tables	20
7.	Material and Property Index	25
8.	List of Figures	29
9.	Curve Fitting Equation Index	34
10.	Flexural Strength (Section I)	I-1
11.	Fracture Toughness (Section II)	II-1
12.	Elastic Properties (Section III)	III-1
13.	Hardness (Section IV)	IV-1
14.	Thermal Properties (Section V)	V-1
15.	References (Section VI)	VI-1

Foreword

The preparation and publication of this databook is sponsored by the U.S. Department of Defense (DoD) as part of the Ceramics Information Analysis Center (CIAC) overall operations.

CIAC, a DoD Information Analysis Center, is sponsored, administratively managed, and funded by the Defense Technical Information Center (DTIC), ATTN: DTIC-AI, 8725 John J. Kingman Road, Suite 0944, Fort Belvoir, VA 22060-6218, and is under the IAC program management of Mr. Ronald E. Hale. It is operated by the Center for Information and Numerical Data Analysis and Synthesis (CINDAS), Purdue University, West Lafayette, IN 47906 under Contract DLA900-90-D-0304. The contract was awarded to Purdue Research Foundation by the Defense Electronics Supply Center (DESC), ATTN: DESC-EACC, Dayton, OH 45444-5181 with Ms. Cheryl A. Montoney as the contracting officer.

CIAC is under the technical direction of Mr. Jerome Persh, Senior Specialist for Materials and Structures, Office of the Director of Defense Research and Engineering (Advanced Technology), ATTN: ODDR&E (AT), The Pentagon, Room 3D1089, Washington, DC 20301-3080.

CIAC is chartered by the DoD as a Full-Service Information Analysis Center and serves as the DoD's central source of engineering and technical data and research and development on all classes of advanced ceramic materials utilized in Defense systems and hardware. It supports the DoD in general and supports in particular the DTIC, the Office of the Secretary of Defense (OSD) program management, and the Joint Logistics Commanders/Joint Directors of Laboratories Technology Panel for Advanced Materials, and provides assistance to or receives guidance from other Defense Programs and Groups as designated by the Technical Monitor.

The authors express their sincere thanks to the many individuals who have assisted in the preparation of this volume. In particular, the authors wish to thank Christopher D. Carpenter and Carolyn S. Dennis of the Ceramics Information Analysis Center for their many hours of hard work and considerable patience. The authors would also like to thank the many Purdue University students who participated in the collection and extraction of the data contained in this volume.

Preface

This databook presents much of the available materials characterization data and information on particulate-reinfõrced ceramic matrix composites collected from the open literauare, company literature, and other sources in a well organized, evaluated, and analyzed manner readily available for engineering calculations, material evaluation and characterization and, to some extent, for design purposes. An exhaustive search was first performed to locate documents containing mechanical, thermal, physical, and other properties required to characterize paraticulate-reinforced ceramic composites. The data were extracted from documents using established techniques, which include transcribing and paraphrasing all the information to fully characterize the material. Data extracted from the literature are stored in "data-sets." Each set of data in the database consists of numerical data points (as a function of temperature and/or other independent variables) and pertinent information on the specification and characterization of the material tested and on the method and conditions of the property measurement. Each "data-set" must include information such as processing techniques, particulate reinforcement content, reinforcement and matrix raw materials, the test method and necessary test environment. Other information was also included if available.

Due to the unique nature of the analysis performed in this databook, the comparison of seemingly unrelated materials may lead to further reinforcement of current understanding of particulate-reinforced ceramic matrix composites and a unique perspective on which to base material selection and application. The authors have done their best to bring insight to the presentation of data within this book. However, it is also the intent that the user community will finally have a single source of data useful for bringing their personal insight to the field of ceramic composites. The data presented in this databook are intended to be a summary of available data on particulate-reinforced ceramic composites for use in materials characterization and materials selection. The data presented in this book are not intended to be design quality data, although to some extent, some of the data may possess the statistical criteria to qualify for design purposes.

The authors express their sincere hope that this volume will advance the awareness of ceramic composites within the materials community and will serve as a valuable source reference for designers, engineers, and scientists in the fields of materials development and in the industrial, military, and aerospace sectors.

1. Introduction

This databook presents much of the available materials characterization data and information extracted from the open literature, company literature, and other sources in a well organized, evaluated, and analyzed manner readily available for engineering calculations, material evaluation and characterization and, to some extent, for design purposes. An exhaustive search was first performed to locate documents containing mechanical, thermal, physical, and other properties required to characterize structural particulate-reinforced ceramic composites. The data were extracted from documents using established techniques, which include transcribing and paraphrasing all the information to fully characterize the material and test method. Data extracted from the literature are stored in "data-sets." Each set of data in the database consists of numerical data points (as a function of temperature and/or other independent variables) and pertinent information on the specification and characterization of the test material and on the method and conditions of the property measurement. Each "data-set" must include the following information to be included in the databook: processing techniques, particulate reinforcement content, reinforcement and matrix raw materials, the test method, and test environment. All tests were assumed to be performed in an air environment unless otherwise specified. Other information was also included if available. The data were extracted from the documents by digitizing the curves and transcribing tables. Depending on the quality or type of figure, or the quality of the reproduction, the accuracy of the digitizer should be within ±5% or better. This is well below the experimental error for flexural strength or fracture toughness test results on particulate-reinforced ceramic composites, which, due to their brittle nature, are statistically controlled by the largest flaw size.

Each data-set includes a dependent variable, which is the mechanical, thermal, or other material property and an independent variable, which is most commonly reinforcement content or test temperature. The independent variable was chosen such that the greatest number of data-sets can be compared directly. After extraction, the data were used to create an exhaustive Paradox™ materials property database. Initial analysis was performed using advanced query functions to compile similar data-sets into "compiled data-sets." The compiled data-sets were then extensively edited, graphed using SigmaPlot™, curved fitted using SigmaPlot™ and TableCurve™, and finally a technical analysis was performed on each compiled data-set.

2. Data Analysis

The data evaluation and technical analysis involve a multi-step process in which all sources of the data for a single compiled data-set are examined. First, the data undergo preliminary evaluation to establish its quality or validity. For a given property of a given material, each data-set from a particular source was compared with all the data on that material and property. Some data-sets were rejected on the basis of inconsistency with the overall values and on evidence concerning measurement methods or materials characterization. Then analysis of the materials behavior is performed. This, at its most rudimentary level, involves trend analysis, and at the most advanced level, involves interpretation of the research article, bringing expert insight to the behavior of the composite, and, if necessary, explanation of important concepts involving the test methods, processing, chemistry, fracture behavior, effect of test environment, and innumerable other concepts that are deemed necessary for comprehension of the data.

In the final production of the databook, the compiled data-sets were combined to create the databook. The book is organized in sections divided by property. This shows a different aspect of the data compared to most research articles which fully test and characterize a single material or a few materials. Since subtle changes in composition and processing makes no two ceramic composites identical, the databook takes the approach to compare a single property of similar composites. Similar composites are those with the same matrix, i.e., silicon nitride, alumina, etc., or similar matrices, i.e., nitrides, carbides, oxides, borides, or glass-ceramics. Similar composites are also considered to be composites that show similar behavior, i.e., composites that demonstrate transformation toughening, or composites that are electrically conductive, etc. Great care is taken to keep each compiled data-set to an appropriate size, large enough to allow for meaningful analysis and not too large to prevent comprehension of the data.

Due to the unique nature of the analysis performed in this databook, the comparison of seemingly unrelated materials may lead to further reinforcement of current understanding of on particulate-reinforced ceramic matrix composites and a unique perspective on which to base new knowledge. For example, the compilation of strength and toughness data for zirconia-alumina composites appears to show general agreement concerning the affect of tetragonal phase content on strength and toughness. The authors have done their best to bring insight to the presentation of data within this book, however, it is also the intent that the user community will finally have a single source of data useful for bringing their personal insight to the field of ceramic composites. The data presented in this databook are intended to be a summary of available data on particulate-reinforced ceramic composites for use in materials characterization and materials selection, and to some extent, for design purposes.

2.1 How to Use This Databook

A schematic of a typical "data-set" used for the presentation of data in this databook is given below. The information needed to describe the composite and test method is presented in a three column format. The first provides information concerning type of reinforcement and range of compositions, processing details, and sintering additives. The second column provides data on the raw materials used to make the ceramic composite. The data include supplier and designation, dimensions, and other properties such as density. Also provided in the second column are data on the reinforcement and matrix properties after composite processing, such as reinforcement grain

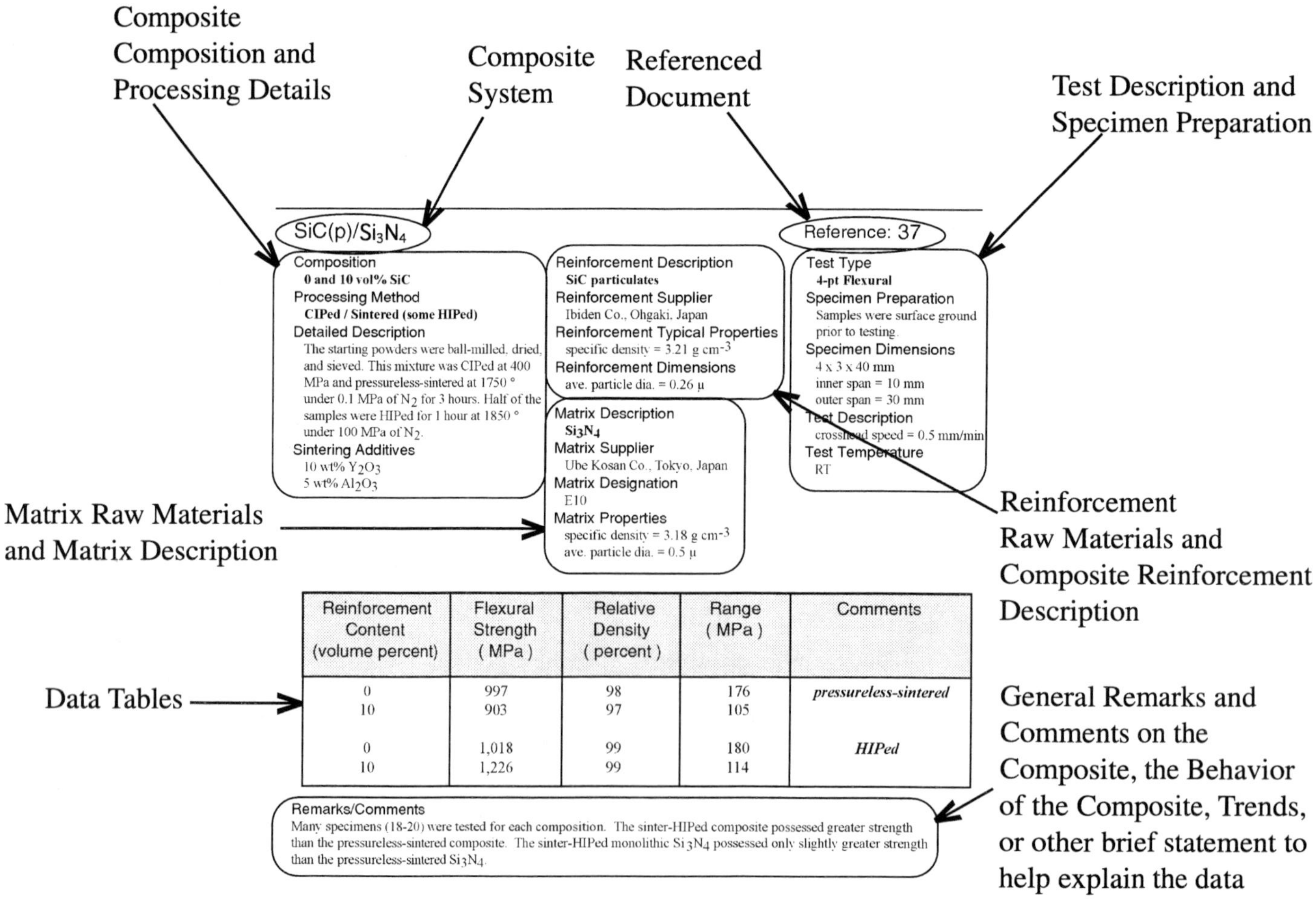

size and matrix grain size. The third column describes the test and specimen preparation procedures used.

The data tables provide the thermal or mechanical data measured along with any other auxiliary variables that would help describe the composite behavior. The first column of the data table usually provides the independent variable, i.e., the property which the thermal or mechanical property is plotted against. The second column provides the dependent variable; it is noteworthy that occasionally there is more than one column providing physical characteristics of the composite such as density, or processing data, processing time, temperature, or pressure. These auxiliary variables may help explain the behavior of the composite. The range of the values for the dependent data is given in another column. Usually, the source does not specify what is meant by the data range. In these instances, it is assumed that the range refers to the maximum and minimum values obtained during the testing of several samples. The range of values is specified as a standard deviation only when it is specified in the reference that the range of values represents one standard deviation. The last column sometimes provides specific comments to describe a grouping of data within the data-set. In this typical example, the comments separate the pressureless sintered materials from the HIPed materials. In other cases, different raw materials may have been used, different test methods may have been used, theoretical values may have been presented as a reference, and so on. In each instance, a blank row separates the data groupings. At the end of each data-set, there is a remarks/comments field that provides additional information given in the reference that does not fit elsewhere in the data-set. Also, some simple analysis is sometimes provided in the remarks/comments field.

3. Materials Overview

The field of structural ceramics is rapidly developing with extensive R&D efforts already or currently being performed at national laboratories, universities and in industry. Many of the results of this research have already found applications as components in military equipment and are slowly penetrating commercial markets and large scale industrial applications. Ceramics have many excellent properties that make their use as structural materials very attractive; however, ceramics are also known as brittle and unreliable materials that will not yield before fracture. In other words, ceramics fail in a "catastrophic" manner when overloaded. This is the characteristic of ceramics that has prevented their use in structural applications in the past, and is the area addressed by much of the research presented in this databook.

The properties of ceramics are the result of the strong covalent and ionic bonding that is characteristic of this class of material. No, or very little, yielding can occur as a result of the strong bonding. This allows large stress concentrations to develop at a crack tip and cause the crack to propagate with little expended energy. The result is a material with low fracture toughness. In contrast, metals have much greater toughness because plastic deformation occurs at the crack tip, which effectively blunts the crack and prevents large stress concentrations.

Monolithic ceramic oxides, carbides, nitrides, and borides have many desirable characteristics that result from the nature of their bonding. Despite low toughness, monolithic ceramics can have very high strengths, although the strength is very dependent on the maximum flaw size within the ceramic. When structural ceramics were first being developed, much work was done to maximize strength by reducing the size of flaws present in a monolithic ceramic. Since flaws can develop at any stage during the lifetime of a ceramic component, i.e., during processing, machining, or service, production of flaw-free ceramics was seen as an impractical route for producing high strength ceramics. Also, eliminating flaws did little to enhance the reliability or toughness of the ceramic throughout the duration of its operational lifetime.

The key to the development of reliable structural ceramics has been to improve the reliability and toughness while retaining other desirable properties. This has been done by creating flaw-tolerant microstructures through techniques such as microstructural control, transformation toughening (e.g., stabilized zirconia), and most importantly, by the development of ceramic matrix composites.

The idea behind particulate-reinforced ceramic matrix composites is a rather simple one in the realm of ceramic matrix composites. In almost all cases, particulate-reinforced CMCs are fabricated by mixing two ceramic powders and processing them in identical manners to those used for monolithic ceramics. The powder mixture may be formed into a complex shape by a green forming process and then pressureless sintered, or poured into a hot-press die and directly hot-pressed, or densified, in a similar manner. Unlike fiber- or whisker-reinforced CMCs, it is not always possible to clearly define the reinforcement and the matrix. The general rule of thumb used in this book is that the minor phase is defined as the reinforcement and the major phase, or the continuous phase, is defined as the matrix. In some cases, when tests were performed on composites with compositions that varied continuously from 100% of one phase to 100% of another phase, the matrix was arbitrarily defined for the sake of consistency.

In comparison with fiber- and whisker-reinforced CMCs, there are several differences that distinguish particulate-reinforced CMCs. First of all, the properties of particulate-reinforced CMCs are mostly isotropic, whereas the properties of fiber- or whisker-reinforced CMCs are quasi-isotropic to anisotropic, depending on reinforcement orientation. The orientation effects of particulate reinforcement are minimal; but some texturing or orientation effects can be caused by hot-pressing. Fiber- and whisker-reinforced CMCs frequently rely on toughening mechanisms, such as fiber or whisker pullout and crack bridging, that require high aspect ratio inclusions in the matrix. Due to the absence of high aspect ratio inclusions, particulate-reinforced CMCs rely on more subtle methods of toughening, and overall toughness of these composites tend to be lower than fiber or whisker reinforcement. Particulate-reinforced CMCs can be thought of as "like monolithic ceramics but better" due to the improved properties and tailorability of these properties.

When it is considered that almost any ceramic could be combined with almost any other ceramic or multiple ceramics to create a particulate CMC, one realizes that there is an almost unlimited number of potential composite systems available, and would take a monumental effort to characterize all potential composite types and present the data in this book. Fortunately, research has focused on a relatively few number of composite systems which, for one reason or another, show distinct advantages over monolithic ceramics. Examples include structural CMCs which demonstrate one or more toughening mechanisms to produce strong and tough ceramic composites, such as zirconia-alumina or zirconia-mullite composites, SiC or TiC reinforced alumina, TiC(p)/SiC, or various particulate-reinforced silicon nitrides and SiAlONs. Particulate reinforcements are also frequently added to a ceramic matrix to optimize thermal or electrical properties. SiC-AlN and SiC(p)/BeO ceramic composites have been designed to optimize thermal conductivity while retaining adequate strength for heat sink applications. One unique material, AlN(p)/BN composite, has been found to have highly anisotropic thermal conductivity for 'heat pipe' applications. Other particulate-reinforced CMCs have been developed to maximize electrical conductivity, particularly for EDM machinable ceramic composites.

Strengthening and Toughening

Like fiber- and whisker-reinforced CMCs, one of the primary objectives for the addition of particulate reinforcement is to control and optimize the microstructure to improve mechanical properties. Tailoring of the microstructure can create toughening mechanisms such as crack deflection, crack branching, microcrack toughening and transformation toughening. All of these mechanisms redistribute stresses at the crack tip and increase the energy needed to propagate a crack through the material, resulting in a more tortuous crack path and rising R-curve behavior. According to the Griffith strength-toughness relationship for brittle materials, increasing toughness should result in a corresponding increase in strength. This relationship is frequently demonstrated in the literature data, however, the literature also shows many examples of the opposite case in which the strength decreased with increasing reinforcement content and toughness. In these cases, the addition of particulate reinforcement may have introduced larger critical flaws in the composite, thereby decreasing its strength. Assuming the strength of the two phases are approximately equal, an increase in strength is generally realized by an increase in toughness or a decrease in critical flaw size. Of course the rule of mixtures plays a significant role in determining the strength of the composite as well.

The following is a summary of the toughening mechanisms present in particulate-reinforced CMCs.

(1) Crack Deflection: The primary method for toughening in most particulate CMCs is crack deflection. A very strong and high modulus particulate can cause a propagating crack to go around the particle instead of propagating through it, increasing the crack length and tortuosity. Factors that affect crack deflection are the particle size and aspect ratio, the chemical compatibility and particulate/matrix bond strength, and CTE mismatch. Many systems are designed to take advantage of thermal expansion mismatch stresses between the reinforcement and matrix. The TiC(p)/SiC system is used as an example [11]. Both TiC and SiC are strong, hard and have high modulus, therefore it may be expected that crack deflection may not play a significant role in this composite. However, the CTE of TiC is nearly double that of SiC, which creates a residual stress field upon cooling to room temperature because the particulate wants to shrink away from the matrix. The stress field deflects a propagating crack as shown in Figure 1. If the matrix has a higher CTE than the particulate inclusion, crack deflection is accomplished by attracting propagating cracks to the particle instead of away from the particle. Thermal mismatch toughening mechanisms, due to their nature, are very sensitive to temperature and decrease with increasing temperature until CTE mismatch stresses equal zero. There is also evidence that pores can alter the crack path and thus increase fracture energy and toughness, however, this effect is limited and usually an increase in porosity results in an decrease in toughness [1].

The size of the particulate inclusions is critical to the effectiveness of crack deflection mechanisms. The magnitude of CTE mismatch stresses is directly dependent on the size of the particle. There is generally an optimum size for toughening: if the particle size is too small, the CTE mismatch stress field is negligible and does not affect the crack path; if the particle size is too large, then CTE mismatch stresses can damage the matrix by extensive microcracking. Liu and Ownby [75] found that the optimum TiB_2 particulate size in an alumina matrix was 3-6 μm, whereas Niihara and Nakahira [77] have achieved excellent results with 'nanocomposites', which are CMCs with a particulate size smaller than 1 μm.

(2) Microcracking: Particulate inclusions, either due to CTE mismatch, a phase transformation, or to its elastic response to the stress field of an approaching crack, can cause microcracking within and around the particle. The proper amount of microcracking can cause a crack to deflect, branch,

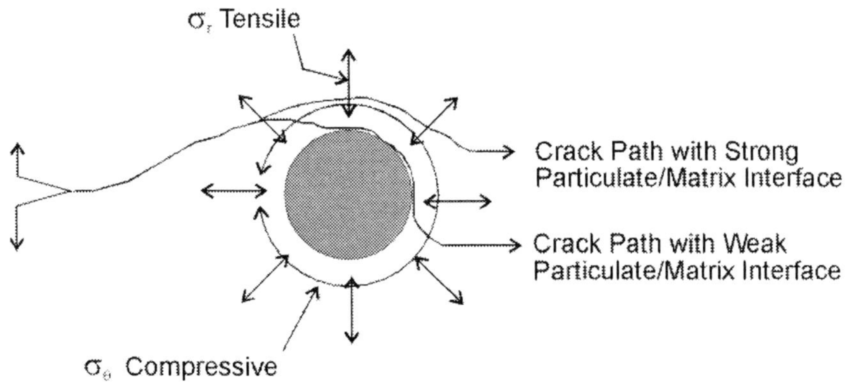

Figure 1. Schematic of a propagating crack being deflected by the stress field around a TiC particle (higher CTE) in a SiC matrix (lower CTE) [11].

and take a tortuous path through the composite, thereby increasing the fracture toughness. It should be warned that microcracking can be detrimental to composite properties as well as beneficial. As microcrack density increases, the microcracks can link up and cause unexpected catastrophic failure of the composite. Proper design of a composite would limit microcrack formation to the "process" zone around the crack tip, whereby the strength limitation imposed by the presence of microcracks can be kept at a minimum. This would allow the composite to retain good strength as well as toughness [2].

(3) Transformation Toughening: Another technique available to toughen a ceramic matrix is the addition of particulate reinforcement which undergoes a phase transformation in the crack tip process zone. A volume expansion is associated with this phase transformation, which imparts a compressive stress on the crack tip and essentially clamps the crack tip and inhibits further crack propagation.

Work in this area has focused almost exclusively on the addition of tetragonal zirconia to a ceramic matrix, where the tetragonal ZrO_2 transforms to the monoclinic phase during fracture. The zirconia is usually stabilized in the zirconia phase by a stabilizing agent, usually yttria, but other stabilizing agents such as magnesia and ceria are also available. Zirconia has a tendency to transform to the monoclinic phase during processing, but some tetragonal ZrO_2 must be preserved in order to retain the transformation toughening effect. Research has demonstrated that the t-ZrO_2 grain size must be larger than a critical size to remain in the tetragonal phase, and this critical size is directly related to the modulus of the matrix. The matrix constrains the ZrO_2 reinforcement to prevent volume expansion, and a higher modulus matrix is better able to constrain the ZrO_2 in tetragonal phase than a lower modulus matrix [7,57].

PSZ and TZP ceramics can be thought of as a simple composite of t-ZrO_2 reinforcing a cubic or monoclinic zirconia matrix. Since the zirconia matrix has a relatively low modulus, the particle size must be < 0.2 μm to prevent spontaneous t-m transformation during processing. Zirconia toughened alumina, or zirconia-alumina composites, are the most extensively studied transformation toughened ceramic matrix composite. The alumina matrix has about twice the stiffness of the zirconia matrix, thus the t-ZrO_2 reinforcement can have a grain size as large as 0.5 μm and still remain in the tetragonal phase [7]. It should be noted that if the matrix modulus is high enough, the zirconia need not be stabilized for transformation toughening. The constraint of the matrix alone can be enough to keep the zirconia reinforcement in the tetragonal phase. Alumina has another great advantage for use as the matrix in a zirconia toughened ceramic: ZrO_2 and Al_2O_3 are chemically compatible and almost completely insoluble in each other. Zirconia-mullite also shows promise as a transformation toughened ceramic composite. Other systems, such as zirconia-cordierite or other zirconia-silicates, have also been studied; however, zirconia can react with the cordierite or silicate to form a detrimental zircon phase ($ZrSiO_4$).

Transformation toughening is not the only toughening mechanism available in zirconia toughened ceramics. The t-m transformation, either during processing or during fracture, can cause matrix microcracking which may be beneficial to composite toughness. Also, zirconia particulates act as inclusions which can deflect cracks. Many researchers have measured the t-ZrO_2 content in the composite before fracture and then measured t-ZrO_2 content at the fracture surface to determine which toughening mechanisms were in effect for that particular composite.

(4) Crack Branching: Crack branching can substantially add to the fracture toughness of composites. The fracture energy and all toughening mechanisms are at least doubled while there are two cracks propagating concurrently. Crack branching may actually be common in particulate-reinforced CMCs, however, it probably never occurs spontaneously by itself. Crack branching is usually the result of the combination of microcracking and crack deflection, see Figure 2 [1].

(5) Crack Bridging: Crack bridging requires high aspect ratio reinforcements and is among the primary toughening mechanisms used in whisker- and fiber-reinforced CMCs. This book deals with

low aspect ratio reinforcements where crack bridging is not a primary toughening mechanism.

Electrical and Thermal Properties

Particulate reinforcement is often added to a ceramic matrix to control the electrical and thermal properties of the ceramic matrix. For example, electrically conductive ceramics, such as TiC, TiB_2, or TiN, are often added to an alumina or silicon nitride matrix to raise the electrical conductivity of the composite. An electrically conductive composite is desirable because it can be machined by EDM. Thermal properties can also be precisely controlled by the type and amount of reinforcement added to a matrix. There have been experiments to add diamond reinforcement to an oxide matrix to increase thermal conductivity. Other composites are designed to be thermal insulators or have directional heat flow.

4. Mechanical Property Measurement Methods

4.1 Flexure Test

The flexure test of monolithic ceramics and ceramic composites is the most popular and perhaps the simplest of all mechanical testing for these materials. A comprehensive review of flexure test methods applied to monolithic ceramics and a discussion of the quality and usefulness of flexure data for design purposes are given elsewhere [3].

ASTM Standardization Efforts

ASTM Committee C-28, Advanced Ceramics, was organized in 1986 for the promotion of knowledge, stimulation of research, and development of standards (classification, specifications, terminology, test methods, guides, and practices) relating to processing, properties, characterization, and performance of advanced ceramic materials. Two ASTM subcommittees are committed to the development of mechanical test standardization of advanced ceramic materials. They are:

- C28.01: Mechanical Testing of Monolithic Ceramics
- C28.07: Mechanical Testing of Ceramic Composites

Subcommittee C28.01 was the one of the first to develop standards and has been the most active of the C28 subcommittees. Two examples of standards that have been developed by ASTM subcommittee C28.01 are:

- C-1161 - Standard test method for flexural strength of advanced ceramics at ambient temperature.
- C-1211 - Standard test method for flexural strength of advanced ceramics at elevated temperatures.

These methods were adaptations of MIL-STD-1942 (MR), Flexural Strength of High Performance Ceramics at Ambient Temperature, published by the U.S. Army.

In November 1983, MIL-STD-1942 allowed several specimen-fixture combinations since it was

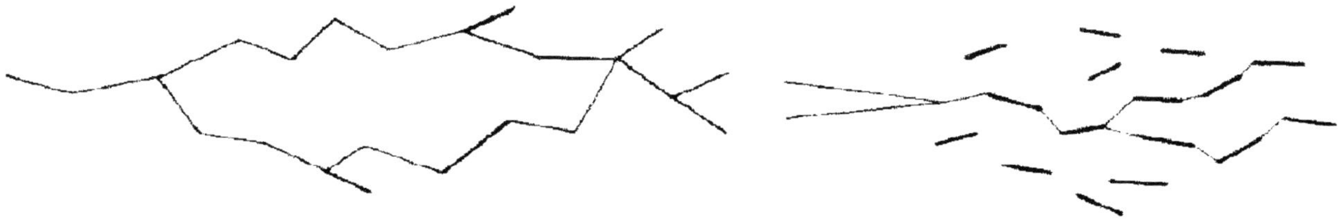

(a) *(b)*

Figure 2. (a) Crack Branching in conjunction with crack deflection, (b) Crack branching in conjunction with microcracking [1].

believed that no one specimen size would meet all the needs of the engineering-ceramic community.

The four-point (1/4 point) and three-point fixture configurations are given in Figure 3.

Specimen cross sections are usually square or rectangular, but sometimes are round, trapezoidal, or triangular.

For the purpose of test method standardization, ceramic composites have been divided into two categories: discontinuously reinforced ceramic composites and continuous fiber-reinforced ceramic composites (CFCCs). These two types of composites require distinct testing schemes due to their differing failure mechanisms and thus require separate sets of test standards.

Discontinuous reinforcements (i.e., whiskers, platelets, particulates, nanoparticulates, and *in situ* reinforcements) are added to a ceramic matrix in order to toughen the ceramic body by increasing the fracture energy required to propagate a crack through the ceramic composite by means of microstructural design and control. Despite the increase in fracture toughness achieved by the addition of discontinuous reinforcements, discontinuously reinforced ceramic composites and monolithic structural ceramics fail in similar manners. Both are prone to catastrophic failure and do not provide flaw tolerant microstructures. It is believed that the test methods for monolithic ceramics, for the time being, are adequate for the characterization of discontinuously reinforced ceramic composites. Thus, ASTM C-1161, C-1211, and the other test methods developed by C28.01 for monolithic ceramics are adaptable to discontinuously reinforced ceramic matrix composites.

Since monolithic ceramic test methods are adaptable to discontinuously reinforced ceramic matrix composites, ASTM C-28.07 has, thus far, made no efforts for test method standardization for this class of material. Instead, C-28.07 is focusing exclusively on test method standardization of continuous fiber-reinforced ceramic composites (CFCCs). Explicit work on test method standardization by C28.07 for discontinuous fiber-reinforced ceramic composites may begin in the future after standardization work has progressed in the areas of monolithic ceramics and CFCCs and if there is a perceived need for standardization.

In addition to the U.S., Japan, Germany, and France have also developed flexure standards. The principal features of these standards are given in Table 1.

There are many similarities between the standards. The specimen and fixture sizes are quite com-

Configuration
A: L=20 mm
B: L=40 mm
C: L=80

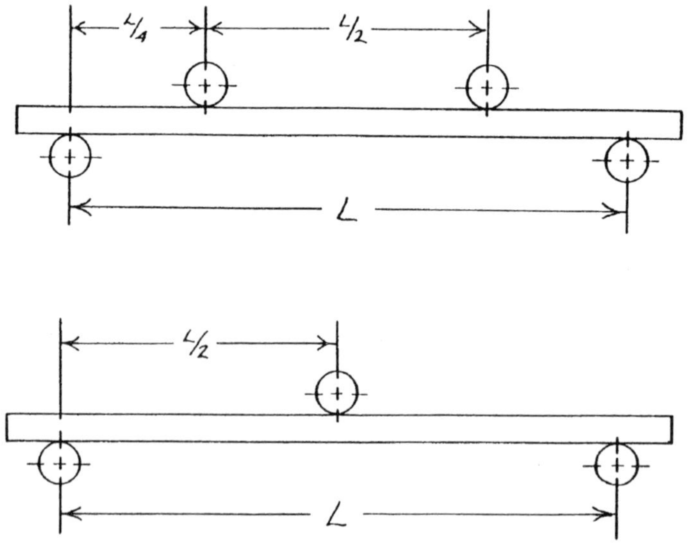

Fig. 3. Schematics of Three- and Four-Point Flexural Test Configurations.

parable and many tolerances and specifications are identical. Nevertheless, there are some differences shown in Table 1 that warrant attention. The ASTM and MIL-STD have more options in the possible testing configurations. The U.S., German, and French standards require the load rollers to be free to rotate to eliminate friction errors that can be present with the JIS R 1601, even with the required polished load points [3]. The chamfer sizes on the Japanese, French, and German standards are liberal, and a 4% error in stress is possible [3].

4.2 Fracture Toughness Test Methods

One of the primary goals for the insertion of whisker reinforcement into a ceramic matrix has been to improve the resistance to crack propagation, or fracture toughness; therefore, there has been a great deal of published and non-published work attempting to characterize the effectiveness of the particulate reinforcement, and much of this data is presented in this book. Although particulate reinforcement can significantly improve the fracture toughness of a ceramic matrix, particulate-reinforced CMCs, like monolithic ceramics, still fail in a brittle manner. There have been several methods devised to measure the fracture toughness of 'brittle' materials such as glasses, monolithic ceramics and high strength metals; these are the same methods that are currently being used to measure the fracture toughness of whisker-reinforced ceramic composites. A complete

Table 1. A Comparison of Flexure Standards [3].

	United States	Japan	Germany	France
Title (year)	MIL-STD 1942 (1983) ASTM C 1161 (1990)	JIS R 1601 (1981)	DIN 51 110 Part 1 (1989)	AFNOR B41-104 (1989, tentative)
Fixture	Three or four point	Three or four point	Four point	Three or four point
Spans (four point)	10 mm x 20 mm 20 mm x 40 mm 40 mm x 80 mm 0.75 in. x 1.5 in.*	10 mm x 30 mm	20 mm x 40 mm	10 mm x 20 mm 20 mm x 40 mm
Fixture	Semiarticulating or fully articulating	Fixed†	Fully articulating	Fully articulating
Fixture load pins	Rotating	Fixed‡	Rotating	Rotating
Specimen sizes	1.5 mm x 2 mm x 25 mm 3 mm x 4 mm x 45 mm 6 mm x 8 mm x 85 mm 0.13 in. x 0.25 in. x 2 in.*	3 mm x 4 mm x 35 mm	3 mm x 4 mm x 45 mm	2 mm x 4 mm x 25 mm 3 mm x 4 mm x 45 mm
Specimen chamfers	Up to 0.15 mm	Up to 0.3 mm§	Up to 0.3 mm§	Up to 0.3 mm§
Specimen preparation	(a) As-fired (b) Application matched (c) Three steps prescribed (wheels, grits, rates, etc.) (d) Other (Most flexible)	(a) As-fired (b) Polished (Final finish may merely conceal damage?)	(a) As-fired (b) Three steps prescribed (wheels, grits, rates, etc.)	(a) Polish deeply (tensile face) (b) Grinding optional (c) Other (Polish at least 100 μm deep to remove damage; roughness of 2 μm?)
Number of specimens (minimum)	10 for mean, 30 for Weibull analysis	10	15 preferably 30	10
Loading rate	0.5 mm/min (for 8 size)	0.5 mm/min	5-10 s	0.2 mm/min 0.5 mm/min (or higher)
Comments	Most flexible: B configuration is most common; tutorial and discussion of flexure testing	Less stringent in some technical details; simple to use	Tolerances similar to MIL STD	Tolerances similar to MIL STD; special inspection required for 20% of specimens; Includes Weibull analysis

*ASTM. †Specimens must be parallel, but no specifications on fixture. ‡5% to 15% error? §4% Stress error.

bibliographic listing on the fracture toughness test methods used to obtain the data in this databook is included in Appendix I.

Examination of fracture toughness data has found that there is considerable variation in the data. Test results can vary according to what material is being tested, what test method is being used, test environment, who is performing the test, random errors, and many other sources of variation. Fracture toughness measurements can vary widely from one test method to another, even when testing an identical material. One reason for the variation between test type is that the different fracture test types may be measuring different aspects of fracture toughness in the same material. For example, indentation microfracture methods measures the surface toughness, while a chevron notched beam method measures the bulk toughness, and an indentation strength in bending technique test measurements are dominated by a materials R-curve characteristics. The Particulate Databook attempts to divide fracture toughness data by test method. The following introduces these fracture toughness test methods.

A round robin was performed using three different test methods on two ceramics in an attempt to gain an understanding of the extent of variation from one test method to another, and the extent of variation of the same test method performed by different people on identical materials. The round robin was performed by the Versailles Advanced Materials and Standards (VAMAS) project, which is an international collaboration for prestandardization research. The participating countries were Canada, France, Germany, Italy, Japan, the United States and the Commission of European Communities. The two materials used in the study were a gas-pressure sintered silicon nitride, grade EC-141*, and a zirconia-alumina composite, grade UTZ-20*. The three test methods examined by the round robin were the indentation microfracture (IM) technique, indentation strength in bending (ISB) technique, and the single-edge precracked beam (SEPB) method. Results from this study are used to introduce these three fracture toughness methods [4].

* *NTK Technical Ceramic, Nagoya, Japan*

Single-Edge Notched Beam or Straight-Edge Notched Beam (SENB)

The SENB test is performed by testing a beam in flexure with a machined notch on the tensile surface. The machined notch is easy to introduce and the size of the notch can be measured accurately, however, the fracture toughness measurements are very sensitive to notch width and depth, and any variation can give markedly different results.

Chevron Notched Beam (CNB)

The test configuration used in Chevron Notched tests are shown in Figure 4. Each of these configurations has a triangular shaped notch machined into the specimen. The notch is loaded in tension during the CNB test. The CNB test has the advantage of producing artificial rising R-curve behavior, i.e., stable crack growth, even for brittle materials with flat R-curves. One disadvantage of the CNB test is the difficulty of machining complex shaped CNB test specimens.

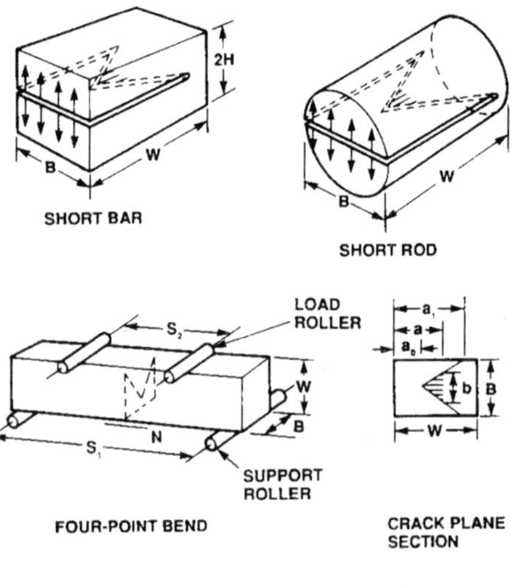

Fig. 4. Chevron Notch Test Configuration for Both Bar and Rod [5]

Vickers Indentation Microfracture (IM) Techniques

The Vickers Indentation microfracture (IM) techniques use a Vickers diamond pyramid indenter to create surface cracks on the sample surface. Figure 5 shows a schematic of the Vickers indentation method [I-16]. The length of crack propagation is measured and used to calculate the toughness of the sample. The primary advantage of the Vickers indentation technique is that it is a simple and very economical test to perform. Not very much material is required to perform the test so that many tests can be run on a single sample. Considering the random and systematic scatter in the data suggests that IM tests can be used to measure absolute fracture toughness with an accuracy level better than 30 to 40% [I-16]. The test is much better when used on a strictly comparative basis. The test is ideally suited to toughness for comparative evaluation between similar materials, such as in a case in which a material is subjected to some process variable.

One drawback to the IM test is that it only measures the surface toughness of the material. In cases where there is reason to believe that the surface toughness may differ from the bulk toughness (which is often the case for ceramic composites), another test method such as the SENB or CNB may be more appropriate because they measure the toughness of the bulk material. The measurement of the indentation crack length is commonly a significant source of error. It is difficult to measure the precise length of the indentation crack; the sample must be prepared to a smooth optical finish to allow for the accurate determination of the crack length. Also, the measurement should be performed quickly and in an inert environment (such as immersed in oil), otherwise slow crack growth will affect the fracture toughness measurements.

The results from the VAMAS round robin indentation microfracture toughness tests are shown in Figure 6. Fracture toughness was calculated from the Miyoshi *et al.* [I-28] equation, which is only one of the many equations used to calculate fracture toughness in this databook. Each line connects the two tests performed at different loads at each laboratory. Results for the silicon nitride material were promising. The U.S. labs were consistently around 5.3-5.5 MPa $m^{1/2}$, four Japanese labs were systematically greater than the U.S. labs by about 0.5 to 1 MPa $m^{1/2}$, while two foreign labs were outliers with results greater than 6-7 MPa $m^{1/2}$. The scatter of the data within a particular lab was typically 0.15-0.3 MPa $m^{1/2}$, so the results were fairly reproducible within a particular laboratory.

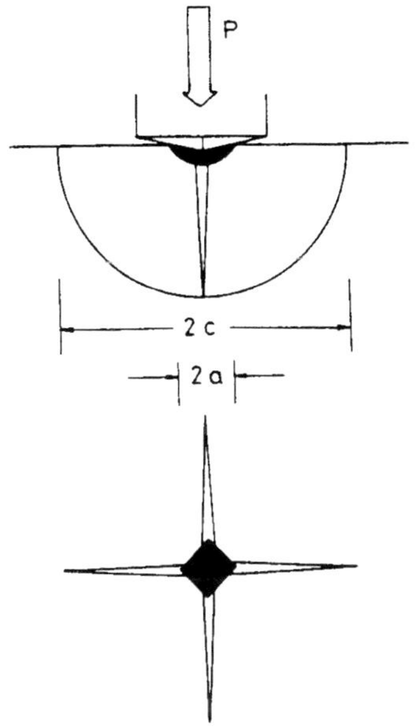

Fig. 5. Schematic of Vickers-Produced Indentation-Fracture System [I-16].

The results from the zirconia-alumina composite tests were not as promising. The fracture toughness data ranged from 5.3 to 9.2 MPa $m^{1/2}$, and the results were widely scattered. Some of the laboratories were internally consistent and showed a variation as low as 0.1 MPa $m^{1/2}$, while others did not perform as well and varied by as much as 1.3 MPa $m^{1/2}$. The primary cause of the wide scatter was the great difficulty in consistently measuring the crack length. Measurement was highly dependent on interpretation, and even on viewing angle. Many of the composites presented in this databook show similar fracture behavior to this alumina-zirconia composite, and based on the results from the VAMAS round robin study, the accuracy of the indentation fracture toughness tests may be suspect.

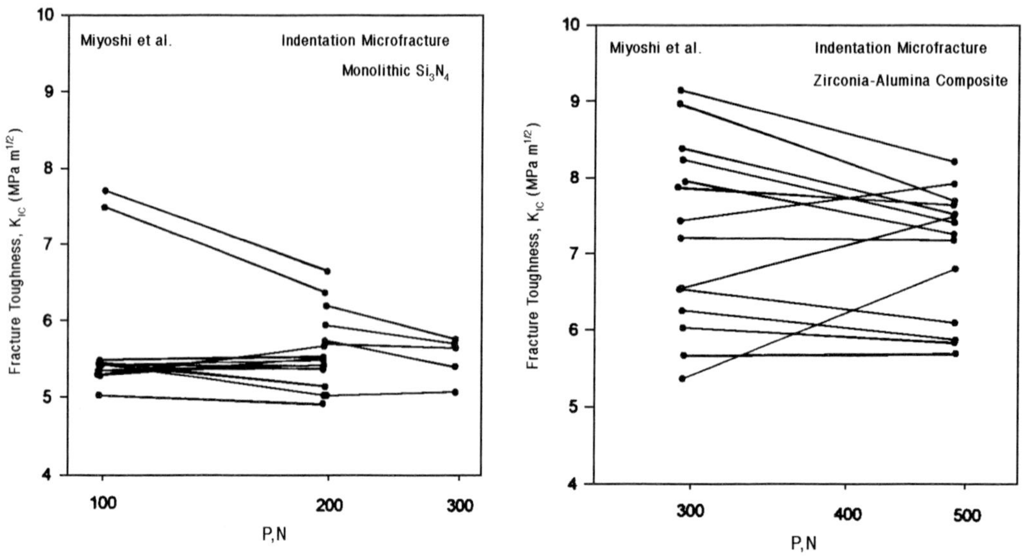

Fig. 6. VAMAS Round Robin Indentation Microfracture Toughness Results [4, I-28].

Indentation Strength in Bending (ISB)

The ISB method is conducted by applying an indentation on the tensile surface of a flexure specimen. The indentor creates a surface flaw of known size, in which the size of the flaw is dependent on the indentation load. Measurement of post-indentation flexural strength is, in essence, a measurement of fracture resistance. Fracture toughness is determined from the post-indentation flexural strength and indentation load by an empirical equation calibrated by test measurement made on specimens of known toughness. One advantage is that no crack measurements need to be made; indentation load replaces crack size as the variable [I-15]. One disadvantage is that all aspects of fracture toughness are embodied in an empirical calibration, rather than based on a theoretical model [4].

Initial evaluation of the ISB method was performed by Chantikul *et al.* [I-15]. ISB fracture toughness tests were performed on several different brittle materials, including WC, zirconia, alumina, silicon nitride, silicon carbide, glass and glass-ceramics. The empirical formula that was obtained for these materials was:

$$K_{IC} = 0.59\,(E/H)^{1/8}\,(\sigma c P^{1/3})^{3/4}$$

From the errors in the data obtained in this initial work, it was estimated that it should be possible to determine K_{IC} to within 30-40%, i.e., with an accuracy comparable to the direct crack measurement method. Results from the VAMAS round robin, shown in Figure 7, suggest that the ISB test can do significantly better than that. The data for the silicon nitride sample (excluding the lab that made the mistake of annealing the sample before testing) showed remarkable consistency. The average of the data at the indentation load of 49 N was 5.8 MPa $m^{1/2}$, and the data were all within the range of 5.5 to 6.0 MPa $m^{1/2}$. The standard deviations within a laboratory were within 0.1 to 0.2 MPa $m^{1/2}$. The result from the zirconia-alumina composite also showed promise. Due to the rising R-curve of these two materials and the large flaw size relative to the specimen cross section, greater confidence should be placed on the tests performed at lower indentation loads.

In summary, the ISB test can be excellent test method on a comparative basis, however, initial test fracture toughness tests should be performed by a test method with a stronger theoretical base.

Multiple Indentation Fracture Toughness Test (MIFS)

A derivative of the ISB method is the Multiple Indentation Flexural Strength (MIFS) test devised by Cook and Lawn [I-10]. Instead of a single indent on the tensile surface, multiple indents are made on

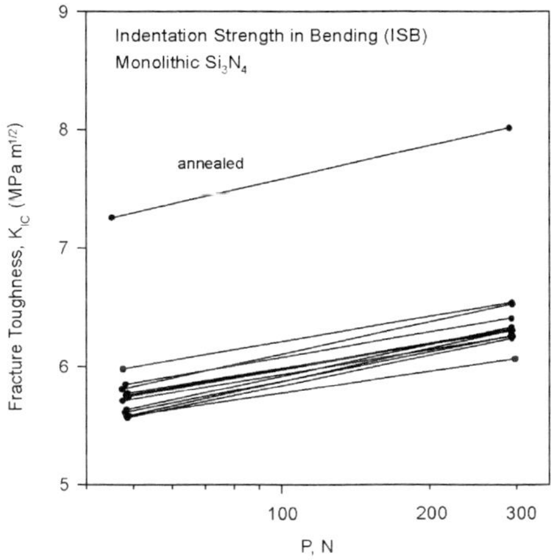

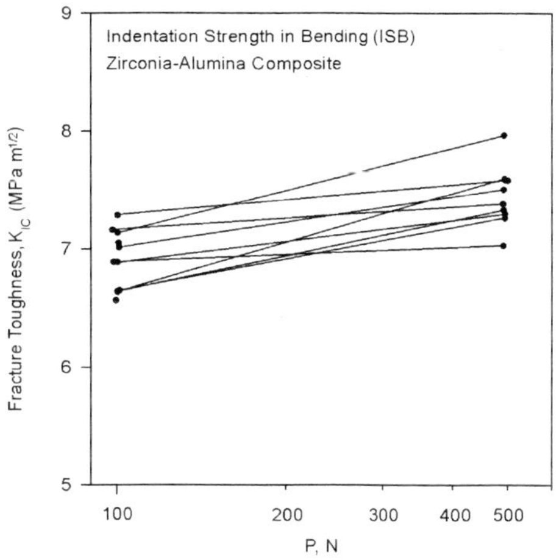

Fig. 7. VAMAS Round Robin Indentation Strength in Bending Results [4].

the tensile surface, as shown in Figure 8. Each indent should experience near identical stress states during the bend test, provided that the indentation fractures do not interact with each other. Slight inhomogeneities cause failure to occur at just one of the indents, leaving the others available for measurement of the critical crack length, i.e., the crack length at failure. Fracture toughness is then calculated from the critical crack length and the flexural strength. In the MIFS technique, direct crack measurement replaces indentation load as a variable used in the ISB method. This introduces the uncertainty of crack length measurements that was encountered in the indentation microfracture technique.

Single-Edge Precracked Beam (SEPB)

The single-edge precracked beam uses a bend specimen with a sharp precrack on the tensile surface of the specimen. The original technique for obtaining the sharp precrack, as-described in ASTM 399 A2, was to fatigue a specimen after indentation. The process is time consuming and it is very difficult to control the fatigue precracking to obtain a crack with precise depth. Nose and Fujii [I-4] proposed a substitute for obtaining a precracked specimen for use in the SEPB technique. Sharp precracks are introduced by bridge-indentation method using a loading fixture similar to the one shown in Figure 9. The sample is placed under compression in an anvil with a central groove. Due to the nature of the loading state, a local tensile field is produced at the Vickers indent. At the 'pop-in sound', the crack propagates into the sample; as the crack propagates, the local tensile stress disappears quickly and crack propagation stops, producing a precrack of known

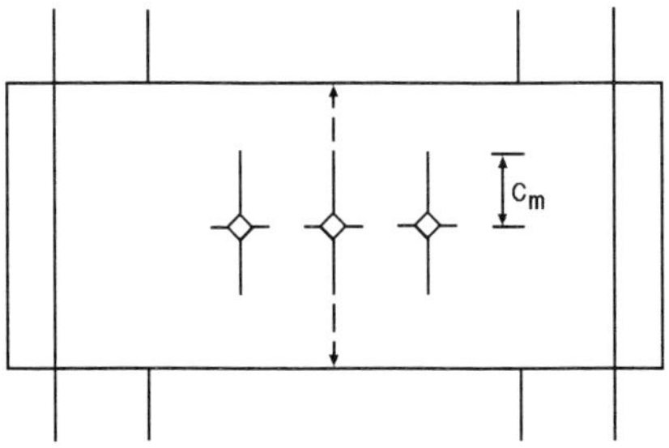

Fig. 8. Schematiac of a MIFS Test Specimen [I-10].

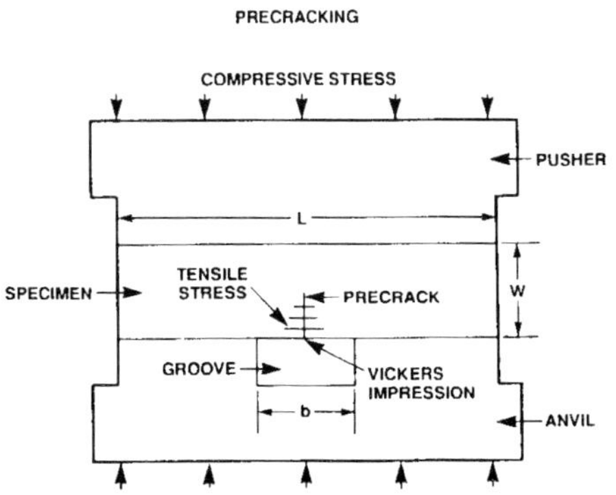

Figure 9. Schematic of the Bridge-Indentation Method [I-4].

depth. Before precracking, the sample must be carefully prepared to a surface finish and edge quality of 0.08 nm (3 mils) or better to eliminate undesirable crack starters.

The VAMAS round robin used SEPB samples precracked by a standard bridge-anvil furnished by the JFCC. The specimens were tested in 3-pt. flexure at two different crosshead rates, permitting an assessment of whether environmental phenomena affected test results. Short (16 mm) spans were used due to the small size of the samples. The results are shown in Figure 10.

Results for the silicon nitride samples from most of the labs showed good consistency around 5.6 MPa $m^{1/2}$ and was not sensitive to rate of loading. The zirconia-alumina composite, on the other hand, did show rate sensitivity, which suggests that environmentally assisted slow crack growth is active. Other experimental observations indicated that R-curve phenomena were active which also could have accounted for the rate effect. The VAMAS trial on the zirconia-alumina composites did raise one significant question concerning the SEPB method: Should the pop-in (initial) precrack length be measured or should the SEPB method use the critical crack length?

In summary, the SEPB method demonstrated consistent results for the silicon nitride material which did not demonstrate environmentally activated slow crack growth. The results from the zirconia-alumina composites were complicated by both slow crack growth and R-curve effects, and results were not as consistent.

4.3 Hardness

Hardness is a material's resistance to deformation, scratching, and erosion. The Vickers hardness test (HV) using a diamond indentor is a common method for ceramic composite hardness measurement. The following guidelines should be followed when performing indentation tests on ceramics [5].

- The indentation must be larger than the microstructural features. This includes matrix grain size and the size of the particulate reinforcement.
- An adequate number of measurements should be made, preferably 10 or more.
- Badly damaged indentations must be ignored.
- The tester must have a way to calibrate the machine, such as a high hardness test block.
- The geometry and condition of the diamond indenter must be checked at intervals, especially for high load tests.
- The errors involved in hardness testing can be as great as 15%.

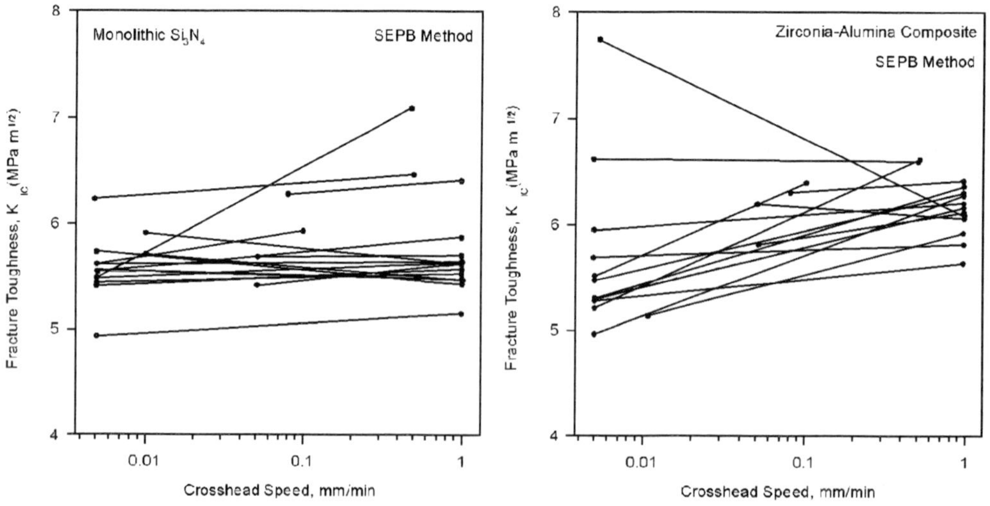

Figure 10. VAMAS Round Robin SEPB Test Results [4].

Vickers hardness values are given in the literature in units of GPa and kgf/mm^2. All data is converted to GPa using the following conversion factor:

$$\text{kgf/mm}^2 * 9.8\text{x}10^{-3} = 1 \text{ GPa}$$

5. Thermophysical Property Measurement Methods

5.1 Thermal Conductivity

There are several techniques of measuring thermal conductivity of materials exhibiting a very wide range of conductivity. These techniques are [6]:

- Axial flow
- Radial flow
- Direct electrical heating
- Guarded hot plate (ASTM C-177)
- Hot-wire method
- Panel test method

The first three methods are classical ones used generally in measuring rigid materials having moderate to high conductivity. The other three methods are generally used in measurements on materials with moderate to low conductivity where the specimens are often nonrigid. Table 2 summarizes the most suitable thermal conductivity measurement techniques for various materials, advantages/disadvantages, temperature range, conductivity range, and measurement uncertainty.

5.2 Thermal Diffusivity

Thermal diffusivity represents a genuine transport property which characterizes thermal phenomena important in many engineering applications and fundamental materials studies. The measurement of thermal diffusivity consists of the accurate recording of the time dependence of temperature due to a transient or periodic thermal perturbation at the specimen boundary as opposed to the measurement of thermal conductivity which involves the measurement of heat fluxes that are difficult to control and measure accurately. Therefore, it is frequently easier to measure thermal diffusivity than thermal conductivity. Since thermal diffusivity is directly related to thermal conductivity in a relationship that involves density and specific heat, which are either known or can be measured or calculated relatively simply and accurately, thermal conductivity can be deduced more easily and accurately.

The most commonly used techniques to measure thermal diffusivity are [6]:

- Step-Intensity Illumination Method
- Mirage-Effect Method
- Photoacoustic-Effect Method
- Laser Flash Diffusivity Technique (ASTM E-1461-91)

Table 3 summarizes some measurement techniques suitable for a variety of materials, advantages/disadvantages, temperature range, diffusivity range, and measurement uncertainty.

5.3 Specific Heat

There are six different calorimetry techniques used for the measurement of specific heat mainly above room temperature and extending up to several thousand degrees. These techniques are [6]:

- Adiabatic
- Drop
- Levitation
- Modulation
- Pulse
- Differential Scanning Calorimetry (ASTM E-1264)

Table 4 summarizes these measurement techniques.

5.4 Thermal Expansion

The most common techniques of measuring thermal expansion of solids are:

- Push-Rod Dilatometer
- Quartz Push-Rod Dilatometer
- Dual Quartz Push-Rod Dilatometer
- Twin Telemicroscope Method
- X-Ray Method
- Interferometric Method
- Linear Thermal Expansion with Interferometry (ASTM E-289)

- Pulse-Heating Interferometric Method
- Graphite Push-Rod Method
- Linear Thermal Expansion by Thermomechanical Analysis (ASTM E-831)
- Linear Thermal Expansion with a Vitreous Silica Dilatometer (ASTM E-228)

For the thermal expansion measurements of ceramic composites presented in this book, the push-rod dilatometer methods appear to be the most commonly used. The push-rod dilatometer methods are simple, reliable, and easy to automate. With these methods, the expansion of the specimen is transferred out of the heated zone to an extensometer by means of rods (or tubes) of some stable material.

When used at temperatures below 1200 K, dilatometers are usually made of vitreous silica. For use at higher temperatures, they are made of alumina, silicon carbide, tantalum, and other refractory materials. Linear variable differential transformers are most frequently used for measuring expansion, but many other types of extensometers have been used. These include dial gauges, optical levers, microscopes, optical gratings, interferometers, and micrometers.

Table 2. Thermal Conductivity Measurement Techniques [6].

Measurement technique	Key features: Advantages	Key features: Disadvantages/limitations	Temperature range (K)	Principal specimen materials	Conductivity range ($W m^{-1} K^{-1}$)	Uncertainty (%)
Axial heat flow	Simultaneous electrical resistivity measurements Temperature coverage (below and above RT) High accuracy	Control of radial heat losses above ~ 500 K	90–1300	Metals and alloys in cylindrical form	10–500	0.5–2.0
Radial heat flow	Accuracy Flexibility at high temperatures	Large specimen size Long measurement time	RT–2600	Solids and powders in cylindrical form	0.01–200	3–15
Direct electrical heating	Very high temperature measurements Multiple property determination including electrical properties Rapid measurements	Sophisticated, often costly equipment Specimen must be an electrical conductor Stable optical properties	400–3000	Wires, rods, tubes of electrical conductors	10–200	2–5
Guarded hot plate	Adaptable to wide range of commercial materials High accuracy	Low conductivity materials Long test times (3–12 h)	80–1500	Thermal insulations "Building" materials	< 1.0	2–5
Hot wire method	Small temperature drop in specimen	Low conductivity materials	RT–1800	Refractory materials including powder and nonrigid forms	0.02–2	5–15
Panel test	Simple apparatus	High-temperature gradient in specimen Long test times	600–1600	Refractory materials in block form	0.05–15	15

Table 3. Thermal Diffusivity Measurement Methods [6].

Measurement technique	Key features: Advantages	Key features: Disadvantages/limitations	Temperature range (K)	Principal specimen materials	Diffusivity range ($m^2 s^{-1}$)	Uncertainty (%)
Pulse method	Wide property range Wide temperature coverage Small and simple specimens Simple and rapid measurement Multilayer specimens Multiple specimen facility	Not well suited for coarse matrix materials Not convenient for translucent materials Sophisticated error analysis	100–3300 (300–2000)	All solids and encapsulated liquid metals Specimens: Disks 6–16 mm in diameter	$10^{-7}-10^{-3}$	1.5–5
Temperature wave method	Wide property range Very wide materials range Suitable for very high pressures Suitable for multiproperty measurement	Quasistationary temperature conditions necessary Complex mathematical treatment Complex error analysis	60–1300 (300–1300)	Solids, liquid metals, liquids, and gases Specimens: rods, cylinders, hollow cylinders, etc.	$10^{-7}-10^{-4}$	1–5 (solids) 1–3 (fluids) 5–9 (liquid metals)
Electron bombardment heat input method	High-temperature coverage Small specimens Applicable to electrical conductors and nonconductors AC techniques applicable Simultaneous specific heat movement possible	Complex experimental apparatus High-vacuum conditions required Arduous mathematical apparatus	330–3200 (1100–2200)	Metals, nonmetals, liquid metals Specimens: disks 6–9 mm, plates 0.25 cm^2 and more, hollow cylinders 6–8 mm i.d. 15–20 mm o.d.	$10^{-7}-5 \times 10^{-5}$	2–10
Monotonic heating regime method	Wide materials range Continuous measurement in very wide temperature range Simple apparatus Simple measurement	Not very suitable for good thermal conductors Somewhat lower precision	4.2–3000	Ceramics, plastics, composites, thermal insulations Specimens: cylinders, plates, paralelepipeds	$10^{-8}-10^{-5}$	2–12

Table 4. Specific Heat Measurement Techniques [6].

Technique	Advantages	Disadvantages	Temp. range[a] (K)	Specimen material	Uncertainty (%)
Adiabatic	Very versatile High sensitivity Solid and liquid specimen	Specimen in container Limitations in high temperatures	4–1300	All	1–3
Drop	Solid and liquid specimen	Specimen in container Slow measurement times Specific heat obtained from enthalpy data	300–2000	All	1–3
Levitation	No container for specimen Solid and liquid specimen	Specimen must be electrical conductor Specific heat obtained from enthalpy data	1000–2500	Electrical conductor (sphere)	2–5
Modulation	No container for specimen Multiproperty measurement capability	Solid specimen	80–3000	All	2–5
Pulse	No container for specimen Measurements at very high temperatures Solid and liquid specimen Multiproperty measurement capability	Specimen must be electrical conductor Sophisticated instrumentation	1000–7000	Electrical conductor (wire, rod, tube)	2–10
Differential scanning	Quick and economical	Specimen in container Limitations in high temperatures	100–1000	All	1–3

[a] For each technique measurements at temperatures higher than that indicated have also been performed. However, the given value indicates the limit of the accurate measurements for that technique.

6. Unit Conversion Tables

SI PREFIXES

FACTOR	PREFIX	SYMBOL
10^{12}	tera	T
10^{9}	giga	G
10^{6}	mega	M
10^{3}	kilo	k
10^{2}	hecto	h
10^{1}	deka	da
10^{-1}	deci	d
10^{-2}	centi	c
10^{-3}	milli	m
10^{-6}	micro	μ
10^{-9}	nano	n
10^{-12}	pico	p

MECHANICAL PROPERTIES

TO CONVERT FROM	TO	MULTIPLY BY
Pressure/Stress		
inch of mercury (32 °F)	Pa	3.386 38 E+03
inch of mercury, tor (60 °F)	Pa	3.376 85 E+03
mm of mercury (torr) (0 °C)	Pa	1.333 22 E+02
inch of water (39.2 °F)	Pa	2.490 82 E+02
inch of water (60 °F)	Pa	2.488 4 E+02
kgf m^{-2}	Pa	9.806 650
kip in^{-2} (ksi)	Pa	6.894 757 E+06
lbf in^{-2} (psi)	Pa	6.894 757 E+03
bar	Pa	1.000 000 E+05
Dyn cm^{-2}	Pa	1.000 000 E-01
atm	Pa	1.013 250 E+05
Fracture Toughness		
ksi $in^{1/2}$	MPa $m^{1/2}$	1.11
kg_f mm^{-2} $mm^{1/2}$	MPa $m^{1/2}$	3.101 140 E-01
Hardness		
kg_f mm^{-2}	GPa	9.806 650 E-03
Fracture Energy		
Btu (international table)/ft^2	J m^{-2}	1.135 653 E+04
erg cm^{-2}	J m^{-2}	1.000 000 E-03

THERMOPHYSICAL PROPERTIES

TO CONVERT FROM	TO	MULTIPLY BY
Thermal Conductivity		
Btu (international table)*ft/(h ft ^{2} °F)	W m^{-1} K^{-1}	1.730 735
Btu (international table)*in/(s ft ^{2} °F)	W m^{-1} K^{-1}	5.192 204 E+02
cal (thermochemical)/(cm s °C)	W m^{-1} K^{-1}	4.184 000 E+02
Thermal Diffusivity		
ft^2/h	m^2/s	2.580 640 E-05
ft^2/s	m^2/s	9.290 340 E-02
in^2/s	m^2/s	6.451 600 E-04
Thermal Expansion		
µin/in	%	1.0 E-04
m in/in	%	1.0 E-01
10^{-6} F^{-1}	10^{-6} K^{-1}	5.555 556 E-01
Heat Capacity		
cal g^{-1} C^{-1}	J kg^{-1} K^{-1}	4.186 800 E+03
Btu lb^{-1} F^{-1}	J kg^{-1} K^{-1}	4.186 800 E+03
Enthalpy		
Btu (international table)/lb	J/kg	2.326 000 E+03
cal (intenational table)/g	J/kg	4.186 800 E+03

MISCELLANEOUS

TO CONVERT FROM	TO	MULTIPLY BY
ACCELERATION		
g, standard acceleration of free fall	m s^{-2}	9.806 650
in s^{-2}	m s^{-2}	2.540 000 E-02
ANGLE		
degree	radian	1.745 329 E-02
minute	radian	2.908 882 E-04
second	radian	4.848 137 E-06
AREA		
ft^2	m^2	9.290 304 E-02
in^2	m^2	6.451 600 E-4
BENDING MOMENT OR TORQUE		
lbf in	N m	1.129 848 E-01
lbf in/in	N m/m	4.448 222
lbf ft	N m	1.355 818
lbf ft/in	N m/m	5.337 866 E+01
ENERGY/WORK		
British Thermal Unit (mean)	J	1.055 870 E+03
calorie (mean)	J	4.190 020
calorie (20 °C)	J	4.181 900
electronvolt	J	1.602 190 E-19
ft lbf	J	1.355 818
kW h	J	3.600 000 E+06
W s	J	1.000 000
ENERGY PER UNIT AREA TIME		
Btu (international table)/(ft^2 s)	W m^{-2}	1.135 653 E+04
cal (thermochemical)/(cm^2 min)	W m^{-2}	6.973 333 E-02
FORCE		
kilogram force	N	9.806 650
kip (1000 lbf)	N	4.448 222 E+03
pound-force (lbf)	N	4.448 222
FORCE PER UNIT LENGTH		
lbf/in	N/m	1.751 268 E+02

TO CONVERT FROM	TO	MULTIPLY BY
LENGTH		
ångström	m	1.000 000 E-10
foot	m	3.048 000 E-01
inch	m	2.540 000 E-02
yard	m	9.144 000 E-01
mile	m	1.609 344 E+03
MASS		
grain	kg	6.479 891 E-05
ounce	kg	2.834 952 E-02
ton (metric)	kg	1.000 000 E+03
MASS PER UNIT LENGTH		
lb/in	kg/m	1.785 797 E+01
MASS PER UNIT AREA		
oz ft^{-2}	kg m^{-2}	3.051 517 E-01
lb ft^{-2}	kg m^{-2}	4.882 428
MASS PER UNIT VOLUME		
oz in^{-3}	kg m^{-3}	1.729 994 E+03
lb in^{-3}	kg m^{-3}	2.767 990 E+04
POWER		
Btu (international table)/s	W	1.055 056 E+03
Btu (thermochemical)/s	W	1.054 350 E+03
cal (thermochemical)/s	W	4.184 000
ft lbf/s	W	1.355 818
horsepower (electric)	W	7.460 000 E+02
TEMPERATURE		
degree Celsius	K	$T_K = t_{oC} + 273.15$
degree Fahrenheit	^oC	$T_C = (t_{oF} - 32)/1.8$
degree Fahrenheit	K	$T_K = (t_{oF} + 459.67)/1.8$
VELOCITY		
ft/h	m/s	8.466 667 E-05
in/s	m/s	2.540 000 E-02
km/h	m/s	2.777 778 E-01
mi/h	km/h	1.609 344
VOLUME PER TIME		
ft^3/min	m^3/s	4.719 474 E-04
in^3/min	m^3/s	2.731 177 E-07

SI DERIVED UNITS WITH SPECIAL NAMES

QUANTITY	NAME	SYMBOL	EXPRESSION IN TERMS OF OTHER UNITS	EXPRESSION IN TERMS OF SI UNITS
force	newton	N		$m\ kg\ s^{-2}$
pressure	pascal	Pa	$N\ m^{-2}$	$m^{-1}\ kg\ s^{-2}$
energy, work, quantity of heat	joule	J	N m	$m^2\ kg\ s^{-2}$
power, radiant flux	watt	W	$J\ s^{-1}$	$m^2\ kg\ s^{-3}$
quantity of electricity, electric charge	coulomb	C		s A
electric potential	volt	V	$W\ A^{-1}$	$m^2\ kg\ s^{-3}\ A^{-1}$
electric resistance	ohm	Ω	$V\ A^{-1}$	$m^2\ kg\ s^{-3}\ A^{-2}$
moment of force		N m		$m^2\ kg\ s^{-2}$
surface tension		$N\ m^{-1}$		$kg\ s^{-1}$
heat capacity, entropy		$J\ K^{-1}$		$m^2\ kg\ s^{-2}\ K^{-1}$
specific heat capacity, specific entropy		$J\ kg^{-1}\ K^{-1}$		$m^2\ s^{-2}\ K^{-1}$
specific energy		$J\ kg^{-1}$		$m^2\ s^{-2}$
thermal conductivity		$W\ m^{-1}\ K^{-1}$		$m\ kg\ s^{-3}\ K^{-1}$
energy density		$J\ m^{-3}$		$m^{-1}\ kg\ s^{-2}$

7. Material and Property Index

Material	*3-pt. Flexural (RT) Section I*	*4-pt. Flexural (RT) Section I*	*Elevated Temperature Flexural Strength Section I*	*Fracture Toughness (Indentation Methods) Section II*	*Fracture Toughness (Precracked Beam Methods) Section II*	*Fracture Toughness (Machined Notched Beam Methods) Section II*
MgO			***4***			
SiC(p)/MgO	1, ***3***		2, ***4***	1, ***2***		
Mullite			***10, 14***	***10, 14***		
SiAlON(p)/Mullite				13, ***14***		
SiC(p)·ZrO_2(p)/Mullite				59, ***60***	59, ***60***	60, ***62***
ZrO_2(p)·ZrO_2(p)/Mullite						15, ***16***
ZrO_2(p)/Mullite		6, ***9***, 12, ***14***	7-8, ***10***	3-4, 6-7, ***8***, ***10***, ***12***		
Cordierite	***14***					
ZrO_2(p)/Cordierite	13, ***14***			5, ***8***		
Al_2O_3	***42***	***29-30***	***22, 46***			
B_4C(p)/Al_2O_3	16-7, ***20***					38, ***42***
Diamond(p)/Al_2O_3						37, ***42***
SiC(p)/Al_2O_3	16, 17, ***20***	23-6, ***27-30***	18-9, ***22***	29, ***32***	20-1, ***22***	31, ***36***
SiC(p) ZrO_2(p)/Al_2O_3				57, ***62***		
TiC(p) ZrO_2(p)/Al_2O_3				58, ***62***		
TiB_2(p)/Al_2O_3	32-4, 37, ***38-9***, ***42***		44-5, ***46***	23, 28-9, 30, ***32***, ***34***		19, ***22***, 38-40, ***41-2***
TiC(p)/ Al_2O_3	31-2, 34-6, ***38-31***		44, ***46***	24-7, 30, ***34-5***		17-8, ***22***
TiN(p)/Al_2O_3			43, 45, ***46***	29-30, ***34***		
ZrO_2(p)/Al_2O_3	47-50, ***51***, ***53-5***	57, 59-61, ***62-3***	58, ***64***	44-9, ***50***, ***52***, ***54***, ***56***	15, ***16***	
ZrO_2·HfO_2(p)/Al_2O_3·Cr_2O_3		65, ***67***			63-4, ***65***	
ZrO_2(p)/$MgCr_2O_4$		66, ***68***				
ZrO_2			***72***			
SiC(p)/ZrO_2	70, ***71***		69, ***72***			66, ***67***

Key: Plain text represents data set number.
Italic bold text represents figure page number.

7. Material and Property Index (continued)

Material	*3-pt. Flexural (RT) Section I*	*4-pt. Flexural (RT) Section I*	*Elevated Temperature Flexural Strength Section I*	*Fracture Toughness (Indentation Methods) Section II*	*Fracture Toughness (Precracked Beam Methods) Section II*	*Fracture Toughness (Machined Notched Beam Methods) Section II*
Si_3N_4			***75***, ***104***		***90***	***98***
Al_2O_3(p)/Si_3N_4	73, ***74***					
BN(p)/Si_3N_4		102, ***105***	99, ***104***			94, 95, ***98***, ***100***
$MoSi_2$(p)/Si_3N_4			73, ***75***			
SiC(p)/Si_3N_4	91, 94, ***96***	80-1, 82, 83-4, ***86-7***	79, ***85***, 92, ***95***	70-1, 73, ***74***, ***76***, 78-9, 83, ***88***	77, 81, 84-5, ***87***	92-3, ***98***, ***100***
TiC(p)/ Si_3N_4		81, ***86***		101, 103, ***104***	80-1, ***87***	94-5, 96-7, ***98***, ***100***
TiN(p)/ Si_3N_4	98, ***103***	100-1, ***103***		86, ***91***, 102-3, ***104***		
Al_2O_3(p)/SiAlON		76, ***77-8***				
SiC(p)/SiAlON			93, ***96***	72, ***74***	82, ***90***	
SiC(p)/SiAlON·Si_3N_4		82-3, ***87***				
AlN			***114***			
BN(p)/AlN	107, ***112***					
SiC(p)/AlN	106, ***111***		110, ***114***			
TiC(p)/AlN	109, ***113***					
TiN(p)/AlN	108, ***113***					
TiO_2(p)/AlN	108, ***113***					
SiC			***135***, ***140***, ***146***		***114***	
Diamond(p)/SiC				121, ***123***		
TiB_2(p)/SiC		143, 145, ***148***	142-3, 144, ***146***	117, ***119***	117-8, ***119***	116, 118, ***119-20***
TiC(p)/SiC	123, 124, ***130***	138, ***139***	123, 125-9, ***132-5***, 136-7, ***140***		107-11, ***112***, ***114***	106-7, ***112***
Al_2O_3(p)/TiB_2				125, ***126***		
B_4C(p)/TiB_2		117, ***119***			125, ***126***	
ZrO_2(p)/TiB_2					124, ***126-7***	
2Y-ZrO_2(p)/TiB_2	115, ***118***					
SiC(p)/ZrB_2	116, ***119***					
SiC(p)/ZrC						68, ***69***
ZrO_2(p)/ZrC		120, ***122***				
Zr(p)·ZrB_2(pl)/ZrC		121, ***122***				
B_4C(p)/Graphite	150, ***151***					

Key: Plain text represents data set number.
Italic bold text represents figure page number.

7. Material and Property Index (continued)

Material	*Elastic Properties (Young's Modulus) Section III*	*Elastic Properties (Shear Modulus) Section III*	*Elastic Properties (Poisson's Ratio) Section III*	*Vickers Hardness Section IV*	*Thermal Properties (Thermal Conductivity / Thermal Diffusivity) Section V*
SiC(p)/MgO					10-11, ***12***, ***16***
Mullite	***6***				
SiC(p)·ZrO$_2$(p)/Mullite					18, ***19***, 32, ***36***
ZrO$_2$(p)/Mullite	3, ***6***				
Cordierite					***4***, ***6***
Diamond(p)/Cordierite					1-3, ***4-6***
Al$_2$O$_3$					***29***, ***36***, ***38***
Diamond(p)/Al$_2$O$_3$					27, ***29***
SiC(p)/Al$_2$O$_3$	7-9, ***10***		9	1-2, ***3***	7-8, ***12***, ***15-16***, 32-3, ***36***
SiC(p)/ ZrO$_2$·Al$_2$O$_3$	12, ***17***			21, ***23-4***	
TiC(p)/ ZrO$_2$·Al$_2$O$_3$	11, ***17***			22, ***23-4***	
TiB$_2$(p)/Al$_2$O$_3$	14-5, ***16***			4, 6, 9, ***10***, ***12***	28, ***29***
TiC(p)/ Al$_2$O$_3$	13, ***16***			4-6, 7, ***10***, ***12-3***	28, ***29***
TiN(p)/Al$_2$O$_3$	14, ***16***			7, 8-9, ***10***, ***12***, ***14***	27, ***29***
ZrO$_2$(p)/Al$_2$O$_3$	1-2, 4, ***5***			15-7, ***18***, ***20***	
ZrO$_2$·HfO$_2$(p)/Al$_2$O$_3$					30-1
ZrO$_2$·HfO$_2$(p)/Al$_2$O$_3$·Cr$_2$O$_3$	18, ***19***				33-5, ***36-8***
BeO					***14***
SiC(p)/BeO					8-10, ***13-4***, ***16***
SiC(p)/ZrO$_2$					31, ***36***, 40-1, ***42***
Al$_2$O$_3$(p)/AlTi$_2$O$_5$					20-2, ***23-4***, ***26***
TiO$_2$(p)/HfO$_2$					40, ***41***

Key: Plain text represents data set number.
Italic bold text represents figure page number.

7. Material and Property Index (continued)

Material	*Elastic Properties (Young's Modulus) Section III*	*Elastic Properties (Shear Modulus) Section III*	*Elastic Properties (Poisson's Ratio) Section III*	*Vickers Hardness Section IV*	*Thermal Properties (Thermal Conductivity / Thermal Diffusivity) Section V*
Si_3N_4	***33, 37***			***33***	
Al_2O_3(p)/Si_3N_4					44, ***47***
BN(p)/Si_3N_4	35, ***37***				
SiC(p)/Si_3N_4	26-31, ***32-4***			25-31, ***32-5***	46, ***47***
TiC(p)/ Si_3N_4	21-2, ***23***			36-7, ***39***	
TiN(p)/ Si_3N_4	36-7, ***38-9***			37-8, ***39***	43, 44-5, ***47***
SiC(p)/SiAlON	20, ***24***				
SiC(p)/AlN					49-50, ***51-3***
TiC(p)/AlN				40, ***41***	
TiN(p)/AlN				40, ***41***	
TiO_2(p)/AlN				41, ***41***	
SiC				***3***	***14, 70***
BN(p)/SiC	40, ***44***				60-1, ***63***, 64-5, ***69***
TiB_2(p)/SiC	41, 43, ***44***		43, ***45***	42, ***44***	67, ***69***
TiC(p)/SiC	42, ***44***		42, ***45***	43, ***44***	
BN					***62***
AlN(p)/BN					46, ***48***, 54-5, ***56-7***
B_4C					***62***
BN(p)/B_4C					58-9, ***61-2***
Al_2O_3(p)/TiB_2					71, ***74***
B_4C(p)/TiB_2				45, ***47***	72, ***74***
ZrO_2(p)/TiB_2				45-6, ***47***	
SiC(p)/ZrB_2	47, ***48***		47, ***49***		71, 72, ***74***, 75, ***76***
Zr(p)·ZrB_2(pl)/ZrC	46, ***48***	46, ***49***			70, ***73***
Graphite					***69***
B_4C(p)/Graphite					66, ***68***

Key: Plain text represents data set number.
Italic bold text represents figure page number.

8. List of Figures

Flexural Strength

Figure 1. 3-pt. Flexural Strength of Hot-Pressed SiC(p)/MgO I-3
Figure 2. 3-pt. Flexural Strength of Hot-Pressed SiC(p)/MgO I-4
Figure 3. 4-pt. Flexural Strength of ZrO_2(p)/Mullite I-9
Figure 4. 4-pt. Flexural Strength vs. Temperature of ZrO_2(p)/Mullite Processed From Sol-Gel Derived Powders I-10
Figure 5. Flexural Strength vs. Sintering Temperature of ZrO_2(p)/Cordierite and ZrO_2(p)/Mullite I-14
Figure 6 3-pt. Flexural Strength of SiC(p)/Al_2O_3 and B_4C(p)/Al_2O_3 .... I-20
Figure 7 3-pt. Flexural Strength vs. Temperature of Hot-Pressed SiC(p)/Al_2O_3 .... I-22
Figure 8. 4-pt. Flexural Strength of SiC(p)/Al_2O_3 .... I-27
Figure 9. 4-pt. Flexural Strength vs. Grain Size of 30 vol% SiC(p)/Al_2O_3 .... I-28
Figure 10. 4-pt. Flexural Strength vs. Applied Load of Hot-Pressed SiC(p)/Al_2O_3 .... I-29
Figure 11. Biaxial Flexural Strength vs. Applied Load of Pressureless Sintered SiC(p)/Al_2O_3 .... I-30
Figure 12a. 3-pt. Flexural Strength of Pressureless Sintered TiB_2(p)/Al_2O_3 and TiC(p)/Al_2O_3 .... I-38
Figure 12b. 3-pt. Flexural Strength of Hot-Pressed TiB_2(p)/Al_2O_3 and TiC(p)/Al_2O_3 .... I-39
Figure 13. 3-pt. Effects of Pressure Processing on Flexural Strength of TiC(p)/Al_2O_3 .... I-40
Figure 14. 3-pt. Flexural Strength vs. Hot-Pressing Temperature of TiC(p)/Al_2O_3 .... I-41
Figure 15. 3-pt. Flexural Strength vs. Sintering Temperature of Sintered TiB_2(p)/Al_2O_3 .... I-42
Figure 16. 4-pt. Flexural Strength vs. Temperature of TiB_2(p)/Al_2O_3, TiC(p)/Al_2O_3, and TiN(p)/Al_2O_3 .... I-46
Figure 17a. 3-pt. Flexural Strength of Sintered ZrO_2(p)/Al_2O_3 .... I-51
Figure 17b. 3-pt. Flexural Strength vs. Sintering Temperature of ZrO_2(p)/Al_2O_3 .... I-53
Figure 18. 3-pt. Flexural Strength of Sintered ZrO_2(p)/Al_2O_3 .... I-54
Figure 19. 3-pt. Flexural Strength vs. Relative Density and Sintering Temperatures of Sintered ZrO_2(p)/Al_2O_3 Sol-Gels I-55
Figure 20. 4-pt. Flexural Strength of ZrO_2(p)/Al_2O_3 .... I-62
Figure 21. 4-pt. Flexural Strength and Relative Density vs. Sintering Temperature of 15 wt% ZrO_2(p)/Al_2O_3 .... I-63
Figure 22. Flexural Strength vs. Temperature of ZrO_2(p)/Al_2O_3 .... I-64
Figure 23. 4-pt. Flexural Strength of Sintered ZrO_2.HfO_2(p)/Al_2O_3.Cr_2O_3 .... I-67
Figure 24. 4-pt. Flexural Strength of Sintered ZrO_2(p)/$MgCr_2O_4$ I-68
Figure 25. 3-pt. Flexural Strength of Hot-Pressed SiC(p)/ZrO_2 .... I-71
Figure 26. 3-pt. Flexural Strength vs. Temperature of Hot-Pressed SiC(p)/YSZ I-72
Figure 27. 3-pt. Flexural Strength of Reaction Bonded Al_2O_3(p)/Si_3N_4 .... I-74
Figure 28. 4-pt. Flexural Strength vs. Temperature of $MoSi_2$(p)/Si_3N_4 .... I-75
Figure 29. 4-pt. Flexural Strength of Al_2O_3(p)/SiAlON I-77
Figure 30. 4-pt. Flexural Strength of Al_2O_3(p)/SiAlON I-78
Figure 31. 4-pt. Flexural Strength of Reaction Bonded SiC(p)/Si_3N_4 .... I-85
Figure 32. 4-pt. Flexural Strength of SiC(p)/Si_3N_4 and TiC(p)/Si_3N_4 .... I-86
Figure 33. 4-pt. Flexural Strength vs. Relative Density of SiC(p)/Si_3N_4 .... I-87
Figure 34. 4-pt. Flexural Strength vs. Sintering Temperature for HIPed SiC(p)/SiAlON.Si_3N_4 .... I-87

Figure 35. Ball-On-Ring Strength of Reaction Bonded SiC(p)/Si_3N_4 .... I-90
Figure 36. 3-pt. Flexural Strength of Reaction Bonded Si_3N_4(p)/SiC I-95
Figure 37. 3-pt. Flexural Strength of Hot-Pressed SiC(p)/Si_3N_4 and Sintered SiC(p)/SiAlON I-96
Figure 38. Flexural Strength of TiN(p)/Si_3N_4 .... I-103
Figure 39. 4-pt. Flexural Strength vs. Temperature of RBSN and 10 vol% BN(p)/RBSN I-104
Figure 40. 3-pt. Flexural Strength vs. Indention Load of Pressureless Sintered BN(p)/Si_3N_4 .... I-105
Figure 41. 3-pt. Flexural Strength of Hot-Pressed SiC(p)/AlN I-111
Figure 42. 3-pt. Flexural Strength vs. Relative Density of 25 wt% Reaction Bonded AlN(p)/BN I-112
Figure 43. 3-pt. Flexural Strength of Hot-Pressed TiN(p), TiO_2(p), and TiC(p) in AlN Matrix I-113
Figure 44. 4-pt. Flexural Strength vs. Temperature of Hot-Pressed SiC(p)/AlN I-114
Figure 45. 3-pt. Flexural Strength of HIPed 2Y-ZrO_2(p)/TiB_2 .... I-118
Figure 46. 4-pt. Flexural Strength of SiC(p)/ZrB_2 and B_4C(p)/TiB_2 .... I-119
Figure 47. 4-pt. Flexural Strength of ZrO_2(p)/ZrC and Zr(p).ZrB_2(pl)/ZrC I-122
Figure 48. 3-pt. Flexural Strength of TiC(p)/SiC I-130
Figure 49. 3-pt. Flexural Strength of TiC(p)/SiC at 1500 °C I-132
Figure 50a. Effect of Sintering Additives on Elevated Temperature Flexural Strength of 10, 18.5, and 30 wt% TiC(p)/SiC I-133
Figure 50b. 3-pt. Flexural Strength vs. Temperature of 50 wt% TiC(p)/SiC With Various Sintering Additives I-134
Figure 51. 3-pt. Flexural Strength vs. Hot-Press Temperature of TiC(p)/SiC I-135
Figure 52. 4-pt. Flexural Strength of Hot-Pressed TiC(p)/SiC I-139
Figure 53. 4-pt. Flexural Strength vs. Temperature of Hot-Pressed TiC(p)/SiC I-140
Figure 54. 4-pt. Flexural Strength vs. Temperature of TiB_2(p)/SiC I-146
Figure 55. 4-pt. Flexural Strength of TiB_2(p)SiC I-148
Figure 56. 3-pt. Flexural Strength of B_4C(p)/Graphite I-151

Fracture Toughness

Figure 57. Indentation Microfracture (IM) Fracture Toughness of SiC(p)/MgO II-2
Figure 58. Fracture Toughness (IM) of ZrO_2(p)/Mullite and Cordierite II-8
Figure 59. Fracture Toughness[I-14] (IM) vs. Relative Density of Zirconia-Mullite II-10
Figure 60. Fracture Toughness[I-23] vs. Stabilizer Content of ZrO_2(p)/Mullite II-12
Figure 61. Fracture Toughness of SiAlON(p)/Mullite II-14
Figure 62. Fracture Toughness vs. Sintering Aid Content of ZrO_2(p)/ZrO_2.Mullite and Al_2O_3(p)/ZrO_2 .... II-16
Figure 63. Fracture Toughness of TiC(p)/Al_2O_3, TiB_2(p)/Al_2O_3, and SiC(p)/Al_2O_3 .... II-22
Figure 64. Indentation Fracture (IM) Fracture Toughness of TiB_2(p)/Al_2O_3 and SiC(p)/Al_2O_3 .... II-32
Figure 65. Fracture Toughness of TiB_2, TiC, and TiN(p)/Al_2O_3 .... II-34
Figure 66. Effect of Processing Conditions on Fracture Toughness of TiC(p)/Al_2O_3 .... II-35
Figure 67. Fracture Toughness vs. Temperature of DIMOX Produced SiC(p)/Al_2O_3 .... II-36
Figure 68. Chevron Notched Short Rod Fracture Toughness[I-7], [I-9] of TiB_2(p)/Al_2O_3 .... II-41
Figure 69. Chevron Notched Short Rod Fracture Toughness of Diamond, B_4C, and TiB_2 Particulate Reinforced Alumina II-42
Figure 70a. Indentation Fracture Toughness[I-23] of 2Y-ZrO_2(p)/Al_2O_3 .... II-50
Figure 70b. Fracture Toughness of ZrO_2(p)/Al_2O_3 .... II-52

Figure 70c. Fracture Toughness of ZrO_2(p)/Al_2O_3 ..... II-54
Figure 71. Effect of Processing Conditions on Indentation Fracture Toughness[I-23] of ZrO_2(p)/Al_2O_3 ..... II-56
Figure 72a. Fracture Toughness of SiC(p).ZrO_2(p)/Mullite II-61
Figure 72b. Fracture Toughness of SiC(p).ZrO_2(p)/Al_2O_3, TiC(p).ZrO_2(p)/Al_2O_3, and SiC(p).ZrO_2(p)/Mullite II-62
Figure 73. Fracture Toughness[I-15] of ZrO_2.HfO_2(p)/Al_2O_3.Cr_2O_3 Ceramics II-65
Figure 74. SENB Fracture Toughness of Hot-Pressed SiC(p)/ZrO_2 ..... II-67
Figure 75. Fracture Toughness[I-20] of ZrO_2(p)/ZrC II-69
Figure 76. Fracture Toughness of SiC(p)/Si_3N_4 and SiAlON II-74
Figure 77. Fracture Toughness of SiC(p)/Si_3N_4 ..... II-76
Figure 78a. Fracture Toughness of SiC(p) and TiC(p)/Si_3N_4 ..... II-87
Figure 78b. Fracture Toughness[I-25] of SiC(p)/Si_3N_4 ..... II-88
Figure 79. Effect of Process Conditions on Fracture Toughness of SiC(p)/SiAlON II-90
Figure 80. Fracture Toughness vs. Grain Size of TiN(p)/Si_3N_4 ..... II-91
Figure 81. Fracture Toughness vs. Temperature of SiC(p)/Si_3N_4, BN(p)/Si_3N_4, and TiC(p)/Si_3N_4 ..... II-98
Figure 82. Fracture Toughness of SiC(p)/Si_3N_4, BN(p)/Si_3N_4, and TiC(p)/Si_3N_4 ..... II-100
Figure 83. Fracture Toughness (IM) of TiC(p) and TiN(p)/Si_3N_4 ..... II-104
Figure 84. Fracture Toughness of TiC(p)/SiC II-112
Figure 85. Fracture Toughness[I-43] vs. Temperature of TiC(p)/SiC II-114
Figure 86. Fracture Toughness of TiB_2(p)/SiC II-119
Figure 87. Fracture Toughness[I-18], [I-26] vs. Temperature of TiB_2(p)/SiC II-120
Figure 88. Fracture Toughness of Diamond(p)/SiC and TiB_2(p)/SiC II-123
Figure 89. Fracture Toughness of YSZ(p)/TiB_2, Al_2O_3(p)/TiB_2, B_4C(p)/TiB_2, SiC(p)/ZrB_2 ..... II-126
Figure 90. Single Edge Pre-Cracked Beam Fracture Toughness vs. Sintering Additive of 20 wt% YSZ(p)/TiB_2 ..... II-127

<u>Elastic Properties</u>

Figure 91. Young's Modulus of ZrO_2(p)/Al_2O_3 ..... III-5
Figure 92. Young's Modulus vs. Relative Density of Sintered ZrO_2(p)/Mullite III-6
Figure 93. Young's Modulus of Hot-Pressed SiC(p)/Al_2O_3 ..... III-10
Figure 94a. Young's Modulus of Hot-Pressed TiB_2, TiC, and TiN Particulate Reinforced Alumina III-16
Figure 94b. Young's Modulus of SiC or TiC Particulate Reinforced Zirconia-Alumina III-17
Figure 95. Young's Modulus vs. ZrO_2 Content of Sintered ZrO_2(p).HfO_2(p)/Al_2O_3.Cr_2O_3 ..... III-19
Figure 96. Young's Modulus of TiC(p)/Si_3N_4 ..... III-23
Figure 97. Young's Modulus vs. HIP Temperature of SiC(p)/SiAlON III-24
Figure 98. Young's Modulus of Reaction Bonded SiC(p)/Si_3N_4 ..... III-32
Figure 99. Young's Modulus vs. Temperature of Reaction Bonded SiC(p)/Si_3N_4 ..... III-33
Figure 100. Young's Modulus of SiC(p)/Si_3N_4 ..... III-34
Figure 101. Young's Modulus vs. Temperature of Reaction Bonded BN(p)/Si_3N_4 ..... III-37
Figure 102. Young's Modulus of Hot-Pressed TiN(p)/Si_3N_4 ..... III-38
Figure 103. Young's Modulus vs. Grain Size of Sintered TiN(p)/Si_3N_4 ..... III-39
Figure 104. Young's Modulus of BN(p), TiB_2(p), and TiC(p) in SiC Matrix III-44
Figure 105. Poisson's Ratio of TiB_2(p) and TiC(p) in SiC Matrix III-45
Figure 106. Young's Modulus of Zr(p).ZrB_2(pl)/ZrC and SiC(p)/ZrB_2 ..... III-48
Figure 107. Shear Modulus of Zr(p).ZrB_2(pl)/ZrC III-49

Figure 108. Poisson's Ratio for $SiC(p)/ZrB_2$ III-49

Hardness

Figure 109. Vickers Hardness vs. Temperature of $SiC(p)/Al_2O_3$ IV-3
Figure 110. Hardness vs. Relative Density of $TiB_2(p)/Al_2O_3$, $TiN(p)/Al_2O_3$, and $TiC(p)/Al_2O_3$ IV-10
Figure 111. Vickers Hardness of $TiC(p)/Al_2O_3$, $TiN(p)/Al_2O_3$, and $TiB_2(p)/Al_2O_3$ IV-12
Figure 112. Effect of Post Sinter HIPing on the Vickers Hardness of $TiC(p)/Al_2O_3$ IV-13
Figure 113. Vickers Hardness of $TiN(p)/Al_2O_3$ IV-14
Figure 114a. Vickers Hardness of $ZrO_2(p)/Al_2O_3$ IV-18
Figure 114b. Vickers Hardness of $ZrO_2(p)/Al_2O_3$ IV-20
Figure 115. Vickers Hardness of $SiC(p)/ZrO_2–Al_2O_3$ and $TiC(p)/ZrO_2-Al_2O_3$ IV-23
Figure 116. The Effects of Processing and Sintering Aids on $TiC(p)/ZrO_2-Al_2O_3$ and $SiC(p)/ZrO_2-Al_2O_3$ IV-24
Figure 117a. Vickers Hardness of $SiC(p)/Si_3N_4$ (Reaction Bonded) IV-32
Figure 117b. Vickers Hardness of $SiC(p)/Si_3N_4$ (Conventional Methods) IV-33
Figure 117c. Vickers Hardness of $SiC(p)/Si_3N_4$ (Reaction Processed and Polymer Derived) IV-34
Figure 118. Vickers Hardness of $SiC(p)/Si3N_4$ Composites from SHS Derived Powders IV-35
Figure 119. Vickers Hardness of $TiN(p)/Si_3N_4$ and $TiC(p)/Si_3N_4$ Composites IV-39
Figure 120. Vickers Hardness of TiN(p)/AlN, $TiO_2(p)/AlN$, and TiC(p)/AlN IV-41
Figure 121. Vickers Hardness of TiC and $TiB_2(p)/SiC$ IV-44
Figure 122. Vickers Hardness of $2Y-ZrO_2(p)/TiB_2$ and $B_4C(p)/TiB_2$ IV-47

Thermal Properties

Figure 123. Diamond Particle Size Dependence of Elevated Temperature Thermal Conductivity of a Diamond(p)/Cordierite Composite V-4
Figure 124. Effect of Diamond Particle Radius on Thermal Conductivity of a 15 vol% Diamond(p)/Cordierite Composite V-5
Figure 125. Specific Heat vs. Temperature of Cordierite and a 15 vol% Diamond(p)/Cordierite Composite V-6
Figure 126. Thermal Diffusivity of SiC(p)/MgO and Al_2O_3 ..... V-12
Figure 127. Thermal Diffusivity of SiC(p)/BeO V-13
Figure 128. Thermal Diffusivity vs. Temperature of SiC(p)/BeO V-14
Figure 129. Thermal Conductivity of $SiC(p)/Al_2O_3$ V-15
Figure 130. Porosity Dependence of the Thermal Conductivity at Room Temperature of SiC Particulate Reinforced Oxides V-16
Figure 131. Temperature Dependence of the Mean Thermal Expansion Coefficient of SiC(p)/ ZrO_2-Mullite V-19
Figure 132. Thermal Expansion Behavior of Alumina-Aluminum Titanate Heated and Cooled from 650 °C and 1250 °C V-23
Figure 133. Thermal Diffusivity of DIMOX Alumina-Aluminum Titanate Composites Heated to and Cooled from A) 600 °C, B) 800 °C, C) 1000 °C, D) 1100 °C, E) 1200 °C V-24
Figure 134. Thermal Diffusivity of a DIMOX Alumina-Aluminum Titanate Composite After Thermal Cycling from RT to 1200 °C V-26
Figure 135. Specific Heat of DIMOX Alumina-Aluminum Titanate Composites V-26

Figure 136. Mean Thermal Expansion Coefficients of Diamond, TiN, TiB_2, and TiC Particulate Reinforced Alumina V-29
Figure 137. Thermal Conductivity vs. Temperature of DIMOX™ $SiC(p)/Al_2O_3$, $SiC(p)/ZrO_2$, $SiC(p)/ZrO_2$-Mullite, and $ZrO_2.HfO_2(p)/Al_2O_3.Cr_2O_3$ V-36
Figure 138. Thermal Diffusivity of Zirconia-Hafnia/Chromia-Alumina Composites vs. Chromia Content V-37
Figure 139. Thermal Diffusivity vs. Temperature of Alumina-Chromia Solid Solutions With or Without ZrO_2 Particulate Reinforcement V-38
Figure 140. Thermal Expansion Coefficient of $TiO_2(p)/HfO_2$ V-41
Figure 141. Temperature Dependence of the Mean Thermal Expansion Coefficient of $SiC(p)/Y-ZrO_2$ V-42
Figure 142. Thermal Expansion Coefficient of Particulate Reinforced Silicon Nitride V-47
Figure 143. Effect of Sintering Additives on the CTE of a Reaction Bonded AlN/BN Composite V-48
Figure 144a. Thermal Diffusivity of SiC/AlN Composites V-51
Figure 144b. Thermal Conductivity of SiC/AlN Composites V-52
Figure 144c. Thermal Diffusivity vs. Temperature of SiC/AlN Composites V-53
Figure 145a. Thermal Conductivity vs. Sintering Aid Content of Hot-Pressed 25 vol% AlN(p)/BN V-56
Figure 145b. Thermal Conductivity vs. Relative Density of Reaction Bonded 25 vol% AlN(p)/BN V-57
Figure 146a. Thermal Conductivity of $BN(p)/B_4C$ V-61
Figure 146b. Thermal Conductivity vs. Temperature of $BN(p)/B_4C$ V-62
Figure 147. Thermal Diffusivity of BN(p)/SiC V-63
Figure 148. Thermal Conductivity vs. Temperature of $B_4C(p)$/Graphite V-68
Figure 149. Thermal Diffusivity vs. Temperature of BN(p)/SiC and $TiB_2(p)/SiC$ V-69
Figure 150. Mean Thermal Expansion Coefficient of $Zr(p).ZrB_2(pl)/ZrC$ V-73
Figure 151. Thermal Expansion Coefficients of Particulate Reinforced TiB_2 and ZrB_2 .... V-74
Figure 152. Thermal Conductivity of $SiC(p)/ZrB_2$, and Diamond(p)/SiC V-76

9. Curve Fitting Equation Index

0004 $y = a + bx^2$

0008 $y = a + be^x$

0014 $y = a + \frac{b}{\ln x}$

0016 $y = a + b\frac{\ln x}{x}$

0024 $\ln y = a + b\sqrt{x}$

0046 $\frac{1}{y} = a + bx^2$

0047 $\frac{1}{y} = a + bx^2 \ln x$

0052 $\frac{1}{y} = a + b(\ln x)^2$

0063 $\frac{1}{y} = a + be^{-x}$

0085 $y^2 = a + bx$

0087 $y^2 = a + bx\sqrt{x}$

0088 $y^2 = a + bx^2$

1025 $y = a + bx\ln x + cx^3$

1091 $y = a + bx^2\sqrt{x} + cx^3$

1147 $y = a + b(\ln x)^2 + c\ln x$

1155 $y = a + b(\ln x)^2 + ce^{-x}$

1301 $\ln y = a + bx^2\sqrt{x} + cx^3$

1342 $\ln y = a + be^x + ce^{-x}$

1368 $\ln y = a + b\frac{x}{\ln x} + \frac{c}{\ln x}$

1494 $\frac{1}{y} = a + bx^2 + ce^{-x}$

1496 $\frac{1}{y} = a + bx^2 \ln x + cx^3$

1511 $\frac{1}{y} = a + bx^2\sqrt{x} + cx^3$

1602 $\frac{1}{y} = a + b\ln x + ce^{-x}$

2656 $y = a + bx^2 \ln x + cx^2\sqrt{x} + d\sqrt{x}$

4173 $y = a + bx + cx\sqrt{x} + dx^2 + ex^2\sqrt{x}$

4189 $y = a + bx + cx\sqrt{x} + dx^2 \ln x + ex^2\sqrt{x}$

4288 $y = a + bx + cx\sqrt{x} + d\sqrt{x} + ee^{-x}$

4341 $y = a + bx + cx^2 + dx^2\sqrt{x} + ex^3$

4492 $y = a + bx + cx^2 \ln x + de^x + e\frac{x}{\ln x}$

6102 $\ln y = a + bx + cx^2 + dx^3 + ex^4$

6131 $y^2 = a + bx + cx^2$

7001 $y^2 = \frac{a + cx}{1 + bx}$

7002 $y = \frac{a + cx}{1 + bx + dx^2}$

7104 $\ln y = \frac{a + cx + ex^2}{1 + bx + dx^2 + fx^3}$

10. Flexural Strength
(Section I)

SiC(p)/MgO

Reference: 77

Composition
0-50 vol% SiC

Processing Method
Hot-Pressed

Detailed Description
The particles were mixed by ball-milling in highly pure toluene with Al_2O_3 balls for 24 hours. The dried mixtures were hot-pressed at 1900 °C and 28 MPa in an N_2 atmosphere.

Reinforcement Description
β-SiC particulates

Reinforcement Supplier
Ibiden Co., Ltd.

Reinforcement Dimensions
ave. particle size < 0.3 μm

Matrix Description
MgO particulates

Matrix Supplier
Ube Industries, Ltd., Japan

Matrix Properties
ave. particle size = 0.1 μm
ave. grain size = 2 - 3 μm

Test Type
3-pt. Flexural

Specimen Preparation
The tensile surface was polished to 1 μm with diamond paste and the tensile edges were beveled.

Specimen Dimensions
3 x 4 x 45 mm
span = 30 mm

Test Description
crosshead speed = 0.5 mm/min

Test Temperature
RT

Reinforcement Content (volume percent)	Flexural Strength (MPa)	Range (MPa)
0	349	+42/-49
5	448	+50/-46
10	487	+24/-31
20	548	+33/-40
30	532	+69/-47
50	632	+44/-41

Remarks/Comments
The strength of MgO was increased with increasing vol% of SiC particulates. The composites were hot-pressed to >99% theoretical density. SiC particles that were usually less than 0.2 μm were predominately dispersed within the MgO grains with some larger SiC particles located at the grain boundaries. High resolution transmission electron microscopy revealed that all the SiC particulates within the MgO grains were irregular shaped.

SiC(p)/MgO

Reference: 77

Composition
0 - 30 vol% SiC

Processing Method
Hot-Pressed

Detailed Description
The particles were mixed by ball-milling in highly pure toluene with Al_2O_3 balls for 24 hours. The dried mixtures were hot-pressed at 1500-1900 °C and 28 MPa in an N_2 atmosphere.

Reinforcement Description
β-SiC particulates

Reinforcement Supplier
Ibiden Co., Ltd.

Reinforcement Dimensions
ave. particle size < 0.3 μm

Matrix Description
MgO particulates

Matrix Supplier
Ube Industries, Ltd., Japan

Matrix Properties
ave. particle size = 0.1 μm
ave. grain size = 2 - 3 μm

Test Type
3-pt. Flexural

Specimen Preparation
The tensile surface was polished to 1 μm with diamond paste and the tensile edges were beveled.

Specimen Dimensions
3 x 4 x 45 mm
span = 30 mm

Test Description
crosshead speed = 0.5 mm/min

Test Temperature
RT-1400 °C

Test Environment
Air

Temperature (°C)	Flexural Strength (MPa)	Reinforcement Content (volume percent)
20	352	0
600	304	0
800	273	0
1,000	219	0
1,200	185	0
1,400	128	0
20	555	10
800	672	10
1,000	733	10
1,200	703	10
1,400	527	10
20	691	30
800	652	30
1,000	747	30
1,200	735	30
1,400	512	30

Remarks/Comments
The strength for the composites increased as the temperature reached 1000 °C. Thereafter, the strength of the composites decreased. SiC particles that were usually less than 0.2 μm were predominately dispersed within the MgO grains with some larger SiC particles located at the grain boundaries. High resolution transmission electron microscopy revealed that all the SiC particulates within the MgO grains were irregular shaped.

Figure 1: 3-pt. Flexural Strength of Hot-Pressed SiC(p)/MgO

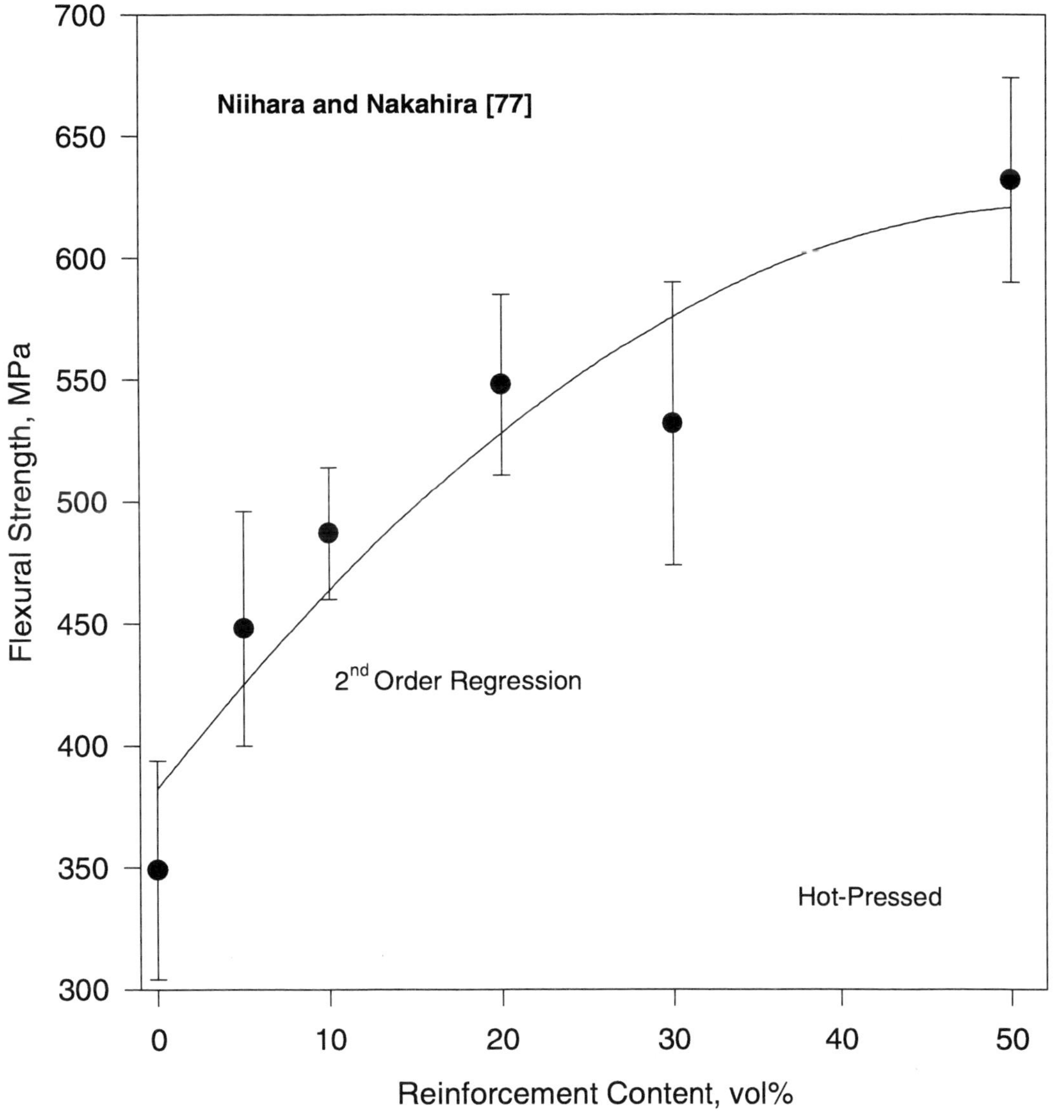

Discussion: The flexural strength of the MgO composite reached over 600 MPa with the addition of 50 vol% SiC particulates. The strength increased with increasing reinforcement content due to the particulates causing crack deflection. This crack deflection, which was caused by hoop stresses of over 1000 MPa around the reinforcement phase, changed the mode of failure from intergranular to transgranular fracture [77]. Large thermal expansion mismatches generated the tensile hoop stress that caused the change in fracture mode. This high tensile hoop stress did not produce microcracks because the SiC particulates were much smaller than the critical particle size required for microcracking [77].

The critical flaw size for the monolithic MgO, about 9 μm, was approximately equal to the critical flaw size for the 10 vol% SiC(p)/MgO composite, about 10 μm. Thus, the increase in strength of the SiC(p)/MgO composite at room temperature was a consequence of the improved toughness by the SiC particles dispersed through the matrix and not a decrease in flaw size.

Figure 2: 3-pt. Flexural Strength of Hot-Pressed SiC(p)/MgO

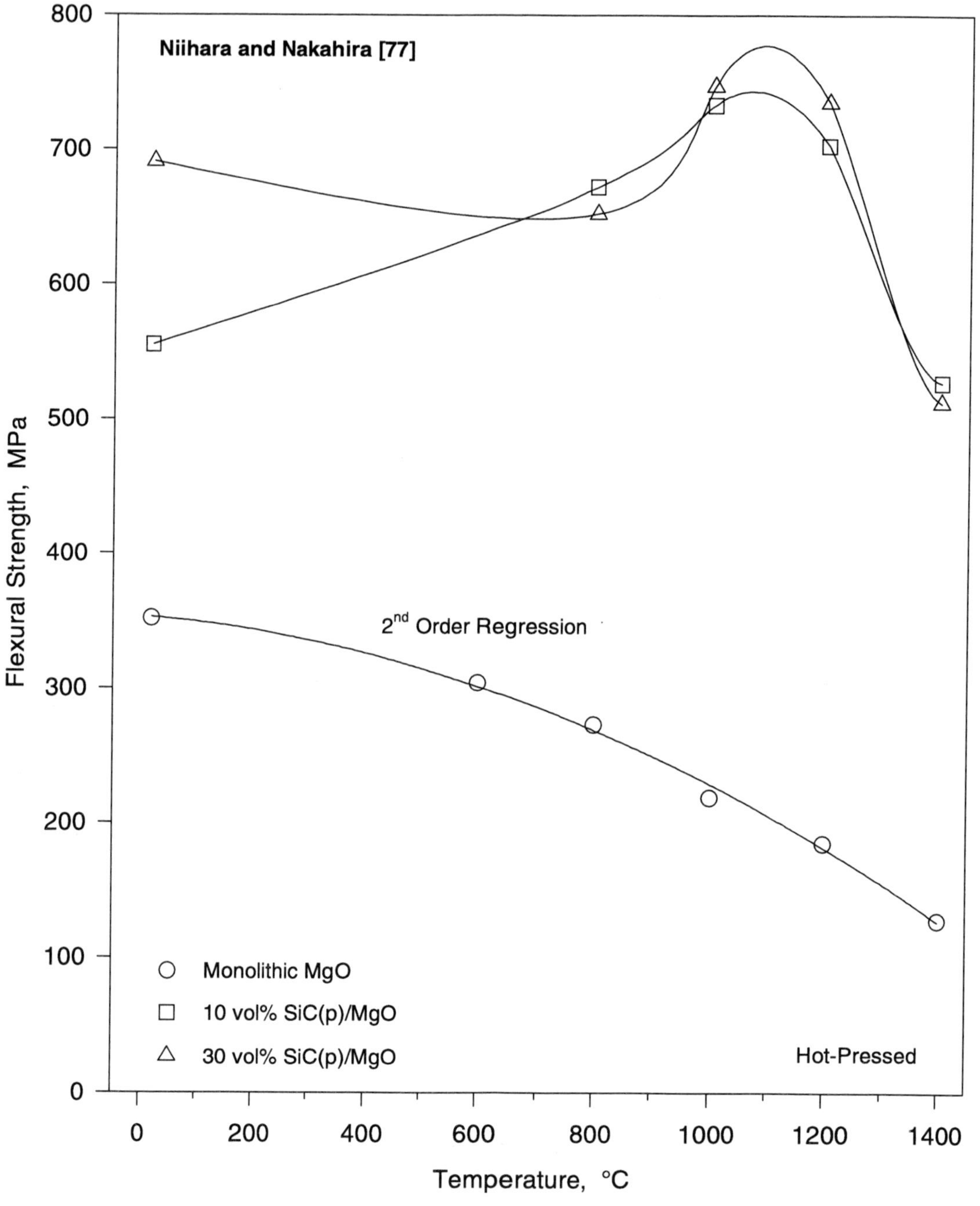

Discussion: Niihara and Nakahira *et al.* [77, 103] tested the elevated temperature flexural strength of SiC(p)/MgO nanocomposites. In the temperature range 1000 to 1400 °C, the 10 and 30 vol% SiC(p)/MgO composites had a considerably higher flexural strength than the monolithic MgO at room temperature [77]. The SiC particulate dispersion within the matrix grains promoted transgranular fracture rather than intergranular fracture at both room temperature and at elevated temperature. The transgranular fracture mode, which resulted in diminished grain boundary impurity effects at elevated temperatures, improved high temperature mechanical properties [103]. The localized tensile stress around the nano-size particles induced the transgranular fracture and then toughening occurred by crack-tip deflection by the hard particles [103]. Thus, it may be concluded that the observed increase in strength at high temperatures was mainly due to the prohibition of grain boundary sliding or cavitation by dispersions within the matrix grains; this promoted transgranular fracture even at high temperatures [77]. Another possible reason for the high temperature strength increase would be the dislocation pinning by the fine SiC dispersions within the matrix grains of the composite. This mechanism was suggested by the fact that the hot-hardness, brittle-to-ductile transition temperature and creep resistance were significantly improved by the nanodispersoids [103]. The nano-size SiC particles pin the dislocation motion at high temperatures and results in improved mechanical properties.

It is possible to explain the decrease in flexural strength at high temperatures in terms of the thermal expansion mismatch. The large thermal expansion mismatch between the matrix and the reinforcement, at a critical temperature, will cause residual stresses in excess of 1000 MPa around the SiC. This will cause crack deflection and initially result in a higher flexural strength up to approximately 1200 °C. However, after this critical temperature, the difference in the thermal expansion caused critical size flaws within the composite that resulted in a decreased strength. In addition, the sudden strength decrease above 1200 °C might also be a result of some grain boundary sliding and/or cavitation at high temperature caused by the softening of grain boundary phases and/or matrix creep. This phenomenon was partially demonstrated by the monolithic MgO which showed a steady decrease in strength with increasing temperature.

The room temperature strengths presented in figure 2 do not correspond to the room temperature strengths presented in figure 1 for the 0, 10, and 30 vol% SiC(p)/MgO composites. The reason for this is not known.

ZrO_2(p)/Mullite

Reference: 30

Composition
0-40 vol% ZrO_2

Composite Architecture
particulate reinforcement

Processing Method
Sol-Gel/Sintered

Detailed Description
Precursor solution with a stoichiometric mullite composition was prepared by partial hydrolysis. Composite powders were calcined at 800 °C for 12 hours and uniaxially pressed at 120 MPa followed by isostatic pressing at 300 MPa. The compacts were sintered at 1400-1700 °C for 2 hours.

Reinforcement Description
zirconia-mullite composite powders

Reinforcement Designation
silica sol: TEOS ($Si(OC_2H_5)_4$)
alumina sol: aluminum i-propoxide ($Al(OC_3H_7)_3$)
zirconia sol: zirconium i-propoxide ($Zr(OC_3H_7)_3$)

Matrix Description
(see reinforcement description)

Test Type
4-pt. Flexural

Specimen Dimensions
3 x 4 x 36 mm

Test Temperature
RT

Reinforcement Content (volume percent)	Flexural Strength (MPa)	Relative Density (percent)	Sintering Temperature (°C)
0	91	79	1,400
0	110	89	1,500
0	116	90	1,600
0	126	92	1,700
10	82	88	1,400
10	62	82	1,600
20	98	83	1,400
20	155	92	1,500
40	153	89	1,400
40	167	92	1,500
40	142	92	1,600

Remarks/Comments
The lower density of the sintered compacts was ascribed to a lower green density of the precursor compact of about 40 %. Crystallization of the tetragonal ZrO_2 and mullite at lower temperatures also hinders densification. This transformation also toughened the composite. The flexural strength increased with increasing reinforcement content.

ZrO_2(p)/Mullite

Reference: 50

Composition

0 and 30 wt% ZrO_2

Processing Method

CIPed/Reaction Sintered

Detailed Description

The sols in the solution (pH ~ 3.0 and 5 wt% in concentration), mixed for one hour, were co-precipitated with diluted aqueous ammonia. All gels were dispersed and filtered in methanol and acetone then subsequently dried in acetone with a rotary evaporator. After vacuum-drying at 60 °C and in an oven at 120 °C, the powders were milled in methanol for 6 hours. The powders were calcined for 30 minutes at 1150 °C for the composites and 1250 °C for the mullite powders. The calcined powders were then milled in methanol for 48 hours in a nylon vibration mill with Si_3N_4-cylinders, dried, and screened (150-mesh) for granulation. All powders were calcined at 800 °C for 1 hour in order to burn out organic inclusions. Green bodies were formed by die-pressing at 30 MPa and subsequent cold isostatically pressing at 200 MPa which were then sintered in air between 1550 to 1650 °C for 2 or 4 hours.

Sintering Additives

0 or 3 mol% Y_2O_3

Reinforcement Description

zirconia-mullite composite powder

Reinforcement Supplier

silica sol: Eka Nobel AB
alumina sol(A), zirconia, and yttria sol: Nyacol Products Inc.
alumina sol(B): Albright & Wilson

Supplier Designation

silica sol: Bindzil 15-500
alumina sol(A): Nyacol AL20
alumina sol(B): Alumina Chlorohydrate

Reinforcement Typical Properties

(A) produced from pure inorganic sols of silica (50 Å), alumina(A) (200 Å), zirconia (75 Å), and yttria (100 Å)
(B) produced from silica and alumina(B) (10 Å) sols with aqueous solution of $ZrO(NO_3)_3$ and $Y(NO_3)_3$

Reinforcement Dimensions

ave. particle size (0 % Y_2O_3) = 1.02 μm
ave. particle size (3 % Y_2O_3) = 0.99 μm

Matrix Description

(see reinforcement description)

Matrix Properties

composition of mullite:
(A) 72.1 wt% Al_2O_3
(B) 72.9 wt% Al_2O_3

Test Type

4-pt. Flexural

Specimen Preparation

The sintered plates were cut and ground with a #400 diamond wheel.

Specimen Dimensions

3 x 4 x 40 mm
inner span = 10 mm
outer span = 30 mm

Test Temperature

RT-1300 °C

4-pt. Flexural Strength

Temperature (°C)	Flexural Strength (MPa)	Range (MPa)	Sintering Temperature (°C)	Relative Density (percent)	Reinforcement Content (weight percent)	Comments
0 mol% Y_2O_3						
22	375	-32/+41	1,650	98.5	0	A, 4 hour sinter
1,100	307	-23/+10	1,650	98.5	0	A, 4 hour sinter
1,200	320	-36/+45	1,650	98.5	0	A, 4 hour sinter
1,300	333		1,650	98.5	0	A, 4 hour sinter
22	311	-26/+29	1,650	96.2	0	A, 2 hour sinter
1,100	306	-63/+55	1,650	96.2	0	A, 2 hour sinter
1,200	304	-41/+43	1,650	96.2	0	A, 2 hour sinter
1,300	312	±30	1,650	96.2	0	A, 2 hour sinter
22	340	±17	1,650	99.2	0	B, 4 hour sinter
1,100	316	±28	1,650	99.2	0	B, 4 hour sinter
1,200	305	±17	1,650	99.2	0	B, 4 hour sinter
1,300	329		1,650	99.2	0	B, 4 hour sinter
22	375		1,650	98.0	0	B, 2 hour sinter
1,200	348		1,650	98.0	0	B, 2 hour sinter
22	335		1,630	97.6	0	B, 4 hour sinter
1,200	300		1,630	97.6	0	B, 4 hour sinter
----	----	----	----	----	----	----
22	344		1,650	98.5	30	A, 2 hour sinter
1,200	266		1,650	98.5	30	A, 2 hour sinter
22	396		1,600	98.2	30	A, 4 hour sinter
1,200	347	-19/ + 44	1,600	98.2	30	A, 4 hour sinter
22	459	±60	1,600	96.2	30	A, 2 hour sinter
1,200	305	±30	1,600	96.2	30	A, 2 hour sinter
22	419	-34/+36	1,650	99.0	30	B, 2 hour sinter
1,100	369	-41/+28	1,650	99.0	30	B, 2 hour sinter
1,200	333	-20/+43	1,650	99.0	30	B, 2 hour sinter
1,300	323		1,650	99.0	30	B, 2 hour sinter
22	408	-43/+38	1,600	97.7	30	B, 4 hour sinter
1,200	340	-14/+8	1,600	97.7	30	B, 4 hour sinter
22	396	-17/+20	1,600	94.2	30	B, 2 hour sinter
1,200	363	-17/+46	1,600	94.2	30	B, 2 hour sinter
3 mol% Y_2O_3						
22	376	-21/+24	1,600	97.5	30	A, 4 hour sinter
1,100	378	±27	1,600	97.5	30	A, 4 hour sinter
1,200	378	±67	1,600	97.5	30	A, 4 hour sinter
1,300	255		1,600	97.5	30	A, 4 hour sinter
22	360	-24/+31	1,600	97.8	30	A, 2 hour sinter
1,100	335	-12/+9	1,600	97.8	30	A, 2 hour sinter
1,200	363		1,600	97.8	30	A, 2 hour sinter
1,300	274	-44/+37	1,600	97.8	30	A, 2 hour sinter
22	406	±17	1,600	97.4	30	B, 4 hour sinter
1,200	414	-27/+17	1,600	97.4	30	B, 4 hour sinter
22	453		1,600	97.9	30	B, 2 hour sinter
1,200	451		1,600	97.9	30	B, 2 hour sinter

22	564	-64/+74	1,550	99.0	30	B, 4 hour sinter
1,100	452	-29/+29	1,550	99.0	30	B, 4 hour sinter
1,200	500	-56/+57	1,550	99.0	30	B, 4 hour sinter
1,300	348		1,550	99.0	30	B, 4 hour sinter

Remarks/Comments

Data shows the variation of the different mullite-based materials fabricated by different processing conditions and tested at different temperatures.

Figure 3: 4-pt. Flexural Strength of ZrO_2(p)/Mullite

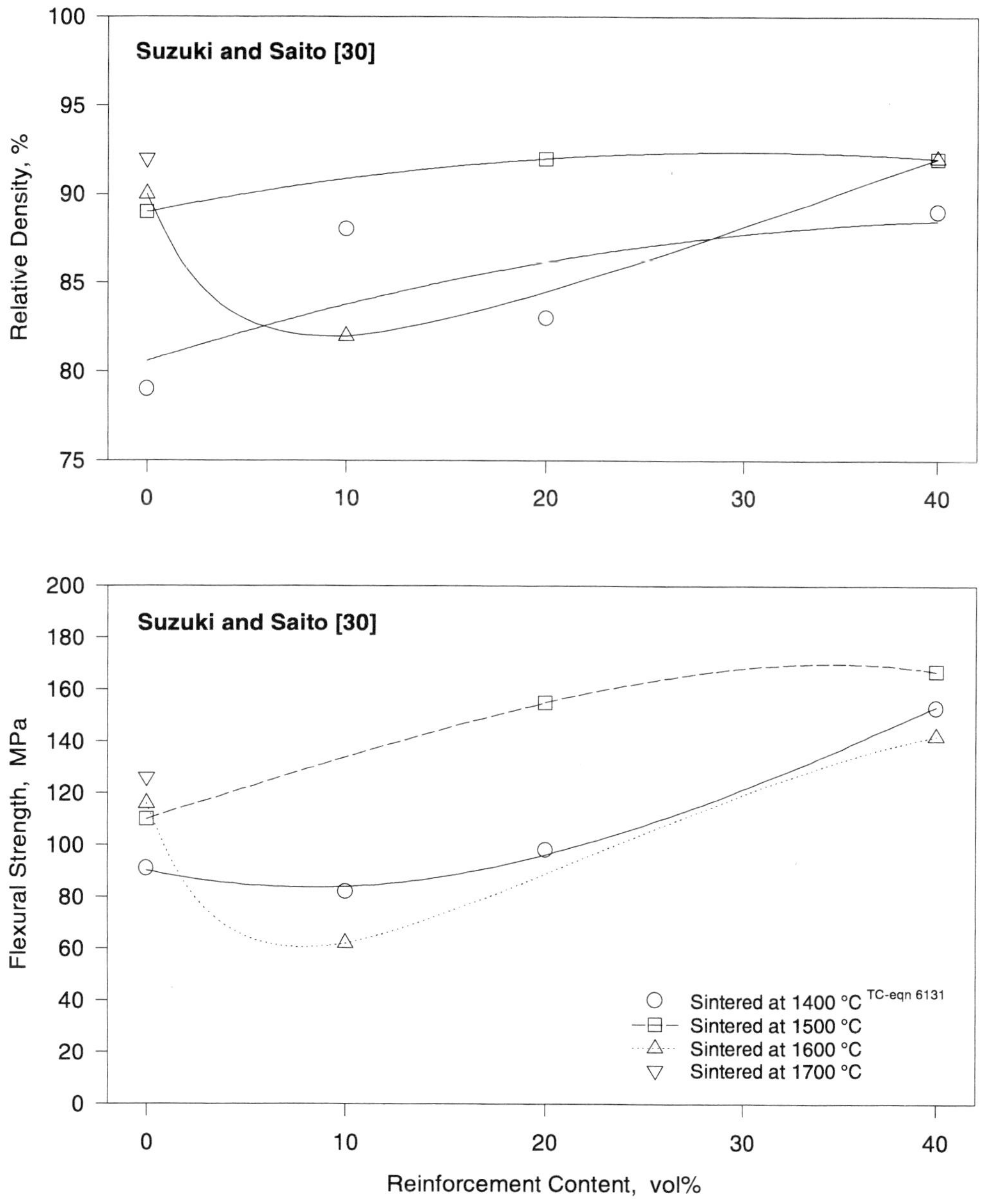

Discussion: Suzuki and Saito [30] tested the flexural strength of zirconia-mullite composites fabricated at different sintering temperatures. Optimal sintering temperature was reached at 1500 °C. Below this temperature, the samples were not dense enough for a high flexural strength. Above this temperature, a glass phase formed along the grain boundaries. This glass phase formation caused a low flexural strength. As the reinforcement content increased for the samples sintered at high temperatures, strengthening started to occur due to the ZrO_2 phase change. Samples were only densified up to 92 % relative density. This also affected the total overall strength. The low relative density was attributed to a low green density caused by large agglomerates formed from the fine powder particles. It was also attributed to the premature crystallization of zirconia and mullite at low temperatures. Both factors hindered the sintering process.

Figure 4: 4-pt. Flexural Strength vs. Temperature of ZrO_2(p)/Mullite Processed From Sol-Gel Derived Powders

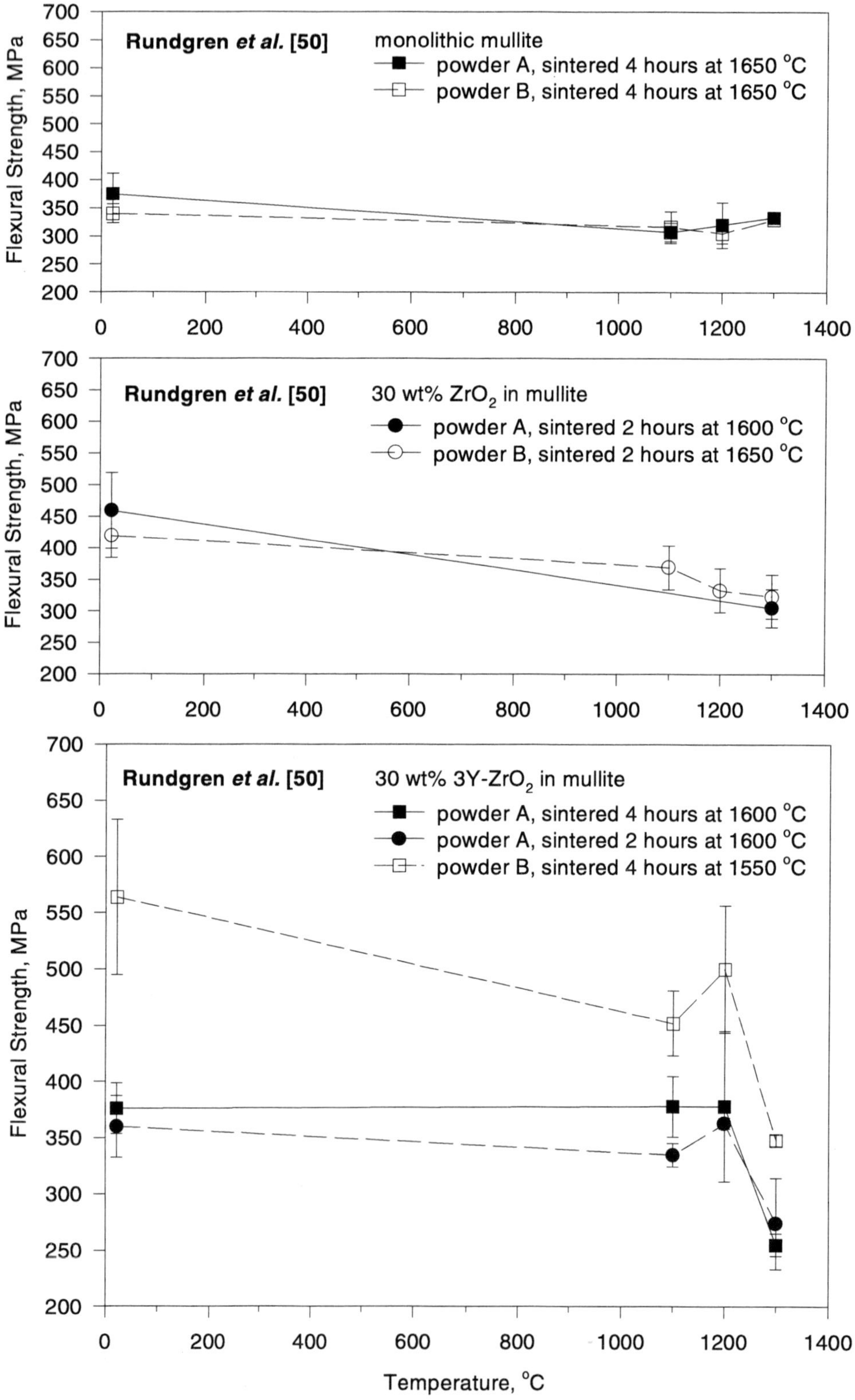

Discussion: Rundgren *et al.* [50] examined the use of powders derived from commercially available sols and salts to make mullite and zirconia-mullite composites. Two types of powders were used. Powder A (colloidal powder) was made from pure inorganic sols of silica, alumina, and yttria. Powder B (polymeric powder) was derived from very fine silica and alumina with aqueous solution of $ZrO(NO_3)_3$ and $Y(NO_3)_3$.

Reaction sintering of sol-gel derived powders generally create fine grain size composites. Microstructural analysis found that the matrix grain size of the powder B composites was much finer than that of the powder A composites. Significant silica-glass was found at the triple junctions of the powder A composite, which accounts for the larger grain growth. The mullite grains in both the A and B composites were slightly elongated, which may have contributed to enhanced strength and toughness of both composites. The zirconia reinforcement in the powder B composite was sharp-edged and irregular in size; the ZrO_2 particulate phase in the powder A composite was large and equiaxed.

The increase or stability in strength as temperature increased for the monolithic mullite samples of either powder A or B is due to the "apparent" strengthening of mullite by the presence of a viscous glassy phase at triple points of the mullite-mullite grain boundaries. Plot two of figure 4 shows the strength versus temperature for composites with 30 wt% ZrO_2 in mullite. Room temperature strength increased for these samples by about 25% in comparison to the monolithic samples. This was due to the zirconia reinforcement and possible transformation toughening. Strengths for these samples decreased with increased test temperature for both powders, probably due to weakening of the grain boundary phase at elevated temperatures. For powder B the stabilization at the higher temperatures was due to an increase in the density of the samples.

Comparing the strength of yttria-containing zirconia-mullite composites, at room temperature it was found that the strength of the powder B composites was nearly 200 MPa stronger than the powder A composites. The powder B composite remained at least 75 MPa stronger than the powder A composite up to 1300 °C. This was due to enhanced toughening by fine particulate reinforcement. At RT, almost all the zirconia (~94%) in the powder B composite was in the tetragonal form due to the small grain size of the ZrO_2 reinforcement. The t-ZrO_2 content in the powder A composites was much less. Both composites retained most of their RT strengths up to 1200 °C, but dropped drastically at 1300 °C. This decrease in strength occurred due to the presence of a zirconium- and yttrium-rich glassy grain boundary phase [50].

ZrO_2(p)/Mullite

Reference: 30

Composition
0-40 vol% ZrO_2

Composite Architecture
particulate reinforcement

Processing Method
Sol-Gel/Sintered

Detailed Description
Precursor solution with a stoichiometric mullite composition was prepared by partial hydrolysis. Composite powders were calcined at 800 °C for 12 hours and uniaxially pressed at 120 MPa followed by isostatic pressing at 300 MPa. The compacts were sintered at 1400-1700 °C for 2 hours.

Reinforcement Description
zirconia-mullite composite powders

Reinforcement Designation
silica sol: TEOS ($Si(OC_2H_5)_4$)
alumina sol: aluminum i-propoxide ($Al(OC_3H_7)_3$)
zirconia sol: zirconium i-propoxide ($Zr(OC_3H_7)_3$)

Matrix Description
(see reinforcement description)

Test Type
4-pt. Flexural

Specimen Dimensions
3 x 4 x 36 mm

Test Temperature
RT

Reinforcement Content (volume percent)	Flexural Strength (MPa)	Relative Density (percent)	Sintering Temperature (°C)
0	91	79	1,400
0	110	89	1,500
0	116	90	1,600
0	126	92	1,700
10	82	88	1,400
10	62	82	1,600
20	98	83	1,400
20	155	92	1,500
40	153	89	1,400
40	167	92	1,500
40	142	92	1,600

Remarks/Comments
The low density of the sintered compacts was ascribed to a low green density of the precursor compact of about 40 vol%. Crystallization of the tetragonal ZrO_2 and mullite at lower temperatures hindered densification. The presence of t-ZrO_2 allowed for transformation toughening of the composite. The flexural strength increased with increasing reinforcement content.

ZrO_2(p)/Cordierite

Reference: 73

Composition
0 or 4.6 vol% ZrO_2

Processing Method
CIPed/Pressureless-Sintered

Detailed Description
Ingredients were mixed in a high density alumina ball mill with water and dried. The batches were isostatically pressed under 140 MPa and then sintered at different temperatures from 1200 to 1350 °C for 4 hours, heating and cooling at 10 °C/min.

Reinforcement Description
m-ZrO_2 particulates

Reinforcement Dimensions
ave. surface area = 12 m^2g^{-1}

Matrix Description
cordierite $2MgO.2Al_2O_3.5SiO_2$

Matrix Supplier
commercial variety

Matrix Properties
ave. particle size = 18 μm

Test Type
3-pt. Flexural

Specimen Dimensions
5 x 7 x 30 mm
span = 20 mm

Test Temperature
RT

Sintering Temperature (°C)	Flexural Strength (MPa)	Range (MPa)	Reinforcement Content (volume percent)
1,200	40	±15	0
1,250	45	±10	0
1,300	55	±5	0
1,320	77	±5	0
1,350	122	±8	0

1,200	108	±5	4.6
1,250	128	±5	4.6
1,300	190	±15	4.6
1,320	165	±30	4.6
1,350	122	±13	4.6

Remarks/Comments
The sintered density was improved with ZrO_2 addition through glass phase formation along grain boundaries. The flexural strength of zirconia reinforced cordierite fabricated at a sintering temperature of 1300 °C was 190 MPa, compared to 55 MPa for pure cordierite. The flexural strength of pure cordierite increased with increasing temperature due to densification. The strength of the composite, with 4.6 vol% ZrO_2 content, increased with sintering temperature up to 1300 °C and then dropped due to the formation of excess glass phase along the grain boundaries.

Figure 5: Flexural Strength vs. Sintering Temperature of ZrO_2(p)/Cordierite and ZrO_2(p)/Mullite

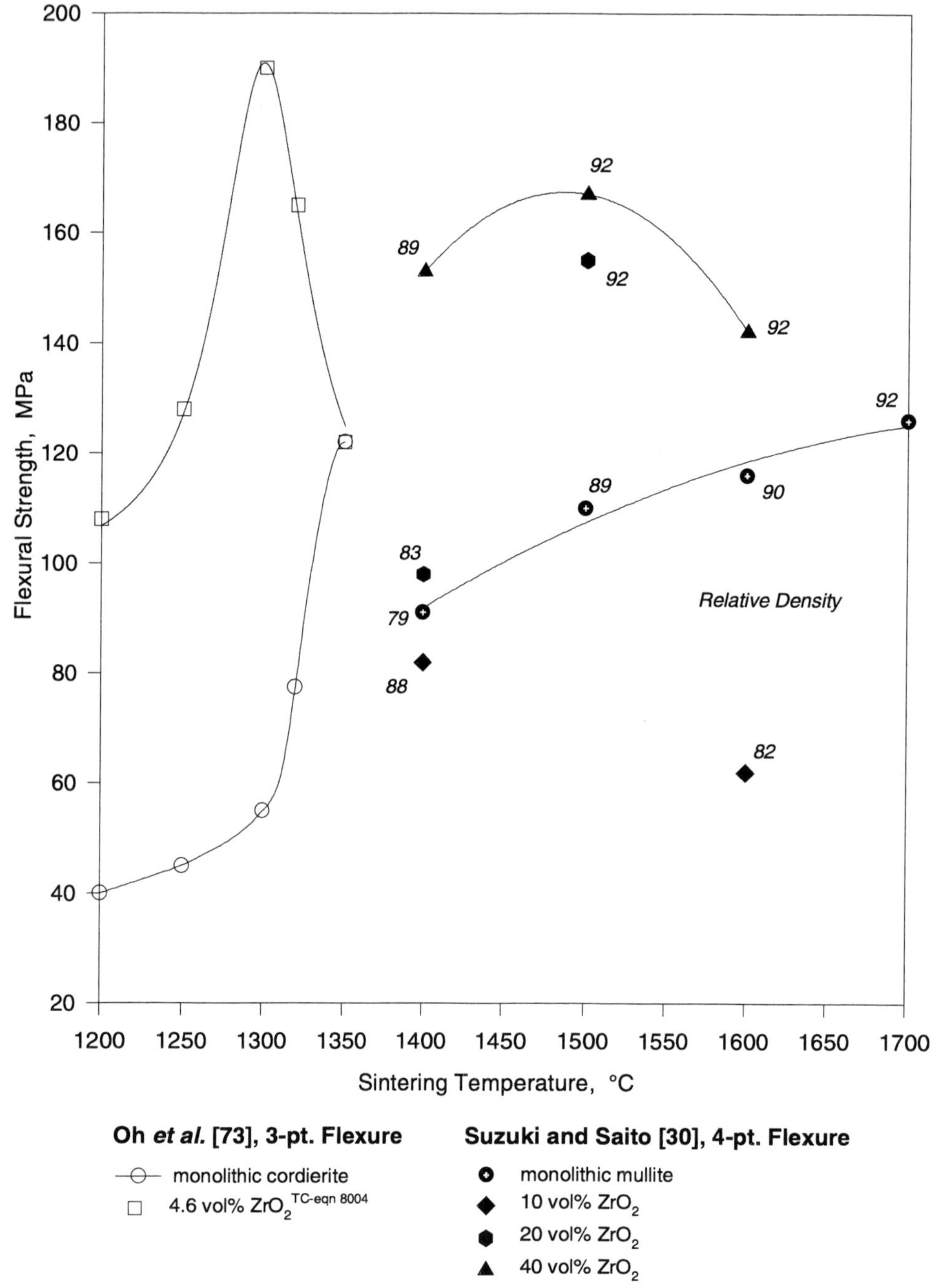

Discussion: Oh *et al.* [73] found that the zirconia in the cordierite matrix completely reacted with SiO_2 to form $ZrSiO_4$ (zircon) at all sintering temperatures. The strength of the composite was considerably greater than the strength of the monolithic cordierite at most sintering temperatures. Crack deflection caused by CTE mismatch stresses between the zircon and cordierite phases strengthened the composite. The strength of cordierite-ZrO_2 increased with sintering temperature up to 1300 °C and then dropped due to the formation of excess glassy phase along the grain boundaries. Conversely, the flexural strength of pure cordierite increased with increasing firing temperature.

Suzuki and Saito [30] tested a mullite and zirconia-mullite composite fabricated by sintering composite powders derived from sol-gel precursors. It was believed that composite precursors have excess free energy which can be utilized to enhance sintering rates. However, full density was not achievable due to a low green density. Strength was highly dependent on density. For similar relative densities, zirconia reinforcement improved strength [30].

$SiC(p)/Al_2O_3$

Reference: 33

Composition
37 vol% SiC (with B_4C)

Processing Method
Conventionally Hot-Pressed or Reaction Hot-Pressed

Detailed Description
Composite powders were mixed by wet ball milling with Si_3N_4 media. Isopropanol, used as the fluid medium, was subsequently removed by drying under vacuum at 50 °C. The dried powders were loaded into graphite dies and hot-pressed under vacuum or Ar at 1600-1800 °C and pressures of 7-28 MPa for up to 1 hour.

Reinforcement Description
B_4C and SiC particulates

Supplier Designation
B_4C: Advanced Ref. Tech.
SiC: Buelher

Reinforcement Dimensions
grain size (SiC):
reaction HP = 2 μm
conventional HP = 3 μm

Matrix Description
Al_2O_3

Matrix Properties
powder particle size ~0.3 μm
grain size:
reaction HP = 4 μm
conventional HP = 6 μm

Test Type
3-pt. Flexural

Specimen Preparation
Specimens were machined either by conventional diamond machining or EDM.

Specimen Dimensions
2 x 3 mm cross-section
span = 12.7 mm

Test Description
crosshead speed = 0.5 mm/min

Test Temperature
RT

Reinforcement Content (volume percent)	Flexural Strength (MPa)	Relative Density (percent)	Comments
37	530	96	*reaction HP*
37	360	99	*conventionally HP*

Remarks/Comments
Fracture strength of 530 MPa for reaction HP specimens vs. 360 MPa for conventionally HP specimens indicate an advantage of the former process. (Because of relatively high resistivities, EDM machining was not attempted on these materials).

$B_4C(p)/Al_2O_3$

Reference: 45

Composition
0 - 20 vol% B_4C

Processing Method
Hot-Pressed

Detailed Description
The powders were mixed in methanol for 2 hours using alumina balls in a plastic jar. The resulting slurries were oven-dried, then hot-pressed in BN coated graphite dies in an argon atmosphere at 1520 °C for 20 minutes.

Reinforcement Description
B_4C particulates

Reinforcement Supplier
Eagle-Picher Industries, Inc.

Reinforcement Dimensions
particle size = 0.2-7 μm

Matrix Description
$\alpha-Al_2O_3$ powders

Matrix Supplier
ALCOA Industrial Chemicals

Matrix Designation
A16SG

Test Type
3-pt. Flexural

Specimen Preparation
Specimens were ground to a 1 μm diamond polish.

Specimen Dimensions
5 x 5 x 25.4 mm

Test Description
crosshead speed = 5 mm/min

Test Temperature
RT

Reinforcement Content (volume percent)	Flexural Strength (MPa)
0	417
5	589
10	598
15	580
20	605

Remarks/Comments

Flexural strength increased by over 100 MPa with the addition of 5 vol% particulates and remained fairly constant with the further addition of B_4C particulates.

$SiC(p)/Al_2O_3$

Reference: 77

Composition

0 - 33 vol% SiC

Processing Method

Hot-Pressed

Detailed Description

The powders were mixed by ball-milling in highly pure toluene with Al_2O_3 balls for 24 hours. The dried mixtures were hot-pressed in N_2 atmosphere at 1600-1800 °C and 28 MPa.

Reinforcement Description

β-SiC particulates

Reinforcement Supplier

Ibiden Co., Ltd.

Reinforcement Dimensions

ave. particle size < 0.3 μm

Matrix Description

α-Al_2O_3 particulates

Matrix Supplier

Asahi Chemical Co.

Matrix Properties

ave. particle size = 0.3 μm

Test Type

3-pt. Flexural

Specimen Preparation

The tensile surfaces were polished to 1 μm with diamond paste, and the edges were beveled.

Specimen Dimensions

3 x 4 x 45 mm
span = 30 mm

Test Description

crosshead speed = 0.5 mm/min

Test Temperature

RT

Reinforcement Content (volume percent)	Flexural Strength (MPa)	Range (MPa)	Hot-Press Temperature (°C)
0.0	355	+88/-90	1600
5.1	1,017	+115/-128	1600
11	931	+90/-96	1700
17	833	+128/-124	1800
33	814	+109/-88	1800

Remarks/Comments

The toughness and strength of Al_2O_3 was improved by incorporating the SiC particles within the matrix. The maximum strength of over 1000 MPa was observed for 5 vol% SiC reinforcement. TEM observations indicated that most of the SiC particles, typically less than 0.2 μm, were predominantly dispersed within the Al_2O_3 grains and some larger particles were at the grain boundaries.

$SiC(p)/Al_2O_3$

Reference: 77

Composition
0 - 17 vol% SiC

Processing Method
Hot-Pressed

Detailed Description
The powders were mixed by ball-milling in highly pure toluene with Al_2O_3 balls for 24 hours. The dried mixtures were hot-pressed in N_2 atmosphere at 1600 or 1800 °C and 28 MPa.

Reinforcement Description
β-SiC particulates

Reinforcement Supplier
Ibiden Co., Ltd.

Reinforcement Dimensions
ave. particle size < 0.3 μm

Matrix Description
α-Al_2O_3 particulates

Matrix Supplier
Asahi Chemical Co.

Matrix Properties
ave. particle size = 0.3 μm

Test Type
3-pt. Flexural

Specimen Preparation
The tensile surfaces were polished to 1 μm with diamond paste, and the edges were beveled.

Specimen Dimensions
3 x 4 x 45 mm
span = 30 mm

Test Description
crosshead speed = 0.5 mm/min

Test Temperature
0-1400 °C

Test Environment
Air

Temperature (°C)	Flexural Strength (MPa)	Hot-Press Temperature (°C)	Reinforcement Content (volume percent)
20	352	1600	0.0
600	356	1600	0.0
800	314	1600	0.0
1,000	309	1600	0.0
1,200	230	1600	0.0
1,400	143	1600	0.0
20	1,050	1600	5.1
900	1,020	1600	5.1
1,000	1,010	1600	5.1
1,100	673	1600	5.1
1,200	538	1600	5.1
1,400	191	1600	5.1
20	893	1800	17
800	870	1800	17
1,000	860	1800	17
1,100	740	1800	17
1,200	675	1800	17
1,300	595	1800	17
1,400	473	1800	17

Remarks/Comments
The strength of the SiC-Al_2O_3 composites remained relatively constant up to about 1000 °C. After 1000 °C, the strength decreased. The strength of the 5.1 vol% SiC composite was greater than that of the 17 vol% SiC composite. Both had greater strengths than the monolithic material.

$SiC(p)/Al_2O_3$

Reference: 112

Composition
~55 vol% SiC

Processing Method
Slip-Cast/DIMOX

Detailed Description
Commercially available $SiC(p)/Al_2O_3$ (Du Pont Lanxide composite Inc., USA) was obtained. This material was prepared using the DIMOX method. An SiC preform was slip-cast and then an Al alloy was infiltrated and oxidized in the furnace, resulting in the composite.

Reinforcement Description
SiC particulates

Reinforcement Properties
ave. particle diameter = 5-20 μm

Matrix Description
aluminum powder

Test Type
4-pt. Flexural

Specimen Dimensions
6 x 3 x 50 mm

Test Temperature
RT-1430 °C

Test Environment
Air

Test Temperature (°C)	Flexural Strength (MPa)	Range (MPa)	Comments
22	503	± 21.5	*7 μm Filler*
380	453	± 21.0	
580	360	± 22.5	
1,000	340	± 19.5	
1,280	310	± 19.5	
1,420	292	± 20.5	
22	186	± 18.5	*64 μm Filler*
570	140	± 21.0	
970	137	± 21.0	
1,430	140	± 21.0	

Figure 6: 3-pt. Flexural Strength of $SiC(p)/Al_2O_3$ and $B_4C(p)/Al_2O_3$

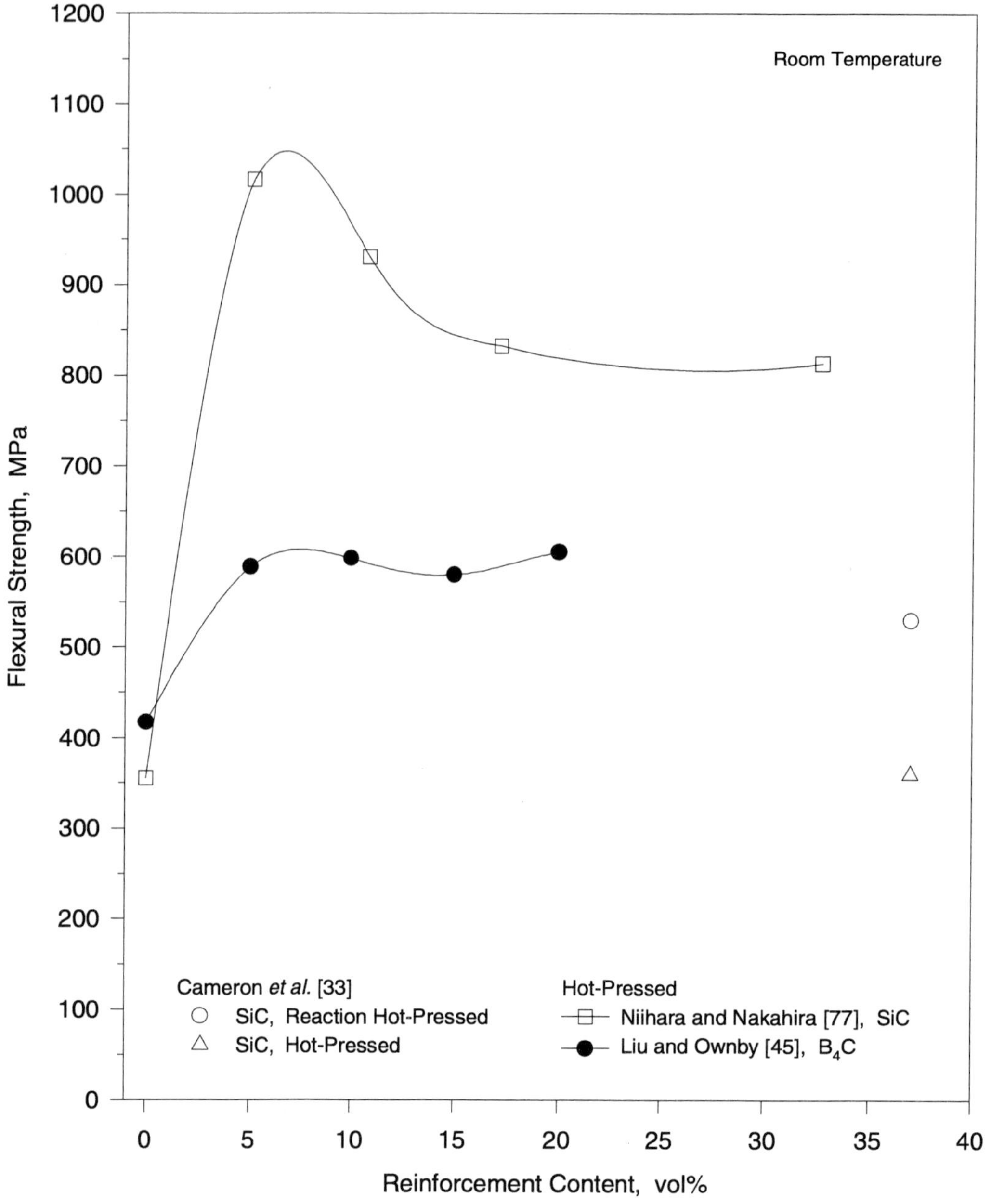

Discussion: The flexural strength of the SiC(p)/Al_2O_3 nanocomposite tested by Niihara and Nakahira [77] increased substantially from 355 MPa to approximately 1000 MPa with the addition of only 5 vol% SiC. This strength improvement was due to the SiC particles distributed throughout the matrix grains. There was also a corresponding decrease in the critical flaw size from approximately 23 to 6 μm for the 5 vol% composite. The flexural strength then decreased to 930 MPa when SiC content was increased to 11 vol%. The flexural strength continued to decrease slightly with the further addition of SiC. The flexural strength for the B_4C(p)/Al_2O_3 composite tested by Liu and Ownby [45] was substantially less than that of the SiC(p)/Al_2O_3 composite for all compositions. However, the flexural strength of the B_4C(p)/Al_2O_3 composite did not decrease with increasing reinforcement content as did the SiC(p)/Al_2O_3 composite. The flexural strength of the SiC(p)/Al_2O_3 composite tested by Cameron *et al.* [33] with approximately 37 vol% SiC was substantially greater (by about 170 MPa) for the reaction hot-pressed composite compared to the conventionally hot-pressed composite.

The SiC-Al_2O_3 nanocomposite tested by Niihara and Nakahira [77] was much stronger than the conventionally hot-pressed composite tested by Cameron *et al.* [33]. The greater strength of the nanocomposite was the result of its fine grain size. The matrix grain size for the SiC(p)/Al_2O_3 nanocomposite from Niihara and Nakahira [77] was one order of magnitude smaller than the grain-size of the composite tested by Cameron *et al.* [33]. The significant reduction in grain size was due to the nano-size SiC particles that restrained the grain-growth of the matrix grains.

Another strengthening mechanism in the nanocomposite was highly localized stresses within and/or around the SiC particles. Thermal expansion mismatch between matrix and SiC during cooling from the fabrication temperature produced these stresses [77]. During processing, the nano-size SiC particles pinned and piled-up the dislocations, and promoted sub-grain boundaries within the matrix grains. This resulted in the refinement of matrix grains for strengthening [77]. Transgranular fracture induced in the SiC(p)/Al_2O_3 nanocomposite was due to localized tensile stress around the nano-size SiC particles. This resulted in strengthening and toughening due to crack tip deflection by the SiC particles [77]. Flexural strength followed the same trend with reinforcement content as fracture toughness, which suggested that the strength was a function of composite fracture toughness.

Figure 7: 3-pt. Flexural Strength vs. Temperature of $SiC(p)/Al_2O_3$

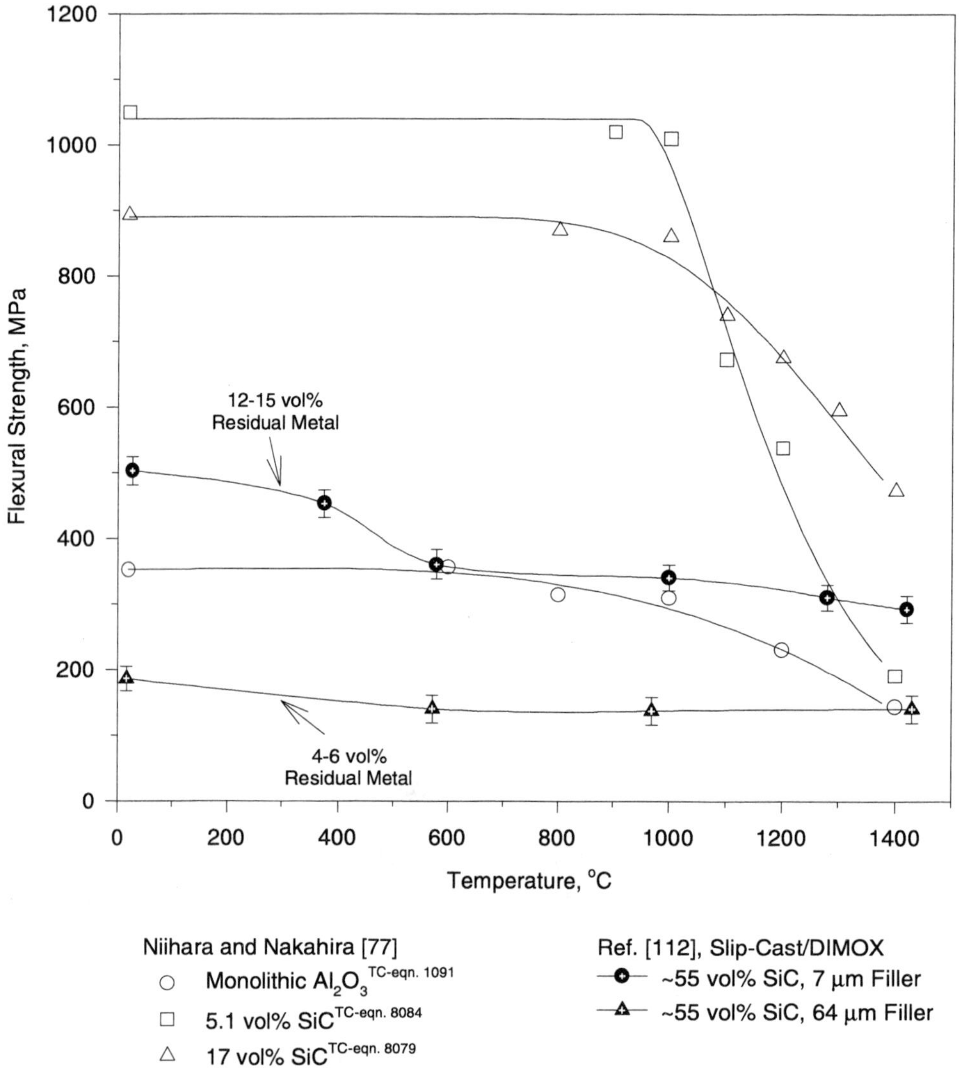

Discussion: Niihara and Nakahira *et al.* [77] tested the flexural strength of $SiC(p)/Al_2O_3$ nano-composites. As shown by figure 6, the addition of 5.1 vol% nano-sized SiC particulates to an alumina matrix remarkably increased the strength of the monolithic Al_2O_3 at room temperature. The monolithic alumina retained its RT strength up to about 1000 °C and then decreased gradually to less than 200 MPa at 1400 °C. The strength of the 5.1 vol% $SiC(p)/Al_2O_3$ composites retained its remarkable RT strength to 1000 °C, followed by a dramatic decrease in strength upon increasing temperature to 1400 °C. This dramatic decrease in strength may have been due to the softening of the matrix phase and the relaxation of thermal residual stress fields as the temperature was increased above 1000 °C. The 17 vol% $SiC(p)/Al_2O_3$ composite was not as strong as the 5.1 vol% composite at room temperature; however, this composite retained its room temperature strength much better at elevated temperatures. The strength of the DIMOX™ composites depended on the size of the SiC filler. The 64 μm filler produced the weaker composite, and the strength of this composite did not significantly change with increasing temperature. The strength of the 7 μm filler composite was at least twice as great as the 64 μm filler at all temperatures; below 600 °C this difference approached 2.5 times. The strength of the 7 μm filler composite decreased in the range of 400 to 600 °C. The reason for this decrease was the high residual aluminum content, which softened and melted in that temperature range.

$SiC(p)/Al_2O_3$

Reference: 24

Composition
20 vol% SiC

Processing Method
Hot-Pressed

Detailed Description
Composite specimens were fabricated by blending Al_2O_3 powders with SiC particulates in hexane and dispersing the mixture with a high-shear homogenizer. The mixtures were dried and hot-pressed in a graphite die at 1750-1850 °C and 41 MPa for 30 minutes.

Reinforcement Description
SiC particulates

Reinforcement Supplier
H.C. Starck, Germany

Supplier Designation
Grade A-1

Matrix Description
Al_2O_3 matrix

Matrix Supplier
Baikowski Corporation, Charlotte, NC

Matrix Designation
Grade CR-10

Test Type
4-pt. Flexural

Specimen Preparation
Specimens had 220-grit diamond ground tensile surfaces.

Specimen Dimensions
2.8 x 3.3 x >26 mm

Test Temperature
RT

Grain Size (μm)	Flexural Strength (MPa)	Range (MPa)
0.6	351	+57/-59
0.9	422	+42/-49
2.3	532	+69/-82
3.2	422	+94/-95
4.0	386	+66/-63
4.4	371	+36/-40
5.0	178	+15/-15
7.6	341	+35/-35
11.8	166	+21/-24
17	107	+10/-9

Remarks/Comments
The results indicate microcracking was not evident in the composites reinforced with SiC particles with sizes up to approximately 2.5 μm. Composites reinforced with SiC with mean diameters of 5 μm or greater exhibited very poor flexural strength, indicating microcracking was present.

SiC(p)/Al_2O_3

Reference: 82

Composition
0 and 5 vol% SiC

Processing Method
Hot-Pressed

Detailed Description
The SiC particles were added to the Al_2O_3 powder, and the mixture was ultrasonically dispersed and then ball-milled in methanol. The slurry was slowly dried to produce a powder uniform in color, which contained only soft agglomerates. The starting powder was calcined in air at 600 °C for 10 hours in order to burn out any residual organics from powder processing. It was then placed into a 76.2 mm diameter graphite die which was lined with graphite foil and a BN coating. The specimens were pressed at 1640 °C and 50 MPa for 60 minutes in flowing nitrogen. The top and bottom surfaces of each disk were removed to eliminate any reaction layer. Hot-pressing was carried out in a graphite-lined furnace using reducing atmospheres to minimize oxidation. Some bars were annealed in argon at 1300 °C for 2 hours before testing.

Reinforcement Description
β-SiC particulates

Reinforcement Supplier
Performance Ceramics Co., Peninsula, OH

Reinforcement Dimensions
ave. particle size = 0.15 μm

Reinforcement Properties
CTE = 4.7 x 10^{-6} °C

Matrix Description
α-Al_2O_3 particulates

Matrix Supplier
Sumitomo, Osaka, Japan

Matrix Designation
AKP-53

Matrix Properties
ave. particle size = 0.2 μm
CTE = 8.8 x 10^{-6} °C

Test Type
4-pt. Flexural

Specimen Preparation
All bars were ground flat on one side, and ground and polished to a 6 μm diamond finish on the other side. Also, the bars were beveled parallel to their lengths.

Specimen Dimensions
3 x 4.4 mm cross-section
inner span = 12.5 mm
outer span = 31.8 mm

Test Temperature
RT

Applied Load (N)	Flexural Strength (MPa)	Reinforcement Content (volume percent)	Comments
0 (Intrinsic Strength)	559 ± 51	0	*annealed*
2	474	0	
5	367	0	
10	271	0	
20	220	0	
50	161	0	
100	140	0	
200	125	0	
0 (Intrinsic Strength)	1001 ± 102	5	*annealed*
2	541	5	
5	401	5	
10	325	5	
20	277	5	
50	202	5	
100	165	5	
200	143	5	
0 (Intrinsic Strength)	760 ± 28		*non-annealed*
2	610	5	
5	502	5	
10	373	5	
20	288	5	
50	214	5	
100	180	5	
200	137	5	

Remarks/Comments

The annealing step increased the intrinsic strength of the composites, but decreased the composites flaw resistance so that the annealing step decreased the strength of the indented composites. Annealing the composite specimens had the effect of diminishing the compressive surface stress while healing surface flaws.

SiC(p)/Al_2O_3

Reference: 82

Composition

0 and 5 vol% SiC

Processing Method

Cold-Pressed/Pressureless Sintered

Detailed Description

The SiC particles were added to the Al_2O_3 powder, and the mixture was ultrasonically dispersed and then ball-milled in methanol. The slurry was slowly dried to produce a powder uniform in color, which contained only soft agglomerates. Disk specimens of the starting powder were uniaxially cold-pressed at 42 MPa, isostatically cold-pressed at 350 MPa, and calcined in air at 600 °C for 10 hours. Sintering was conducted in flowing nitrogen at 1775 °C for 4 hours. For calcining and sintering, the disks were surrounded by SiC powder in a covered Al_2O_3 crucible. The sintered disks were approximately 25 mm in diameter and 2.8 mm thick. Pressureless sintering was carried out in a graphite-lined furnace using reducing atmospheres to minimize oxidation.

Reinforcement Description

β-SiC particulates

Reinforcement Supplier

Performance Ceramics Co., Peninsula, OH

Reinforcement Dimensions

ave. particle size = 0.15 μm

Reinforcement Properties

CTE = 4.7 x 10^{-6} °C

Matrix Description

α-Al_2O_3 particulates

Matrix Supplier

Sumitomo, Osaka, Japan

Matrix Designation

AKP-53

Matrix Properties

ave. particle size = 0.2 μm

CTE = 8.8 x 10^{-6} °C

Test Type

Post-Indentation Biaxial Flexure

Specimen Preparation

All disks were ground flat on one side, and ground and polished to a 6 μm diamond finish on the other side.

Specimen Dimensions

dia. = 25 mm

thickness = 2.8 mm

Type of Specimen

Disk

Test Description

radius flat = 2.47 mm

three-ball support radius = 9.37 mm

Test Temperature

RT

Applied Load (N)	Flexural Strength (MPa)	Reinforcement Content (volume percent)
0 (Intrinsic Strength)	569 ± 72	0
2	552	0
5	386	0
10	281	0
20	214	0
50	161	0
100	131	0
200	112	0
0 (Intrinsic Strength)	586 ± 72	5
2	448	5
5	333	5
10	250	5
20	196	5
50	159	5
100	130	5
200	108	5

Remarks/Comments

The composite showed a slightly greater intrinsic strength than the monolithic alumina, however, the strength of the composite declined more than the alumina after indentation.

$SiC(p)/Al_2O_3$

Reference: 88

Composition
0 - 30 vol% SiC

Processing Method
Hot-Pressed

Detailed Description
The alumina was ball-milled for 24 hours to break up the agglomerates in a deflocculated, water-based slurry (Darvan 821-A). The slurry was passed through a 325-mesh sieve to remove any large agglomerates. The SiC was added, and the mixture was milled for one hour. The material was then dried, dry pressed at 12 MPa, and then pre-fired to 600 °C for one hour to remove any residual volatile species. The pre-fired discs were then inductively hot-pressed in a nitrogen atmosphere for 1 hour at 1600 °C. Pressure was applied beginning at 400 °C and was gradually increased to 15 MPa and held.

Reinforcement Description
SiC particulates

Supplier Designation
Lonza LS-13

Reinforcement Typical Properties
density = 3.2 g cm^{-3}
CTE = 4.3 $x10^{-6}$ $°C^{-1}$
E = 470 GPa
fracture toughness = 3-5 MPa $m^{1/2}$

Reinforcement Dimensions
ave. size = 13 μm

Matrix Description
Al_2O_3

Matrix Designation
Ferro A-310

Matrix Properties
ave. grain size = 0.35 μm
density = 3.96 g cm^{-3}
CTE = 7.7-8.3 $x10^{-6}$ $°C^{-1}$
E = 380 GPa
fracture toughness = 2-3 MPa $m^{1/2}$

Test Type
4-pt. Flexural

Specimen Preparation
The sample was polished to one μm surface finish, and the tensile edges were chamfered. The tensile side was perpendicular to the hot-pressing direction.

Specimen Dimensions
3 x 4 x 50 mm

Test Temperature
RT

Reinforcement Content (volume percent)	Flexural Strength (MPa)	Range (MPa)
0	346	±30
10	271	±10
20	199	±5
30	243	±15

Remarks/Comments
The strength of monolithic alumina was greater than that of the composites. 20 vol% additions of particulates caused the strength to drop further. Further increasing the volume fraction of particulates to 30 vol% caused the strength to increase.

Figure 8: 4-pt. Flexural Strength of $SiC(p)/Al_2O_3$

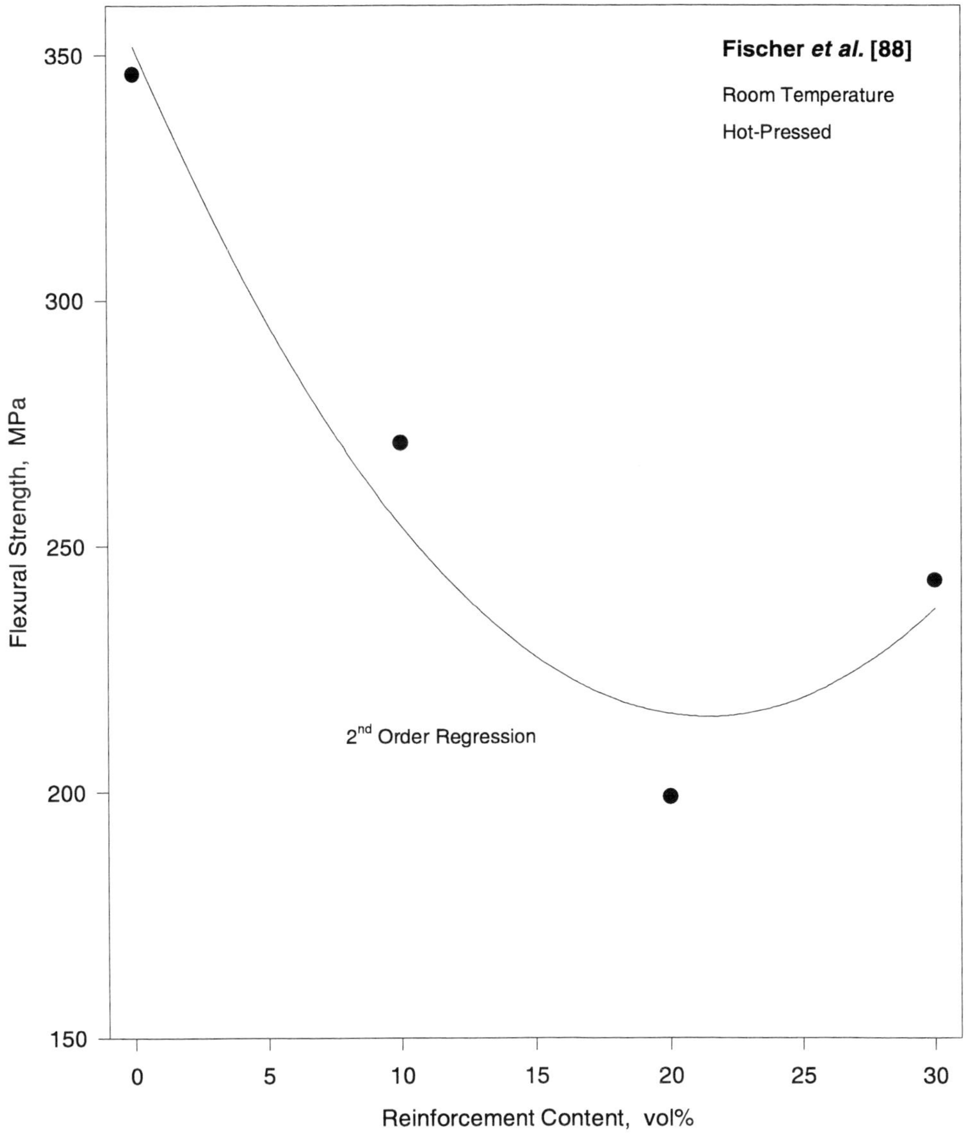

Discussion: The flexural strength decreased with the addition of SiC particulates to an alumina matrix. A decrease in flexural strength found in particulate composites often results from defects (microcracks) caused by high internal stresses, or processing defects. Both increase the critical flaw size of the composite and are associated with the second phase [88]. The strength of monolithic alumina was 350 MPa and dropped with the addition of SiC reinforcement until it reached a minimum of 200 MPa at 20 vol% SiC. Strength then increased by ~40 MPa with the further increase of SiC reinforcement to 30 vol%. This increase in strength could be a result of a decrease in the mean free path between SiC particulates that could potentially limit the size of the radial crack associated with the SiC reinforcement [88].

Figure 9: 4-pt. Flexural Strength vs. Grain Size of $SiC(p)/Al_2O_3$

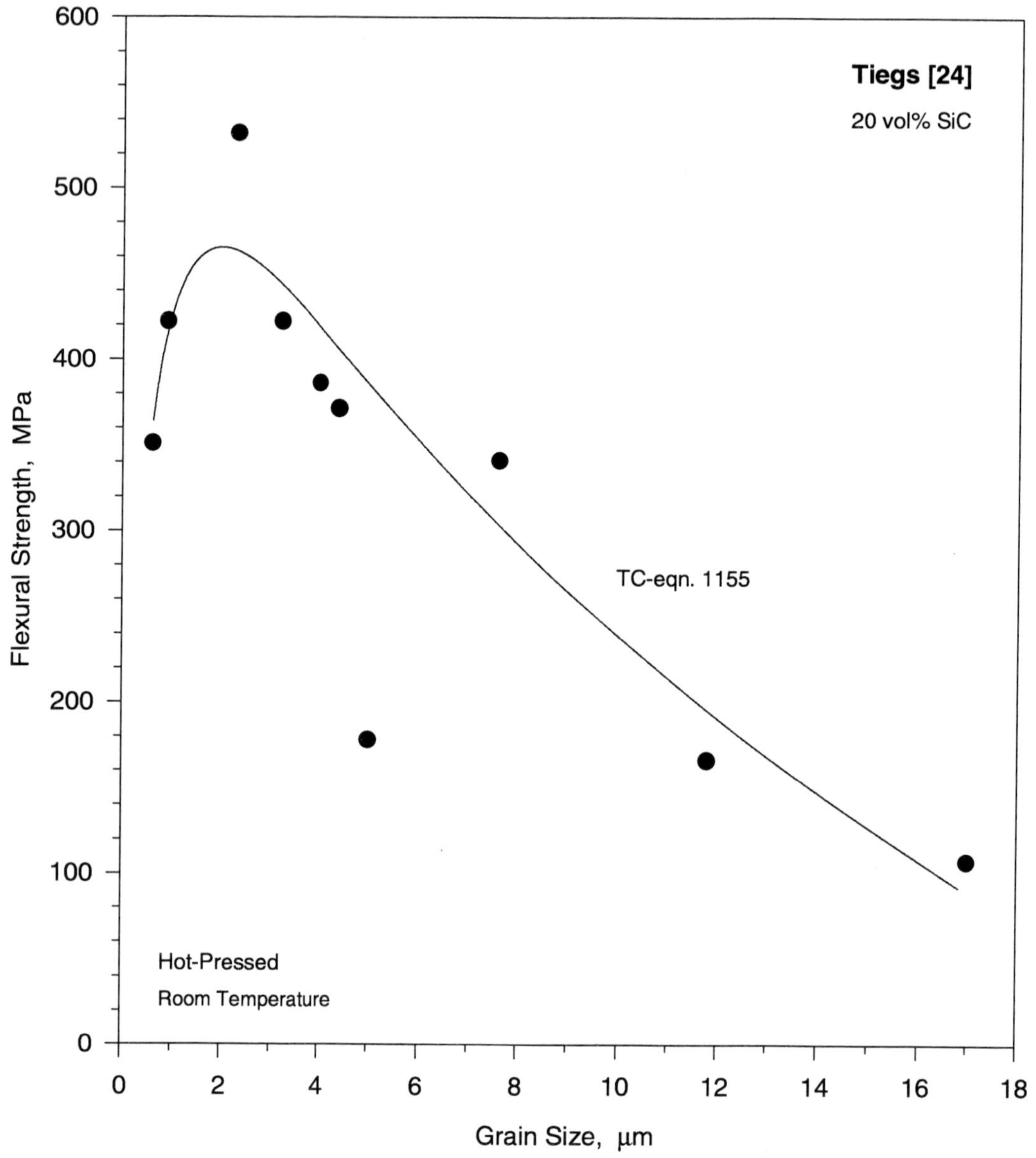

Discussion: Tiegs [24] showed that grain size affects the magnitude of thermal stresses and the extent of matrix microcracking; and strength can be used as an indicator of the extent of microcracking present in the composite. A corresponding loss of strength generally occurs with the occurrence of microcracking. The increase in strength observed up to 2.5 µm indicated that no microcracking was present. Because of the limited data gathered it was not apparent whether microcracking occurred between the grain sizes of 2.5 and 4 µm [24]. After 4 µm, however, microcracking was apparent and possibly increased the fracture toughness of the composite even as it led to a deterioration in strength.

Figure 10: 4-pt. Flexural Strength vs. Applied Load of Hot-Pressed $SiC(p)/Al_2O_3$

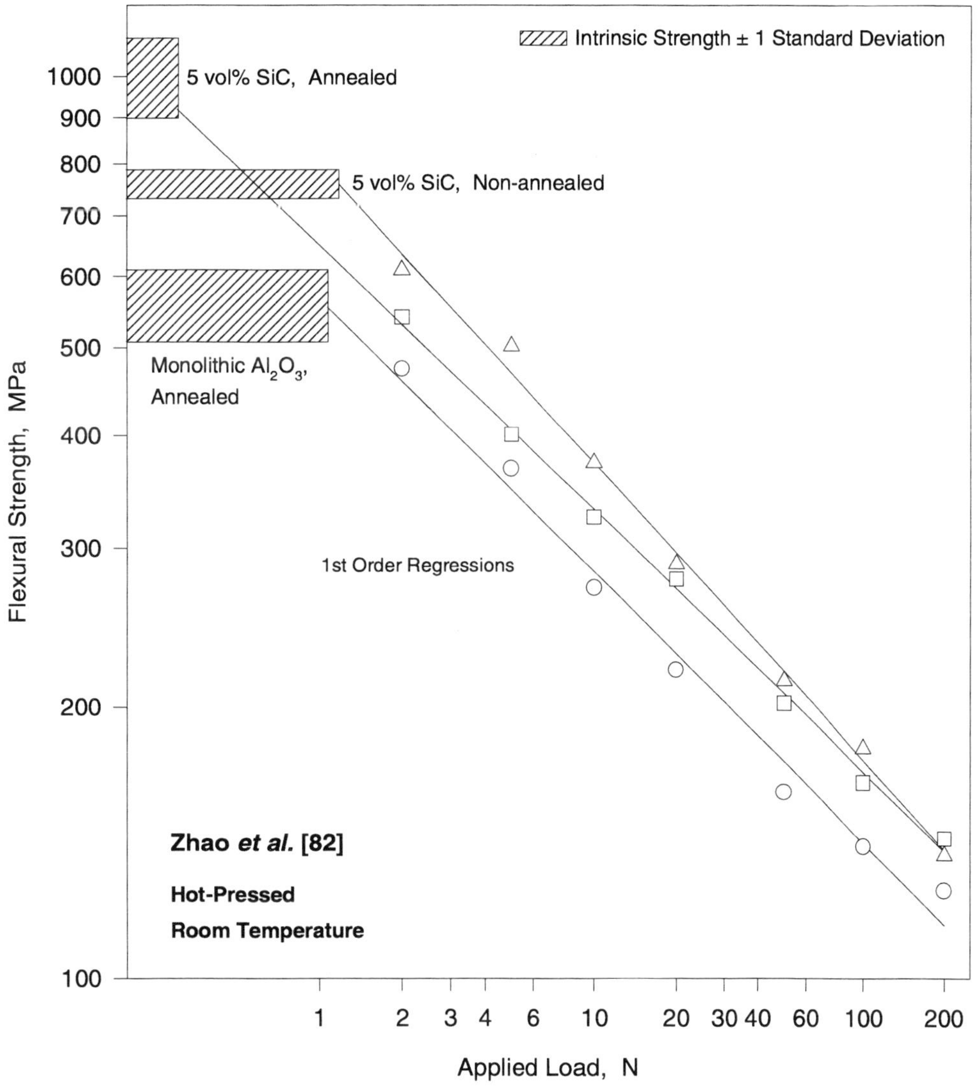

Discussion: Flexural strength of ceramics and ceramic composites is determined primarily by the flaw tolerance (fracture toughness) of the material and the maximum flaw size. In the indentation flexural strength test, a flaw is introduced of known size; this eliminates flaw size as a variable. Zhao *et al.* [82] performed 4-pt. flexural strength tests on indented samples to measure the flaw tolerance of annealed monolithic alumina, and as-processed or annealed 5 vol% $SiC(p)/Al_2O_3$ composites.

Figure 10 shows data from tests performed on hot-pressed samples. Annealing had two effects on the as-hot-pressed samples. First, the annealing step healed surface flaws from the machining step. Tests on the unindented composites showed that strength increased from 760 before annealing to 1000 MPa after annealing. Second, the annealing step diminished the compressive surface layer, which diminished the flaw tolerance and fracture toughness of the composite. It was likely that the compressive surface layer was introduced by the machining operation and that the layer extended to a depth of tens of microns [82].

The effect of the diminishing surface layer was quite evident in the indented samples. Since the as-hot-pressed and annealed samples had approximately the same maximum flaw size (the indentation microcracks), the strength of these composites was primarily due to flaw tolerance and fracture toughness of the composite. Test results showed that the strength of the indented samples decreased after annealing due to diminished compressive surface stresses [82].

Figure 11: Biaxial Flexural Strength vs. Applied Load of Pressureless Sintered SiC(p)/Al_2O_3

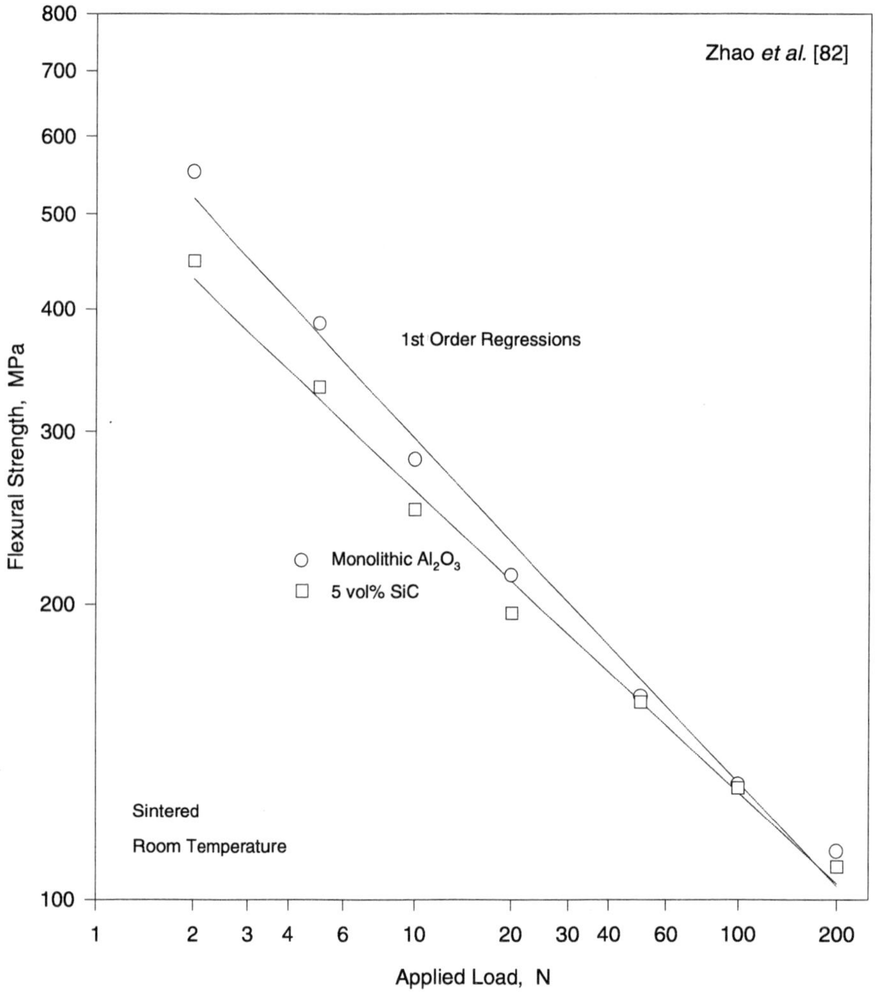

Discussion: Figure 11 shows results of the biaxial indentation fracture tests performed on sintered samples, Zhao *et al.* [82]. The flexural strength of the pressureless sintered SiC(p)/Al_2O_3 composite was less than the monolithic Al_2O_3 composite for all the applied loads. This result was opposite to that of both the hot-pressed and the hot-pressed and annealed SiC(p)/Al_2O_3 composites (see figure 10). The SiC particulates inhibited densification in the sintering process because pressureless sintering does not supply sufficient driving force for densification. This was not a problem with the hot-pressing technique since the high applied pressure was able to overcome SiC inhibited densification. The lower strength of the sintered composite compared to the hot-pressed composites may have been due to three different factors: (1) the different test methods, (2) greater porosity in the sintered composite, and (3) the absence of a compressive surface layer in the sintered composites.

TiC(p)/Al_2O_3

Reference: 8

Composition
0 - 35 wt% TiC

Processing Method
Hot-Pressed

Detailed Description
Composites were supplied by Feldmuhle AG, Plochingen, Germany.

Reinforcement Description
TiC particulates

Matrix Description
Al_2O_3 powder

Test Type
3-pt. Flexural

Specimen Preparation
Samples were surface ground with a D30 diamond wheel or tension side polished to $R_t = 0.8$.

Specimen Dimensions
2 x 4 x 32 mm
span = 30 mm

Test Description
crosshead speed = 0.1 mm/min
The relation $\sigma_B = 3Pl/2bw^2$ was used to find the flexural strength, where:
P = load
l = span
b = breadth
w = width

Test Equipment
Instron universal testing machine

Reinforcement Content (volume percent)	Flexural Strength (MPa)	Range (MPa)	Porosity (percent)	Comments
0.0	332		0.5	*Ground Specimen*
4.1	465		0.87	
8.3	536		0.69	
17	318		1.93	
26	400		1.29	
35	352		1.28	
0.0	305	+62/-68	0.5	*Polished Specimen*
4.1	462	+21/-21	0.87	
8.3	502	+26/-28	0.69	
17	530	+51/-65	1.93	
26	635	+89/-91	1.29	
35	670	+74/-89	1.28	

Remarks/Comments
A general trend of increasing flexural strength with increasing TiC content was observed for the polished specimen. For the ground specimen, the strengthening due to TiC particulates was counteracted by surface flaws introduced during grinding.

$TiC(p)/Al_2O_3$

Reference: 23

Composition
~ 25 vol% TiC

Processing Method
CIPed/Pressureless Sintered

Detailed Description
The powders were mixed by ball-milling in ethanol for 24 hours. After drying, the mixtures were uniaxially pressed then isostatically pressed at 200 MPa. Sintering of the compact was done in a radio-frequency induction heating furnace. Specimens were set in an SiC crucible to prevent reaction with the graphite susceptor and were fired at 1800 °C for 1 hour under an argon gas flow at atmospheric pressure with Al_2O_3 encapsulation.

Sintering Additives
MgO (0.5 wt% of matrix)

Reinforcement Description
TiC particulates

Reinforcement Supplier
H.C. Stark

Supplier Designation
c.a.s. grade

Reinforcement Dimensions
ave. grain size = 1.5 μm

Matrix Description
Al_2O_3

Matrix Supplier
Shinkosha

Matrix Properties
ave. grain size = 0.2 μm

Test Type
Flexural Test

Type of Specimen
Beam

Test Temperature
RT

Reinforcement Content (volume percent)	Flexural Strength (MPa)
25	530

Remarks/Comments
Pressureless-sintered TiC/Al_2O_3 achieved a strength increase over monolithic Al_2O_3, which has a strength of 300-400 MPa. Control of vaporization and grain growth allowed mechanical properties comparable to these prepared by hot-pressing.

$TiB_2(p)/Al_2O_3$

Reference: 25

Composition
0 - 37 wt% TiB_2

Processing Method
CIPed/Pressureless-Sintered

Detailed Description
The powders were mixed in acetone in an alumina ball-mill. The mixture was then isostatically pressed at 100 MPa into cylinders of 50 to 55 mm diameter and 105 to 115 mm length. These cylindrical bars were pre-sintered under an Ar atmosphere at 1100 °C for 2 hours. Upon cooling, specimens were diamond-cut from these blocks. Ten specimens were then sintered at 1400-1800 °C under Ar.

Reinforcement Description
TiB_2 particulates

Reinforcement Supplier
H. C. Starck, Gosler, Germany

Supplier Designation
Grade A

Reinforcement Dimensions
ave. dia. = 1.75 μm

Matrix Description
Al_2O_3 powder

Matrix Supplier
ALCOA, USA

Matrix Designation
CT 3000 SG

Matrix Properties
ave. dia. = 0.5 μm

Test Type
3-pt. Flexural

Specimen Dimensions
3 x 4 x 30 mm
span = 20 mm

Test Description
loading rate = 12 N/min.

Sintering Temperature (°C)	Flexural Strength (MPa)	Reinforcement Content (weight percent)	Weibull's Modulus (dimensionless)	Relative Density (percent)
1,400	99	0.0	9.7	
1,500	160	0.0	18.0	96.0
1,600	298	0.0	8.5	97.5
1,700	357	0.0	9.9	97.9
1,800	319	0.0	9.4	98.0
1,400	168	4.4	9.6	
1,500	192	4.4	10.9	93.4
1,600	292	4.4	5.5	95.5
1,700	265	4.4	3.4	95.7
1,800	272	4.4	12.0	95.9
1,400	140	8.9	11.0	
1,500	169	8.9	8.2	91.1
1,600	297	8.9	9.1	93.7
1,700	308	8.9	17.1	94.9
1,800	368	8.9	9.3	95.9
1,400	107	18	7.6	
1,500	134	18	12.1	88.6
1,600	265	18	14.8	92.1
1,700	293	18	5.6	92.5
1,800	290	18	15.7	93.0
1,400	70	27	11.6	
1,500	79	27	8.2	87.1
1,600	181	27	6.3	89.7
1,700	229	27	12.2	91.4
1,800	206	27	9.5	91.8
1,400	79	37	10.0	
1,500	78	37	9.7	86.0
1,600	199	37	9.0	89.0
1,700	193	37	11.6	90.0
1,800	205	37	12.9	90.3

Remarks/Comments

The addition of TiB_2 to Al_2O_3 generally resulted in a reduction in flexural strength. An increase in sintering temperature increased the flexural strength, but changes in the flexural strength for sintering temperature changes of 1600 to 1800 °C were nearly within experimental error.

$TiB_2(p)/Al_2O_3$

Reference: 43

Composition
0 - 20 vol% TiB_2

Processing Method
Hot-Pressed

Detailed Description
The powders were mixed in a ball-mill in ethanol for 2 hours using alumina balls in a plastic jar. The resulting slurries were oven dried. The Al_2O_3-TiB_2 powders were hot-pressed in BN coated graphite dies at 1470 °C for 20 minutes.

Reinforcement Description
TiB_2 powder

Reinforcement Dimensions
ave. particle size = 3 μm
range = 1-15 μm

Matrix Description
Al_2O_3 powder

Matrix Supplier
ALCOA Industrial Chemicals, Bauxite, AR

Matrix Designation
A-16SG

Test Type
3-pt. Flexural

Specimen Preparation
Specimens were diamond cut.

Specimen Dimensions
5 x 5 x 25.4 mm
span = 19 mm

Test Description
crosshead speed = 0.5 mm/min
applied pressure = 32 MPa

Test Temperature
RT

Reinforcement Content (volume percent)	Flexural Strength (MPa)
0	416
5	641
10	629
15	600
20	715

Remarks/Comments
Each point is the average of 5 to 7 specimens.

$TiC(p)/Al_2O_3$

Reference: 47

Composition
0 - 76 vol% TiC

Processing Method
Hot-Pressed

Detailed Description
The composites were wet-mixed for 20 hours in a ball-mill. Specimens were hot-pressed in a graphite die at 1500-1750 °C and 20 MPa for 60 minutes. Applied pressure for the temperature range from room temperature to the hold temperature was one-half the final pressure. Hot-pressed specimens were removed from the die after cooling to room temperature.

Sintering Additives
TiO_2 and MgO powder

Reinforcement Description
TiC powder

Reinforcement Typical Properties
S < 400 ppm, Fe < 200 ppm, Ni < 20 ppm, Cr < 10 ppm

Reinforcement Dimensions
ave. particle size = 1.0 μm

Matrix Description
α–Al_2O_3 powder

Matrix Properties
Si < 50 ppm, Na < 10 ppm, Mg < 10 ppm, Cu < 10ppm, Fe < 20 ppm
particle size = 0.8 μm

Test Type
3-pt. Flexural

Specimen Preparation
Specimens were diamond cut from hot-pressed products. Surfaces were ground, and the edges of the tensile surface were beveled with grinding paper.

Specimen Dimensions
36 x 4 x 3 mm

Test Description
crosshead speed = 0.5 mm/min

Test Temperature
RT

Reinforcement Content (volume percent)	Flexural Strength (MPa)	Range (MPa)
0.0	637	+3/-10
8.3	579	+6/-6
17	588	+5/-8
26	628	+7/-5
35	647	+7/-5
55	765	+3/-5
76	716	+4/-4

Remarks/Comments

Flexural strength of each composite showed a large scatter and a rising trend with the increase of additional TiC. This data has been averaged. Referenced values were converted from kg/mm^2 to MPa.

$TiC(p)/Al_2O_3$

Reference: 58

Composition

26 wt% TiC

Processing Method

Cold Pressed / Sintered (some HIPed) or Hot-Pressed

Detailed Description

Al_2O_3, TiC, and TiH_2 powders were dispersed and milled with hexane in a 15 cm diameter stainless steel mill for 24 hours with a ball to charge ratio of 10:1. Larger batches were attrition milled in deionized water, and a binder was added prior to spray drying. Powders were dry screened to 80-mesh and wet screened to 325-mesh. Powders were uniaxially pressed and sintered in Ar or N_2 at 1840 or 1875 °C for 10 minutes. HIPing was performed for some specimens by stacking the parts on a graphite setter plate and heating for 1 hour at 1300-1700 °C while increasing the argon pressure and holding the temperature for 15-30 minutes. Some specimens were only hot-pressed at 1650 °C for 10 minutes in argon.

Sintering Additives

3.7 wt% TiH_2 (TI 353, Atlantic Equipment Engineers, Bergenfield, NJ)

Reinforcement Description

TiC powder

Reinforcement Supplier

Hermann C. Starck Inc., New York, NY

Supplier Designation

Grade CA

Matrix Description

Al_2O_3 powder

Matrix Supplier

ALCOA Aluminum Co., Pittsburgh, PA

Matrix Designation

A-16 Superground

Test Type

3-pt. Flexural

Specimen Preparation

30 μm surface finish

Specimen Dimensions

4 x 5 x 50 mm
inner span = 20 mm
outer span = 40 mm

Test Description

crosshead speed = 0.5 mm/min.

Test Equipment

Model 1123 Universal Testing Machine (Instron Corporation, Canton, MA)

Relative Density (percent)	Flexural Strength (MPa)	Range (MPa)	HIPed Temperature (°C)	Sintering Temperature (°C)	Comments
94.9	480	±94		1875	*Sintered in argon*
97.5	415	±55		1840	*Sintered in N_2*
97.9	555	±58	1300	1840	*Sintered in N_2 and HIPed*
100	681	±28	1600	1840	
99.5	717	±10	1700	1840	
99.5	638	±127	1600	1875	*Sintered in argon and HIPed*
98.6	583	±73	---	---	*Hot-Pressed*

Remarks/Comments

Maximum strength was achieved by sintering in N_2 and HIPing at 1700 °C

$TiC(p)/Al_2O_3$

Reference: 68

Composition
30 wt% TiC

Processing Method
CIPed/Sintered (some HIPed) or Hot-Pressed

Detailed Description
The powders were milled in acetone. The milled slurry was dried, uniaxially pressed at 50 MPa, and isostatically pressed at 200 MPa. Some compacts were sintered using a heating rate of 5 °C/min. under helium with a 10 minute hold at 1870 °C. Some samples were also embedded with C powder to prevent chemical reactions and vaporization from the specimen. Final densification for some of the sintered specimens was done by HIPing at 1550 °C for 30 minutes under 150 MPa pressure of argon. For comparison, other samples were prepared by hot-pressing at 1650 °C and 40 MPa in a vacuum for 30 minutes.

Sintering Additives
0.01 wt% MgO

Reinforcement Description
TiC powder

Reinforcement Supplier
H.C. Starck, West Berlin, Germany

Reinforcement Dimensions
particle size = 1-3 μm

Matrix Description
Al_2O_3 powder

Matrix Supplier
Sumitomo Chemical Co., Osaka, Japan

Matrix Properties
particle size = 0.4 μm

Test Type
3-pt. Flexural

Specimen Preparation
Surfaces and edges were polished with an 800-grit diamond wheel and 1 μm diamond paste.

Specimen Dimensions
3 x 4 x 25 mm

Test Description
indentor load = 100 N

Test Temperature
RT

Relative Density (percent)	Flexural Strength (MPa)	Range (MPa)	Comments
96.7	625	±121	*sintered at 1870 °C in He for 10 minutes with embedding*
99.8	686	±95	*sintered and HIPed at 1550 °C and 150 MPa in Ar for 30 minutes*
99.9	721	±84	*hot-pressed at 1650 °C and 40 MPa in vacuum for 30 minutes*

Remarks/Comments
Flexural strength was taken as the average of 5 samples. Typical flexural strength of Al_2O_3-30 wt% TiC composites which were sintered with embedding powder was 625 MPa; compared to 690 MPa for the composites that were sintered with embedding powder and hot isostatically pressed. Hence, the increase of sintered density by HIPing resulted in improved flexural strength and better reliability.

Al_2O_3(p)/TiB_2

Reference: 100

Composition
0 and 30 vol% particulates

Processing Method
Hot-Pressed

Detailed Description
Monolithic TiB_2 was hot-pressed at 1850 °C and 30 MPa for 1 hour. TiB_2-Al_2O_3, 30 vol% TiB_2, was produced by hot-pressing at 1600 °C and 30 MPa for 30 minutes.

Reinforcement Description
Al_2O_3 particulates

Reinforcement Supplier
ALCOA

Supplier Designation
A16 S.G.

Matrix Description
TiB_2 powder

Matrix Supplier
Starck

Matrix Designation
Grade F

Test Type
Flexural Test

Reinforcement Content (volume percent)	Flexural Strength (MPa)	Range (MPa)
0	473	±19
30	711	±9

Remarks/Comments
Flexural strength increased with the addition of 30 vol% Al_2O_3.

Figure 12a: 3-pt. Flexural Strength of Pressureless Sintered $TiB_2(p)/Al_2O_3$ and $TiC(p)/Al_2O_3$

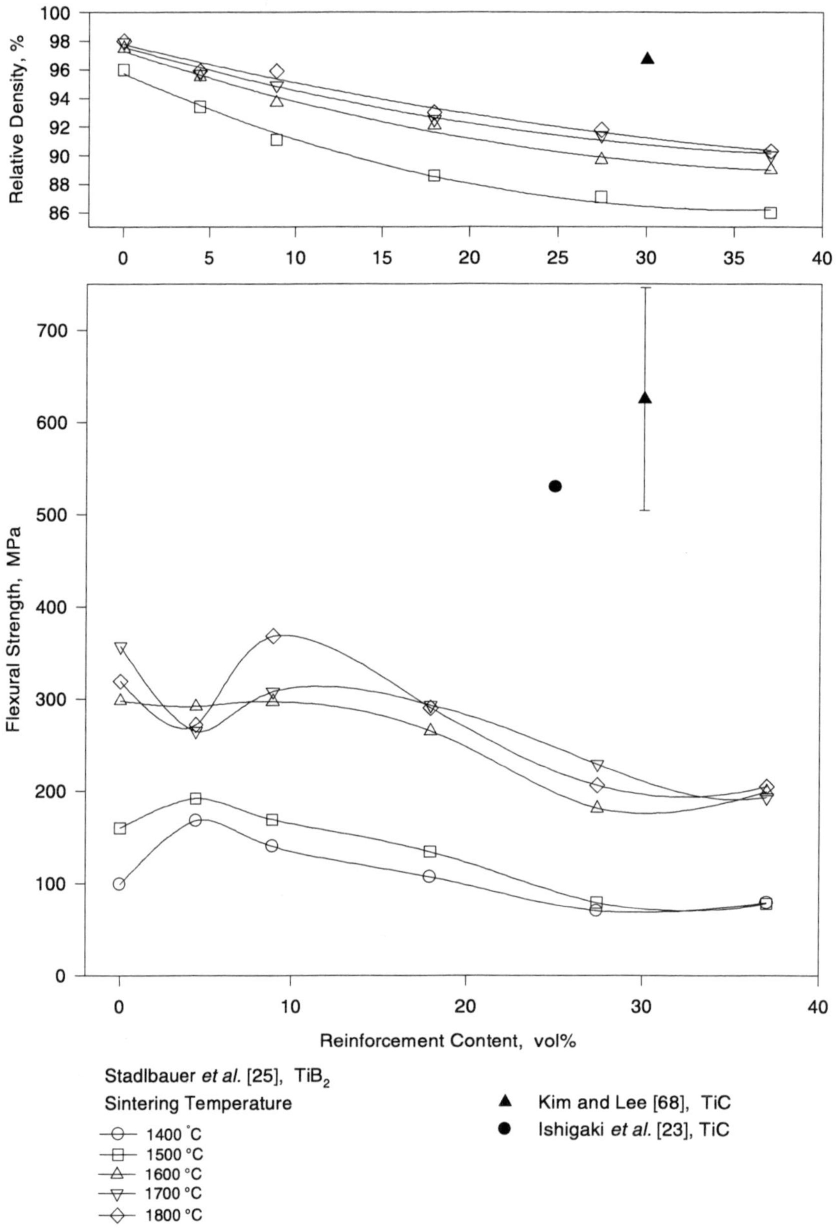

Discussion: Figure 12a shows the flexural strength of pressureless-sintered TiB_2 and TiC particulate reinforced alumina. Stadlbauer *et al.* [25] found that increasing the sintering temperature increased the densification of the $TiB_2(p)/Al_2O_3$ composite. For a given sintering temperature, the addition of TiB_2 particulates inhibited densification, and relative density decreased as TiB_2 content increased. This increase in porosity with increasing TiB_2 content caused the strength of the composite to decrease. The $TiC(p)/Al_2O_3$ composites showed a considerably greater flexural strength at an equivalent reinforcement content. The composite tested by Kim and Lee [68] had the highest strength of over 600 MPa despite being less than 97% dense. The composite tested by Ishigaki *et al.* [23] showed a slightly lower strength of 530 MPa, but was still well above monolithic Al_2O_3, which had a strength of around 300 to 400 MPa.

Figure 12b: 3-pt. Flexural Strength of Hot-Pressed $TiB_2(p)/Al_2O_3$ and $TiC(p)/Al_2O_3$

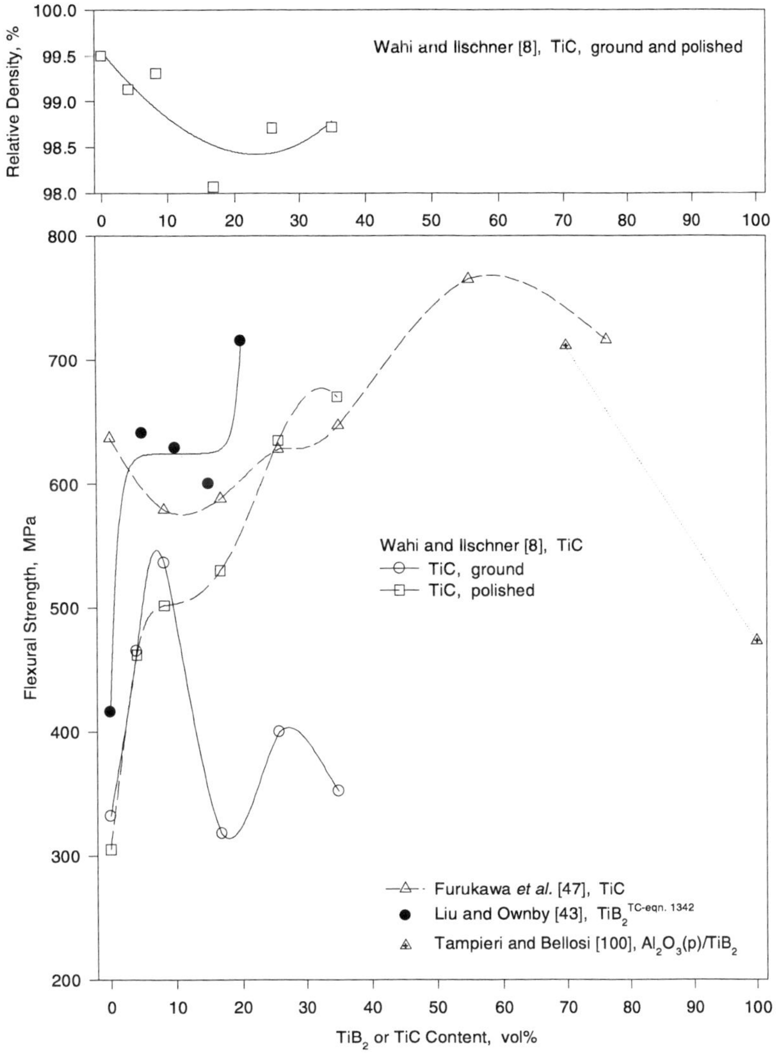

Discussion: Wahi and Ilschner [8] examined the effect of sample preparation on flexural strength. The grinding-induced surface defects were large enough to compensate for the smaller increase in strength due to the addition of TiC particulates. Polishing the samples brought this effect to a minimum influence. At 10 % reinforcement content, flexural strength reached a maximum for the ground sample. This was due to the density at that location reaching a maximum. Liu and Ownby [43] found that the TiB_2 composites had an increasing strength with a maxima of 712 MPa at 20 vol% TiB_2. This effect was much like the way the ground sample from ref. [8] behaved. Tampieri and Bellosi [100] found that the addition of 30 vol% Al_2O_3 to a TiB_2 matrix increased the strength by over 200 MPa.

The fracture toughness of the composites tested by Furukawa *et al.* [47] increased linearly with the addition of TiC from 3 MPa $m^{1/2}$ for monolithic alumina to 6.9 MPa $m^{1/2}$ for the 55 vol% $TiC(p)/Al_2O_3$ composite shown in figure 65. This fracture toughness data strongly suggests that the increase in strength was due to the increase in toughness. The slight decrease in strength at 10 and 20 wt% TiC was due to the grinding induced surface defects that were large enough to more than compensate for the increase in toughness due to the addition of TiC [47].

Figure 13: 3-pt. Effects of Pressure Processing on Flexural Strength of $TiC(p)/Al_2O_3$

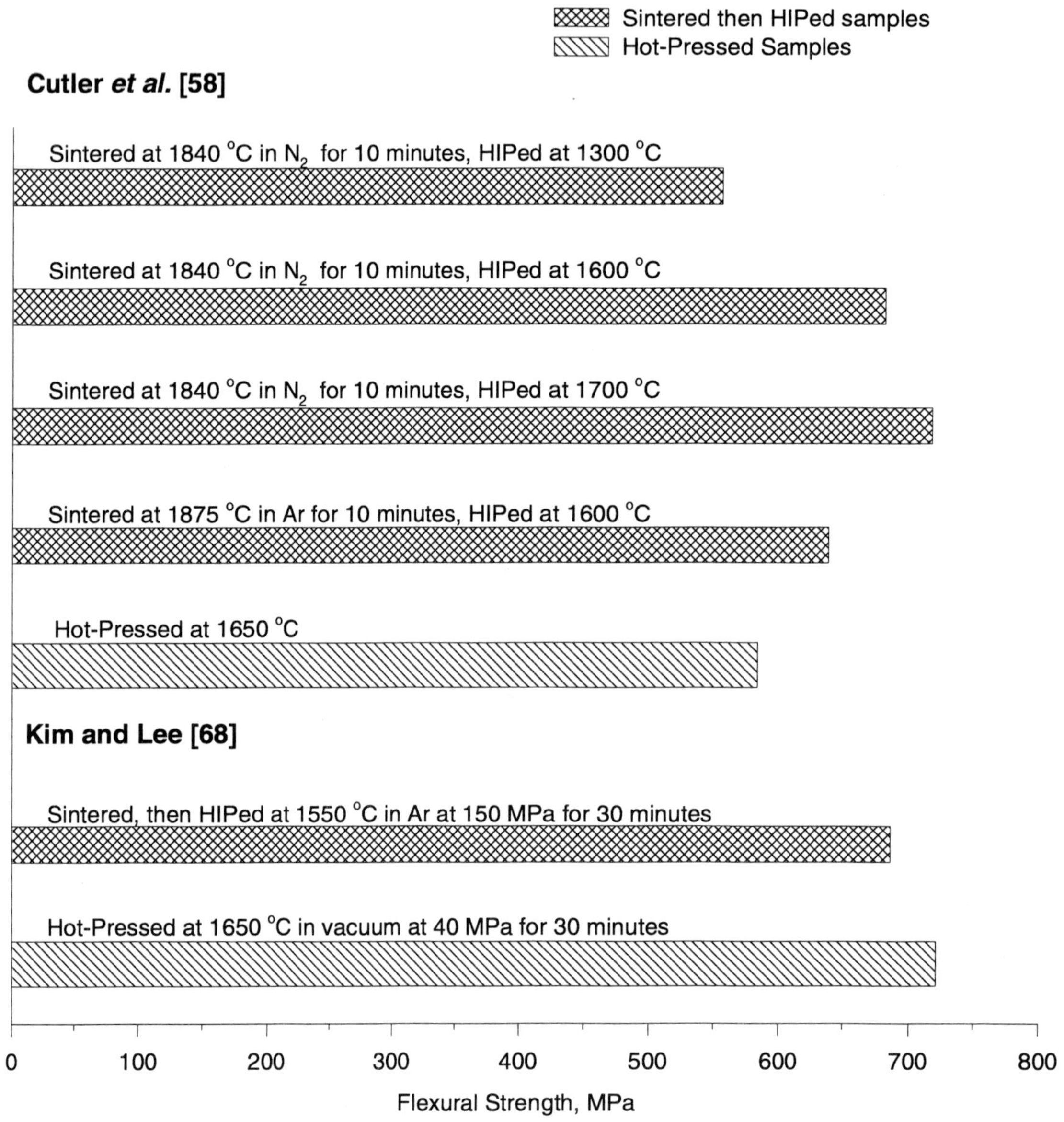

Discussion: References [58] and [68] studied the change in strength as a function of processing method. Cutler *et al.* [58] examined the effect of adjusting the HIP temperature and other processing parameters. For the samples sintered at 1840 °C (the top three samples shown in figure 13), it was found that increasing the HIP temperature increased the composite strength. For the sample sintered in Ar at 1875 °C, strength was lower than another sample sintered with the same HIPing conditions that were sintered at 1840 °C in N_2. Cutler *et al.* [58] found that the strength of the hot-pressed composite was lower than sintered-HIPed composites which were HIPed at or above 1600 °C

The sinter/HIPed samples from Kim and Lee [68] corresponded to the samples of ref. [58] with like processing (sintered in Ar). The hot-pressed composite was slightly stronger than the sinter/HIPed composite.

Figure 14: 3-pt. Flexural Strength vs. Hot-Pressing Temperature of TiC(p)/Al_2O_3

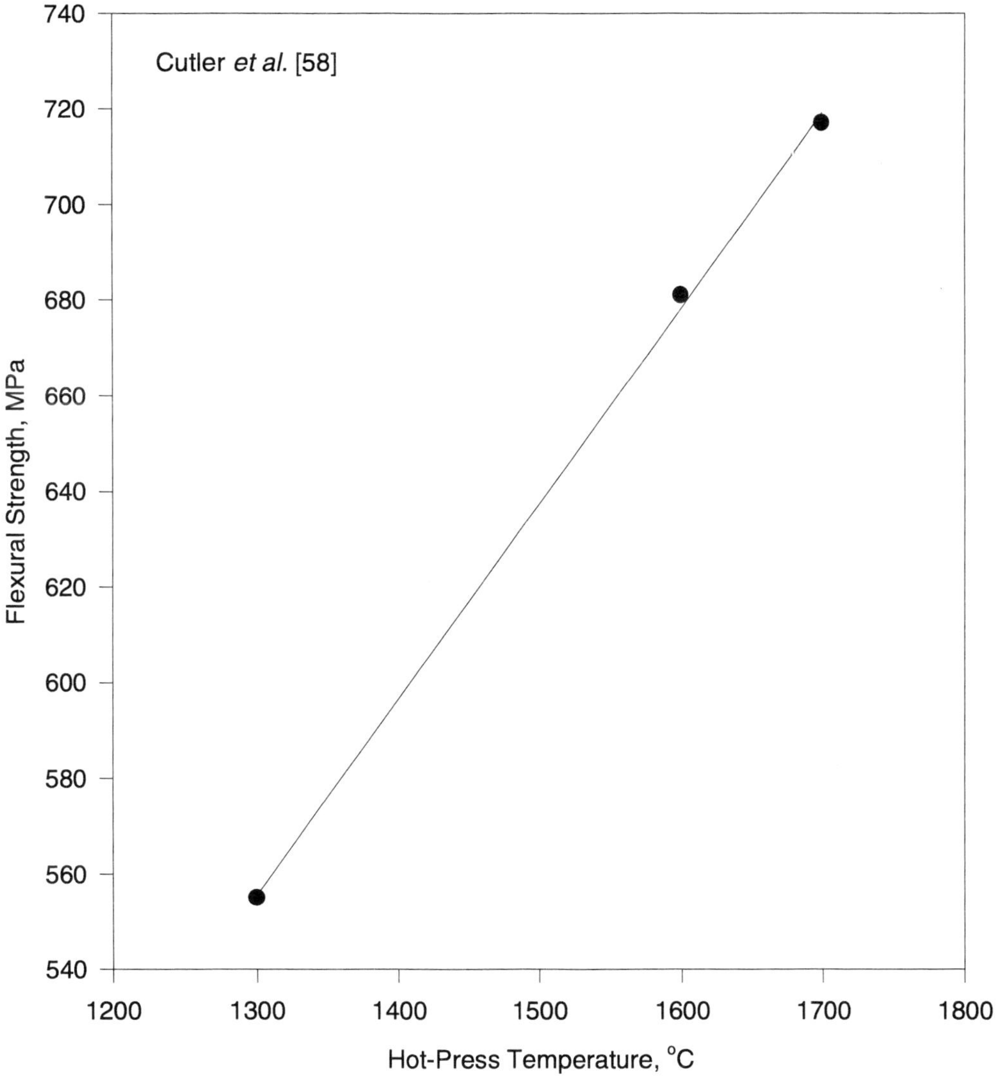

Discussion: Flexural strength increased with hot-pressing temperature due to a sintering boost given by the Al_2O_3-Al_4O_4C eutectic liquid formed from the small amount of free carbon that existed in the TiC powders. Sintering the powders in excess of 1925 °C lowered the density of the composite owing to the reaction between Al_2O_3 and carbon [58].

Figure 15: 3-pt. Flexural Strength vs. Sintering Temperature of Sintered $TiB_2(p)/Al_2O_3$

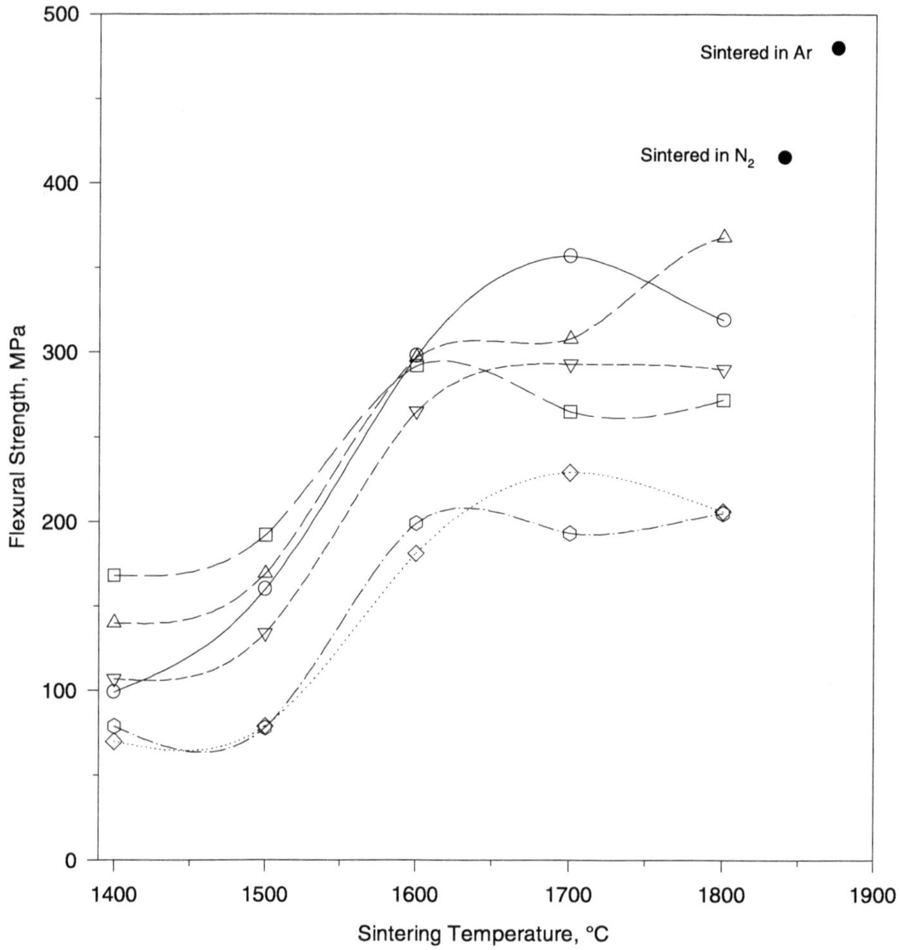

Stadlbauer *et al.* [25]
- Monolithic Al_2O_3
- 4.4 vol% TiB_2
- 8.9 vol% TiB_2
- 18.0 vol% TiB_2
- 27.4 vol% TiB_2
- 37.0 vol% TiB_2

Cutler *et al.* [58]
- 20 vol% TiC

Discussion: Stadlbauer *et al.* [25] found that density increased with increasing sintering temperature; and for a given sintering temperature, density decreased with an increase in TiB_2 concentration. This increase in density led to an increase in flexural strength, while an increase in TiB_2 concentration led to a decrease in flexural strength. The samples of Cutler *et al.* [58], both of 20 vol% TiC, showed that sintering in Ar at a higher temperature, ~65 °C higher, gave a greater strength than sintering in N_2.

TiN(p)/Al_2O_3

Reference: 84

Composition
0 and 30 vol% TiN

Processing Method
Hot-Pressed

Detailed Description
Al_2O_3 powders were homogenized with 30 vol% TiN in plastic jars with alumina balls and isobuthyl alcohol. The composites were hot-pressed at 1600-1650 °C and 30 MPa in a vacuum.

Reinforcement Description
TiN particulates

Matrix Description
Al_2O_3 powder

Test Type
4-pt. Flexural

Specimen Dimensions
3 x 3 x 30 mm
inner span = 13 mm
outer span = 26 mm

Test Description
crosshead speed = 0.5 mm/min

Test Temperature
RT, 700, and 800 °C

Test Environment
Air

Temperature (°C)	Flexural Strength (MPa)	Range (MPa)	Relative Density (percent)	Reinforcement Content (volume percent)
20	436	±35	100	0
700	346	±35	100	0
800	364	±19	100	0
20	729	±53	99.3	30
700	521	±44	99.3	30
800	448	±16	99.3	30

Remarks/Comments
The flexural strength increased for the composites compared to the base Al_2O_3 matrix. A progressive decrease in flexural strength was observed at high temperatures. This is linked to the defect population and it is known that this population changes unpredictably with temperature.

$TiC(p)/Al_2O_3$

Reference: 84

Composition
0 and 30 vol% TiC

Processing Method
Hot-Pressed

Detailed Description
Al_2O_3 powders were homogenized with 30 vol% TiC in plastic jars with alumina balls and isobuthyl alcohol. The composites were hot-pressed at 1600-1650 °C and 30 MPa in a vacuum.

Reinforcement Description
TiC particulates

Matrix Description
Al_2O_3 powder

Test Type
4-pt. Flexural

Specimen Dimensions
3 x 3 x 30 mm
inner span = 13 mm
outer span = 26 mm

Test Description
crosshead speed = 0.5 mm/min

Test Temperature
RT, 700, and 800 °C

Test Environment
Air

Temperature (°C)	Flexural Strength (MPa)	Range (MPa)	Relative Density (percent)	Reinforcement Content (volume percent)
20	436	±35	100	0
700	346	±35	100	0
800	364	±19	100	0

20	785	±86	99.4	30
700	565	±52	99.4	30
800	491	±12	99.4	30

Remarks/Comments
The flexural strength increased for the composites compared to the base Al_2O_3 matrix. A progressive decrease in flexural strength was observed at high temperatures. This is linked to the defect population and it is known that this population changes unpredictably with temperature.

$TiB_2(p)/Al_2O_3$

Reference: 84

Composition
0 and 30 vol% TiB_2

Processing Method
Hot-Pressed

Detailed Description
Al_2O_3 powders were homogenized with 30 vol% TiB_2 in plastic jars with alumina balls and isobuthyl alcohol. The composites were hot-pressed at 1600-1650 °C and 30 MPa in a vacuum.

Reinforcement Description
TiB_2 particulates

Matrix Description
Al_2O_3 powder

Test Type
4-pt. Flexural

Specimen Dimensions
3 x 3 x 30 mm
inner span = 13 mm
outer span = 26 mm

Test Description
crosshead speed = 0.5 mm/min

Test Temperature
RT, 700, and 800 °C

Test Environment
Air

Temperature (°C)	Flexural Strength (MPa)	Range (MPa)	Relative Density (percent)	Reinforcement Content (volume percent)
20	436	±35	100	0
700	346	±35	100	0
800	364	±19	100	0
20	711	±9	98.6	30
700	487	±23	98.6	30
800	437	±14	98.6	30

Remarks/Comments

The flexural strength increased for the composites compared to the base Al_2O_3 matrix. A progressive decrease in flexural strength was observed at high temperatures. This is linked to the defect population and it is known that this population changes unpredictably with the temperature.

TiN(p)/Al_2O_3

Reference: 94

Composition

50 wt% TiN

Processing Method

CIPed/Reaction Sintered or Reaction Hot-Pressed

Detailed Description

When the two powders react, they produce a mixture of Al_2O_3 and TiN according to the reaction:
$xTiO_2 + 2AlN \rightarrow Al_2O_3 + xTiN + (1\text{-}x/2)N_2 + (x\text{-}3/2)O_2$.
The powders were ball-milled with acetone and were reacted in an alumina crucible under a 10 mm overpressure of nitrogen at 1600 °C. The weight loss varied from 5-5.5 %. 50 x 20 x 20 mm sintered billets were made by initial CIPing, followed by sintering at 1600 °C in N_2 under a cover of AlN and with a 10 mm overpressure of N_2. For other specimens, hot-pressing was done on 70 mm discs at 1550 °C and 30 MPa in a BN coated graphite die and punch and using 10 mm N_2 overpressure.

Reinforcement Description

TiO_2

Reinforcement Supplier

Glaxo Labs, India

Supplier Designation

Laboratory grade

Matrix Description

AlN

Matrix Supplier

C.S. Starck, Berlin, Goslav, Germany

Matrix Designation

Grade A

Test Type

4-pt. Flexural

Specimen Dimensions

inner span = 7 mm
outer span = 20 mm

Test Description

crosshead speed = 0.1 mm/min

Test Temperature

RT, 700, and 800 °C

Test Environment

Air

Temperature (°C)	Flexural Strength (MPa)	Std. Dev. (MPa)	Comments
20	421	16	*reaction hot-pressed*
700	420	15	
800	391	20	
20	358	13	*reaction-sintered*
700	358	20	
800	338	19	

Remarks/Comments

The data points are the average of four samples. The strength values were very well retained up to 700 °C, falling only by 6 % at 800 °C. For the hot-pressed samples, the load was applied parallel to the hot-pressing direction. The hot-pressed samples showed higher strengths than the pressureless-sintered samples.

Figure 16: 4-pt. Flexural Strength vs. Temperature of TiB_2(p)/Al_2O_3, TiC(p)/Al_2O_3, and TiN(p)/Al_2O_3

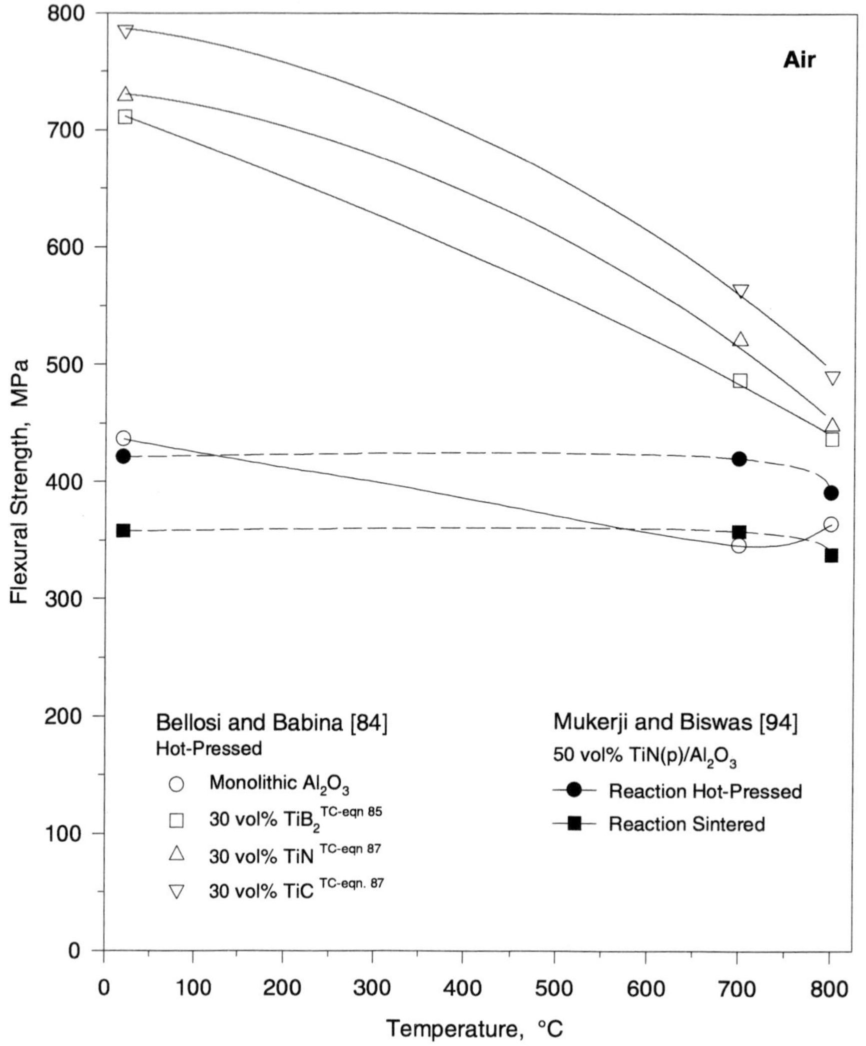

Discussion: Bellosi and Babina [84] found that the composites tended to decrease in strength as temperature increased. Toughening and reinforcing mechanisms were due to crack interactions with the particulate reinforcement. The progressive decrease in strength observed at high temperatures was due to the oxidation of the reactive second phase particles. The large volume expansion during oxidation created microcracks and weak points in the matrix that caused a decrease in strength at a much lower temperature than would normally be expected without the effect of oxidation. The reaction hot-pressed composite tested by Mukerji and Biswas [94] exhibited a greater flexural strength than the reaction sintered composite because of its higher relative density.

$ZrO_2(p)/Al_2O_3$

Reference: 16

Composition
0-30 wt% ZrO_2

Processing Method
CIPed/Sintered

Detailed Description
The appropriate mixtures of Al_2O_3 and ZrO_2 were ground for 6 hours in acetone. Grinding media, as well as crucibles, were made of alumina. The mixed powders were isostatically pressed at 150 MPa at room temperature into cylindrical bars of 200 x 8 mm dimensions and then pre-sintered in air at 900 °C. From these, specimens were diamond cut to standard sizes and then sintered at temperatures from 1150 °C to 1680 °C. The heating rate up to 1150 °C was 3 °C/min and from 1150 °C to the final temperature, 1 °C/min. The sintering time was 3 hours, and the cooling rate was 2 °C/min.

Reinforcement Description
0Y-ZrO_2 and 5Y-ZrO_2

Reinforcement Supplier
Dynamit Nobel

Supplier Designation
Dynazirkon F (monoclinic unstabilized), Degussa (pyrogenic unstabilized), and Dynazirkon F5Y (Y_2O_3 stabilized)

Reinforcement Dimensions
mean particle size = 0.7 µm (Dynazirkon F), 0.01 µm (Degussa), or 0.35 µm (Dynazirkon F5Y)

Matrix Description
Al_2O_3 powder

Matrix Supplier
ALCOA

Matrix Designation
A16SG

Matrix Properties
mean particle size = 0.5 µm

Test Type
3-pt. Flexural

Specimen Dimensions
35 x 4.5 x 3.5 mm
span = 25 mm

Test Description
crosshead speed = 0.1 mm/min

Test Equipment
Zwick 147 (Zwick, Germany)

Reinforcement Content (weight percent)	Flexural Strength (MPa)	Sintering Temperature (°C)	Comments
0	351	1,510	*0Y-ZrO_2*
0	350	1,540	
0	331	1,570	
0	299	1,600	
0	239	1,640	
5	334	1,540	*0Y-ZrO_2*
5	394	1,600	
5	283	1,640	
5	370	1,540	*5Y-ZrO_2*
5	378	1,600	
5	438	1,640	
5	403	1,680	
15	263	1,540	*0Y-ZrO_2*
15	300	1,600	
15	140	1,640	
15	427	1,540	*5Y-ZrO_2*
15	474	1,600	
15	528	1,640	
15	442	1,680	
30	208	1,540	*0Y-ZrO_2*
30	190	1,600	
30	130	1,640	
30	442	1,540	*5Y-ZrO_2*
30	550	1,600	
30	575	1,640	
30	480	1,680	

Remarks/Comments
The flexural strength was found to depend on the sintering temperature. The maximum strength was found at around 1640 °C sintering temperature. Samples with the Y_2O_3 stabilizer tended to have a higher flexural strength over those samples without any stabilizer.

$ZrO_2(p)/Al_2O_3$

Reference: 22

Composition
0-100 wt% ZrO_2

Processing Method
Slip-Cast/Sintered

Detailed Description
Powders were milled for 3 days with 1 % dispersant (Celuna D-305) and formed by slip casting in a rectangular mold 4 x 10 x 80 mm. Cast samples were sintered at 1550 °C for 3 hours.

Reinforcement Description
(1) 3Y-ZrO_2
(2) 0Y-ZrO_2

Reinforcement Supplier
(1) TSK Co., Sinnanyo, Yamaguchi, 746, Japan
(2) Daiichi Kigenso Chemicals Co., Osaka, Japan

Supplier Designation
(1) TZ-3Y
(2) EP-ZrO_2

Matrix Description
α-Al_2O_3

Matrix Supplier
ALCOA

Matrix Designation
A-16

Matrix Properties
ave. particle size = 0.3 μm

Test Type
3-pt. Flexural

Specimen Preparation
Specimens were surface polished.

Specimen Dimensions
span = 30 mm

Test Description
crosshead speed = 102 mm/min

Test Temperature
RT

Reinforcement Content (weight percent)	Flexural Strength (MPa)	Comments
0	263	*unstablilized ZrO_2*
10	429	
20	364	
30	165	
40	178	
50	156	
60	179	
70	138	
80	133	
90	69	
0	383	*3 mol% Y_2O_3*
10	540	
20	575	
30	629	
40	575	
50	734	
60	767	
70	747	
80	999	
90	1,030	
100	700	

Remarks/Comments
Transformation toughening caused an increase in strength with the addition of stabilized zirconia. The addition of unstabilized zirconia increased critical flaw size and caused macrocracking, thus resulted in a decrease in strength.

$ZrO_2(p)/Al_2O_3$

Reference: 57

Composition
0-15 vol% ZrO_2

Processing Method
Pressure-Assisted Slip Cast/Pressureless-Sintered

Detailed Description
Slip preparation was done using an ultrasonic probe for 10 minutes. Alumina/zirconyl chloride dispersions were sprayed into 1 mol/dm^3 NH_4OH solution and stirred with a magnetic stirrer. Centrifuging was done at 3000 rpm for 1 hour, the supernatant was discarded, and the sediment was redispersed in a minimum amount of distilled water at pH = 5.5. Slip was pressure-casted into small discs then dried in an oven at 80 °C for 24 hours. Two sets of differently processed and sintered zirconia/alumina composites were then prepared from the pressure cast and dried green samples.

(1) The first set of dry green samples were sintered at 1600 °C for 2 hours then cooled. To minimize thermal stresses due to the release of water during heating, the green samples were first heated at the rate of 2 °C/min. until the furnace reached 200 °C when most of the water was released, baked at 200 °C for 2 hours, and ramped to 500 °C at 2 °C/min. The samples were heat treated at 500 °C for 2 hours. Later, these samples were heated at the rate of 3 °C/min. until the furnace reached the sintering temperature of 1600 °C.

(2) The second set of green samples were heated directly to 500 °C (calcining temperature) at the rate of 2 °C/min. After cooling, these green bodies were dry pressed at 50 MPa and sintered at 1600 °C.

Reinforcement Description
ZrO_2 powder

Reinforcement Typical Properties
density = 4.29 g cm^{-3} for as-synthesied zirconia hydrate

Reinforcement Dimensions
surface area = 82 m^2 g^{-1}
ave. dia. < 0.2 μm
crystallite size = 17 nanometer for zirconia hydrate as-synthesized

Matrix Description
Al_2O_3 powder

Matrix Supplier
ALCOA Aluminum Co., Pittsburgh, PA

Matrix Designation
A16SG

Matrix Properties
density = 3.98 g cm^{-3}
surface area = 10.2 m^2 g^{-1}
ave. dia. = 2.0 μm

Test Type
3-pt. Flexural

Specimen Preparation
Specimens were polished prior to testing.

Reinforcement Content (volume percent)	Flexural Strength (MPa)	Comments
0.0	365	*(1) sintered at 1600 °C with drying*
2.5	393	
5.0	401	
7.5	409	
10	422	
13	470	
15	491	
5.0	454	*(2) presintered at 500 °C and sintered at 1600 °C without drying*
7.5	523	
10	580	
13	531	
15	511	

Remarks/Comments
The strength of the green samples calcined at 500 °C and dry-pressed was 50 % higher at 10 vol% ZrO_2 than those that were directly sintered.

$Al_2O_3(p)/ZrO_2$

Reference: 108

Composition
10 and 82 wt% Al_2O_3

Processing Method
Sol-Gel/Sintered

Detailed Description
Sol gels were aged at room temperature for one day then pre-heated at 500, 700, or 900 °C. All powders were ground in a zirconia automortar to reduce aggregation during pretreatments. The powders were pressed under 295 MPa pressure at room temperature with 1 minute hold. Pellets were formed with a diameter of 13 mm and a thickness between 3.5-4 mm. The pellets were dried at 110 °C for 24 hours and fired as follows: from 100 to 500 °C with a heating rate of 120 °C/hour with a 2 hour hold, then to a selected temperature at 1000 °C/h with a 3 hour hold.
Series A: 85.36 wt% ZrO_2 + 10.44 wt% Al_2O_3 + 4.2 wt% Y_2O_3
Series B: 17.28 wt% ZrO_2 + 82.0 wt% Al_2O_3 + 0.72 wt% Y_2O_3

Reinforcement Description
Al_2O_3 particulates

Matrix Description
ZrO_2

Test Type
3-pt. Flexural

Specimen Dimensions
span = 9 mm

Test Description
The test was conducted at a selected temperature at a heating rate of 240 °C/min. with 3 hours isothermal dwell time.

Test Temperature
RT

Sintering Temperature (°C)	Flexural Strength (MPa)	Relative Density (percent)	Comments
1,400	40.1	73.2	*pre-heated at 500 °C, Series A*
1,500	152	86.4	
1,575	348	93.9	
1,650	230	88.5	
1,400	86.3	84.8	*pre-heated at 700 °C, Series A*
1,500	224	87.4	
1,575	568	95.6	
1,650	621	96.9	
1,400	80.8	76.8	*pre-heated at 900 °C, Series A*
1,500	165	84.4	
1,575	451	93.6	
1,650	381	98.1	
1,400	51.5	59.0	*pre-heated at 500 °C, Series B*
1,500	67.0	69.9	
1,575	204	91.2	
1,650	276	97.5	
1,400	26.1	61.4	*pre-heated at 700 °C, Series B*
1,500	48.1	68.4	
1,575	208	71.3	
1,650	320	96.9	
1,400	76.5	58.9	*pre-heated at 900 °C, Series B*
1,500	78.5	68.4	
1,575	118	83.9	
1,650	250	69.5	

Remarks/Comments
Flexural strength depended on composition, preparation, and heat treatments. The best pre-heating temperature was 700 °C in order to achieve the highest flexural strength. At 900 °C, densification was slow, and the particles coarsened rapidly. At 500 °C, the composite still retained some porosity, thus lowering its flexural strength. Series A precursors resulted in much greater strength values than the Series B precursors.

Figure 17a: 3-pt. Flexural Strength of Sintered ZrO_2(p)/Al_2O_3

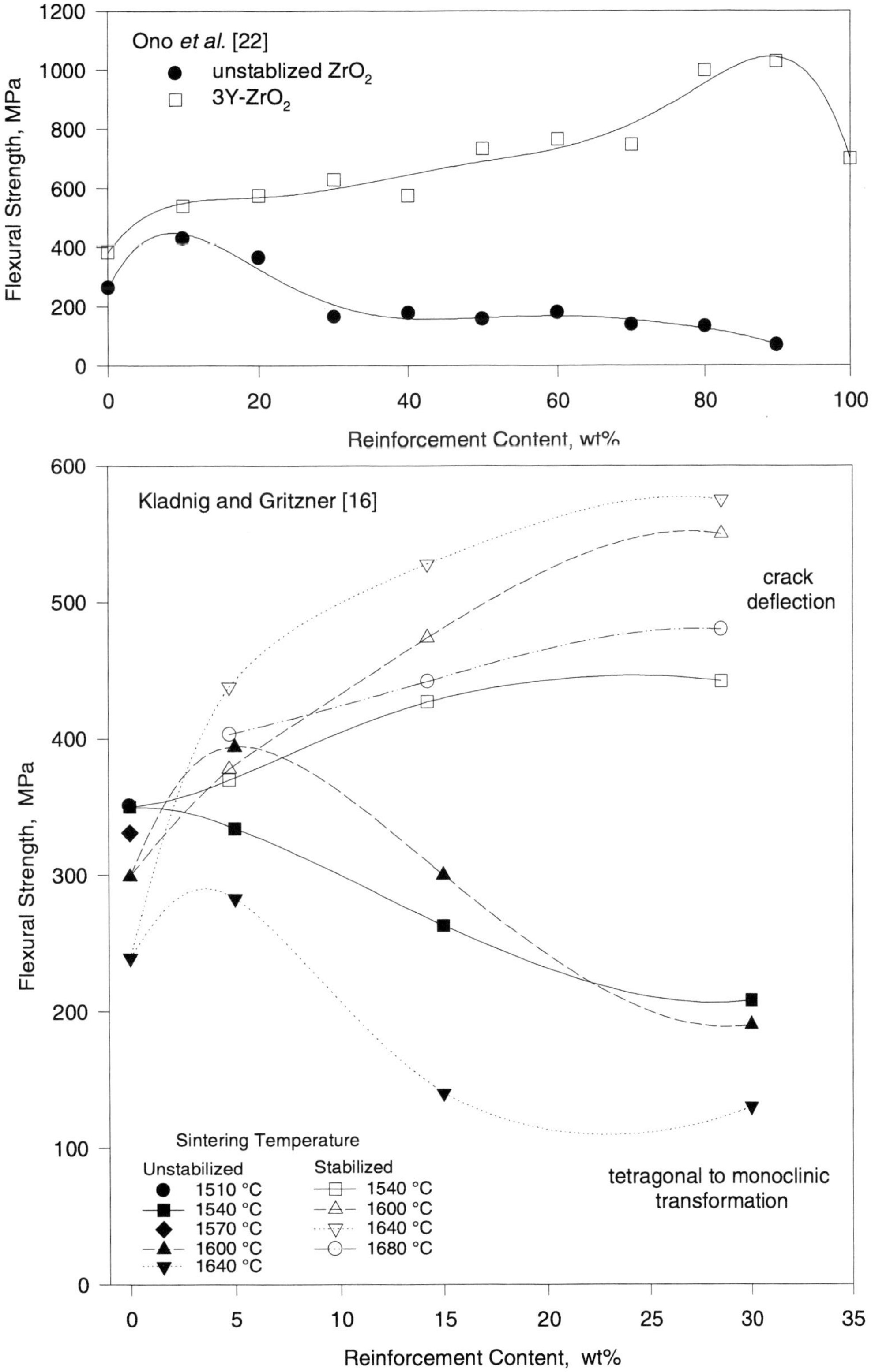

Discussion: Ono *et al.* [22] found that the amount and type of ZrO_2 added to the Al_2O_3 greatly affected the flexural strength of the samples (figure 17a); when more than 5 wt% unstabilized ZrO_2 was added to the composite, a large increase of the monoclinic phase occurred in the fracture area. The larger monoclinic size caused a coefficient of thermal expansion mismatch that resulted in macrocracks. For samples with less than 5 wt% ZrO_2, the phase change had a mismatch that the matrix absorbed without the need for macrocracks to relieve stress. In these cases there was a decrease in the percentage of tetragonal ZrO_2 on the fracture surface, indicating a phase change had impeded crack propagation by the t→m transformation. Fracture toughness of both the stabilized and unstabilized zirconia composites increased with increasing ZrO_2 content and reached a maximum around 40-80 wt% ZrO_2. At the same time the toughness increased, flexural strength decreased with the addition of unstabilized zirconia. This suggests an increasing critical flaw size with the addition of unstabilized zirconia. See the fracture toughness analysis presented with figures 70a-c for a more detailed analysis of fracture mechanisms and the effect of zirconia phase transformations.

Kladnig and Gritzner [16] found that the small size of the powder particles, at least at concentrations below 15% ZrO_2, caused most of the ZrO_2 to remain tetragonal. For the most part, only the tetragonal phase was present when the Y_2O_3 stabilized the ZrO_2 phase. This indicated that the increase in bend strength was due mainly to particulate crack deflection and only slightly to a phase change. The bend strength dropped well below the values for the pure Al_2O_3 samples whenever a substantial amount of monoclinic phase was present; the reason for this was that the t→m phase change caused macrocracking. Only when most of the ZrO_2 remained in the tetragonal phase were the bend strengths of pure Al_2O_3 reached [16].

For Figure 17b, all samples' strengths had a strong dependence on the sintering temperature. Maximum strength for the monolithic Al_2O_3 was 350 MPa. Maximum strength for the stabilized ZrO_2 composite was able to reach a strength in excess of the monolithic Al_2O_3 when sintered around 1650 °C. For the unstabilized ZrO_2, the maximum strength reached at sintering temperatures around 1600 °C was not as high as the monolithic Al_2O_3. Maximums in bend strength for the unstabilized ZrO_2 composites were noticeable, despite the fact that the densities of all samples increased with increasing sintering temperature [16]. Decreases in strength with increasing sintering temperature were reportedly due to the growth of ZrO_2 grains. The highest bend strength found in these studies was the composite with 30 % 5Y-ZrO_2 content. The value of 575 MPa was 225 MPa higher than the best observed value for pure Al_2O_3, and higher than the usual values recorded for composites of this nature. As shown in figure 17a, strength increased with increasing stabilized ZrO_2 content above the strength of monolithic Al_2O_3, and decreased below monolithic Al_2O_3 with increasing unstabilized ZrO_2 content.

Figure 17b: 3-pt. Flexural Strength vs. Sintering Temperature of $ZrO_2(p)/Al_2O_3$

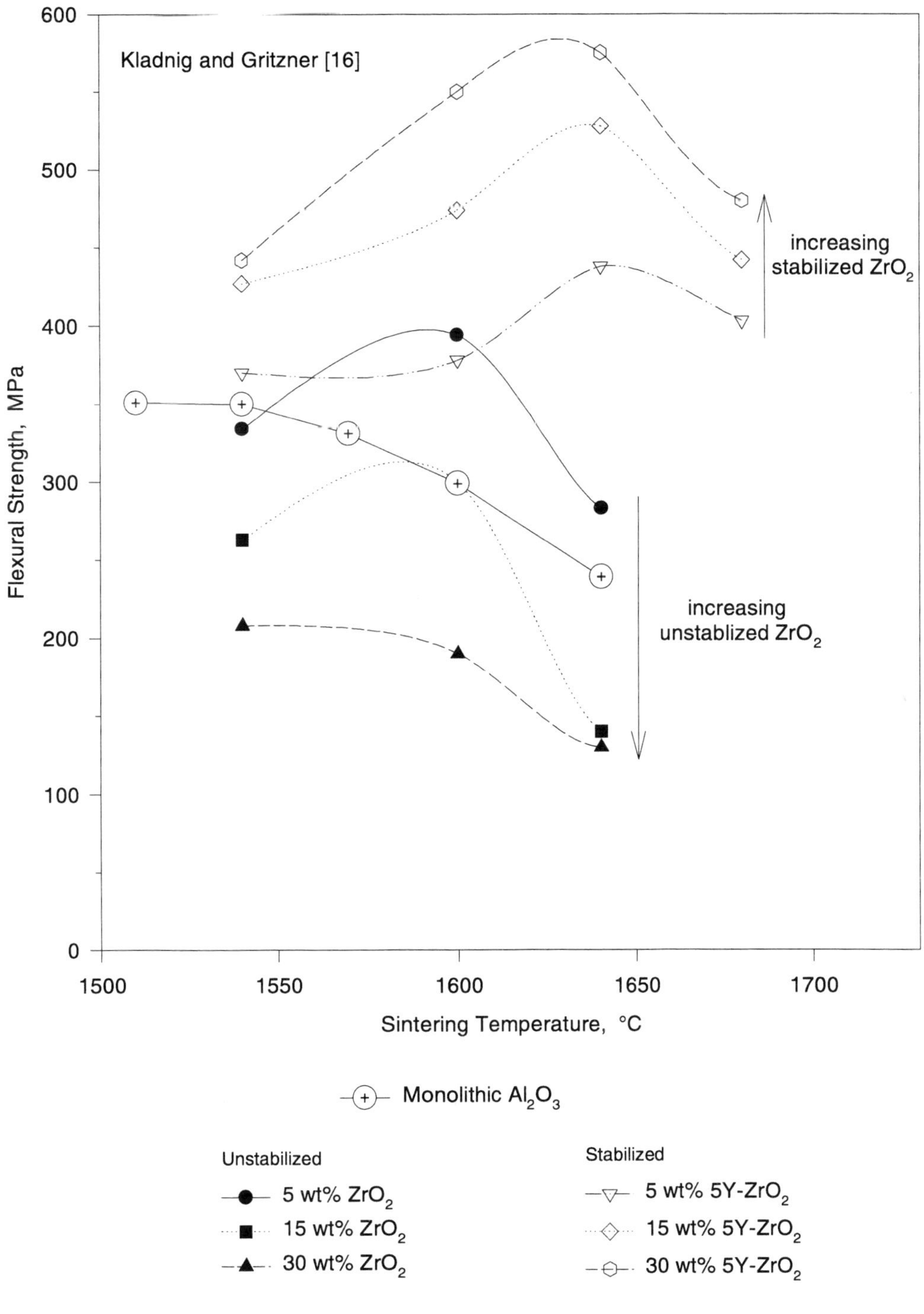

Figure 18: 3-pt. Flexural Strength of Sintered ZrO_2(p)/Al_2O_3

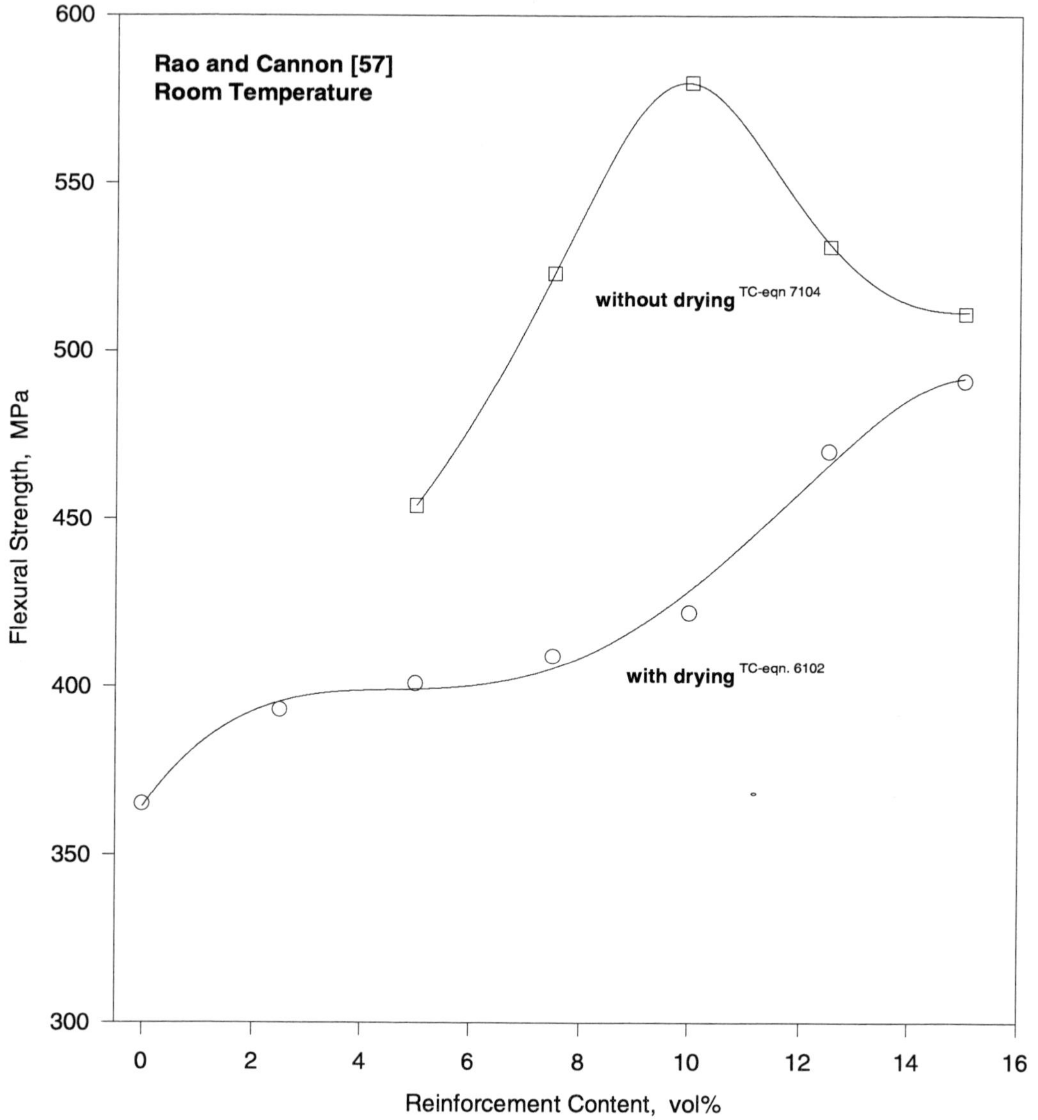

Discussion: Rao and Cannon [57] tested a zirconia-alumina composite slip-cast from sol precursors in a slurry. The slip-cast green bodies were either slowly dried or directly calcined and sintered without a drying step. The flexural strengths in both processes increased with the addition of up to 10 vol% zirconia due to several phenomena. First, fine zirconia particles coat the larger alumina particles before sintering. During sintering these fine particles coalesce into larger particles at different rates depending on orientation and concentration. If there is enough zirconia present, it will coalesce into particles of tetragonal zirconia, which are necessary for transformation toughening and strengthening. The particles must be larger than 0.1-0.2 μm to allow the t→m transformation. It can be observed from experimental results that the composite should contain at least 7.5 vol% zirconia [57]. For zirconia concentrations less than this amount, the particles do not migrate far enough to aggregate and coalesce. A presintering step (without drying) allowed for coalescence of the zirconia at a much lower zirconia content, which was demonstrated by the higher strength of the these composites. Slow drying of the sol-precursors may have limited migration of the zirconia. This suggests that pre-sintering conditions play a large role in the sintering thermodynamics, and can make it easier for zirconia particles to migrate for increased strength.

Figure 19: 3-pt. Flexural Strength vs. Relative Density and Sintering Temperatures of Sintered ZrO_2(p)/Al_2O_3 Sol-Gels

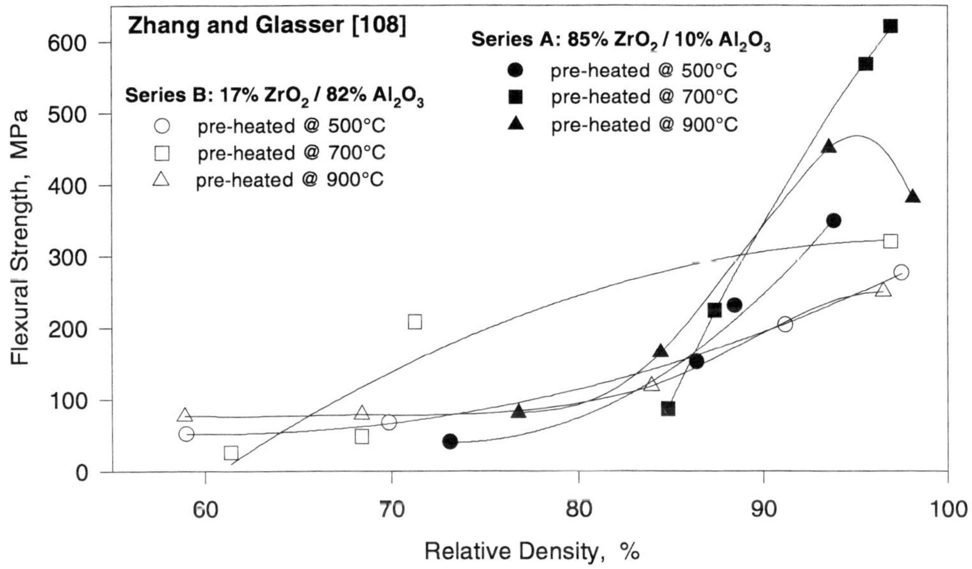

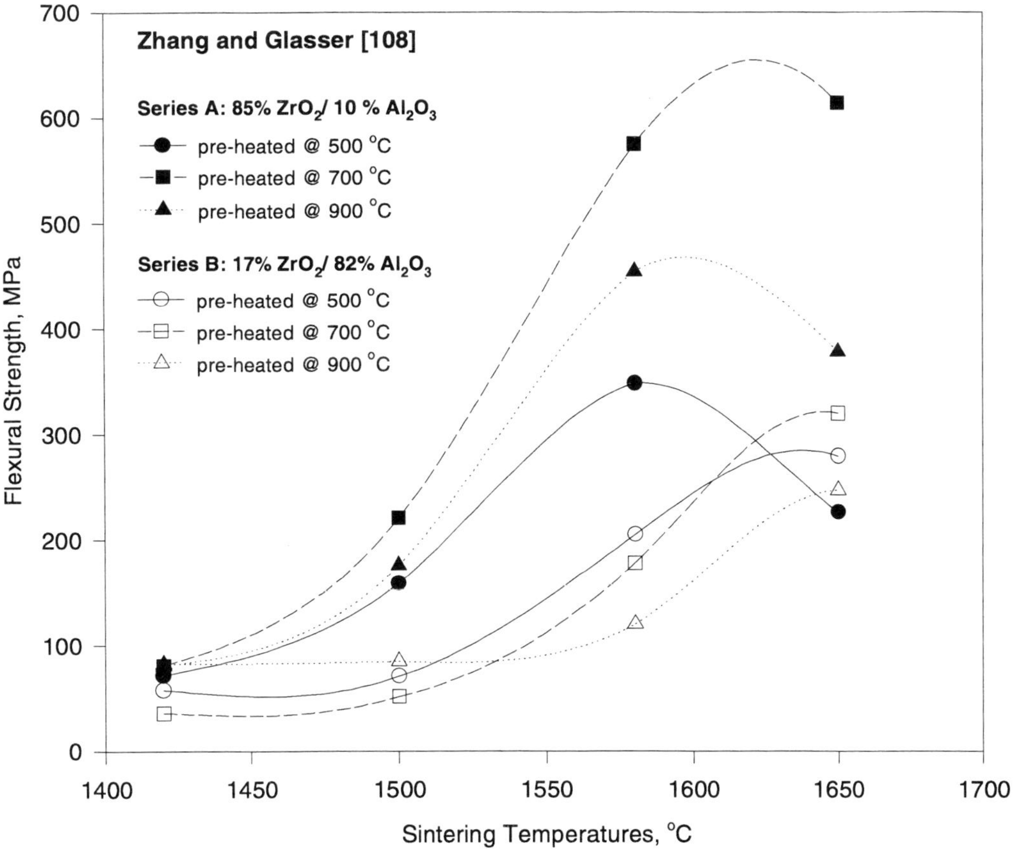

Discussion: The sol-gel precursors have many properties that make them preferable to other composite processing routes. One is an increased sintering rate. This is due to the gels containing smaller particles or pores, and gel densification during sintering is achieved by viscous flow instead of the usual diffusion process found in crystalline materials. The driving force for this viscous flow densification is the decrease of surface energy. Because of higher driving forces, sintering can be performed at relatively low temperatures. During higher temperature sintering water loss is progressive, and the resulting condensation reaction increases crosslink density while structural relaxation increases the skeletal density. The disadvantages of higher temperature sintering are that pore closure may trap gases that later expand and the rapid escape of volatiles may cause cracking [108]. So, this data explores the different heating rates to find an optimum heating schedule for sintering.

The materials in series A showed that the best preheating temperature to obtain the highest flexural strength was 700 °C. The 900 °C preheated samples were slow to densify and crystalline particles coarsened rapidly during the high temperature sintering. For the 500 °C treatment, rapid densification occurred because of the high specific area due to low pretreatment temperatures. After sintering, the 500 °C samples retained pores due to initial low green body density. In a comparison between series A and B, series B needed higher temperatures or longer sintering times to increase density for higher strengths. When sintered at the highest temperature (1650 °C), the 700 °C preheat cycle gave the best results for the series B composites.

The flexural strengths of series A were generally much higher than that of series B because series A contained much more stabilized tetragonal ZrO_2 [108]. The close grouping in series A and series B was due to the amount of ZrO_2. Series B did not have enough zirconia to undergo a phase change from tetragonal to monoclinic for inhibiting and clamping cracks. This resulted in overall lower strengths. Series A, on the other hand, did have a large enough zirconia concentration to do this, so it had higher strengths than series B. The large difference in maximum values was due to the different heating rates of the gels during sintering. The flexural strength of these composites were probably not optimal because of the relatively low green densities. Yet the flexural strengths were higher than that of typical ZrO_2/Al_2O_3 composites, which lie in the range of 400-550 MPa [108].

Zhang and Glasser [108] cited possible errors in the measurement of relative density, which may have had the effect of overestimating the relative density.

$ZrO_2(p)/Al_2O_3$

Reference: 7

Composition
0-30 vol% ZrO_2

Processing Method
Hot-Pressed

Detailed Description
Submicron composite powders, including yttrium nitrate (a Y_2O_3 precursor), were mixed by ball-milling with methanol and Al_2O_3 balls in a plastic container. All powders were dried and calcined at 400 °C for 4 hours. The composites were hot-pressed at 1600 °C for 2 hours. The pure Al_2O_3 end-member was hot-pressed at 1400 °C to achieve a grain size comparable to the two-phase materials.

Reinforcement Description
2Y-ZrO_2 particulates

Reinforcement Supplier
Zircas Corp.

Matrix Description
Al_2O_3 powder

Matrix Supplier
Lindy B, Union Carbide Corp.

Test Type
4-pt. Flexural

Specimen Preparation
Samples were polished with a 220 grit diamond wheel. Some samples were annealed at 1300 °C prior to testing.

Specimen Dimensions
3.2 x 3.2 mm cross-section
inner span = 12.2 mm
outer span = 25.4 mm

Type of Specimen
Bar

Test Temperature
RT

Reinforcement Content (volume percent)	Flexural Strength (MPa)	Range (MPa)	Comments
0	599	+29/-32	*Non-annealed*
6	767	+87/-151	
12	749	+38/-64	
18	1,070	+39/-29	
24	856	+52/-94	
30	1,130	+23/-23	
30	951	+90/-11	*Annealed at 1300 °C*

Remarks/Comments
The flexural strength of the composite was found to increase proportionally to the increase in fracture toughness. Additional strengthening can be obtained by compressive stressing the surface through abrasion. The specimen that was annealed at 1300 °C to remove the transformed layer had a flexural strength closer to what was predicted by theoretical modeling.

ZrO_2(p)/Al_2O_3

Reference: 7

Composition
24 vol% ZrO_2

Processing Method
Hot-Pressed

Detailed Description
Submicron composite powders, including yttrium nitrate (a Y_2O_3 precursor), were mixed by ball-milling with methanol and Al_2O_3 balls in a plastic container. All powders were dried and calcined at 400 °C for 4 hours. The composites were hot-pressed at 1600 °C for 2 hours.

Reinforcement Description
2Y-ZrO_2 particulates

Reinforcement Supplier
Zircas Corp.

Matrix Description
Al_2O_3 powder

Matrix Supplier
Lindy B, Union Carbide Corp.

Test Type
4-pt. Flexural

Specimen Preparation
Specimens were polished with a 220-grit diamond wheel.

Specimen Dimensions
3.2 x 3.2 mm cross-section
inner span = 12.2 mm
outer span = 25.4 mm

Type of Specimen
Bar

Test Temperature
-180-1000 °C

Test Environment
Air

Temperature (°C)	Flexural Strength (MPa)
-180	1,220
-75	1,220
25	880
35	911
200	720
300	765
400	618
600	669
700	610
800	527
1,000	305

Remarks/Comments
The flexural strength values corresponded well with the theoretical values, except below 25 °C. The low temperature fracture behavior was not well understood.

$Al_2O_3(p)/ZrO_2$

Reference: 12

Composition
0-7.5 wt% Al_2O_3

Processing Method
Cold-Pressed/Sintered

Detailed Description
The YSZ was washed with distilled water to remove chlorine. Fifty gram batches were prepared by adding the appropriate amount of Al_2O_3 to the washed YSZ and ball-milling for 12 hours. Milling was done in polypropylene jars using zirconia balls in a fish oil dispersing agent with distilled water and isopropanol. A binder mixture was added and milled for an additional 1.5 hours. Next, the mixtures were spray dried then cold pressed into pellets, 1.6 cm in diameter and ~1.2 mm thick, and bars 3.1 x 0.4 x 0.4 cm at 220 MPa and 280 MPa, respectively. The specimens were fired in a $MoSi_2$ resistance furnace on ZrO_2 setters in air. Sintering was for 8 hours at 1300 °C.

Reinforcement Description
Al_2O_3 particulates

Reinforcement Supplier
Apache Chemicals Inc., Seward, IL

Reinforcement Dimensions
ave. particle size = 0.03 μm

Matrix Description
4.5Y-ZrO_2

Matrix Supplier
Zircor Corp., Florida, NY

Matrix Properties
ave. particle size = 0.03 μm
contains 8.0 wt% Y_2O_3

Test Type
4-pt. Flexural

Specimen Preparation
Samples were polished with 45, 30, 15, 6, 1, and 0.25 μm diamond paste after initial grinding on a 70 μm diamond wheel.

Specimen Dimensions
25 x 4 x 2 mm
inner span = 12 mm
outer span = 23 mm

Test Description
crosshead speed = 0.1 mm/min

Test Equipment
Instron model 1125 strength tester

Reinforcement Content (weight percent)	Flexural Strength (MPa)	Std. Dev. (MPa)
0.00	272	19
0.33	317	50
0.65	304	34
1.30	272	37
3.25	318	36
5.00	422	58
7.50	332	48

Remarks/Comments
The flexural strength increased with increasing volume fraction up to only 0.325 wt% Al_2O_3 then decreased while maintaining values above the monolith. At 1.3 wt% Al_2O_3, the composite began to increase strength up to 5 wt% reinforcement. Beyond 5 wt%, strength decreased.

$ZrO_2(p)/Al_2O_3$

Reference: 26

Composition
15 wt% ZrO_2

Processing Method
CIPed/Pressureless-Sintered

Detailed Description
A 0.333 wt% water solution of polyvinyl alcohol was filtered through a 10-20 μm glass frit. 400 ml of the filtered solution was poured into a polypropylene bottle, and 40 g of an appropriate mixture of alumina, zirconia, and zirconia-yttria powders were then added. The bottle's contents were adjusted to pH 2-3 with 0.3 ml of concentrated nitric acid and ultrasonically agitated at high power for 8 minutes. This agitation warmed the slurry. The bottle containing the slurry was resealed, and the contents were allowed to gravitationally settle for 24 hours. The middle 300 ml of the supernatant was removed using a cannulation procedure at a rate of about 5 ml/sec. Filtered air was used as the cover gas during cannulation. The cannulated supernatant was homogenized before further use. It was then flocculated with 375 ml of filtered reagent-grade acetone. The flocculated slurry was allowed to stand for 15 minutes and was then poured onto a 100-mesh stainless steel screen. The retained flocs were scooped into a colloid press where they were consolidated at a pressure of 35 MPa over a period of 15-30 minutes. The sample was held in the press for 30 minutes after the press came to equilibrium, then was ejected from the press, dried at 60-80 °C in laboratory air, and CIPed at 280 MPa. Binder burnout and bisque firing were accomplished by ramping the specimen at 2 °C/min. from room temperature to 600 °C and at 5 °C/min. from 600 to 1200 °C.

Reinforcement Description
ZrO_2 and 3Y-ZrO_2 particulates

Reinforcement Supplier
Toyo Soda USA, Atlanta, GA

Supplier Designation
ZrO_2: TZ-0Y
Y-ZrO_2: TZ-3Y

Matrix Description
Al_2O_3 powder

Matrix Supplier
Sumitomo Chemical America, New York

Matrix Designation
AKP-HP

Test Type
4-pt. Flexural

Specimen Dimensions
2 x 1.5 x 25 mm
inner span = 10 mm
outer span = 20 mm

Test Description
crosshead speed = 0.25 mm/min
stress rate = 50 MPa/sec

Test Temperature
RT

Sintering Temperature (°C)	Flexural Strength (MPa)	Weibull's Modulus (dimensionless)	Relative Density (percent)
1,300	508	9.1	71.7
1,350	734	9.0	92.1
1,400	860	8.2	98.1
1,450	943	7.7	99.2
1,500	1,048	15.8	100
1,550	939	8.1	98.8
1,600	838	8.5	97.8

Remarks/Comments
Ten bars were tested for each experiment. Flexural strength increased with sintering temperature up to 1500 °C. After 1500 °C, flexural strength decreased. Weibull modulus decreased slightly with increasing sintering temperature until 1450 °C was reached. At 1500 °C, Weibull modulus experienced a large increase. Thereafter, it decreased with increasing sintering temperature.

Al_2O_3(p)/ZrO_2

Reference: 72

Composition
20 wt% Al_2O_3

Processing Method
Sinter/HIPed

Detailed Description
Composite billets were formed by a sinter/ HIP process. Primary sintering was done at 1400 °C followed by HIPing at 1500 °C.

Sintering Additives
4.2 wt% Y_2O_3

Matrix Description
3Y-Al_2O_3.ZrO_2 powder (4.2 wt% Y_2O_3)

Matrix Supplier
TOSOH Corp. of Tokyo

Matrix Designation
TOSOH TZ-3$Y_2$0A ("Super Z")

Test Type
4-pt. Flexural

Applicable Test Standard
MIL-STD-1942 A

Specimen Dimensions
3 x 4 x 50 mm
inner span = 20 mm
outer span = 40 mm

Test Description
crosshead speed = 0.5 mm/min
Half of the bars were heat treated for 100 hours at 1000 °C.

Test Temperature
RT

Reinforcement Content (weight percent)	Flexural Strength (MPa)	St. Dev. (MPa)	Weibull's Modulus (dimensionless)	Comments
20	1,840	308	6.0	***as-received***
20	1,490	59	24.0	***after heat treatment***

Remarks/Comments
Data taken was the average of five measurements. Density was 5.5 g cm^{-3}.

Figure 20: 4-pt. Flexural Strength of ZrO_2(p)/Al_2O_3

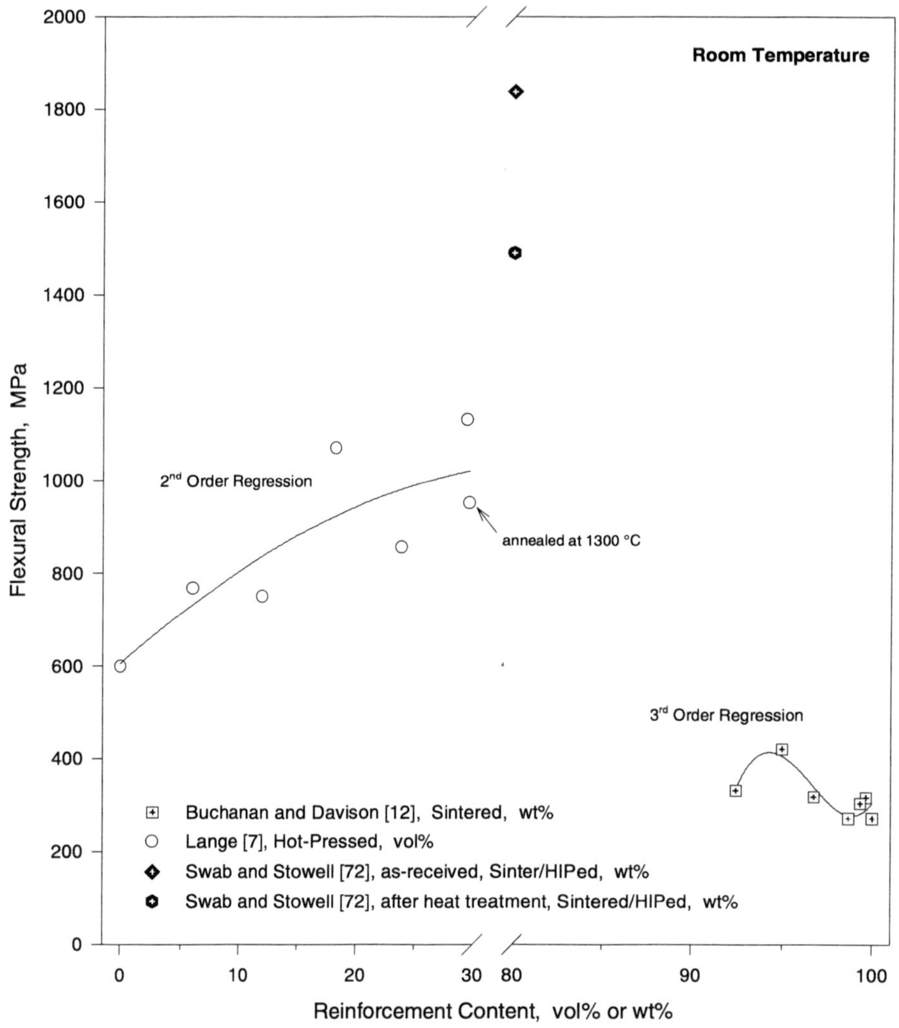

Discussion: Lange [7] found that the increase in strength was due to an increase in fracture toughness, which in turn was influenced by the amount of t-ZrO_2 present in the sample. A minimum amount of ZrO_2 was necessary in a composite of this type to inhibit crack growth by t→m transformation. According to Swab and Stowell [72], the strength discrepancy between the as-received strength and the strength after heat treatment was due to a change in the surface compressive stresses that were introduced during machining. Machining caused transformation (tetragonal to monoclinic) of the surface zirconia, which induced stresses on the surface that led to a higher flexural strength. The heat treatment reduced these stresses and thus decreased the flexural strength of the sample tested after heat treatment. The high strength of the composite after heat treatment was due to the alumina. The alumina enhanced the rigidity of the matrix (due to the higher modulus of Al_2O_3) which increased the stress necessary to initiate a tetragonal to monoclinic phase transformation, thus increasing overall strength [72].

Figure 21: 4-pt. Flexural Strength and Relative Density vs. Sintering Temperature of 15 wt% ZrO_2(p)/Al_2O_3

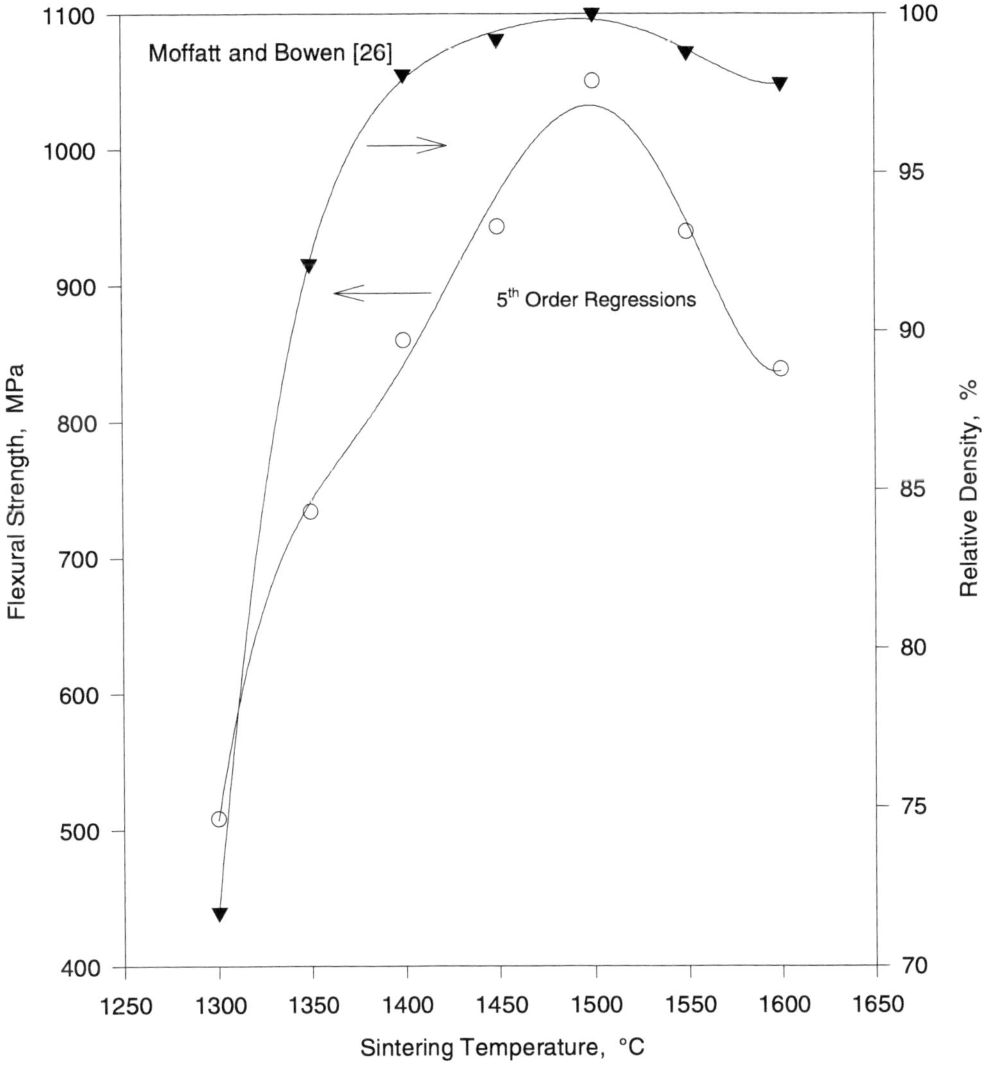

Discussion: Maximum strength was reached in excess of 1000 MPa at a sintering temperature of 1500 °C. Samples fired either above or below this optimum temperature tended to a dependence of strength on firing temperature, Moffatt and Bowen [26]. Below this optimum temperature the samples were not fully densified, and above this temperature the samples did not reach theoretical density due to rapid grain growth. The overall strengths shown are still much higher than those found in the studies on pure alumina. This may be due to the small grain size of the alumina-zirconia composites. The initial rise in strength at high fired densities are consistent with the reported dependence of alumina strength on porosity. It seems that the dependence of fracture strength on porosity arose from largely material-independent geometrical considerations for this composite [26].

Figure 22: Flexural Strength vs. Temperature of ZrO_2(p)/Al_2O_3

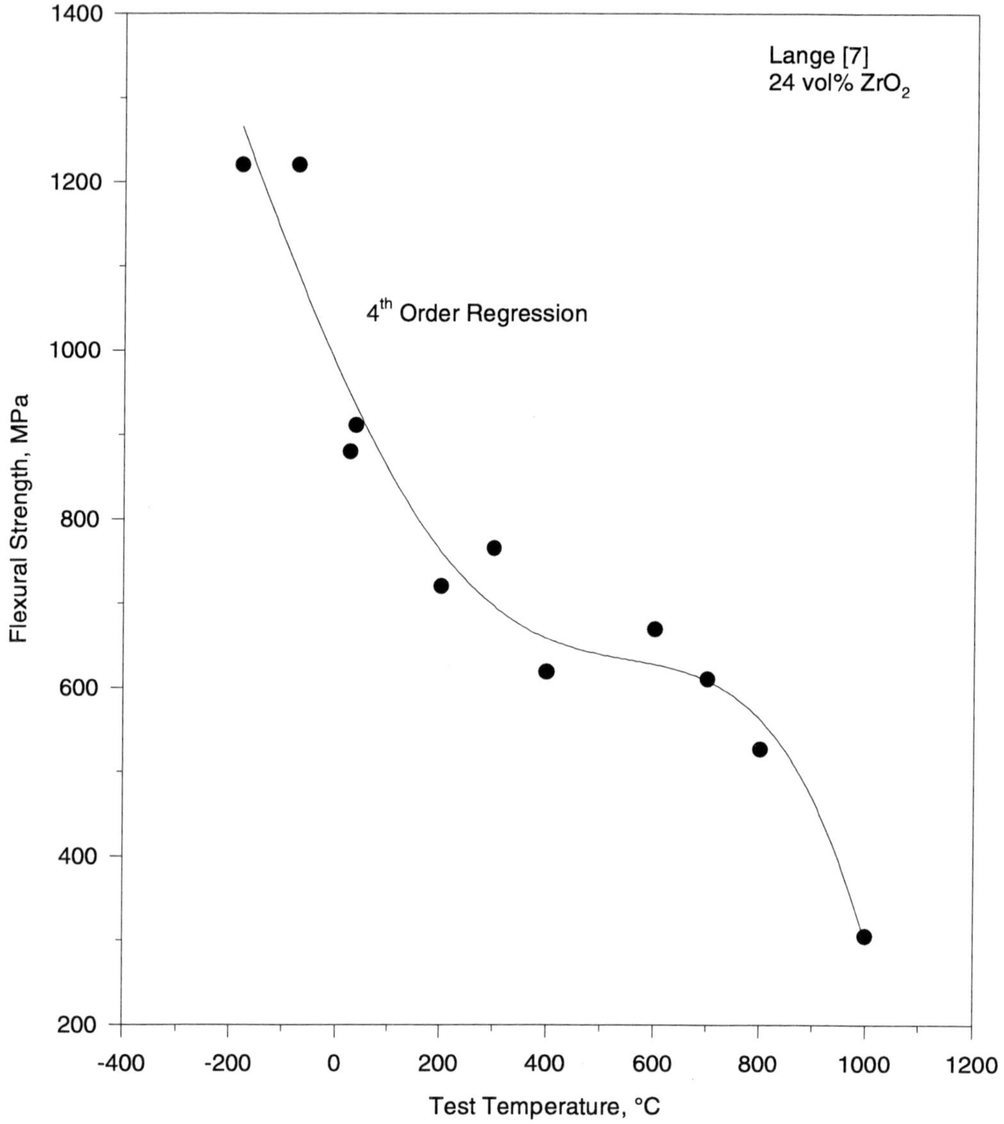

Discussion: As temperature increased the flexural strength decreased. This was as expected taking Young's modulus into consideration. Strength data above room temperature was well documented, but below this temperature data was not well understood. The relationship at this low temperature may be due to an increase in process zone size [7].

$ZrO_2.HfO_2(p)/Al_2O_3.Cr_2O_3$

Reference: 13

Composition
0 - 15 vol% ZrO_2 (containing 10 and 30 mol% HfO_2)

Processing Method
Dry Pressed/Sintered

Detailed Description
Particulate and matrix powders were attrition milled separately in proponal. These powders were then mixed by ball-milling with water. The mixture was dried, CIPed at 28 MPa, and sintered at 1550 °C for 2 hours in argon.

Reinforcement Description
$ZrO_2.HfO_2$ particulates

Matrix Description
$Al_2O_3.Cr_2O_3$ (Cr_2O_3-5,20 mol%)

Test Type
4-pt. Flexural

Specimen Dimensions
2.06 x 2.80 x 50 mm
inner span = 9.5 mm
outer span = 19.0 mm

Test Temperature
RT

ZrO_2 Reinforcement Content (volume percent)	Flexural Strength (MPa)	Comments
0.0	433	*5 mol% Cr_2O_3 in Al_2O_3*
2.5	480	*30 mol% HfO_2 in ZrO_2*
5.0	490	
7.5	489	
10.0	478	
12.5	454	
15.0	426	
0.0	433	*5 mol% Cr_2O_3 in Al_2O_3*
2.5	480	*10 mol% HfO_2 in ZrO_2*
5.0	490	
7.5	489	
10.0	477	
12.5	434	
15.0	354	
0.0	426	*20 mol% Cr_2O_3 in Al_2O_3*
2.5	435	*30 mol% HfO_2 in ZrO_2*
5.0	434	
7.5	427	
10.0	414	
12.5	383	
15.0	333	
0.0	426	*20 mol% Cr_2O_3 in Al_2O_3*
2.5	435	*10 mol% HfO_2 in ZrO_2*
5.0	434	
7.5	418	
10.0	364	
12.5	282	
15.0	168	

Remarks/Comments
Hafnia was added to increase the transformation temperature of the zirconia. Chromia was added to reduce the thermal conductivity of the alumina.

ZrO_2(p)/$MgCr_2O_4$

Reference: 111

Composition
0 - 21 vol% ZrO_2

Processing Method
Cold Pressed/Sintered

Detailed Description
$MgCr_2O_4$ powder was made by mixing $MgCO_3$ and Cr_2O_3 powders and then calcining the mixture at 1200 °C for 4 h in air. The $MgCr_2O_4$ was mixed with ZrO_2, and ball-milled in methanol for 16 hours. The mixture was dried in air and then mixed with a 5 % acryloid-stearic acid binder system dissolved in methanol. The methanol was slowly evaporated by heating. The dry mix was sieved through a 30-mesh screen, pressed in a steel die at 276 MPa, and sintered at 1650 °C for 1.3 hours at an oxygen partial pressure of 9e-13 atmospheres

Reinforcement Description
ZrO_2 particulates

Matrix Description
$MgCr_2O_4$

Test Type
4-pt. Flexural

Specimen Dimensions
inner span = 22 mm
outer span = 38 mm

Test Description
crosshead speed = 1.3 mm/min

Test Temperature
RT

Reinforcement Content (volume percent)	Flexural Strength (MPa)	Range (MPa)
0	65	±9.2
7	107	±14.6
10	134	±12.8
14	142	±33.4
16	156	±27.3
21	137	±31.2

Remarks/Comments
Microcracking was the predominant strengthening mechanism. Little change was noticed before 7 vol%. At 7 vol%, microcracking was evident. Microcracks developed into macrocracks at 16.5 vol% particulate reinforcement. Maximum strength occurred at 16.5 vol% ZrO_2, at which point the macrocracks developed.

Figure 23: 4-pt. Flexural Strength of Sintered $ZrO_2.HfO_2(p)/Al_2O_3.Cr_2O_3$

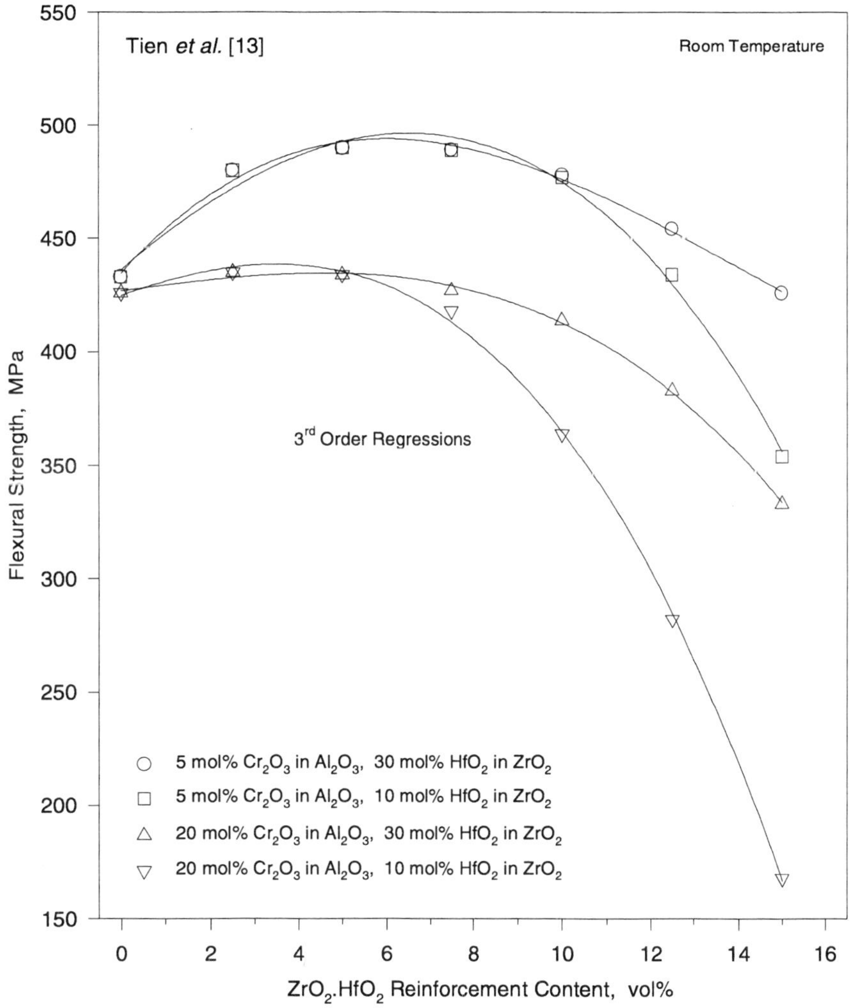

Discussion: The strength of the composites containing 5 mol% Cr_2O_3 in the alumina matrix reached a maximum at 6 - 8 vol% $ZrO_2.HfO_2$ content. The strength of the composites containing 20 mol% Cr_2O_3 did not show an increase up to 8 vol% reinforcement, but showed a similar drop in strength as the 5 mol% Cr_2O_3 composites above 8 vol% reinforcement. The addition of 30 mol% HfO_2 in ZrO_2, in contrast to the addition of only 10 vol% HfO_2 in ZrO_2, allowed for greater strengths at high (from 10 to 16 vol%) reinforcement content. Due to the small volume change of the dispersed particles during the t—>m transformation, samples which had a high HfO_2 content had a lower microcrack density. As a result of the fewer microcracks with high HfO_2 content, the samples' flexural strengths were greater. Analysis of the effect of microcracking on fracture toughness for these composites is presented in figure 73.

Figure 24: 4-pt. Flexural Strength of Sintered ZrO_2(p)/$MgCr_2O_4$

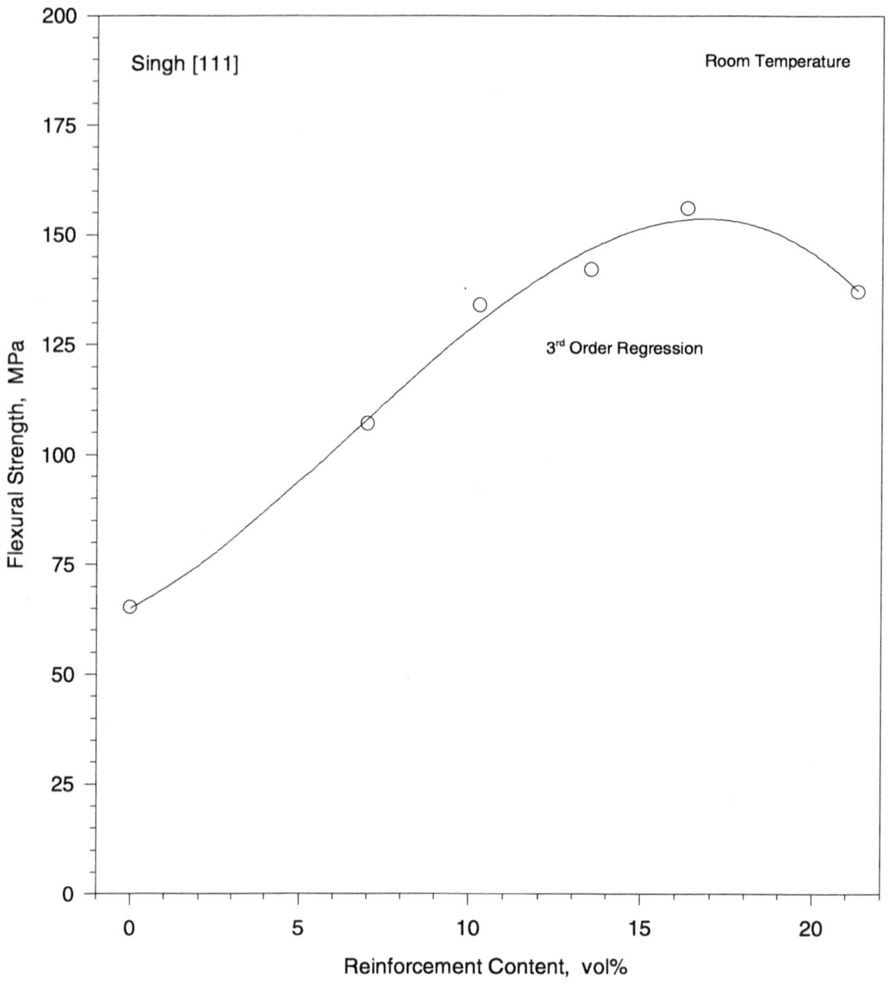

Discussion: The flexural strength increased as the ZrO_2 reinforcement content increased. At a critical reinforcement content the ZrO_2 particles began to interact with each other through overlapping stress fields. This led to the relaxation of stresses within the composite through the propagation of microcracks. As the reinforcement content increased past the critical amount, microcrack density increased and the flexural strength decreased.

SiC(p)/ZrO_2

Reference: 85

Composition
0 and 15 wt% SiC

Composite Architecture
particulate reinforcement

Processing Method
Sol-Gel/Hot-Pressed

Detailed Description
ZrO_2Cl_2-$8H_2O$, $Y(NO_3)_3$-$6H_2O$, and α–SiC were used as starting materials. Y-TZP and SiC(p)/Y-TZP were prepared by the coprecipitation method from the starting solution. The mixture was hot-pressed in a boron-coated graphite mold at 1600 °C and 25 MPa for 1 hour.

Reinforcement Description
α–SiC particulates

Matrix Description
2.8Y-TZP

Test Type
3-pt. Flexural

Test Description
crosshead speed = 0.05 mm/min.

Test Temperature
RT-1000 °C

Test Environment
Air

Test Equipment
Instron 1195 testing machine

Temperature (°C)	Flexural Strength (MPa)	Range (MPa)	Reinforcement Content (weight percent)
20	1,230	+49/-52	0
200	1,060	+51/-60	0
300	931	+25/-45	0
400	796	+47/-55	0
600	487	+36/-57	0
800	283	+24/-49	0
1,000	214	+24/-21	0
20	1,050	+23/-23	15
300	1,020	+19/-21	15
500	847	+41/-42	15
800	648	+20/-24	15
1,000	264	+16/-12	15

Remarks/Comments
Flexural strength of composite and monolith was inversely proportional to temperature. At 200 °C and 800 °C, the strength of the composite was equal to and double the strength of the monolithic ZrO_2, respectively. The differences in strength were due to the loss of transformation toughening in the monolith at high temperature. The composite remains strong due to the particle dispersion effect which is almost independent of temperature.

SiC(p)/ZrO_2

Reference: 85

Composition
0 - 30 wt% SiC

Composite Architecture
particulate reinforcement

Processing Method
Sol-Gel/Hot-Pressed

Detailed Description
ZrO_2Cl_2-$8H_2O$, $Y(NO_3)_3$-$6H_2O$, and α–SiC were used as starting materials. Y-TZP and SiC(p)/Y-TZP were prepared by the coprecipitation method from the starting solution. The mixture was hot-pressed in a boron-coated graphite mold at 1600 °C and 25 MPa for 1 hour.

Reinforcement Description
α–SiC particulates

Matrix Description
2.8Y-TZP

Test Type
3-pt. Flexural

Specimen Dimensions
5 x 2.5 x 26 mm
span = 20 mm

Test Description
crosshead speed = 0.05 mm/min.

Test Temperature
RT

Test Equipment
Instron 1195 testing machine

Reinforcement Content (weight percent)	Flexural Strength (MPa)	Range (MPa)	Relative Density (percent)
0	1,270	+45/-55	100
5	1,140	+27/-29	99.2
10	1,130	+24/-17	98.8
15	1,080	+24/-29	98.4
20	1,050	+25/-44	97.2
30	188	+13/-14	88.4

Remarks/Comments
Flexural strength decreased with increasing reinforcement content, but was fairly constant from 6 wt% to 20 wt% SiC. The large decrease in strength above 20 wt% SiC is due to poor sinterability of the composite.

Figure 25: 3-pt. Flexural Strength of Hot-Pressed SiC(p)/ZrO_2

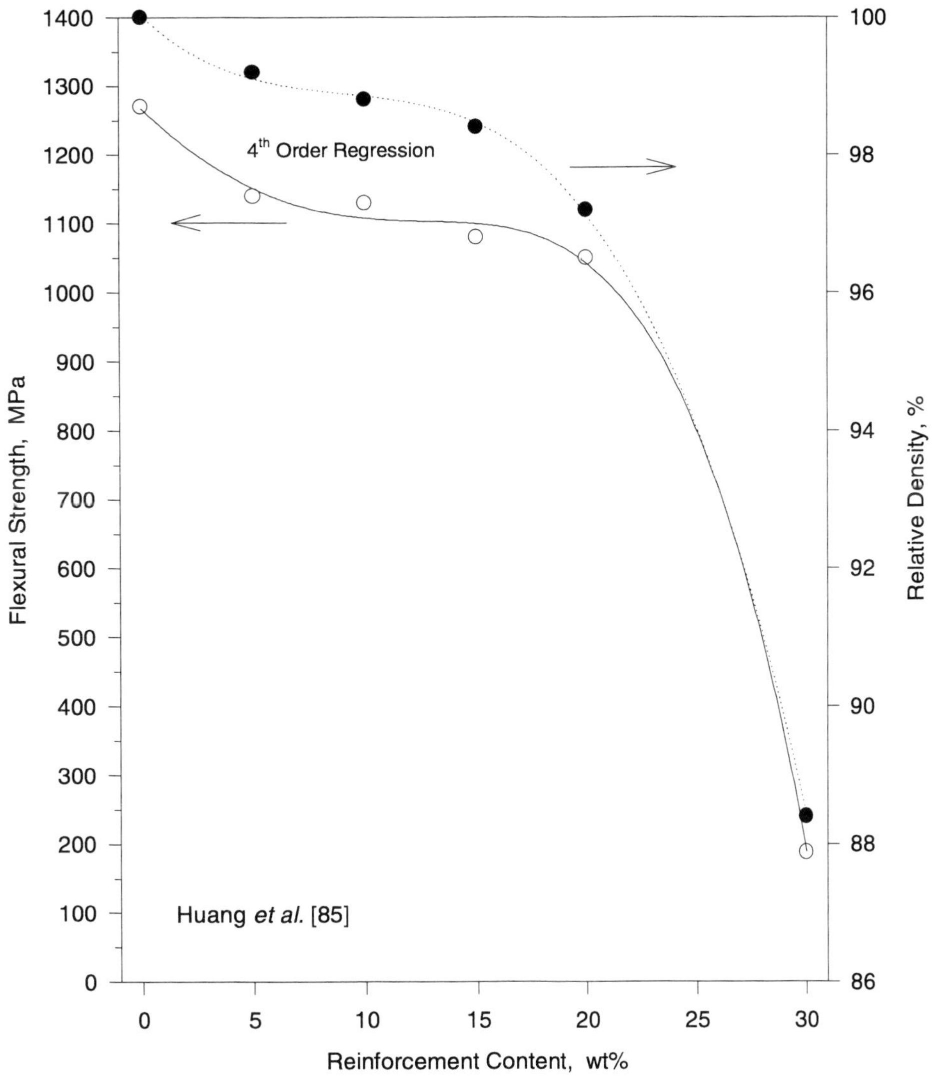

Discussion: Huang *et al.* [85] found that the strength of a SiC(p)/YSZ composite was highly dependent on fracture toughness and relative density. Strength decreased with the addition of SiC particulates to a YSZ matrix. Strength decreased considerably with the addition of a small quantity of SiC reinforcement (below 6 wt%) because the SiC reinforcement decreased the amount of transformable t-zirconia in the matrix. There was a sharp decrease in toughness in this range. The t-zirconia content continued to decrease with further addition of SiC particulates from 6 to 15 wt%. However, strengthening and toughening mechanisms created by the SiC reinforcement compensated for the decrease in tetragonal phase. The result was that there was no change in strength as SiC content increased from 6 to 15 wt%. Further addition of SiC reinforcement dramatically decreased relative density, resulting in a sharp decrease in strength.

Figure 26: 3-pt. Flexural Strength vs. Temperature of Hot-Pressed SiC(p)/YSZ

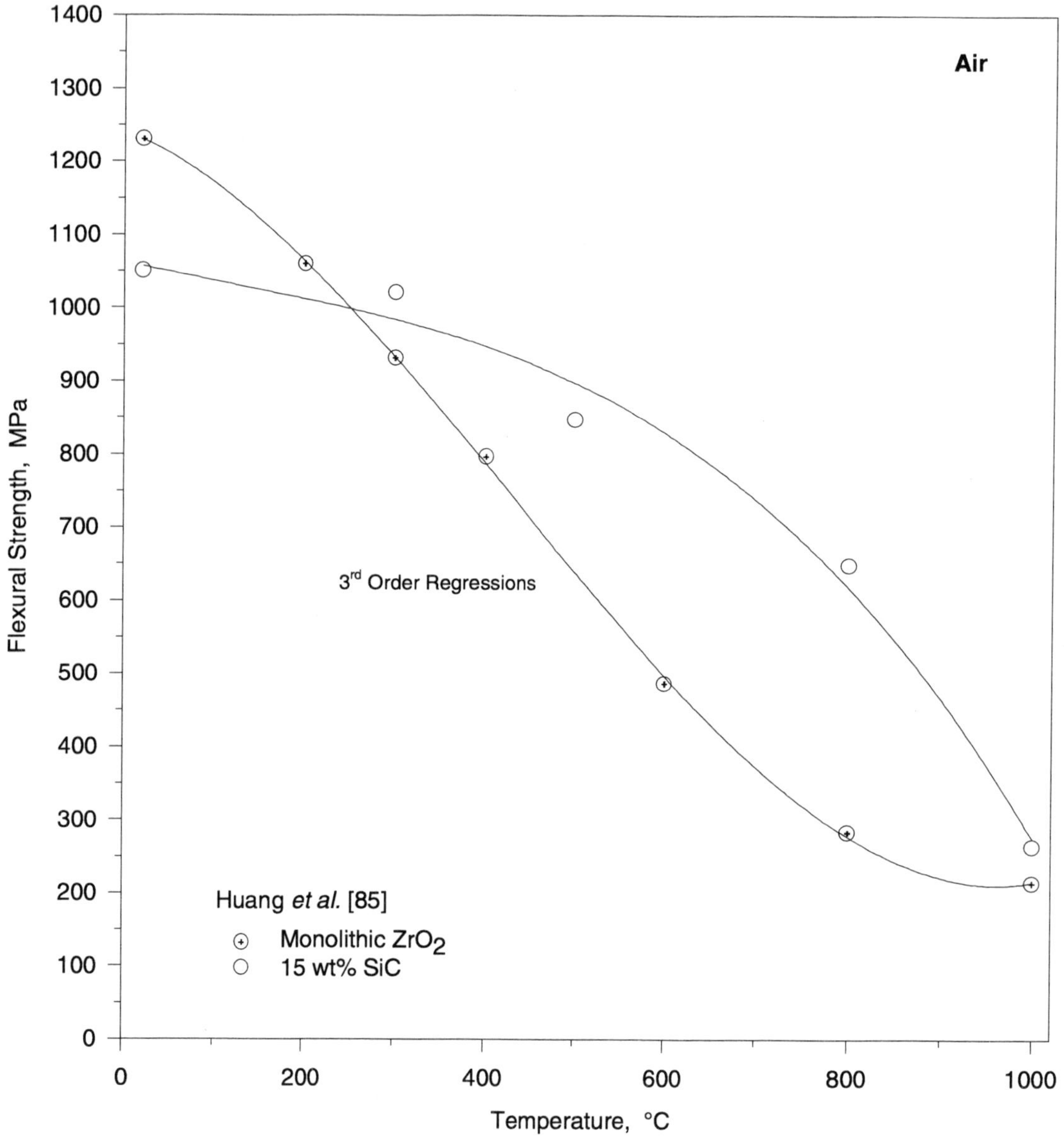

Discussion: Strength of the YSZ ceramic decreased rapidly with increasing temperature because transformation toughening and strengthening disappeared at elevated temperatures. The strength of the 15 wt% SiC(p)/YSZ composite was not as dependent on transformation toughening mechanisms for its high strength. As a result, the composite retained higher strength in the range of 300 to 1000 °C.

$Al_2O_3(p)/Si_3N_4$

Reference: 36

Composition
20 - 70 vol% Al_2O_3

Processing Method
Reaction Bonded

Detailed Description
Suspended in alcohol, these particles were mixed for 24 hours in a pot-mill with Si_3N_4 balls. The mixtures were dried and kneaded with a thermoplastic resin binder, and then cooled and crushed. The crushed materials were heated to 160 °C in a die and pressed into a disk shape under 98 MPa. The samples were dewaxed and heated stepwise to 1350 °C in a nitrogen atmosphere for nitridation.

Reinforcement Description
Al_2O_3 particulates

Reinforcement Typical Properties
particle size = 0.5 µm

Matrix Description
Si powder

Matrix Properties
particle size = 0.9 µm

Test Type
3-pt. Flexural

Specimen Dimensions
dia. = 40 mm
thickness = 10 mm

Type of Specimen
Disk

Test Temperature
RT

Reinforcement Content (volume percent)	Flexural Strength (MPa)	Relative Density (percent)
20	192	89
30	194	85
40	106	81
50	106	73
60	96	72
70	89	68

Remarks/Comments
Flexural strength decreased as the alumina content increased. Strength correlated to the relative density.

$MoSi_2(p)/Si_3N_4$

Reference: 41

Composition
0 and 20 vol% $MoSi_2$

Processing Method
Hot-Pressed

Detailed Description
The starting powders were mixed, then hot-pressed at 1750 °C and 30 MPa in argon.

Sintering Additives
5 wt% MgO

Reinforcement Description
$MoSi_2$ particulates

Reinforcement Supplier
Alfa Products, Danvers, MA

Matrix Description
Si_3N_4

Matrix Supplier
H. C. Starck, Berlin, Goslar, FRG

Matrix Designation
Grade LC-12

Test Type
4-pt. Flexural

Specimen Dimensions
2.5 x 5.1 x 25.4 mm
inner span = 9.5 mm
outer span = 19 mm

Test Temperature
1200, 1400, and 1500 °C

Test Environment
Air

Test Equipment
Instron mechanical testing unit

Temperature (°C)	Flexural Strength (MPa)	Reinforcement Content (volume percent)
1,200	435	0
1,400	94	0
1,500	58	0
1,200	284	20
1,400	122	20
1,500	115	20

Remarks/Comments
The composite had a lower flexural strength value at 1200 °C than the monolithic Si_3N_4. However, at 1500 °C, the composite had nearly twice the strength of the monolithic Si_3N_4. The strength data presented at 1400 and 1500 °C were yield strengths.

Figure 27: 3-pt. Flexural Strength of Reaction Bonded Al_2O_3(p)/Si_3N_4

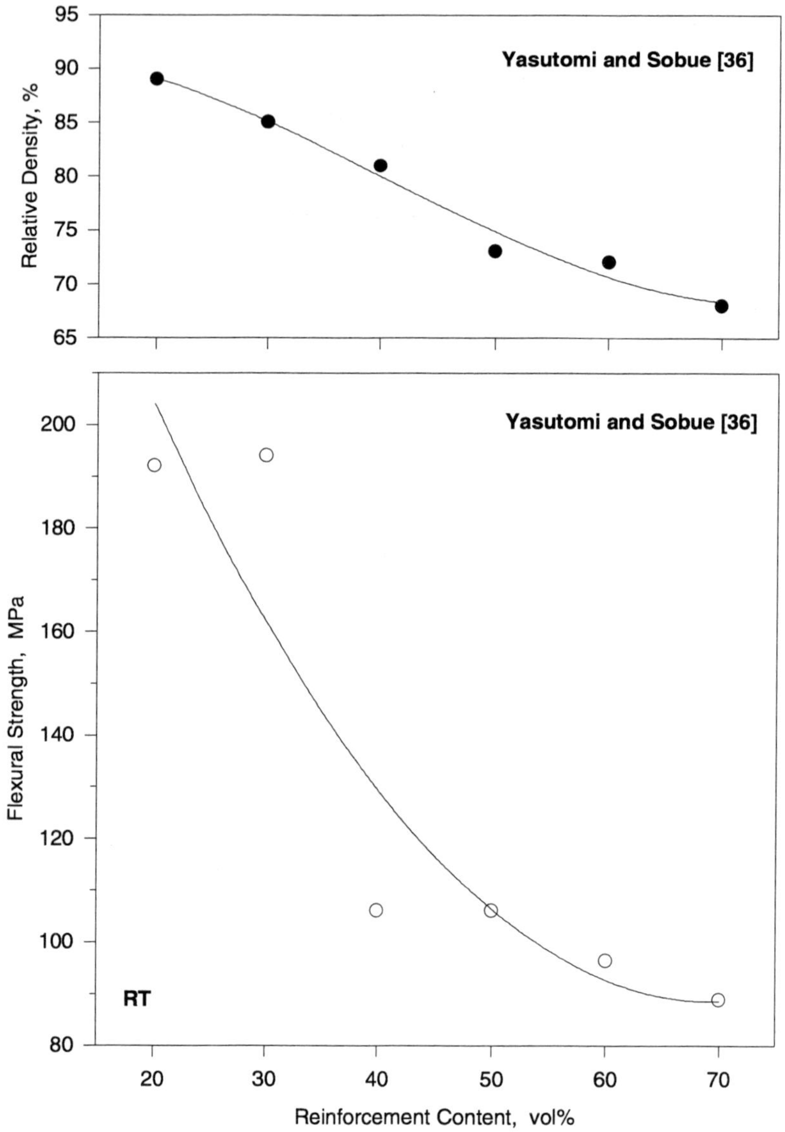

Discussion: The flexural strength of the reaction bonded composites tested by Yasutomi and Sobue [36] was a maximum at 20 and 30 vol% alumina content, and significantly decreased when the alumina content was increased above 30 vol%. The decrease in strength from 30-70 vol% reinforcement content was primarily due to the decrease in relative density.

Figure 28: 4-pt. Flexural Strength vs. Temperature of $MoSi_2(p)/Si_3N_4$

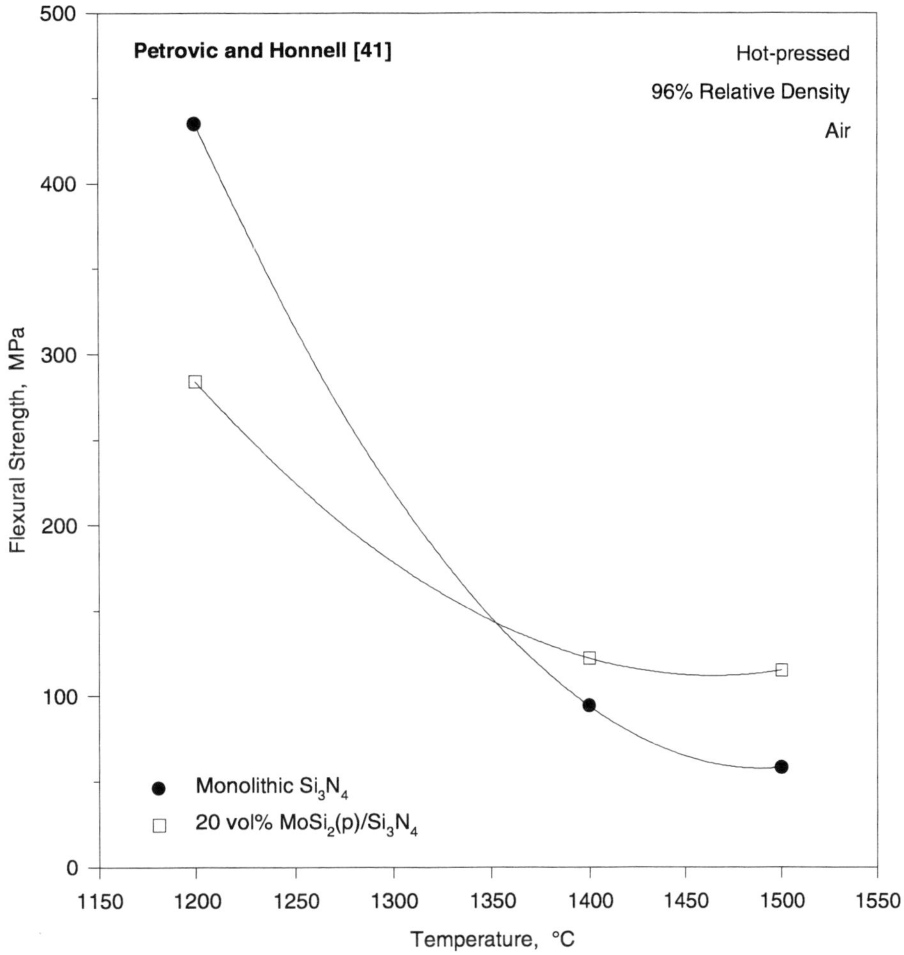

Discussion: At 1200 °C, both the monolithic and the composite materials were brittle. At this temperature the monolithic silicon nitride was 150 MPa stronger than the $MoSi_2(p)/Si_3N_4$ composite. At 1400 °C to 1550 °C, plasticity of both materials became apparent in the yielding behavior prior to ultimate fracture [41]. The strength data presented at these temperatures were yield strengths. The yield stress in the four-point bend tests was taken as the stress at a small (0.05 mm) plastic offset deviation from elastic loading. At 1500 °C, the composite was twice as strong as the monolithic material. Since the composite retained its strength at elevated temperatures better than the monolithic Si_3N_4, and was twice as strong as the monolithic material at 1500 °C, the $MoSi_2$ particles appeared to have a beneficial influence on the mechanical properties of Si_3N_4 at elevated temperatures [41].

Al_2O_3(p)/SiAlON

Reference: 87

Composition
4 - 21 wt% Al_2O_3

Processing Method
Sintered or Sintered then HIPed

Detailed Description
The mixed powders were milled in ethanol in a planetary type alumina ball-mill for 3 hours. Samples were dry pressed and sintered in a graphite resistance furnace at 1730 °C for 3 hours and embedded in Si_3N_4 powder. The presintered bending bars were post-HIPed in the graphite-heated hot isostatic press. Specimens were placed in the HIP furnace on BN-coated graphite plates and were heated at a rate of 10 °C/min. to 300 °C under a vacuum to remove all moisture from the samples. The furnace was heated to 1730 °C at 40 °C/min. with a constant nitrogen pressure ramp up to 10 MPa. After holding for 15 minutes, the samples underwent further pressurization with argon to 200 MPa in 45 minutes at constant temperature. After holding for 30 minutes, the heat was turned off. With natural cooling and depressurization, the specimens were withdrawn from the HIP after 1 hour.

Composition (wt%) :
sample 1: 21 Al_2O_3, 69 Si_3N_4, 6 AlN
sample 2: 12 Al_2O_3, 69 Si_3N_4, 15 AlN
sample 3: 6 Al_2O_3, 69 Si_3N_4, 21 AlN
sample 4: 4 Al_2O_3, 84 Si_3N_4, 8 AlN

Sintering Additives
4 wt% Y_2O_3
6-21 wt% AlN

Reinforcement Description
Al_2O_3 particulates
AlN powders

Matrix Description
Si_3N_4 (Sialon) powder

Test Type
4-pt. Flexural

Specimen Preparation
Sample surface was polished.

Specimen Dimensions
inner span = 20 mm
outer span = 40 mm

Test Temperature
RT

Reinforcement Content (weight percent)	Flexural Strength (MPa)	Density (g cm^{-3})	$\frac{\beta'}{\alpha'+\beta'}$	Comments
21	131	3.04	1.00	*sintered; sample 1*
12	329	3.05	0.81	*sample 2*
6	411	3.10	0.41	*sample 3*
4	259	3.12	0.98	*sample 4*
21	118	3.18	0.98	*post HIP; sample 1*
12	330	3.24	0.70	*sample 2*
6	459	3.27	0.87	*sample 3*
4	253	3.23	0.87	*sample 4*

Remarks/Comments
Flexural strength did not improve substantially with post-HIP treatment.

Figure 29: 4-pt. Flexural Strength of Al_2O_3(p)/SiAlON

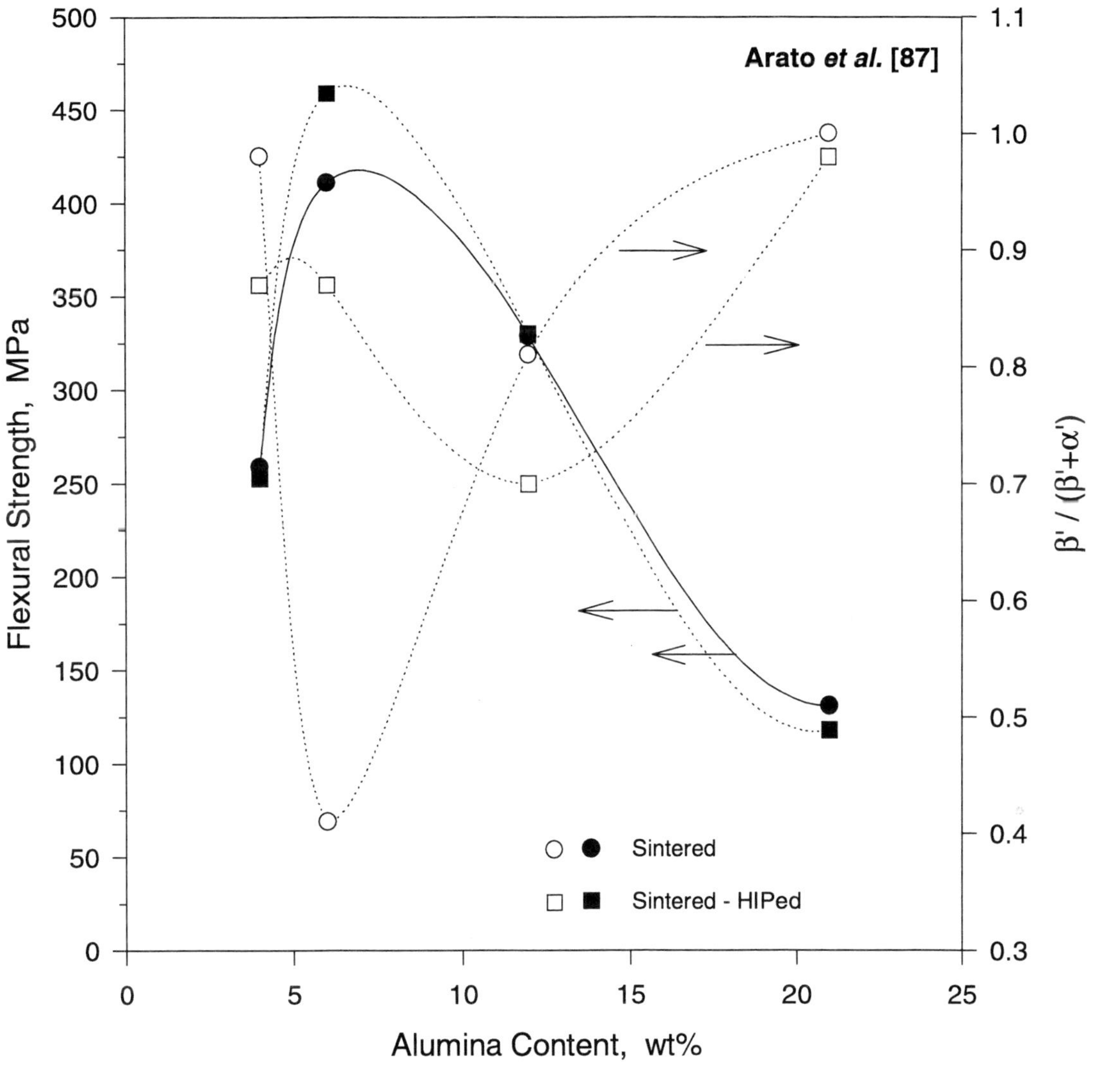

Discussion: Figure 29 shows the flexural strength and phase content of the sialon samples tested by Arato *et al.* [87] plotted against alumina content. Optimum strength of over 400 MPa was achieved by sample 3, which was produced from 6 wt% Al_2O_3 and 21 wt% AlN powders. A post-sintering HIPing process changed the phase content of the as-sintered composite; however, the change in phase content after HIPing did not affect strength of any of the samples except sample 3. In sample 3, the β' content was much greater in the HIPed condition than the as-sintered condition. This increase in the β' phase led to an increase in strength of 48 MPa after HIPing.

Figure 30: 4-pt. Flexural Strength of Al_2O_3(p)/SiAlON

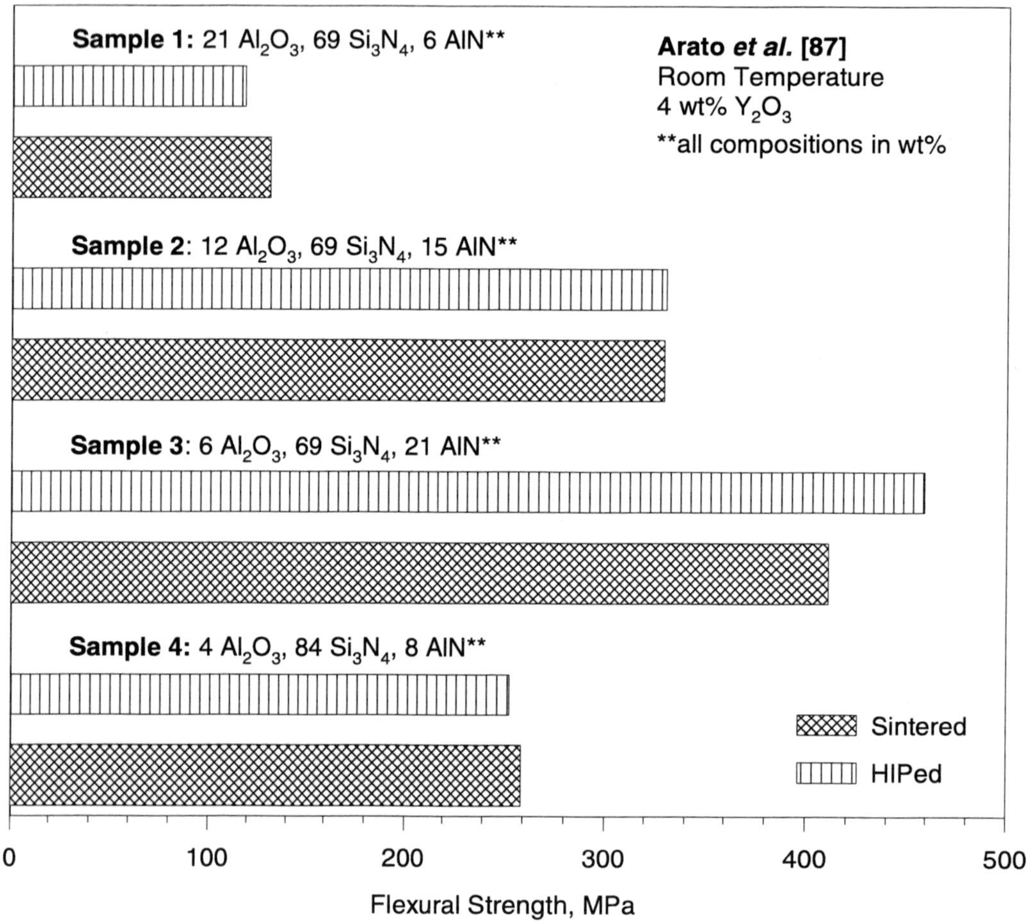

Discussion: Figure 30 shows the same data presented in figure 29.

$SiC(p)/Si_3N_4$

Reference: 27

Composition
0 - 20 vol% SiC

Processing Method
CIPed/Reaction Bonded

Detailed Description
Composites were formed by the nitridation of isostatically pressed green billets of silicon powder in which SiC had been incorporated prior to compaction. Nitridation was done in a dry N_2 atmosphere in the temperature range of 1350-1380 °C.

Reinforcement Description
α-SiC particulates

Reinforcement Supplier
Grindwell Norton, India

Matrix Description
Si_3N_4

Test Type
4-pt. Flexural

Specimen Dimensions
45 x 4.5 x 3.5 mm
inner span = 10.04 mm
outer span = 40.1 mm

Test Description
loading rate = 1.25 N/min

Test Temperature
RT-1400 °C

Test Environment
Air

Test Equipment
High Temperature Bending Strength Tester Model 422S, Netzsch, Germany

Temperature (°C)	Flexural Strength (MPa)	Range (MPa)	Reinforcement Content (volume percent)	Relative Density (percent)
30	63	±5	0	72
1,200	55	±6	0	72
1,400	66	±5	0	72
30	131	± 6	5	77
1,000	156	±12	5	77
1,200	181	±15	5	77
1,400	165	±5	5	77
30	102	±4	10	81
1,000	96		10	81
1,200	95		10	81
1,400	98		10	81
30	84	±8	15	73
1,000	96		15	73
1,200	94		15	73
1,400	75		15	73
30	89	±18	20	76
1,000	98		20	76
1,200	83		20	76
1,400	108		20	76

Remarks/Comments
The greatest improvement in strength at ambient temperature occurred at 5 vol% particulates.

SiC(p)/Si_3N_4

Reference: 31

Composition
0 - 60 vol% SiC

Processing Method
CIPed/Pressureless-Sintered (Some HIPed)

Detailed Description
All powders were mixed together in an attrition mill. This mixture was then CIPed at 630 MPa. Next, sintering was performed in a graphite element furnace at 0.1 MPa in nitrogen. Some specimens were then HIPed at 2000 °C and 100 MPa in nitrogen. The HIPed specimens were presintered at 1850 °C without encapsulation.

Sintering Additives
10.5 wt% Al_2O_3 (ALCOA, A16)
4.5 wt% Y_2O_3 (Ventron, 4N)

Reinforcement Description
β-SiC powder

Reinforcement Supplier
H. C. Starck

Supplier Designation
B 10

Reinforcement Dimensions
ave. grain size = 1.0 μm

Matrix Description
α-Si_3N_4 powder

Matrix Supplier
H. C. Starck

Matrix Designation
LC 10

Matrix Properties
ave. grain size = 0.5 μm

Test Type
4-pt. Flexural

Specimen Dimensions
inner span = 20 mm
outer span = 40 mm

Test Description
crosshead speed = 0.1 mm/min

Test Temperature
RT

Reinforcement Content (volume percent)	Flexural Strength (MPa)	Relative Density (percent)	Range (MPa)	Comments
0	348	96.0	±108	*pressureless sintered*
5	405	95.3	±28	
10	439	94.9	±31	
20	273	92.0	±56	
40	208	91.7	±57	
60	291	87.0	±78	
0	352	96.9	±33	*HIPed*
10	491	96.6	±101	
40	561	96.6	± 38	

Remarks/Comments
The strength was found to increase with the SiC dispersion up to SiC fraction of 0.1. Samples with a fraction of 0.2 or more showed lower strength, which can be attributed to the increasing residual porosity after sintering.

SiC(p)/Si_3N_4

Reference: 37

Composition
0 and 10 vol% SiC

Processing Method
CIPed / Sintered (some HIPed)

Detailed Description
The starting powders were ball-milled, dried, and sieved. This mixture was CIPed at 400 MPa and pressureless-sintered at 1750 °C under 0.1 MPa of N_2 for 3 hours. Half of the samples were HIPed for 1 hour at 1850 °C under 100 MPa of N_2.

Sintering Additives
10 wt% Y_2O_3
5 wt% Al_2O_3

Reinforcement Description
SiC particulates

Reinforcement Supplier
Ibiden Co., Ohgaki, Japan

Reinforcement Typical Properties
specific density = 3.21 g cm^{-3}

Reinforcement Dimensions
ave. particle dia. = 0.26 μm

Matrix Description
Si_3N_4

Matrix Supplier
Ube Kosan Co., Tokyo, Japan

Matrix Designation
E10

Matrix Properties
specific density = 3.18 g cm^{-3}
ave. particle dia. = 0.5 μm

Test Type
4-pt. Flexural

Specimen Preparation
Samples were surface ground prior to testing.

Specimen Dimensions
4 x 3 x 40 mm
inner span = 10 mm
outer span = 30 mm

Test Description
crosshead speed = 0.5 mm/min

Test Temperature
RT

Reinforcement Content (volume percent)	Flexural Strength (MPa)	Relative Density (percent)	Range. (MPa)	Comments
0	997	98	±176	***pressureless-sintered***
10	903	97	±105	
0	1,018	99	±180	***HIPed***
10	1,226	99	±114	

Remarks/Comments

Many specimens (18-20) were tested for each composition. The sinter-HIPed composite possessed greater strength than the pressureless-sintered composite. The sinter-HIPed monolithic Si_3N_4 possessed only slightly greater strength than the pressureless-sintered Si_3N_4.

TiC(p)/Si_3N_4

Reference: 38

Composition
0 and 15 vol% TiC

Processing Method
Hot-Pressed

Detailed Description
The starting powders were homogenized and mixed together. The mixtures were filter pressed at 2.07 MPa for ~45 minutes over a 0.02 µm polymer filter and were dried for 2 weeks. The resulting green bodies were hot-pressed at 1800 °C and 41 MPa for 2 hours.

Sintering Additives
6 wt% CeO_2 (Cerac Inc., Milwaukee, WI)

Reinforcement Description
TiC particulates

Reinforcement Supplier
Hermann C. Starck, Inc., New York, NY

Supplier Designation
c.a.s. Grade

Reinforcement Dimensions
ave. particle size = 1.25 µm

Matrix Description
α-Si_3N_4

Matrix Supplier
Hermann C. Starck, Inc., New York, NY

Matrix Designation
Grade LC12

Matrix Properties
ave. particle size = 0.55 µm

Test Type
4-pt. Flexural

Specimen Dimensions
50.8 x 6.35 x 3.18 mm
inner span = 22.2 mm
outer span = 44.4 mm

Test Description
crosshead speed = 30.6 mm/min

Test Temperature
RT

Reinforcement Content (volume percent)	Flexural Strength (MPa)
0	490
15	578

Remarks/Comments

Five samples were tested for each reinforcement content.

SiC(p)/Si_3N_4

Reference: 40

Composition
0 and 10 vol% SiC

Processing Method
Hot-Pressed

Detailed Description
The starting powders were mixed, dried, and sieved. The powders were hot-pressed at 30 MPa under 0.1 MPa nitrogen at 1700 °C and 1780 °C.

Sintering Additives
5 wt% Y_2O_3 (in addition to the Y_2O_3 and Al_2O_3 premixed in matrix powder)

Reinforcement Description
β-SiC particulates

Reinforcement Supplier
Ibiden Co. Ltd., Oogaki-city, Gifu-prefecture, Japan

Reinforcement Typical Properties
specific density = 3.21 g cm^{-3}

Reinforcement Dimensions
ave. particle dia. = 0.26 μm
surface area = 15.1 $m^2 g^{-1}$

Matrix Description
Si_3N_4

Matrix Supplier
Ube Industries, Ltd., Akasaka, Tokyo, Japan

Matrix Designation
COA

Matrix Properties
specific density = 3.18 g cm^{-3}
ave. particle dia. = 0.5 μm
surface area = 10.3 $m^2 g^{-1}$
contains 5 wt% ea. Y_2O_3 and Al_2O_3
composition (wt%):
N > 38, O = 1.27, °C < 0.3,
Al < 0.05, Ca < 0.05, Fe < 0.01

Test Type
4-pt. Flexural

Specimen Preparation
Specimens were polished with 6 and 3 μm diamond paste.

Specimen Dimensions
6 x 3 x 50 mm
inner span = 10 mm
outer span = 30 mm

Test Description
crosshead speed = 0.5 mm/min

Test Temperature
RT

Reinforcement Content (volume percent)	Flexural Strength (MPa)	Weibull's Modulus (dimensionless)
0	966	10.9
10	953	13.0

Remarks/Comments
Data is the average of ten specimens. The flexural strength decreased with the addition of particulates.

SiC(p)/SiAlON.Si_3N_4

Reference: 44

Composition
10 vol% SiC

Processing Method
Hot-Pressed/HIPed

Detailed Description
The mixed powders were ball-milled for 24 hours in ethanol. The slurry was dried in a rotary evaporator and sieved to a particle size of <210 μm. The mixed powders were then hot-pressed at 1720 °C and 24.5 MPa for 30 minutes in 0.1 MPa N_2 and HIPed at 1850-1950 °C in 10-100 MPa N_2.

Reinforcement Description
β-SiC particulates

Reinforcement Supplier
Ibiden Co., Ltd.

Supplier Designation
UF30

Reinforcement Dimensions
ave. particle size = 0.28 μm

Matrix Description
α-SiAlON and α-Si_3N_4 powders

Matrix Supplier
Ube Kosan Co., Ltd.

Matrix Properties
60 wt% Si_3N_4
40 wt% SiAlON

Test Type
4-pt. Flexural

Specimen Dimensions
4 x 3 x 40 mm
inner span = 10 mm
outer span = 30 mm

Test Description
crosshead speed = 0.5 mm/min

Test Temperature
RT

HIP Temperature (°C)	Flexural Strength (MPa)	Density (g cm^{-3})	Range (MPa)	Comments
1,720	762	3.20	±109 MPa	***as-Hot-Pressed***
1,850	781	3.27	±83 MPa	***HIPed at 10 MPa for 1 hour***
1,850	824	3.27	±65 MPa	***HIPed at 100 MPa for 1 hour***
1,950	746	3.27	±49 MPa	***HIPed at 100 MPa for 1 hour***

Remarks/Comments

The composite HIPed at 1850 °C in 100 MPa N_2 exhibited the highest strength at room temperature.

SiC(p)/Si_3N_4

Reference: 67

Composition

24 wt% SiC

Processing Method

CIPed/ Pressureless-Sintered (some Polymer Pyrolyzed)

Detailed Description

Polymethylsilazane was dissolved in tetrahydrofuran (THF) in an argon atmosphere. 10.5 wt% Al_2O_3 and 4.5 wt% Y_2O_3 were added. The THF was evaporated in a vacuum at 0.1 Pa and room temperature. Dried residue was held at 550 °C for 1 hour, and then at 1000 °C for 1 hour. It was then attrition milled for 3 hours at 1000 rpm, and then cold isostatically pressed at 630 MPa into green compacts of 35 x 17 x 15 mm. Reference samples were made from α–Si_3N_4, β–SiC, 10.5 wt% Al_2O_3, and 4.5 wt% Y_2O_3. They were attrition milled, dried, sintered, and hot-pressed in the same manner as the pyrolyzed sample. Both the as-pyrolyzed sample and the reference sample were sintered in a graphite resistivity heated furnace at 1750 and 1850 °C under 0.1 MPa N_2.

Sintering Additives

10.5 wt% Al_2O_3 (A16, ALCOA, Pittsburgh, PA)
4.5 wt% Y_2O_3 (Ventron Alta Produkte, Karlsruhe, FRG)

Reinforcement Description

β-SiC particulates

Reinforcement Supplier

H.C. Stark, Berlin, Germany

Supplier Designation

B 10

Matrix Description

β-Si_3N_4 powder

Matrix Supplier

H. C. Stark, Berlin, Germany

Supplier Designation

LC 10

Matrix Properties

Polymethylsilazane powder A
(7.1 wt% C, 22.5 wt% N, 10.4 wt% O)
Reference powder B
(9 wt% C, 20.6 wt% N, 9.2 wt% O)

Test Type

4-pt. Flexural

Specimen Dimensions

25 x 2.5 x 3.0 mm

Test Description

crosshead speed = 0.1 mm/min

Test Temperature

RT

Reinforcement Content (weight percent)	Flexural Strength (MPa)	Relative Density (percent)	Range (MPa)	Comments
24	256	95	±17	***polymer derived powders***
24	228	93	±32	***conventional***

Remarks/Comments

The pyrolyzed sample demonstrated improved sintering over the reference sample, which resulted in greater flexural strength.

$SiC(p)/Si_3N_4$

Reference: 81

Composition
0 - 25 vol% SiC

Processing Method
Hot-Pressed

Detailed Description
Materials were mixed together by ball-milling in a polypropylene jar with cyclohexane and Si_3N_4 media for 3 hours. The composite was mixed for another hour after pre-wetted SiC was added. After ball-milling, the slurry was dried in a thin film rotating evaporator, granulated, and pre-fired to burn off residual organics. Monolithic silicon nitride and all composites were hot-pressed at 1725 °C and 26 MPa in an N_2 atmosphere.

Sintering Additives
5 wt% Y_2O_3
2 wt% Al_2O_3

Reinforcement Description
SiC particulates

Reinforcement Supplier
Lonza Ltd.

Supplier Designation
LS-13

Reinforcement Dimensions
ave. size = 13 µm

Matrix Description
Si_3N_4 powder

Matrix Supplier
Ube Industries Ltd.

Matrix Designation
SN-E10

Test Type
4-pt. Flexural

Specimen Preparation
Test bars were cut from bulk material. Samples were 1 µm finish polished. The edges on the tensile side of the bars were chamfered.

Specimen Dimensions
3 x 4 x 50 mm

Test Temperature
RT

Reinforcement Content (volume percent)	Flexural Strength (MPa)
0	855
0	822
0	800
0	767
5	804
5	660
5	643
5	609
15	710
15	686
15	644
15	634
25	677
25	658
25	618
25	578

Remarks/Comments
Strength decreased with the addition of particulates.

Figure 31: 4-pt. Flexural Strength of Reaction Bonded SiC(p)/Si_3N_4

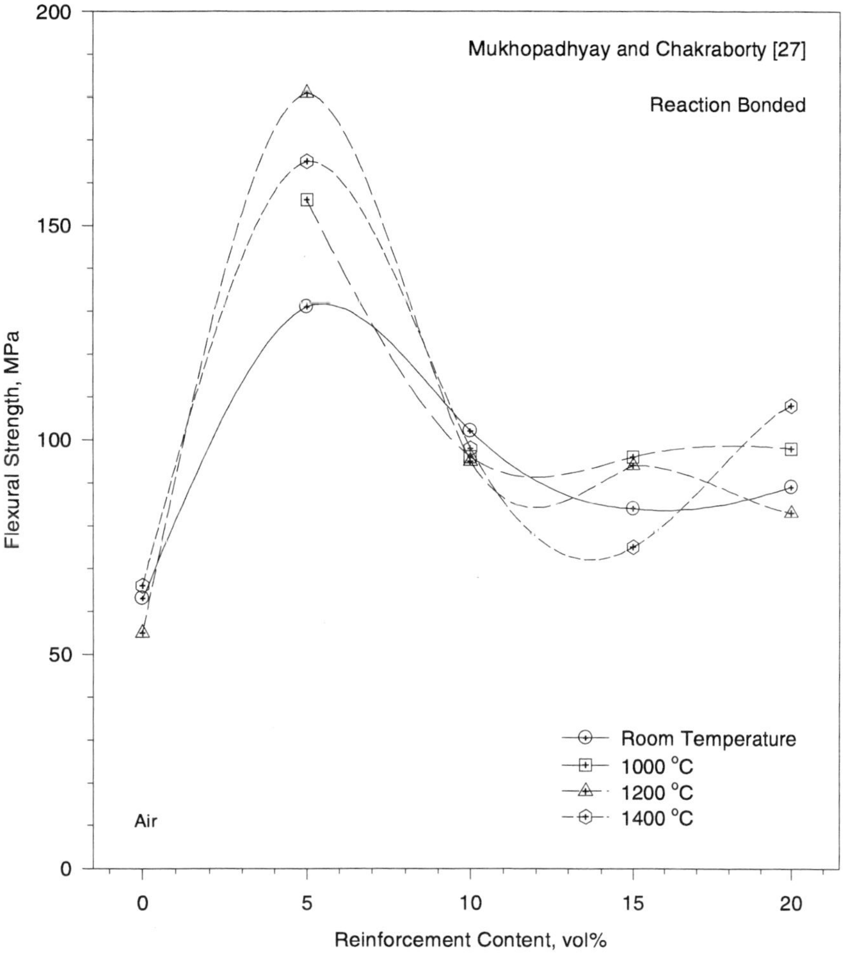

Discussion: The flexural strength of the SiC(p)/Si_3N_4 composites was relatively low due to the process of reaction bonding resulting in low densities. The flexural strength increased substantially with the addition of 5 vol% SiC to the monolithic Si_3N_4 composite for all four test temperatures. As the SiC content increased from 5 to 10 vol%, the flexural strength of all composites decreased. The strength of the composites when tested at RT, 1000 °C, and 1200 °C did not significantly change as the SiC content was increased from 10 to 20 vol%. When tested at 1400 °C, the composite did show a slight increase in strength as the SiC content was increased from 15 to 20 vol%.

It was observed that the elevated temperature strength of the 5 vol% SiC(p)/RBSN composite was greater than the room temperature strength. The increase in strength as test temperature increased was a result of Si_3N_4 having a woolly phase and a hard, dense phase. As the woolly phase converted to a more stable phase at high test temperatures and under pressure, the flexural strength increased.

Figure 32: 4-pt. Flexural Strength of SiC(p)/Si_3N_4 and TiC(p)/Si_3N_4

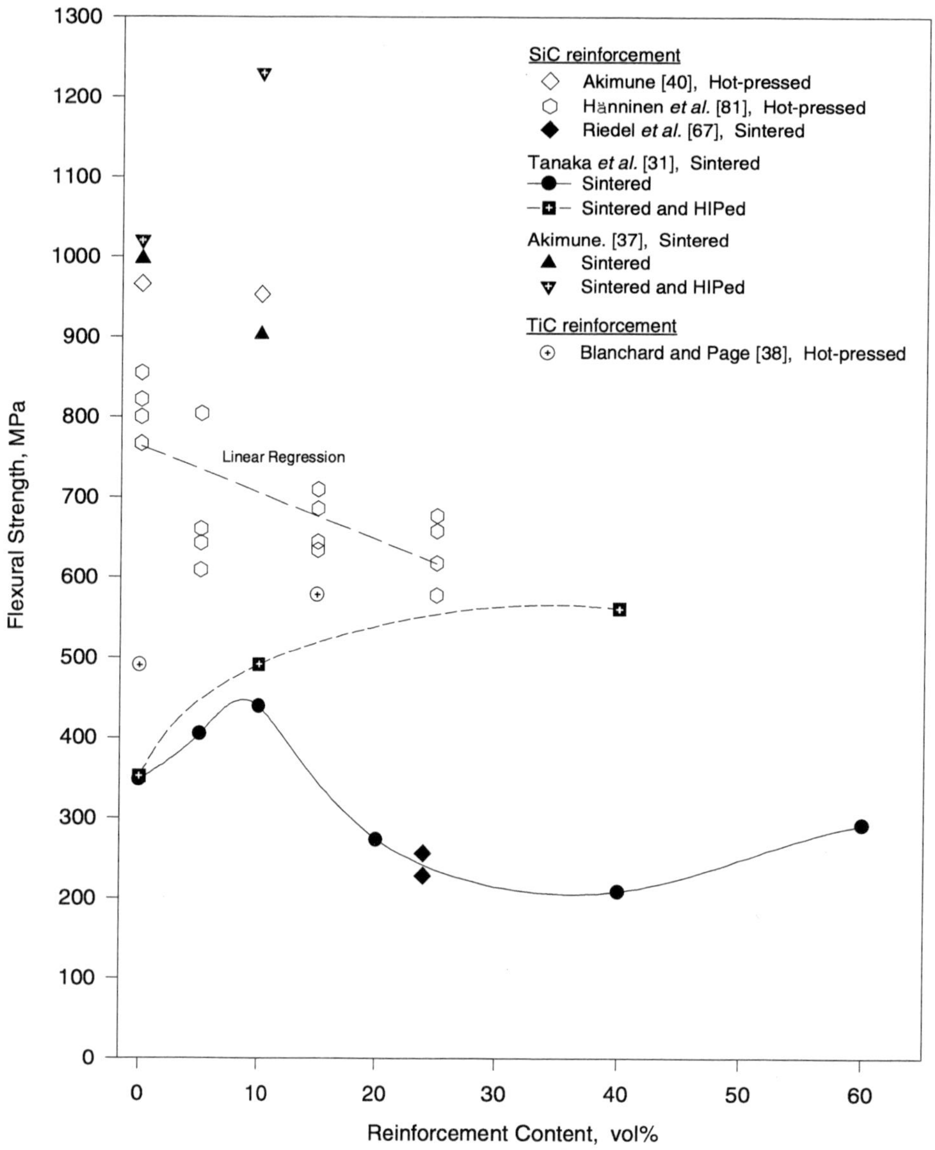

Discussion: For the sintered sample tested by Tanaka *et al.* [31], the flexural strength increased up to a SiC volume fraction of 0.1. As the SiC reinforcement content was increased above 10 vol%, the strength decreased due to the increased residual porosity after sintering. Some samples were HIPed to improve the strength. The HIPing reduced the porosity with a SiC fraction of more than 0.2. Due to the increased density the flexural strength also improved. General trends include a decrease in flexural strength verses reinforcement content for all hot-pressed and sintered composites. This, like ref. [31], was due to residual porosity after the densification process. For all HIPed samples an increase in strength occurred as the reinforcement content increased. This was due to a densification of the pores that was evident in the subsequent samples. The flexural strength of the hot-pressed titanium carbide particulate reinforced Si_3N_4 tested by Blanchard and Page [38] was a little less than similar SiC(p)/Si_3N_4 composites tested by Akimune [40] and Hänninen *et al.* [81].

Figure 33: 4-pt. Flexural Strength vs. Relative Density of $SiC(p)/Si_3N_4$

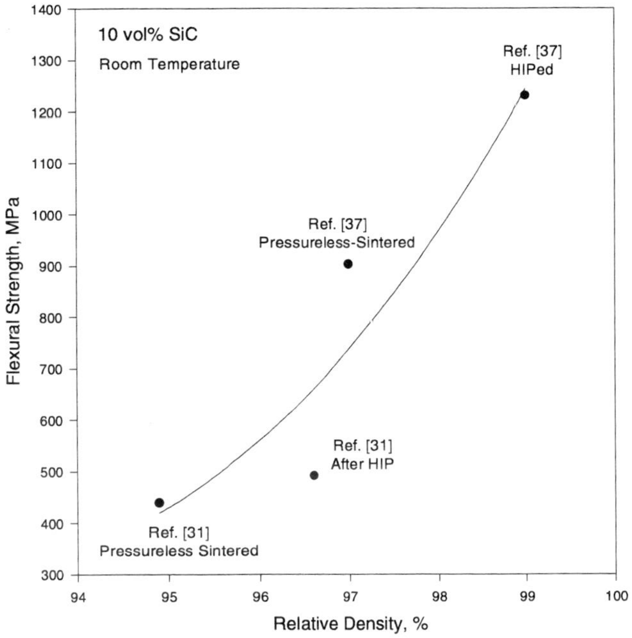

Discussion: This graph shows the relationship between the sintering process and densification. For the 10 wt% SiC sample, HIP processing by both Akimune [37] and Tanaka *et al.* [31] produced greater densification than pressureless sintering alone. A higher densification always produced a higher strength, but the samples from ref. [37], at a comparable density to the ref. [31] sample, were stronger.

Figure 34: 4-pt. Flexural Strength vs. Sintering Temperature for HIPed $SiC(p)/SiAlON.Si_3N_4$

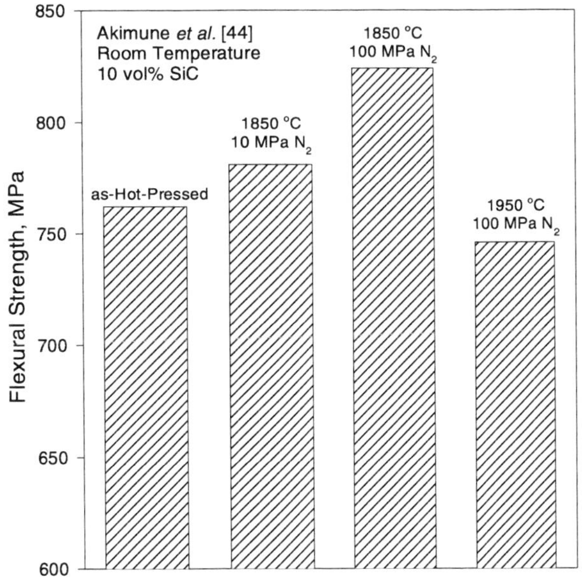

Discussion: Three of the composites showed higher density levels resulting from the HIPing process. SiC(p)/SiAlON composites produced by HIPing at 1850 °C under 100 MPa nitrogen pressure exhibited the highest strength at room temperature. This was because SiC-particles and sintering conditions inhibited the growth of SiAlON grains, and had the effect of reducing the size of the fracture origin, Akimune *et al.* [44].

$SiC(p)/Si_3N_4$

Reference: 35

Composition
12 wt% Si-rich SiC
10 and 20 wt% C-rich SiC

Processing Method
Colloidal Pressed / Reaction Bonded

Detailed Description
Four sample groups were fabricated. The first group, the monolithic RBSN was fabricated either in methanol with CIPing or in octanol without CIPing. The remaining RBSN composite samples were fabricated in octanol without CIPing. Octanol was substituted for methanol to reduce the probability of differential sedimentation between the dissimilar Si and SiC powders before samples were pressed. A high-intensity ultra-sonic probe was used to break up agglomerates in the dispersed powder mixture rather than the low-intensity ultrasonic bath used for the methanol processed baseline group. Si and SiC powders were ultrasonically dispersed separately, mixed together by hand to visual uniformity, and again sonicated. Total sonication time was 5-8 minutes at up to 40 W. Oxygen exposure was avoided except during pressing. Samples were colloidally pressed at 17 MPa to yield 1.27 diameter x 0.2 cm thick samples. A 0.4 °C/min. ramp to 200 °C followed by a 24 hour soak in N_2 was used to dry the samples. Nitridation was accomplished in ultra-high purity N_2 at atmospheric pressure. The firing schedule for the samples was:

(A) RBSN (methanol):
1 °C/min. to 1200 °C
0.5 °C/min. to 1400 °C; 1 hour hold

(B) RBSN (octanol):
1 °C/min. to 1250 °C; 4 hour hold
0.5 °C/min. to 1400 °C; 7 hour hold

(C) RBSN + 12 % Si-rich SiC:
2 hours at 1250-1400 °C (controller failed); 4 hour hold
10 hours at > 1400 °C (max: 1540 °C)

(D) RBSN + 10 % C-rich SiC:
3 °C/min. to 1250 °C; 4 hour hold
0.5 °C/min. to 1400 °C; 10 hour hold

(E) RBSN + 20 % C-rich SiC:
3 °C/min. to 1250 °C; 4 hour hold
0.5 °C/min. to 1400 °C; 10 hour hold

Reinforcement Description
SiC particulates

Reinforcement Dimensions
ave. Si dia. = 0.23 μm
ave. C-rich SiC dia. = 0.04 μm
ave. Si-rich SiC dia. = 0.05 μm

Matrix Description
Si powder

Matrix Designation
F93 S

Test Type
Biaxial Ball-On-Ring Method

Specimen Preparation
Samples were ground and polished to 1 μm on the tensile surface.

Specimen Dimensions
dia. = 12.43-12.67 mm
thickness = 1 mm

Type of Specimen
Disk

Test Temperature
RT

Reinforcement Content (weight percent)	Flexural Strength (MPa)	Relative Density (percent)	Range (MPa)	Comments
0	441	77	± 129	*A, optimized*
0	401	77	± 107	*A, baseline*
0	592	76	± 148	*B*
10	517	76	± 130	*D*
12	438	73	± 54	*C*
20	313	73	± 66	*E*

Remarks/Comments

The strength of the RBSN plus 10 or 12 wt% SiC were both unusually high compared with conventional RBSN, which typically exhibits strengths in the 75-150 MPa range. The high strengths of the RSBN plus 10-12 wt% SiC samples indicate that highly perfect microstructures were achieved in green bodies made from very dissimilar powders. Also, the strength of the octanol-pressed RSBN was higher than that of methanol pressed RBSN.

SiC(p)/Si_3N_4

Reference: 74

Composition
0 - 22 vol% SiC

Processing Method
Colloidal Pressed/Reaction Bonded

Detailed Description
Si and SiC powders were colloidally pressed from an octanol dispersion. The bodies were then reaction bonded to form the composite.

Reinforcement Description
SiC particulates

Reinforcement Dimensions
dia. = 15 μm (type A)
dia. = 50 μm (type B)

Matrix Description
Si powder

Matrix Properties
dia. = 0.2-0.3 μm

Test Type
Ball-On-Ring Method

Specimen Preparation
One side was polished to 9 μm surface finish.

Specimen Dimensions
dia. = 12 mm
thickness = 1.5 mm

Type of Specimen
Disk

Test Temperature
RT

Reinforcement Content (volume percent)	Flexural Strength (MPa)	Range (MPa)	Comments
0	295	± 138	*type A*
22	273	± 32	
0	457	± 111	*type B*
21	171	± 6	

Remarks/Comments

Type B was processed with a higher solids loading in the colloidal slurry to prevent segregation of SiC particulates. A large degree of segregation was found in the type A composite, and some agglomeration was seen in the type B composite.

Figure 35: Ball-On-Ring Strength of Reaction Bonded $SiC(p)/Si_3N_4$

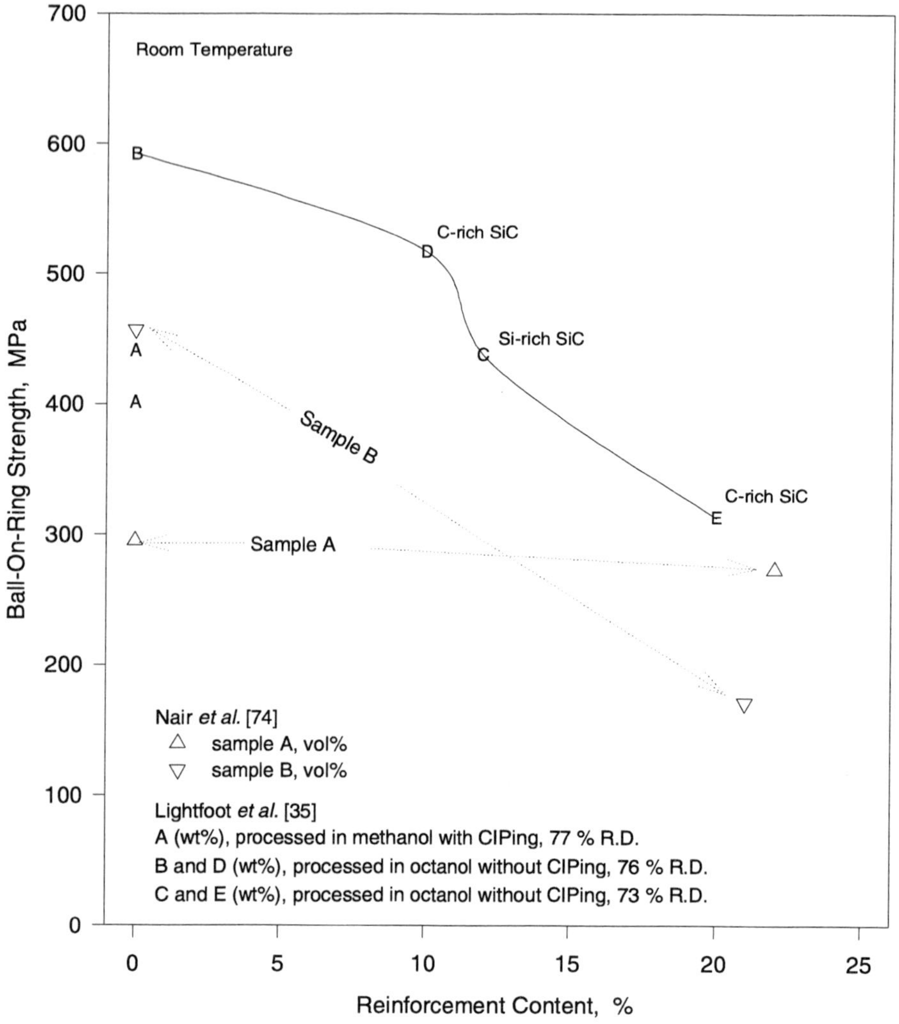

Discussion: Nair *et al.* [74] found that the decrease in flexural strength verses reinforcement content was due to a decrease in densification. A large degree of segregation had occurred in the sample A composite; whereas some agglomeration was found in the sample B composite.

Lightfoot *et al.* [35] tested RBSN ceramics processed in octanol and methanol. The octanol processed RBSN was almost 200 MPa stronger than the methanol processed RBSN samples. In fact, the octanol processed RBSN sample (B) was about 600 MPa, which was very high for a ceramic that was only 76 % dense. This was significantly higher than the strength that had previously been achieved by any RBSN ceramic; which generally have flexural strengths in the neighborhood of 75 to 150 MPa. Also, reaction bonded SiC(p)/RBSN composites were processed in octanol. In all cases the SiC was either slightly C-rich or Si-rich; the effect of this was unclear. Compared to monolithic RBSN (B), the SiC reinforcement decreased the strength of the composites. The addition of very fine SiC powders into the Si powder lowered the green packing densities, and thus lowered the final density. This suggested that the voids between Si particles were not filled effectively by the SiC particles. Fracture analysis showed that the sample set was relatively free of processing flaws, which was the primary reason for the overall high strength. The CIPing step had little effect on density or strength.

SiC(p)/Si_3N_4

Reference: 46

Composition
0 - 52 vol% SiC

Processing Method
Hot-Pressed

Detailed Description
An amorphous Si-C-N powder was prepared by introducing vaporized $((CH_3)_3Si)_2NH$ and NH_3 gas into an Al_2O_3 tube. Various powders with different C:N ratios were prepared by controlling the volume of NH_3 gas. These composite powders were stabilized by heat treatment in N_2 gas. The composite powder was added to 6 wt% Y_2O_3 and 2 wt% Al_2O_3. This powder was hot-pressed for 2 hours at 34 MPa and 1800 °C.

Sintering Additives
6 wt% Y_2O_3
2 wt% Al_2O_3

Reinforcement Description
SiC particulates

Matrix Description
Si_3N_4 powder

Test Type
3-pt. Flexural

Specimen Dimensions
span = 20 mm

Test Description
crosshead speed = 0.5 mm/min

Test Temperature
RT

Reinforcement Content (volume percent)	Flexural Strength (MPa)	Range (MPa)
0	1,033	±100
3	1,004	±64
10	1,225	±107
17	1,124	±83
24	1,184	±110
34	1,152	±95
52	1,146	±60

Remarks/Comments
Bending strength increased up to 10 vol% because of the growth of tree-shaped crystals, and it was almost constant in the region beyond 20 vol%.

Si_3N_4(p)/SiC

Reference: 56

Composition
10 - 50 % Si_3N_4

Processing Method
Cold Pressed/Reaction Bonded

Detailed Description
SiC and Si were mixed and pressed under 96.6 MPa. 1% of polyvinyl alcohol was added as a binder. The green bodies were dried in a furnace at 200 °C for 2 hours. The green shapes were nitrided at 1350-1450 °C.

Reinforcement Description
Si powder

Matrix Description
SiC

Test Type
3-pt. Flexural

Specimen Dimensions
152 x 19 x 13 mm

Test Description
load rate = 50 kg/min.
The equation $W = 3PL/(2bd^2)$ where:
W = MOR
P = applied load of failure
L = span width
b = width of the specimen
d = depth of specimen

Test Temperature
RT-1400 °C

Test Environment
Air

Test Equipment
Instron Universal Testing Machine model 1195

Temperature (°C)	Flexural Strength (MPa)	Reinforcement Content (percent)	Relative Density (percent)
22	27	10	72.8
1,250	36	10	72.8
1,350	45	10	72.8
1,400	48	10	72.8
22	40	20	76.0
1,250	47	20	76.0
1,350	61	20	76.0
1,400	68	20	76.0
22	46	30	80.4
1,250	59	30	80.4
1,350	68	30	80.4
1,400	73	30	80.4
22	51	40	83.2
1,250	61	40	83.2
1,350	72	40	83.2
1,400	77	40	83.2
22	58	50	84.5
1,250	68	50	84.5
1,350	82	50	84.5
1,400	86	50	84.5

Remarks/Comments
The flexural strength of the composites increased with increasing temperature and increasing SiC particulate content.

SiC(p)/SiAlON

Reference: 78

Composition
0 - 12 wt% SiC

Processing Method
CIPed / Pressureless-Sintered

Detailed Description
β-SiAlON powder was prepared by carbo-thermal reduction at 1450 °C for 5 hours in N_2. Raw materials included high purity SiO_2 powder, carbon powder, and $Al(OH)_3$. Ingredients totaling 100 g of the above powder, along with SiC powder and Y_2O_3, were milled with 250 ml of ethanol for 24 hours using a pot mill with silicon nitride beads, and they were then dried under diminished pressure. Composites were formed into green bodies 50 mm diameter x 5 mm thick at 33 MPa and CIPed at 98 MPa. The green bodies were pressureless-sintered in a carbon crucible at 1780 °C for 3 hours in N_2.

Sintering Additives
5 wt% Y_2O_3

Reinforcement Description
β-SiC particulates

Reinforcement Dimensions
grain size = 0.2-0.3 μm

Matrix Description
β-SiAlON

Matrix Properties
$Si_{5.5}Al_{0.5}O_{0.5}N_{7.5}$

Test Type
3-pt. Flexural

Specimen Preparation
Specimens were machined from sintered bodies.

Specimen Dimensions
40 x 4 x 3 mm
span = 30 mm

Test Description
crosshead speed = 0.5 mm/min

Test Temperature
RT and 1200 °C

Test Environment
Air

Reinforcement Content (weight percent)	Flexural Strength (MPa)	Temperature (°C)
0	785	20
3	730	20
6	800	20
9	740	20
12	750	20
0	475	1200
3	495	1200
6	480	1200
9	540	1200
12	520	1200

Remarks/Comments
The strength remained relatively constant at 800 MPa for composites containing up to 12 wt% β-SiC at room temperature. At 1200 °C, strength increased for composites containing up to 9 wt% β-SiC content.

SiC(p)/Si_3N_4

Reference: 93

Composition
0 - 20 vol% SiC

Processing Method
Hot-Pressed

Detailed Description
The powders were mixed in an Al_2O_3 ball-mill in an ethanol solvent for 77.8 hours in a polyethylene vessel. Hot-pressing was performed at 1800 °C and 30 MPa for 1 hour in an N_2 atmosphere.

Sintering Additives
8 wt% Y_2O_3 (Nippon Yttrium Co.)

Reinforcement Description
β-SiC powders

Reinforcement Supplier
Ibiden Co., Ltd.

Supplier Designation
UF grade

Reinforcement Dimensions
ave. particle size = 0.3 μm

Matrix Description
α-Si_3N_4 powder

Matrix Supplier
Ube Industries Co., Ltd.

Matrix Designation
E10 grade

Matrix Properties
ave. particle size = 0.2 μm

Test Type
3-pt. Flexural

Specimen Dimensions
3 x 4 x 40 mm
span = 30 mm

Test Description
crosshead speed = 0.5 mm/min

Test Temperature
RT

Reinforcement Content (volume percent)	Flexural Strength (MPa)	Relative Density (percent)	Range (MPa)
0	764	99.9	±46
5	854	99.9	±59
10	940	98.4	±121
20	444	98.7	±32

Remarks/Comments
The bending strength reached a maximum at 10 vol% SiC, and then fell rapidly as the SiC content increased. Increasing SiC content inhibited sintering and caused a fine, equiaxed Si_3N_4 grain structure.

Figure 36: 3-pt. Flexural Strength of Reaction Bonded Si_3N_4(p)/SiC

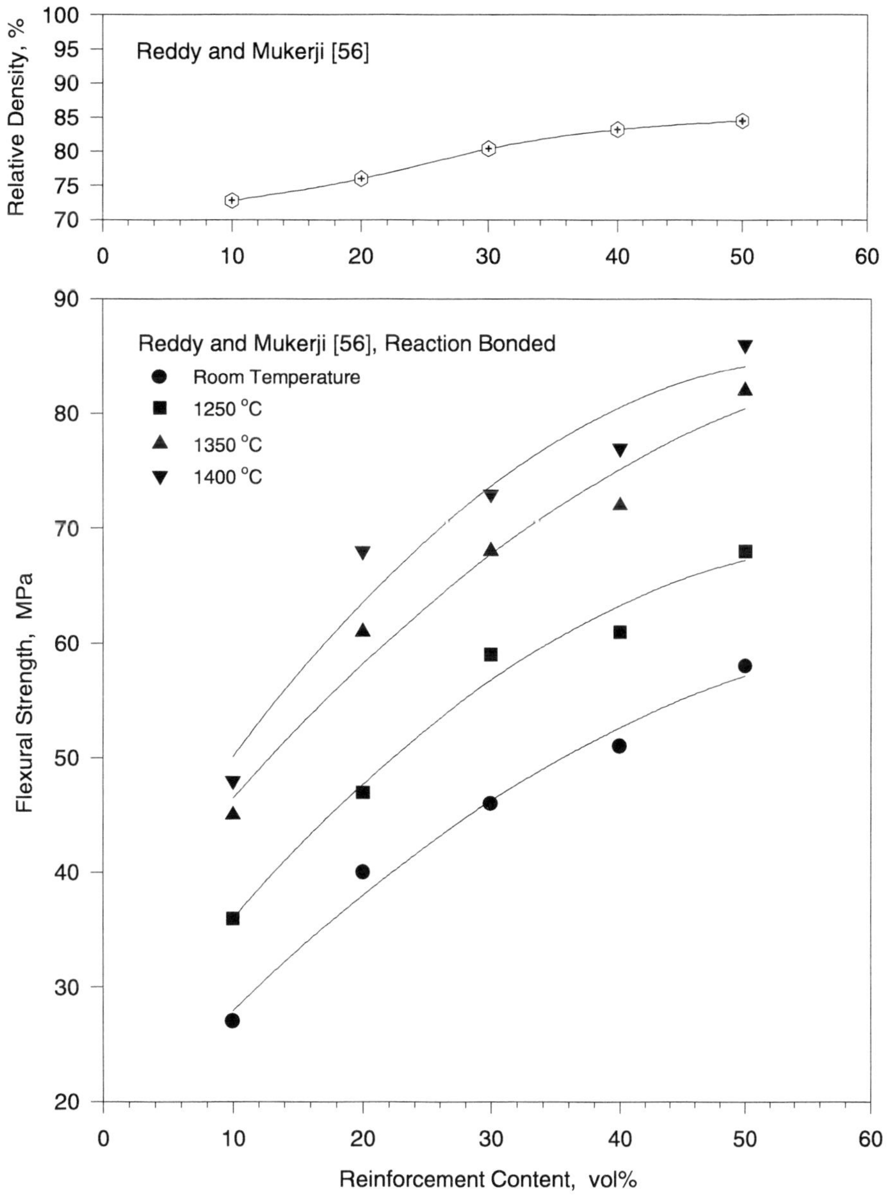

Discussion: Overall, the flexural strengths of the Si_3N_4(p)/SiC composites were relatively low due to the reaction bonding that resulted in low composite densities. This was particularly a problem because of the low silicon nitride content, and because the nitridation temperatures were much lower than the temperatures required to sinter silicon carbide. Relative density and flexural strength increased with the addition of silicon to the silicon carbide powder since the reaction bonded silicon nitride was better able to fill in the pores and bond the SiC powder at higher RBSN content. In addition, the flexural strength increased as the test temperature increased for each given vol% RBSN content. This was due to the RBSN having two phases, a dense and a woolly phase. As the woolly phase converted to the more consolidated phase at high temperatures and under pressure, the flexural strength increased.

Figure 37: 3-pt. Flexural Strength of Hot-Pressed SiC(p)/Si_3N_4 and Sintered SiC(p)/SiAlON

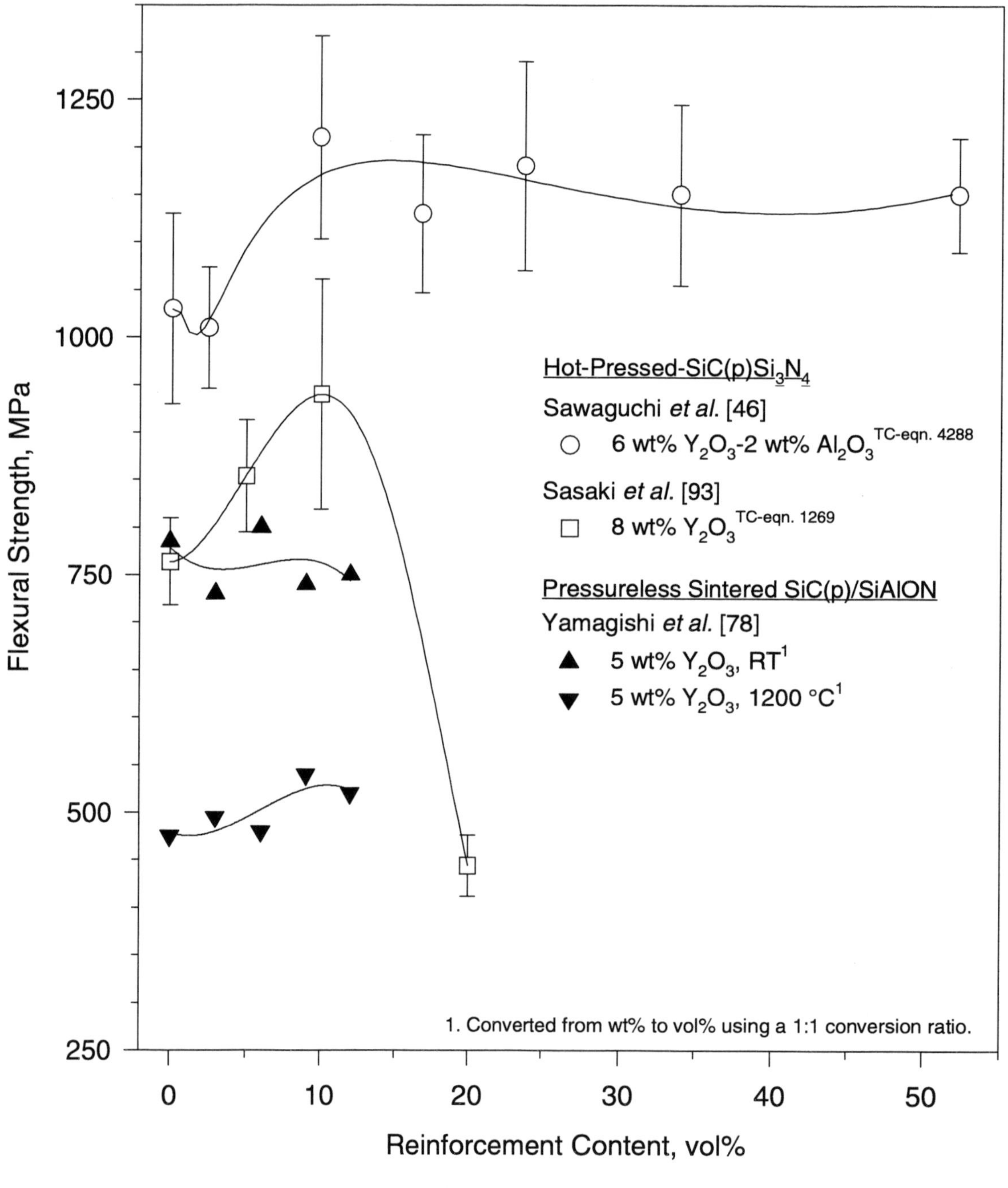

Discussion: The SiC(p)/Si_3N_4 nanocomposite tested by Sawaguchi *et al.* [46] was very strong; the 3-pt. flexural strength was greater than 1000 MPa at all SiC contents. Strength initially increased as a function of the SiC content up to 10 vol% SiC due to the growth of tree shaped crystals. The flexural strength remained approximately constant with reinforcement content after the initial increase in strength.

The SiC(p)/Si_3N_4 composite tested by Sasaki *et al.* [93] showed a substantial increase in strength as the SiC content was increased from 0 to 10 vol%. This increase in strength was due to the additive toughening effects of an elongated Si_3N_4 grain structure combined with crack deflection by the SiC particulates. The addition of 20 vol% SiC caused a dramatic decrease in toughness and strength. This decrease in composite toughening was due to the higher SiC content causing the silicon nitride grain structure to become equiaxed, as well as causing a slight decrease in relative density.

Yamagishi *et al.* [78] found that the flexural strength for the β-SiAlON-β-SiC composite prepared by pressureless sintering was substantially greater, by 200 to 250 MPa, for the composite tested at room temperature than the composite tested at 1200 °C for all SiC contents. The flexural strength increased only a very slight amount with the addition of SiC for the composite tested at 1200 °C. The flexural strength for the pressureless sintered composite tested at room temperature remained approximately constant with increasing SiC content from 0 to 12 wt%. Contrary to this behavior, fracture toughness increased by 0.5-1 MPa $m^{1/2}$ to a maximum at 6 wt% SiC content. This increase in toughness should result in an increase in strength. However, this was not observed. The lack of strengthening by the SiC particulates suggested that the critical flaw size also increased with the addition of SiC particulates.

$TiN(p)/Si_3N_4$

Reference: 19

Composition
0 - 65 wt% TiN

Processing Method
Slip Cast/Sintered

Detailed Description
Powders were ball-milled in a liquid medium. Green bodies, produced by slip casting, were then sintered under a nitrogen atmosphere at 1750 °C.

Sintering Additives
Y_2O_3
Al_2O_3

Reinforcement Description
TiN particulates

Supplier Designation
grade A and grade B

Reinforcement Dimensions
grade A: grain size = 0.8 μm
grade B: grain size = 1.2 μm

Matrix Description
Si_3N_4

Test Type
3-pt. Flexural

Specimen Preparation
Samples were machined by EDM wire.
speed = 80 mm/s
cutting rate = 3-5 mm/min
final roughness = 1.6-0.6 Ra μm

Test Temperature
RT

Reinforcement Content (weight percent)	Flexural Strength (MPa)	Comments
50	971	*grade A*
0.0	860	*grade B*
30	837	
40	822	
50	822	
55	862	
60	825	
65	605	

Remarks/Comments
Densities greater than 97% T.D. were achieved for up to 65 wt% TiN. When fine TiN powders (grade A) were used, a strengthening effect was observed for a TiN content of 50 wt%, possibly due to thermal expansion mismatch. Grade B TiN powders produced a composite with fairly constant strength in the range of 0 to 60 wt% TiN, and then the strength dropped above 60 wt% as the strengthening effect of elongated β–Si_3N_4 grains becomes ineffective.

$BN(p)/Si_3N_4$

Reference: 27

Composition
0 and 10 vol% BN

Processing Method
CIPed / Reaction Bonded

Detailed Description
SiC and Si_3N_4 powders were isostatically pressed and then nitrided in a dry N_2 atmosphere at 1350-1380 °C.

Reinforcement Description
BN particulates

Reinforcement Supplier
H.C. Starck

Matrix Description
Si_3N_4

Test Type
4-pt. Flexural

Specimen Dimensions
45 x 4.5 x 3.5 mm
inner span = 10.04 mm
outer span = 40.1 mm

Test Description
loading rate = 1.25 N/min.

Test Temperature
RT-1400 °C

Test Environment
Air

Test Equipment
High Temperature Bending Strength Tester Model 422S, Germany

Temperature (°C)	Flexural Strength (MPa)	Reinforcement Content (volume percent)	Relative Density (percent)
30	111	10	80
1,000	89	10	80
1,200	95	10	80
1,400	98	10	80
30	64.6	0	72
1,200	54.8	0	72
1,400	67.7	0	72

Remarks/Comments
The strength improvements were 76 %, 74 %, and 49 % at 30 °C, 1200 °C, and 1400 °C when compared with monolithic reaction bonded Si_3N_4. The composite density was 2.21 g cm^{-3} with a porosity of 20 vol%.

$TiN(p)/Si_3N_4$

Reference: 36

Composition
20 - 70 vol% TiN

Processing Method
Reaction Bonded

Detailed Description
Suspended in alcohol, these particles were mixed for 24 hours in a pot mill together with Si_3N_4 balls. The mixtures were dried and kneaded with a thermoplastic resin binder, and then cooled and crushed. The crushed materials were heated to 160 °C in a die and pressed into a disk shape under 98 MPa. The samples were dewaxed and heated stepwise to 1350 °C in an N_2 atmosphere.

Reinforcement Description
TiN particulates

Reinforcement Typical Properties
particle size = 2.0 μm

Matrix Description
Si powder

Matrix Properties
particle size = 0.9 μm

Test Type
Flexural Test

Specimen Dimensions
dia. = 40 mm

thickness = 10 mm

Type of Specimen
Disk

Test Temperature
RT

Reinforcement Content (volume percent)	Flexural Strength (MPa)	Relative Density (percent)
20	457	87.2
40	274	79.7
50	199	77.5
60	167	74.7
70	64	69.6

Remarks/Comments
$TiN-Si_3N_4$ showed a maximum value of 460 MPa when it contained 20 vol% of TiN.

TiN(p)/Si_3N_4

Reference: 54

Composition
0 - 40 vol% TiN

Processing Method
Hot-Pressed

Detailed Description
Si_3N_4 powder and sintering aids were homogenized in water for 48 hours and freeze dried. The second phase (TiN) was added to the matrix mix and mixed in isobutyl alcohol for 48 hours. The materials were then uniaxially hot-pressed at 1800 °C and 30 MPa for 10 minutes to nearly full density.

Sintering Additives
3 wt% Al_2O_3
8 wt% Y_2O_3

Reinforcement Description
TiN particulates

Reinforcement Supplier
Starck, Germany

Supplier Designation
Grade C

Reinforcement Dimensions
ave. particle size = 0.3-3 μm

Matrix Description
Si_3N_4 powder

Matrix Supplier
Starck, Germany

Matrix Designation
LC12

Test Type
4-pt. Flexural

Specimen Dimensions
3 x 3 x 30 mm
inner span = 13 mm
outer span = 26 mm

Test Description
crosshead speed = 0.5 mm/min

Test Temperature
RT

Reinforcement Content (volume percent)	Flexural Strength (MPa)	Relative Density (percent)	Range (MPa)
0	815	100	±35
20	545	99.4	±172
30	613	99.6	±134
40	529	99.4	±61

Remarks/Comments
The flexural strength decreased as 20 vol% TiN was added, and then increased to a peak as 30 vol% TiN was added. Thereafter, the flexural strength decreased. Overall, strength values were lower for composites than for monolithic Si_3N_4.

$BN(p)/Si_3N_4$

Reference: 66

Composition
10 - 40 wt% BN

Processing Method
Slip-Cast / Pressureless-Sintered

Detailed Description
BN and Si_3N_4 powders were mixed, and then pressureless sintered.

Reinforcement Description
BN powder

Reinforcement Supplier
Kawasaki Steel, Tokyo, Japan

Reinforcement Typical Properties
CTE:
parallel to c-axis = 40×10^{-6} °C^{-1}
transverse to c-axis < 0.1×10^{-6} °C^{-1}

Matrix Description
Si_3N_4 powder

Test Type
Indentation Strength in Bending Test (ISB)

Specimen Preparation
One side of each sample was indented using various loads between 0.2 and 1000 N.

Specimen Dimensions
5.2-6.2 x 2.5-3.1 x 32 mm

Test Description
load = 0.2-1000 N

Test Temperature
RT

Applied Load (N)	Flexural Strength (MPa)	Reinforcement Content (weight percent)	Relative Density (percent)
0.2	48.0	40	62
4.6	48.0	40	62
10	38.7	40	62
45	31.6	40	62
100	31.0	40	62
215	18.3	40	62
0.2	64.0	30	66
4.6	64.0	30	66
10	62.4	30	66
45	50.8	30	66
100	44.4	30	66
215	33.8	30	66
465	21.0	30	66
0.2	89.0	20	74
10	89.0	20	74
45	87.3	20	74
100	76.2	20	74
215	62.2	20	74
465	50.8	20	74
1,000	38.7	20	74
0.2	212.0	10	82
10	212.0	10	82
45	172.0	10	82
100	150.0	10	82
215	107.0	10	82
465	87.3	10	82

Remarks/Comments
Strength values were measured in three-point bend loading before subjecting to thermal shocks.

Figure 38: Flexural Strength of TiN(p)/Si_3N_4

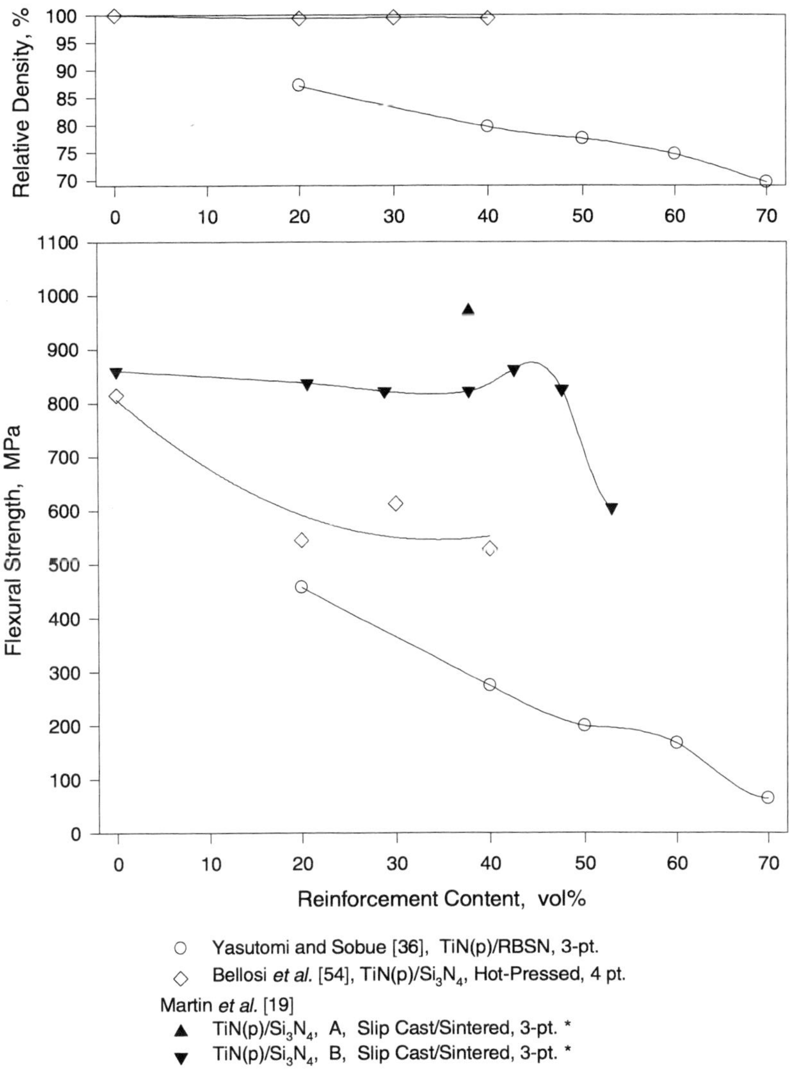

Discussion: For the TiN(p)/Si_3N_4 composite tested by Martin *et al.* [19], up to 48 vol% (60 wt%) of coarse TiN particulate (grade B) can be added to the Si_3N_4 matrix without modification of the room-temperature strength with respect to the properties of the matrix itself. As the TiN particulate content increased to greater than 50 vol%, the flexural strength of the TiN(p)/Si_3N_4 composite decreased. The TiN(p)/Si_3N_4 composite with fine TiN particulates was ~ 150 MPa greater than the equivalent composite with coarse (type B) TiN reinforcement [19]. This difference in strengthening effect between the coarse and the fine powders was likely due to the magnitude of the thermal mismatch stresses. Both TiN powders strengthened the composite by crack deflection. The coarse TiN particulates caused greater microcracking, which may not have been beneficial to the strength of the composite.

The reaction bonded TiN(p)/Si_3N_4 composite tested by Yasutomi and Sobue [36] had a maximum strength of ~ 460 MPa at 20 vol% TiN particulates. RBSN composites typically result in lower strength values compared to sintered composites due to the lower density of the RBSN composites. The flexural strength of the reaction bonded TiN(p)/Si_3N_4 composite decreased with increasing amounts of TiN (from 20 vol%). The unreinforced silicon nitride tested by Bellosi *et al.* [54] possessed a flexural strength of over 800 MPa. The addition of 20, 30, or 40 vol% TiN to the silicon nitride matrix decreased its strength to the range of 520 to 620 MPa.

Figure 39: 4-pt. Flexural Strength vs. Temperature of RBSN and 10 vol% BN(p)/RBSN

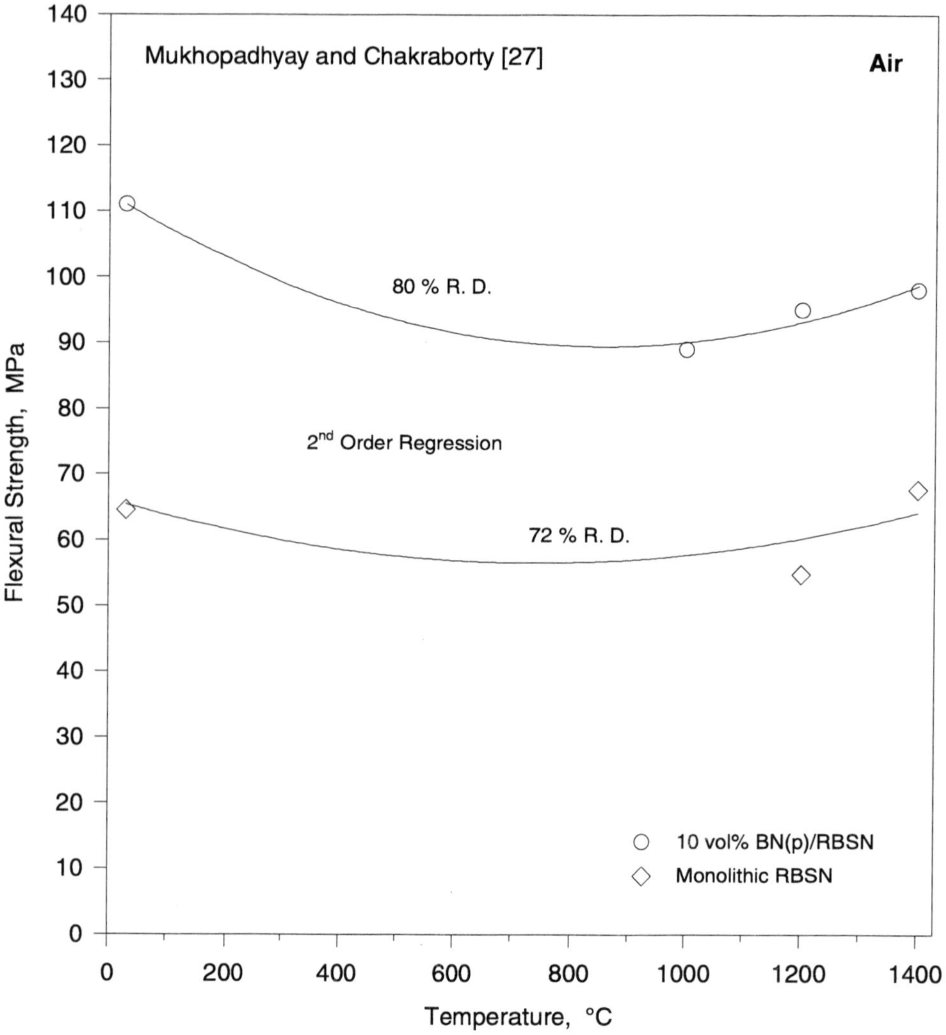

Discussion: The low overall strength of both samples was due to the low relative density, which is typical for reaction bonded materials. The strength of the monolithic RBSN and 10 vol% BN(p)/RBSN composite remained fairly constant with increasing temperature up to 1400 °C. The flexural strength for the 10 vol% BN(p)/RBSN composite was substantially greater than the monolithic RBSN for all temperatures. This was at least partially due to the higher relative density of the composite.

Figure 40: 3-pt. Flexural Strength vs. Indention Load of Pressureless Sintered BN(p)/Si_3N_4

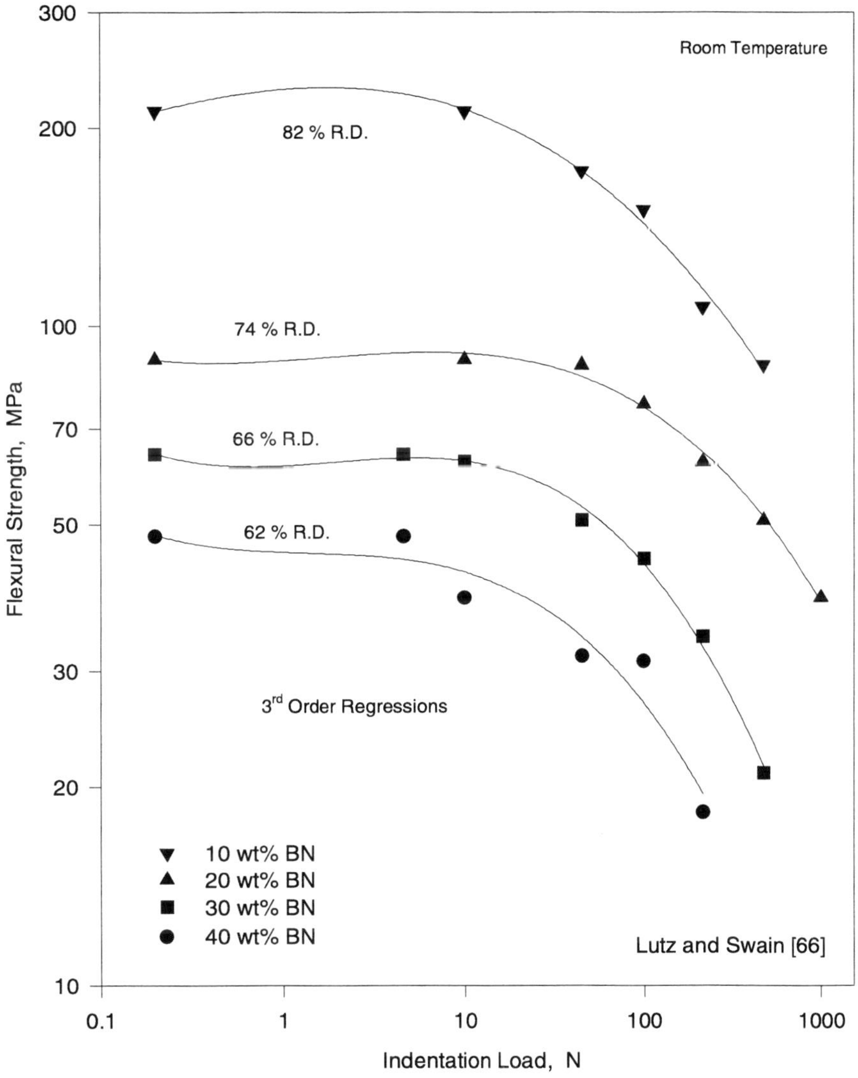

Discussion: The flexural strength for the pressureless sintered BN(p)/Si_3N_4 composite decreased as the indentation load increased for each given vol% BN reinforcement, and the strength decreased with the addition of BN reinforcement [66]. These tests indicated that the fracture resistance lowered significantly with the addition of BN. Boron nitride inhibited densification, and it was likely that the low overall strength and fracture resistance was due to low relative density.

SiC(p)/AlN

Reference: 63

Composition
30 - 90 mol% SiC

Processing Method
Hot-Pressed

Detailed Description
A deflocculent, used to get better powder dispersion, was first dissolved in the solvent (anhydrous ethyl alcohol), then both powders were added to the solvent. Before each addition, magnetic stirring was used. The slurries were ball-mixed for 3 days, then dried at 100 °C. The powder mixtures were poured into a graphite die lined with carbon foil. The sample was uniaxially pressed at room temperature and 35 MPa, and then hot-pressed at 2150 °C and 40 MPa in a flowing N_2 atmosphere. Two different hot-pressing times were selected, 45 minutes and 15 minutes.

Reinforcement Description
β-SiC particles

Reinforcement Supplier
Herman C. Starck, Goslar, Germany

Supplier Designation
Grade B10

Reinforcement Dimensions
ave. particle size < 0.8 μm

Matrix Description
AlN powder

Matrix Supplier
Herman C. Starck, Goslar, Germany

Matrix Designation
Grade B

Matrix Properties
particle size = 1.3-2.0 μm

Test Type
3-pt. Flexural

Specimen Preparation
Specimens were diamond ground, and the edges were chamfered.

Specimen Dimensions
17 x 4 x 3 mm
span = 15 mm

Test Description
crosshead speed = 0.1 mm/min

Test Temperature
RT

Reinforcement Content (mol %)	Flexural Strength (MPa)	Relative Density (percent)	Comments
30	450	>99	*45 min. soaking time*
75	1,000	>99	
90	800	97.7	
75	700	>99	*15 min. soaking time*

Remarks/Comments
The grain size and size distribution of these fully dense solid solutions affected the mechanical properties. The material, which consists of uniform fine grains around 1 μm in size, possessed very high flexural strength (as high as 1000 MPa). A relatively low strength of 450 MPa was obtained with coarse grains, 5 μm in size.

AlN(p)/BN

Reference: 79

Composition
25 wt% AlN

Processing Method
CIPed/Reaction Bonded

Detailed Description
MBN-050(80 wt%), MBN-250(20 wt%), atomized Al powder, and the sintering and nitriding aids were mixed at a weight ratio of BN-AlN = 3:1 and then ball-milled. The mixed powder was dried, uniaxially pressed at 10 MPa, and isostatically pressed at 700 MPa until a 50 x 50 x 4 mm green body was obtained. The nitridation of aluminum was carried out in a mixed gas atmosphere N_2:H_2 = 3:1 in vol%. The temperature was initially kept at 150 °C for approximately 30 minutes, then increased at the rate of 6 °C/min up to 500 °C, and increased at the rate of 3 °C/min up to 600 °C. The temperature was held at 600 °C for 4 hours, then increased at the rate of 6 °C/min up to 1900 °C with an atmosphere change to pure N_2 at 1100 °C. The final densification was done at 1900 °C for 4 hours in a N_2 atmosphere, and then cooled at the rate of 10 °C/min.

Sintering Additives
Nitridation aid:
- CaF_2 = 1 wt%

Sintering Aids:
- Y_2O_3 = 8 wt% or
- 2 $CaO.3B_2O_3$ = 4 wt% or
- CaC_2 = 4 wt% or
- CaF_2 = 2 wt%

Reinforcement Description
Al powder

Reinforcement Supplier
Toyo Aluminum K.K.

Reinforcement Typical Properties
purity >= 99.3 %
grain size = 7-8 μm

Matrix Description
h-BN

Matrix Supplier
Mitsui Toatsu Chemicals Inc.

Matrix Designation
MBN-050
MBN-250

Matrix Properties
MBN-050:
- purity 97.0 %
- grain size = 0.1-0.2 μm

MBN-250:
- purity 92.8 %
- ave. grain size = 0.01 μm

Test Type
3-pt. Flexural

Applicable Test Standard
JIS-R1601

Specimen Dimensions
50 x 50 x 4 mm

Test Temperature
RT

Relative Density (percent)	Flexural Strength (MPa)	Sintering Aids
87.1	98	8 wt% Y_2O_3
84.7	87	4 wt% 2 $CaO.3B_2O_3$
86.3	89	4 wt% CaC_2
85.9	91	2 wt% CaF_2

Remarks/Comments
The effect of various sintering additives on reaction bonded AlN(p)/BN composites was examined. The three-point bending strength was about 90 MPa for all composites with sintering aids, and was lower for reaction bonded BN (49 MPa) and BN-AlN (69 MPa) without sintering aids.

TiN(p)/AlN

Reference: 96

Composition
0 - 20 wt% TiN

Processing Method
Hot-Pressed

Detailed Description
The powders were milled in a planetary mill and then hot-pressed in a graphite mold.

Reinforcement Description
TiN powder

Matrix Description
AlN powder

Matrix Properties
ave. particle size = 4 μm
composition (wt%): 64.6 Al, 30.9 N
0.7 Fe, 0.7 O
balance: Mg, Mn, Ti, Cr

Test Type
3-pt. Flexural

Test Temperature
RT

Reinforcement Content (weight percent)	Flexural Strength (MPa)
0	200
5	409
20	459

Remarks/Comments
Self-reinforcement occurred by the intragranular crystal-oriented transition "2H to multilayer polytypes" to produce plate-like grains.

TiO_2(p)/AlN

Reference: 96

Composition
0 - 20 wt% TiO_2

Processing Method
Hot-Pressed

Detailed Description
The powders were milled in a planetary mill and then hot-pressed in a graphite mold.

Reinforcement Description
TiO_2 powder

Matrix Description
AlN powder

Matrix Properties
ave. particle size = 4 μm
composition (wt%): 64.6 Al, 30.9 N
0.7 Fe, 0.7 O
balance: Mg, Mn, Ti, Cr

Test Type
3-pt. Flexural

Test Temperature
RT

Reinforcement Content (weight percent)	Flexural Strength (MPa)
0	200
5	342
20	338

Remarks/Comments
The introduction of 5 wt% TiO_2 greatly increased the flexural strength; no further increase was seen at 20 wt% TiO_2. The strength increase may be due to the formation of plate-shaped interlayers in the grains of these materials.

TiC(p)/AlN

Reference: 96

Composition
0 or 5 wt% TiC

Processing Method
Hot-Pressed

Detailed Description
The powders were milled in a planetary mill and then hot-pressed in a graphite mold.

Sintering Additives
2 wt% Al

Reinforcement Description
TiC powder

Matrix Description
AlN powder

Matrix Properties
ave. particle size = 4 μm
composition (wt%): 64.6 Al, 30.9 N, 0.7 Fe, 0.7 O
balance: Mg, Mn, Ti, Cr

Test Type
3-pt. Flexural

Test Temperature
RT

Reinforcement Content (weight percent)	Flexural Strength (MPa)
0	200
5	408

Remarks/Comments
Self-reinforcement occurred by the intragranular crystal-oriented transition "2H to multilayer polytypes" to produce plate-like grains.

SiC(p)/AlN

Reference: 98

Composition
0 - 100 wt% SiC

Processing Method
Hot-Pressed

Detailed Description
The materials were fabricated by Cercom, Inc. under the license from FMC Corp. under processing parameters which would prevent or minimize the formation of a SiC-AlN solid solution. The composite was fabricated by mechanical mixing of the SiC and AlN powders and hot-pressing below 1900 °C.

Reinforcement Description
SiC particulates

Reinforcement Dimensions
ave. grain size:
7.0 μm (45 wt% SiC)
25.0 μm (15 wt% SiC)

Matrix Description
AlN particulates

Matrix Supplier
Cercom, Inc., Vista, CA

Test Type
4-pt. Flexural

Applicable Test Standard
MIL-STD 1942

Specimen Dimensions
3 x 4 x 40 mm

Test Description
crosshead speed = 0.5 mm/min
The specimen was preloaded to 6 N to hold it in place during ramping.

Test Temperature
RT-1050 °C

Test Equipment
ATS 3201 Series high temperature furnace, Instron 8502 load frame, 5 kN load cell

Temperature (°C)	Flexural Strength (MPa)	Reinforcement Content (weight percent)	Std. Dev. (MPa)
25	390	0	±26.0
700	375	0	±9.5
1,050	267	0	±40.8
25	477	15	±40.1
700	322	15	±36.4
900	240	15	±15.3
25	347	45	±49.4
600	369	45	±42.3
800	252	45	±45.6
900	182	45	±8.6
25	398	100	±29.2
800	359	100	±35.4
1,000	306	100	±17.8

Remarks/Comments
The monolithic AlN material exhibited a slightly lower ambient temperature strength and toughness when compared to the composites, but retained its strength and toughness at elevated temperatures. The diminished potency of the toughening mechanisms in the SiC(p)/AlN composites contributed to the elevated temperature strength degradation at moderate temperature regimes, but alternative damage mechanisms such as oxidation could have played a more significant role at higher temperatures.

Figure 41: 3-pt. Flexural Strength of Hot-Pressed SiC(p)/AlN

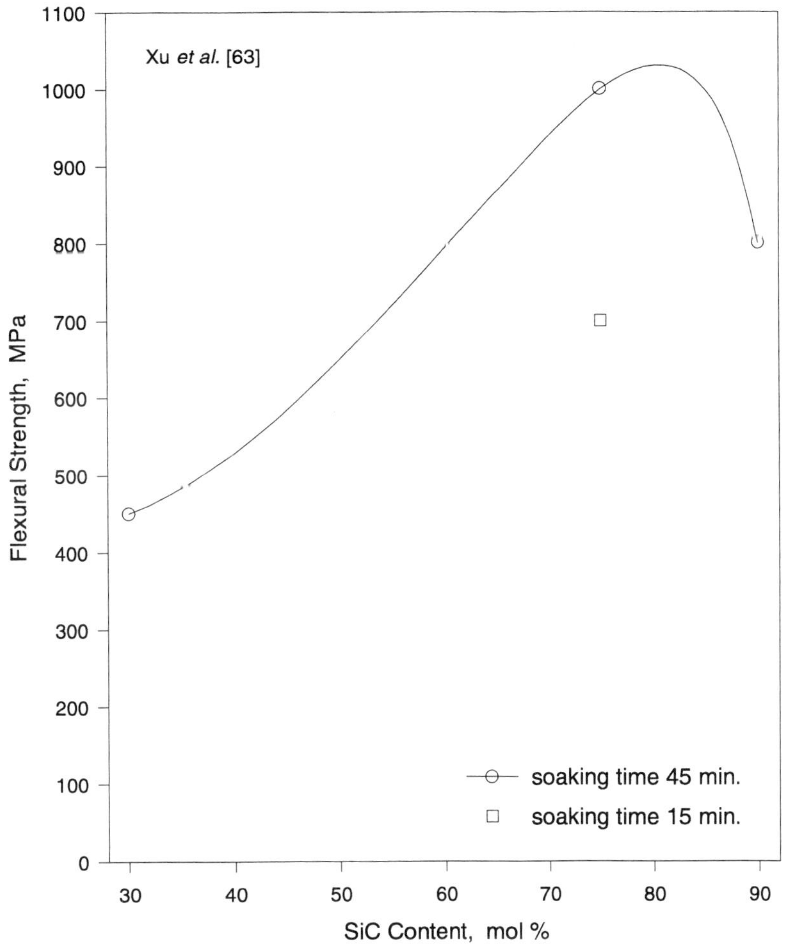

Discussion: The grain size and distribution affected the flexural strength, i.e., smaller grain size resulted in an increase in strength. The SiC/AlN ceramic with approximately 75 mol% SiC consisted of uniform fine grains around 1 μm in size and corresponded to flexural strengths in excess of 1000 MPa. A lower flexural strength of 450 MPa was determined for a 30 mol% SiC/AlN ceramic, which resulted in a structure with coarse grains approximately 5 μm in size. The two dominant mechanisms that assisted in controlling the grain size were: (1) phase-transformation-related grain multiplication and (2) nucleation and growth of Wurtzite-type 2H(d) grains [63]. In addition, a third possible mechanism was grain refinement through formation and migration of subgrain boundaries, a process similar to recrystalization [63]. Alumina, graphite and WSi_2 inclusions were observed in the grain boundaries of the composites. These impurities had little effect on the room temperature strength; however, they can degrade the high temperature properties. A decrease in the soaking time from 45 minutes to 15 minutes resulted in a 300 MPa decrease in flexural strength for the 70 mol% SiC composite. This was due to the sample with the 45 minute hot-pressing time being able to go completely into a solid solution. The resulting solid solution was harder and stronger than either of the two phases that were present in the sample that was hot-pressed for 15 minutes [63].

Figure 42: 3-pt. Flexural Strength vs. Relative Density of Reaction Bonded AlN(p)/BN

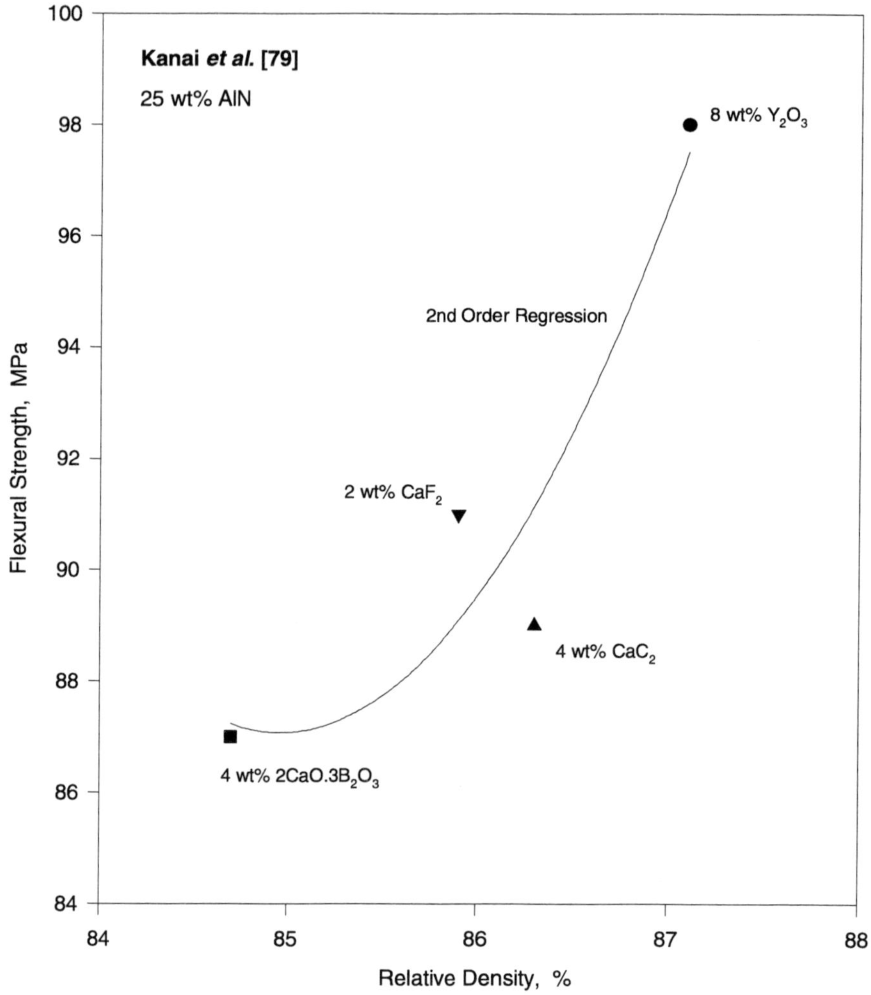

Discussion: The overall strength and density of the reaction bonded AlN(p)/BN composites were low, but flexural strength was increased as the relative density increased. The yttria sintering additive achieved the highest densification, thus, the highest flexural strength. The 4 wt% $2CaO.3B_2O_3$ was the least effective sintering additive and resulted in the lowest strength.

Figure 43: 3-pt. Flexural Strength of Hot-Pressed TiN(p), TiO_2(p), and TiC(p) in an AlN Matrix

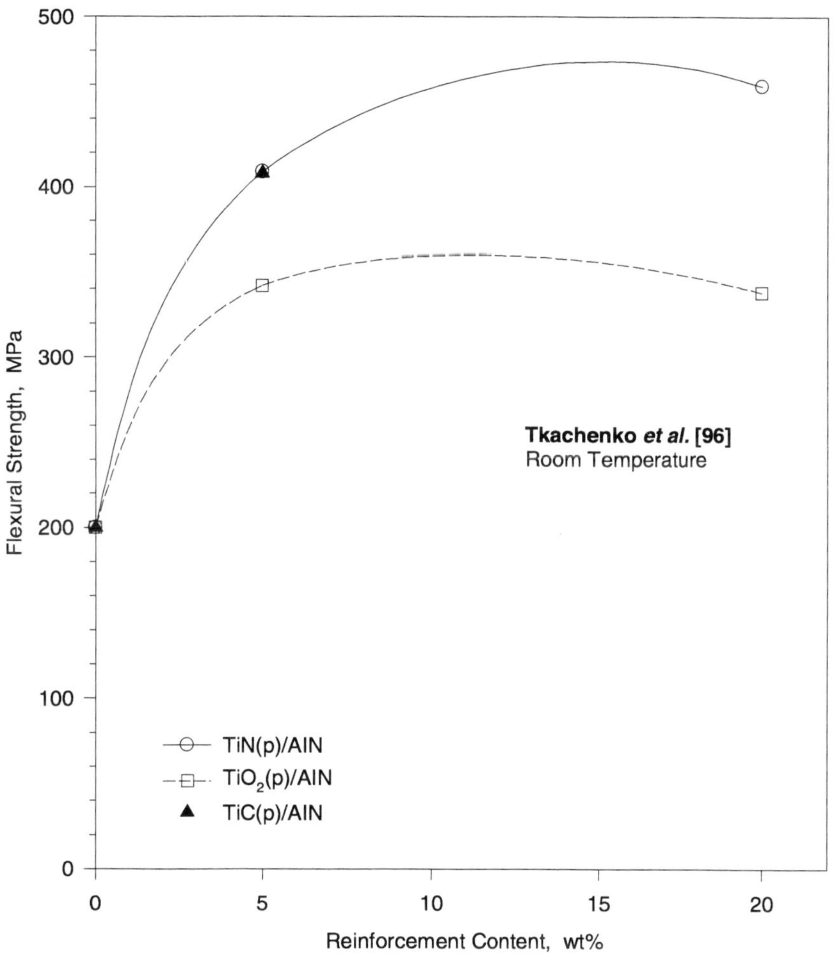

Discussion: TiN, TiO_2, and TiC formed plate shaped particles with different morphologies in the AlN matrix due to precipitates of Al and Ti, Mo, or Si. The precipitates' geometry, regular polyhedrons, indicated the occurrence of diffusion mass transfer at the precipitate-AlN matrix boundary [96]. The hardening of the composite material was due to the content of these plate-shaped precipitates and the dislocation clusters that formed around the precipitates.

Figure 44: 4-pt. Flexural Strength vs. Temperature of Hot-Pressed SiC(p)/AlN

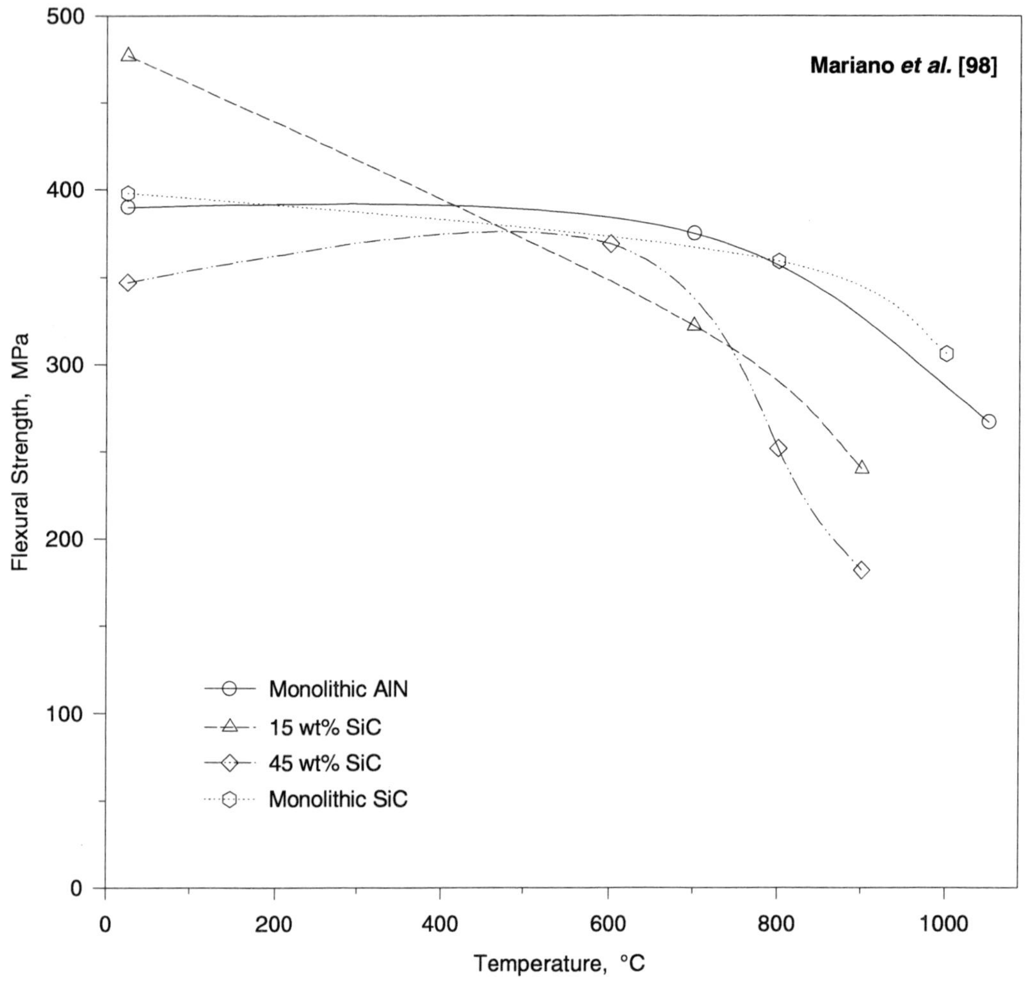

Discussion: The decrease of flexural strength at high temperatures could have been due to several mechanisms such as oxidation, creep, flaw generation, and microstructural instability due to the amorphous phases at grain boundaries degrading with temperature [98]. It was supposed that initial strength decreased due to the diminished effect of room temperature toughening mechanisms such as crack deflection. At high temperatures, the decrease in strength was accredited to oxidation of the samples [98].

$2Y\text{-}ZrO_2(p)/TiB_2$

Reference: 34

Composition
20 wt% 2Y-ZrO_2

Processing Method
CIPed/HIPed

Detailed Description
Starting powders were ball-milled in ethanol. The mixed powder was dried, uniaxially pressed under 9.8 MPa, and CIPed at 300 MPa. The green bodies were HIPed at 1600 °C under 200 MPa for 2 hours in Ar gas by the glass-encapsulation method.

Sintering Additives
0-5 wt% SiC

Reinforcement Description
2Y-ZrO_2 particulates

Reinforcement Dimensions
particle dia. = 1-3 μm

Matrix Description
TiB_2

Matrix Properties
particle dia. = 2-4 μm

Test Type
3-pt. Flexural

Specimen Dimensions
3 x 4 x 36 mm

Test Description
crosshead speed = 0.5 mm/min

Sintering Additive (weight percent)	Flexural Strength (MPa)	Range (MPa)
0.0	1,200	+15/-16
2.5	1,280	+79/-60
5.0	1,290	+45/-32

Remarks/Comments
The addition of SiC to the composite improved the flexural strength for up to 5 wt% SiC due to grain size reduction.

$2Y\text{-}ZrO_2(p)/TiB_2$

Reference: 34

Composition
0 - 40 wt% 2Y-ZrO_2

Processing Method
CIPed/HIPed

Detailed Description
Starting powders were ball-milled in ethanol. The mixed powder was dried, uniaxially pressed under 9.8 MPa, and CIPed at 300 MPa. The green bodies were HIPed at 1600 °C under 200 MPa for 2 hours in Ar gas by the glass-encapsulation method.

Reinforcement Description
2Y-ZrO_2 particulates

Reinforcement Dimensions
particle dia. = 1-3 μm

Matrix Description
TiB_2

Matrix Properties
particle dia. = 2-4 μm

Test Type
3-pt. Flexural

Specimen Dimensions
3 x 4 x 36 mm

Test Description
crosshead speed = 0.5 mm/min

Test Temperature
RT

Reinforcement Content (weight percent)	Flexural Strength (MPa)	Range (MPa)
0	448	±21
5	869	±88
10	1,170	±143
20	1,200	±36
40	865	±147

Remarks/Comments
Maximum flexural strength was observed at 20 wt% particulates.

SiC(p)/ZrB_2

Reference: 109

Composition
10 wt% SiC

Processing Method
Hot-Pressed

Reinforcement Description
SiC particulates

Matrix Description
ZrB_2

Test Type
3-pt. Flexural

Specimen Preparation
Test specimens were produced by wire EDM

Test Temperature
RT and 1200 °C

Temperature (°C)	Flexural Strength (MPa)
22	539
1,200	470

Remarks/Comments
Room temperature strength dropped by only 13% at 1200 °C.

SiC(p)/ZrB_2

Reference: 92

Composition
0 and 20 vol% SiC

Processing Method
Hot-Pressed

Detailed Description
All materials were fabricated by hot-pressing and contain some degree of anisotropy.

Reinforcement Description
SiC particulates

Reinforcement Supplier
Superior Graphite

Reinforcement Dimensions
ave. size = 3 μm

Matrix Description
ZrB_2 powder

Test Type
4-pt. Flexural

Test Description
Mechanical properties were tested in the transverse direction.

Reinforcement Content (volume percent)	Flexural Strength (MPa)	Density ($g\ cm^{-3}$)
0	305	6.10
20	408	5.52

Remarks/Comments
Flexural strength was higher for the particulate reinforced composite.

$B_4C(p)/TiB_2$

Reference: 101

Composition
0 - 30 vol% B_4C

Processing Method
Cold-Pressed/Pressureless-Sintered

Detailed Description
Powder compacts formed by cold-pressing were fabricated by pressureless-sintering at 1600 °C in an argon atmosphere.

Sintering Additives
8.6 mol% Ni
1.4 mol% C

Reinforcement Description
B_4C powder

Matrix Description
TiB_2 powder

Test Type
4-pt. Flexural

Specimen Dimensions
3 x 4 mm cross-section
inner span = 10 mm
outer span = 30 mm

Test Description
crosshead speed = 0.5 mm/min

Test Temperature
RT

Reinforcement Content (volume percent)	Flexural Strength (MPa)	Range (MPa)	Relative Density (percent)
0	379	+21/-20	94.3
10	502	+79/-75	94.7
20	533	+76/-77	93.7
30	401	+59/-81	88.3

Remarks/Comments
The flexural strength increased to a maximum of approximately 533 MPa with 20 vol% B_4C. Flexural strength decreased with further addition of B_4C reinforcement.

Figure 45: 3-pt. Flexural Strength of HIPed 2Y-ZrO_2(p)/TiB_2

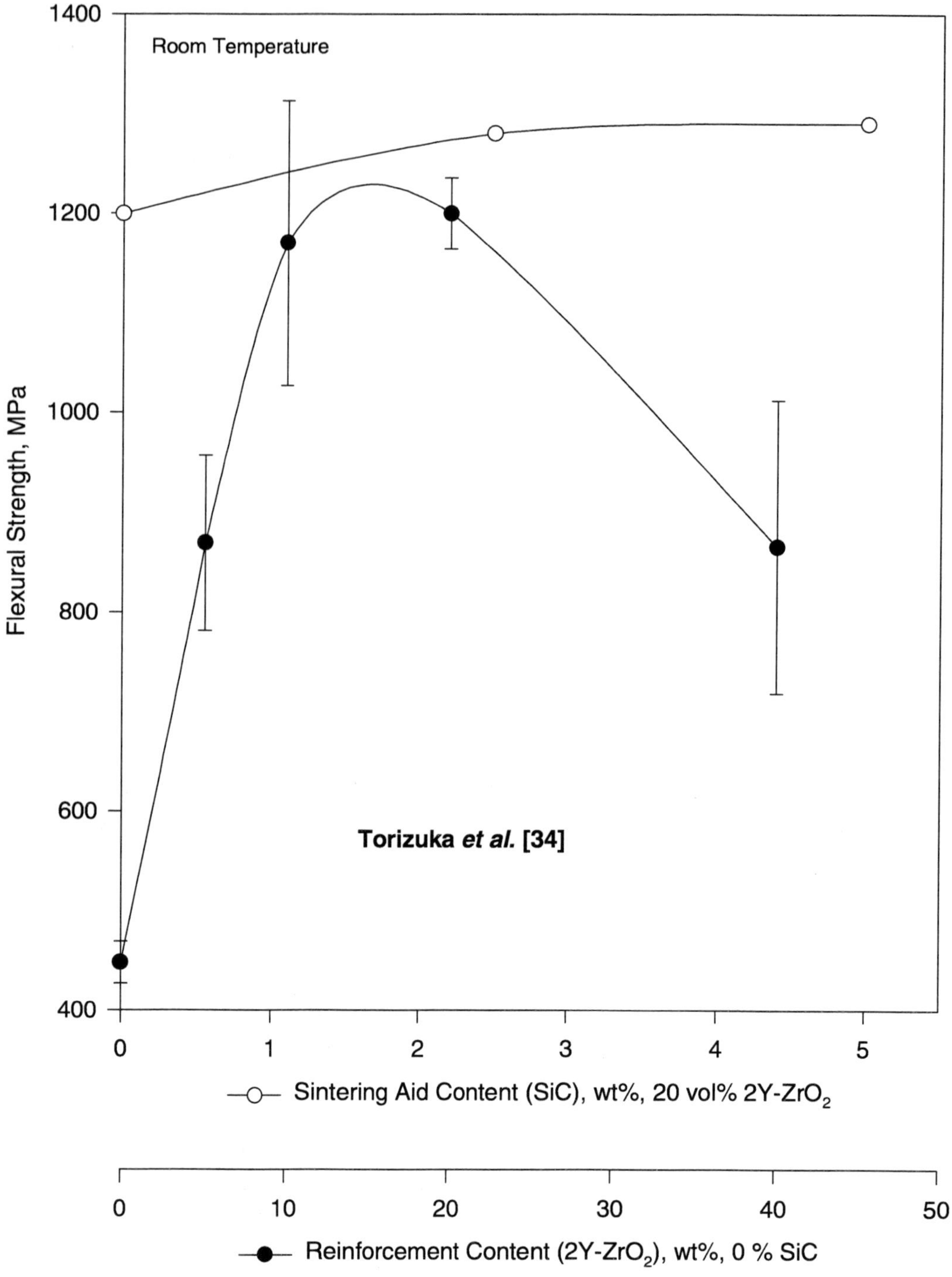

Discussion: The very high flexural strength of 2Y-ZrO_2(p)/TiB_2 increased by a small amount with the addition of silicon carbide sintering additives to the composite. SiC restrained the grain growth of the ZrO_2 phase for increased sinterability. The result was a highly dense, fine-grain sample. The addition of up to 20 wt% 2Y-ZrO_2 reinforcement increased the flexural strength of the composite. Additions above 20 wt% reinforcement caused a decrease in the strength of the 2Y-ZrO_2(p)/TiB_2 composite.

Figure 46: 4-pt. Flexural Strength of SiC(p)/ZrB_2 and B_4C(p)/TiB_2

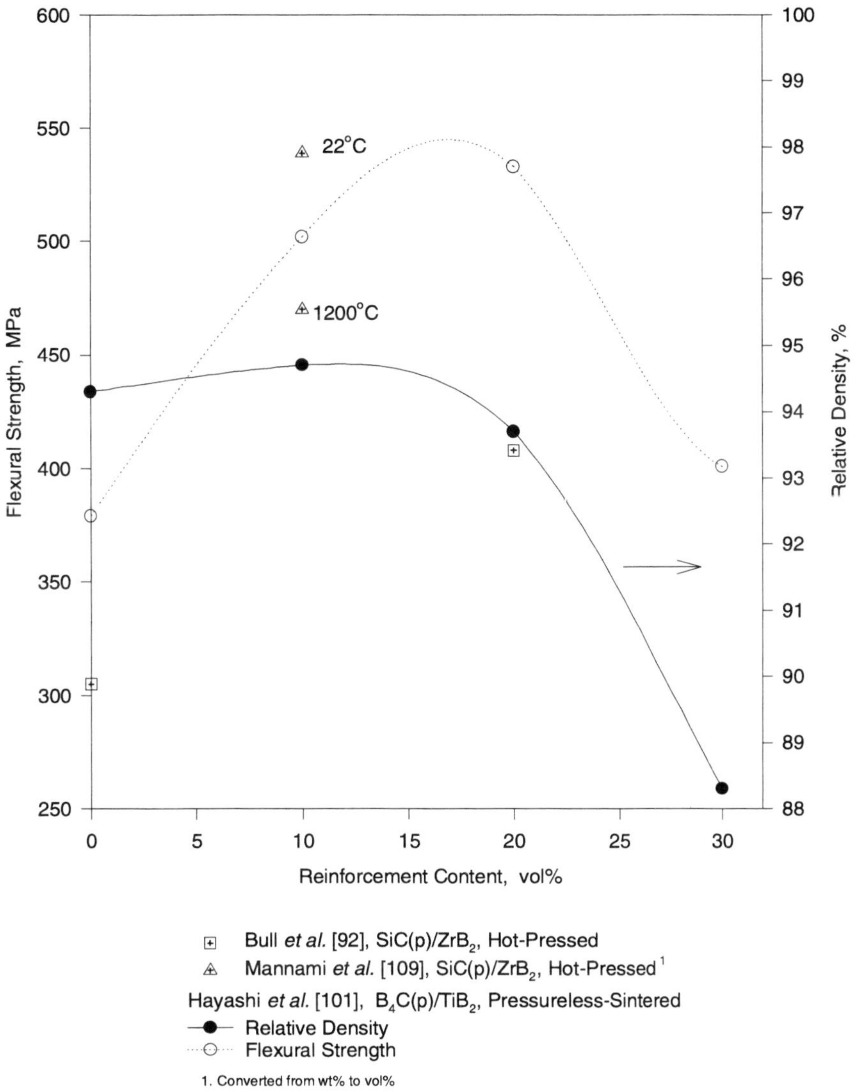

Discussion: Crack deflection by the B_4C and SiC particulates was the main strengthening mechanism for all the composites presented above. Mannami *et al.* [109] showed significant strength retention for the hot-pressed SiC(p)/ZrB_2 composite; its RT strength decreased by only 13% at 1200 °C. Bull *et al.* [92] found that the strength of monolithic ZrB_2 was increased by over 100 MPa by the addition of 20 vol% SiC. Hayashi *et al.* [101] found that the B_4C(p)/TiB_2 composite increased strength with the addition of up to 20 vol% B_4C; the composite reached a maximum strength of ~525 MPa, then flexural strength decreased. This correlated with the decrease in the relative density at 30 vol% B_4C reinforcement.

ZrO_2(p)/ZrC

Reference: 18

Composition
0-40 wt% ZrO_2

Processing Method
Sintered

Detailed Description
Type A zirconia was a pre-reacted stabilized material. Particles below 1 μm were extracted from the powders by sedimentation. ZrC was dispersed in the aqueous solution. Type B zirconia was a physical mixture of ZrO_2 and Y_2O_3. ZrC and Type B ZrO_2 mixtures were ball-milled in polyethylene mills using ZrO_2 balls for up to 16 hours. Test blocks of each material were slip-cast and allowed to go to completion. The side molds used were removed, and the specimens were air dried for 12 hours. The specimens were finally dried at 50 °C for 48 hours, then cut and hand sanded to size. A graphite tube furnace with an Ar atmosphere was used for sintering. Samples were set in a graphite crucible and packed in loose powder of the same ZrC-ZrO_2 composition. The samples were heated at a rate of 30 °C/min. to the desired temperature and held for one hour.

Reinforcement Description
3Y-ZrO_2 (type A) and 2.7Y-ZrO_2 (type B) particulates

Reinforcement Supplier
Type A: Toyo Soda Manufacturing Co., Tokyo, Japan
Type B: Wah Chang Corp., Albany, OR

Matrix Description
ZrC powder

Matrix Supplier
Furrchi Chemical Co., Tokyo, Japan

Test Type
4-pt. Flexural

Specimen Preparation
The surface was finished with a 400-grit diamond wheel and chamfered on the corners with a 15 μm diamond wheel.

Specimen Dimensions
3.5 x 40 x 3.0 mm
inner span = 12.7 mm
outer span = 25.4 mm

Test Description
crosshead speed = 0.2 mm/min

Test Equipment
Instron testing machine

Reinforcement Content (weight percent)	Flexural Strength (MPa)	Sintering Temperature (°C)	Range (MPa)	Comments
0	219	2,400	+20/-15	*Monolithic ZrC*
20	280	2,000	+11/-10	*Type A ZrO_2*
30	229	2,000		
40	308	2,000		
5	225	2,200	+25/-23	*Type B ZrO_2*
6	230	2,200	+21/-18	
7	243	2,200	+12/-11	
35	290	2,400	+24/-19	

Remarks/Comments
The flexural strength increased marginally with increasing ZrO_2 content. This may be attributed to the presence of tetragonal ZrO_2 and/or residual stress in the two-phase microstructure.

$Zr(p).ZrB_2(pl)/ZrC$

Reference: 28

Composition
2.5 - 24 vol% Zr particles, ZrB_2/ZrC ratio = 58:42 by volume

Composite Architecture
multi-reinforcement

Processing Method
Lanxide Reaction Bonded

Detailed Description
The material was formed by the direct reaction of molten zirconium with a shaped preform of boron carbide particulates, which were substantially consumed in the process. These materials are currently being commercialized as the Lanxide NX-3400 series platelet-reinforced ceramics.

Reinforcement Description
Zr particulates
ZrB_2 platelets

Matrix Description
ZrC

Test Type
4-pt. Flexural

Applicable Test Standard
MIL-STD-1942(MR)

Specimen Dimensions
3 x 6 x 50 mm
inner span = 20 mm
outer span = 40 mm

Test Description
crosshead speed = 0.5 mm/min

Test Temperature
RT

Residual Zirconia Content (volume percent)	Flexural Strength (MPa)
2.5	870
3.1	842
24	888

Remarks/Comments
The high flexural strength resulted from the presence of the ZrB_2 platelets and metal, as well as the narrow distribution of processing flaw sizes.

Figure 47: 4-pt. Flexural Strength of ZrO_2(p)/ZrC and Zr(p).ZrB_2(pl)/ZrC

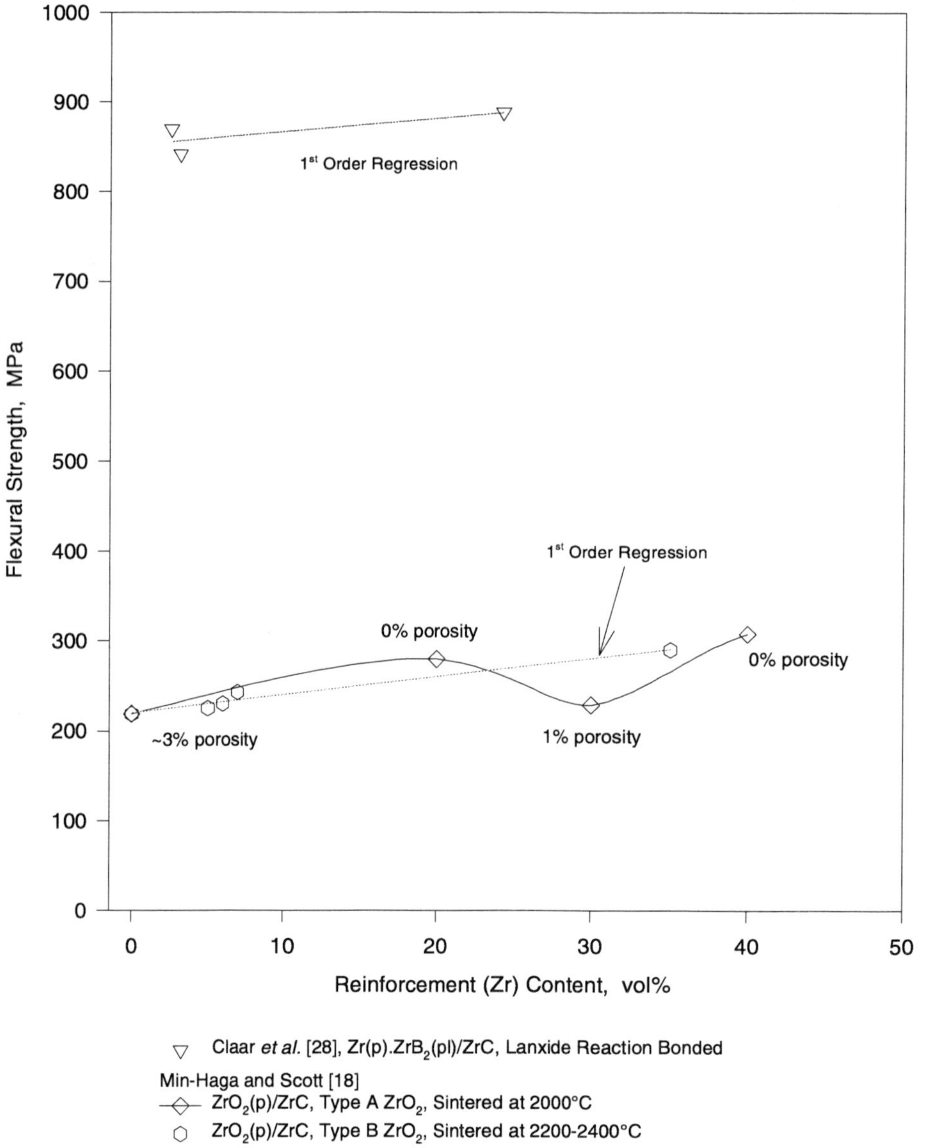

Discussion: In the ZrO_2(p)/ZrC composite, there was a slight increase in strength with increasing ZrO_2 content [18]. The observed increase in flexural strength may be attributed to the presence of tetragonal ZrO_2. This would have caused residual stresses in the two-phase microstructure from the transformation of ZrO_2. The phase change would have caused ZrO_2 to be in compression and ZrC to be in tension. Porosity had an effect on strength; increased strength was dependent on decreasing porosity.

The composite tested by Claar *et al.* [28] was very strong due to unique processing and microstructure. Strength was nearly independent of residual metal content; however, the residual zirconia did significantly increase fracture toughness of the composite.

TiC(p)/SiC

Reference: 20

Composition
0 - 25 vol% TiC

Processing Method
Hot-Pressed

Detailed Description
Powders and sintering aids were mixed together. Samples were hot-pressed at 2000-2100 °C and 40 MPa.

Sintering Additives
1.5 wt% Al_2O_3
1 wt% C

Reinforcement Description
TiC particulates

Reinforcement Dimensions
ave. particle size = 7.2 μm (coarse)
ave. particle size = 0.82 μm (fine)

Matrix Description
α-SiC powder

Matrix Properties
ave. particle size = 2.0 μm

Test Type
3-pt. Flexural

Specimen Dimensions
3 x 4 x 26 mm

Test Description
crosshead speed = 0.5 mm/min

Test Temperature
RT

Reinforcement Content (volume percent)	Flexural Strength (MPa)	Relative Density (percent)	Comments
0	487	98.6	*coarse TiC*
7	463	96.7	
13	497	98.1	
25	418	100	
0	487	98.6	*fine TiC*
7	455	95.5	
13	497	98.4	
25	586	99.7	

Remarks/Comments
The coarse and fine TiC powders gave nearly identical results except at 25 vol% TiC content where the strength was much greater for the composite fabricated with fine TiC powders.

TiC(p)/SiC

Reference: 20

Composition
13 vol% TiC

Processing Method
Hot-Pressed

Detailed Description
Powders and sintering aids were mixed together. Samples were hot-pressed at 2000-2100 °C and 40 MPa.

Sintering Additives
1.5 wt% Al_2O_3
1 wt% C

Reinforcement Description
TiC particulates

Reinforcement Dimensions
ave. particle size = 7.2 μm (coarse)

Matrix Description
α-SiC powder

Matrix Properties
ave. particle size = 2.0 μm

Test Type
3-pt. Flexural

Specimen Dimensions
3 x 4 x 26 mm

Test Description
crosshead speed = 0.5 mm/min

Test Temperature
RT-1250 °C

Test Environment
Air

Temperature (°C)	Flexural Strength (MPa)	Range (MPa)
20	498	±54
1,100	403	±38
1,250	345	±106

Remarks/Comments
Flexural strength decreased with increasing temperature.

TiC(p)/SiC

Reference: 53

Composition
0 - 100 wt% TiC

Processing Method
Hot-Pressed

Detailed Description
β–SiC powders were mixed with TiC together with one of the sintering aids. Well-mixed powders were obtained by ball-milling in n-hexane for 24 hours using polyamide media. After drying, the powders were loaded into a graphite die with an inner diameter of 90 mm and hot-pressed at 1800-2150 °C for 2 hours under 40 MPa in an Ar gas atmosphere. Hot-pressed discs were machined into bars.

Sintering Additives
composite: 0.5 wt% B and 1 wt% C
SiC monolith: 1 wt% B and 1 wt% C

Reinforcement Description
TiC particulates

Reinforcement Supplier
Kojundo Chem. Lab. Co., Ltd.

Reinforcement Typical Properties
ave. particle size = 1.2 μm
purity > 99 %

Matrix Description
β-SiC powder

Matrix Supplier
Ibiden Co., Ltd.

Matrix Designation
Ultrafine

Matrix Properties
ave. particle size = 0.3 μm
purity > 98 %

Test Type
3-pt. Flexural

Applicable Test Standard
JIS R-1601

Specimen Dimensions
3 x 4 x 38 mm

Test Temperature
RT

Hot-Press Temperature (°C)	Flexural Strength (MPa)	Reinforcement Content (weight percent)	Relative Density (percent)
1,800	418	0	84.1
1,900	391	0	92.6
2,000	585	0	98.5
2,150	598	0	99.2
1,800	334	50	81.1
1,900	477	50	94.4
2,000	699	50	98.9
2,150	706	50	99.2
1,800	662	100	98.9
2,000	457	100	98.8
2,150	409	100	98.7

Remarks/Comments
The flexural strength at room temperature increased with the hot-pressing temperature in accordance with the increase in the density for SiC and SiC-50 wt% TiC. The highest strengths recorded were 620 and 710 MPa, respectively. On the other hand, the TiC monolith which was hot-pressed at 1800 °C exhibited a high strength (662 MPa), but the strength decreased monotonically to 409 MPa with increasing hot-pressing temperature.

TiC(p)/SiC

Reference: 53

Composition
5-90 wt% TiC

Processing Method
Hot-Pressed

Detailed Description
Powders of β–SiC were mixed with TiC. Well-mixed powders were obtained by ball-milling in n-hexane for 24 hours using polyamide media. After drying, the powders were loaded into a graphite die with an inner diameter of 90 mm, and hot-pressed at 2150 °C for 2 hours under 40 MPa in an Ar gas atmosphere. Hot-pressed discs were machined into bars.

Reinforcement Description
TiC particulates

Reinforcement Supplier
Kojundo Chem. Lab. Co., Ltd.

Reinforcement Typical Properties
ave. particle size = 1.2 μm
purity > 99 %

Matrix Description
β-SiC powder

Matrix Supplier
Ibiden Co., Ltd.

Matrix Designation
Ultrafine

Matrix Properties
ave. particle size = 0.3 μm
purity > 98 %

Test Type
3-pt. Flexural

Applicable Test Standard
JIS R-1601 and JIS R-1604 (for RT and elevated temperatures, respectively)

Specimen Dimensions
3 x 4 x 38 mm

Test Temperature
RT and 1500 °C

Test Environment
Ar

Reinforcement Content (weight percent)	Flexural Strength (MPa)	Temperature (°C)	Relative Density (%)
5	302	25	76.6
10	376	25	81.5
20	510	25	89.8
22	625	25	94.6
23	698	25	96.0
25	682	25	98.2
30	681	25	98.2
50	658	25	99.6
70	512	25	99.4
90	443	25	99.1
5	378	1,500	73.6
10	498	1,500	81.5
20	592	1,500	89.8
22	644	1,500	94.6
23	718	1,500	96.0
25	883	1,500	98.2
30	933	1,500	98.2
50	512	1,500	99.6
70	257	1,500	99.4
90	218	1,500	99.1

Remarks/Comments
Flexural strengths of SiC-TiC composites without sintering aids at both room temperature and 1500 °C (in an Ar atmosphere) are shown by the data as a function of the TiC content. In the composition with greater than 25 wt% TiC, the room temperature flexural strengths were almost equivalent to those of the composites with boron and carbon additives. However, both the room and elevated temperature strengths decreased more drastically for those composites with less than 25 wt% TiC content, compared with those of composites containing boron and carbon due to the difficulty in sintering SiC without sintering additives. The high temperature strengths were lower than those of SiC-TiC composites with boron and carbon. Therefore, it may be suggested that additions of B and C will affect the fracture behavior of the composites at elevated temperatures.

TiC(p)/SiC

Reference: 53

Composition
10 - 50 wt% TiC

Processing Method
Hot-Pressed

Detailed Description
β–SiC powders were mixed with TiC and sintering aids. Well-mixed powders were obtained by ball-milling in n-hexane for 24 hours using polyamide media. After drying, the powders were loaded into a graphite die with an inner diameter of 90 mm and hot-pressed at 2150 °C for 2 hours under 40 MPa in an Ar gas atmosphere. Hot-pressed discs were machined into bars.

Sintering Additives
some with 0.5 wt% B and 1 wt% C or
2 wt% Al_2O_3

Reinforcement Description
TiC particulates

Reinforcement Supplier
Kojundo Chem. Lab. Co., Ltd.

Reinforcement Typical Properties
ave. particle size = 1.2 μm
purity > 99 %

Matrix Description
β-SiC powder

Matrix Supplier
Ibiden Co., Ltd.

Matrix Designation
Ultrafine

Matrix Properties
ave. particle size = 0.3 μm
purity > 98 %

Test Type
3-pt. Flexural

Applicable Test Standard
JIS R-1601 and JIS R-1604 (for RT and elevated temperatures, respectively)

Specimen Dimensions
3 x 4 x 38 mm

Test Temperature
RT - 1500 °C

Test Environment
Ar

Temperature (°C)	Flexural Strength (MPa)	Reinforcement Content (weight percent)	Relative Density (percent)	Comments
25	380	10	81.1	*no sintering additives*
1,200	400	10	81.1	
1,500	500	10	81.1	
25	690	30	97.8	
1,200	800	30	97.8	
1,500	930	30	97.8	
25	660	50	99.6	
1,200	580	50	99.6	
1,500	510	50	99.6	
25	610	10	98.0	*B and C additives*
1,200	830	10	98.0	
1,500	970	10	98.0	
25	750	30	98.9	
1,200	800	30	98.9	
1,500	930	30	98.9	
25	710	50	98.8	
1,200	830	50	98.8	
1,500	800	50	98.8	
25	580	10	99.9	*Al_2O_3 additive*
1,200	350	10	99.9	
1,500	180	10	99.9	
25	570	30	99.6	
1,200	390	30	99.6	
1,500	230	30	99.6	
25	550	50	99.2	
1,200	360	50	99.2	
1,500	220	50	99.2	

Remarks/Comments

In the case of Al_2O_3 additions, sintered bodies with densities higher than 99 % TD could be attained. However, the flexural strength was lower than for those composites with B and C, especially the high temperature strength. The composites with B and C added appeared to have the highest strengths when compared to the composites without sintering aids or with Al_2O_3 added.

TiC(p)/SiC

Reference: 53

Composition
50 wt% TiC

Processing Method
Hot-Pressed

Detailed Description
Powders of either α– or β–SiC were mixed with TiC together with one of the sintering aids. Well-mixed powders were obtained by ball-milling in n-hexane for 24 hours using polyamide media. After drying, the powders were loaded into a graphite die with an inner diameter of 90 mm and hot-pressed at 2150 °C for 2 hours under 40 MPa in an Ar gas atmosphere. Hot-pressed discs were machined into bars.

Sintering Additives
0.5 wt% B and 1 wt% C

Reinforcement Description
TiC particulates

Reinforcement Supplier
Kojundo Chem. Lab. Co., Ltd.

Reinforcement Typical Properties
ave. particle size = 1.2 μm
purity > 99 %

Matrix Description
SiC powder (β– , α– and β– (1:1 ratio), or α–)

Matrix Supplier
Ibiden Co., Ltd. (β–)
Showa Denko K.K. (α–)

Matrix Designation
Ultrafine (β–)
DU A-1 (α–)

Matrix Properties
β–:
ave. particle size = 0.3 μm
purity > 98 %
α–:
ave. particle size = 0.4 μm
purity > 98 %

Test Type
3-pt. Flexural

Applicable Test Standard
JIS R-1601 and JIS R-1604 (for RT and elevated temperatures, respectively)

Specimen Dimensions
3 x 4 x 38 mm

Test Temperature
RT and 1500 °C

Test Environment
Ar

Temperature (°C)	Flexural Strength (MPa)	Relative Density (percent)	Comments
25	710	98.8	*β– only*
1,500	800	98.8	
25	660	98.4	*α– and β–*
1,500	330	98.4	
25	560	98.5	*α– only*
1,500	280	98.5	

Remarks/Comments
The composite made with β-SiC had higher flexural strengths at room temperature and much greater flexural strength at 1500 °C (710 and 800 MPa, respectively). Also, the flexural strength increased for the β-SiC composite with the increase in temperature, but decreased for the composites made with α– and β– SiC or α– only SiC.

TiC(p)/SiC

Reference: 70

Composition
0 - 100 wt% TiC

Processing Method
Hot-Pressed

Detailed Description
Powders of β-SiC and TiC were mixed with B_4C powder and carbon black as sintering aids for SiC. Well-mixed powders were attained by ball-milling in n-hexane for 24 hours using a polyamide media. After drying, the powders were loaded into a graphite die with id = 90 mm and hot-pressed at 2150 °C for 2 hours under 40 MPa in an Ar gas atmosphere. Hot-pressed discs of SiC-TiC composites were machined into bars.

Sintering Additives
B_4C and C (carbon black)

Reinforcement Description
TiC powders

Reinforcement Supplier
Kojundo Chem. Lab. Co., Ltd

Reinforcement Typical Properties
ave. particle size = 0.4 μm
purity > 99 %

Matrix Description
β-SiC powders

Matrix Supplier
Ibiden Co. Ltd.

Matrix Designation
Ultrafine

Matrix Properties
ave. particle size = 0.3 μm
purity > 98 %

Test Type
3-pt. Flexural

Applicable Test Standard
JIS R-1601 and R-1604

Specimen Dimensions
3 x 4 x 38 mm

Test Temperature
RT and 1500 °C

Test Environment
Ar

Reinforcement Content (weight percent)	Flexural Strength (MPa)	Range (MPa)	Temperature (°C)
0	608	±64	22
10	608	±3	22
30	753	±22	22
50	710	±34	22
70	551	±70	22
90	649	±60	22
100	427	±90	22
0	1,020	±35	1,500
10	971	±80	1,500
30	931	±150	1,500
50	807	±65	1,500
70	424	±50	1,500
90	361	±40	1,500
100	315		1,500

Remarks/Comments
The flexural strengths for the hot-pressed SiC and TiC at room temperature were ~620 and ~420 MPa, respectively. For the composites, strengths higher than 750 MPa were recorded for 30-50 wt% TiC composites. These results agreed well with the results on the composition dependence of the fracture toughness. It can be concluded that strengthening was achieved through toughening, at least at room temperature. The strength at 1500 °C decreased monotonically with increasing TiC content from ~1000 MPa (for monolithic SiC) to ~300 MPa (for monolithic TiC).

Figure 48: 3-pt. Flexural Strength of TiC(p)/SiC

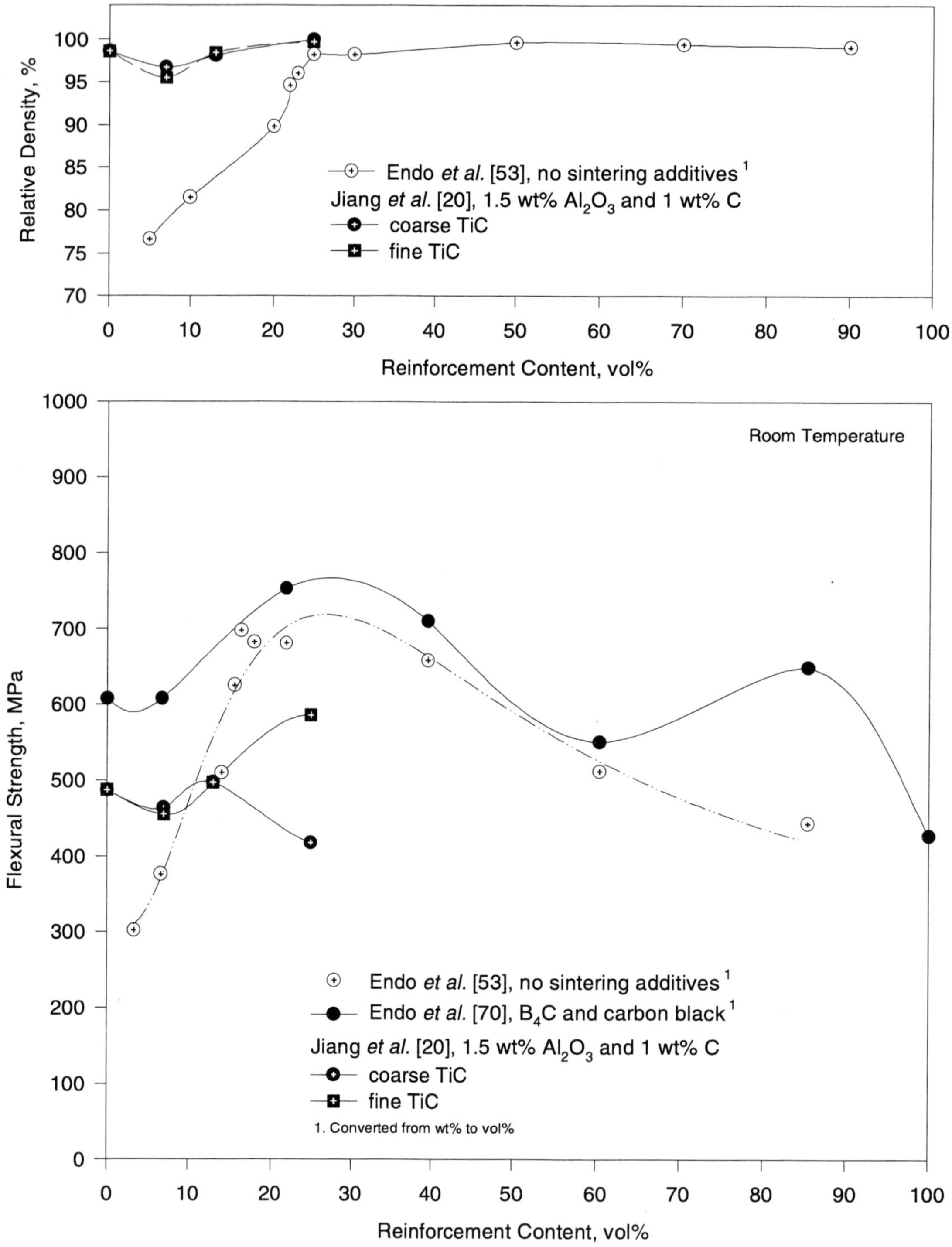

Discussion: Endo *et al.* [53, 70] tested TiC(p)/SiC composites fabricated with or without B_4C and carbon black sintering additives. The strength of the TiC(p)/SiC composite fabricated without sintering additives was fairly low at TiC content below 25 wt% due to low density. The low density of these composites at less than 30 wt% (~25 vol%) reinforcement content was due to the poor sinterability of SiC. TiC additions significantly enhanced the densification of the SiC matrix. The TiC(p)/SiC composites reached near full density when the TiC content was 30 wt% or greater. The strength of the TiC(p)/SiC composite reached a maximum, close to 700 MPa, at 25 wt% TiC. Strength decreased as the TiC content increased from 25 to 90 wt%. This decrease in strength may have been due to the lower strength of TiC, microcracking, and/or grain growth.

The B_4C and carbon sintering additives allowed near full density to be reached at all compositions. Consequently, relative density was not a significant factor in determining the strength of the composite fabricated with sintering additives [70]. The shape of the curve for this composite may have been due to the less abundant phase, TiC on the left of the graph and SiC on the right of the graph, impeding dislocation motion. The impeded dislocations strengthened the sample. The samples were weaker at 0, 60 and 100 vol% TiC because grains were able to grow and the larger grains were not as effective in pinning dislocations.

From Jiang *et al.* [20], the coarse and fine TiC composites had a high strength due to the high relative density obtained through the use of sintering additives. Both composites had identical strength at 0, 7, and 13 vol% TiC reinforcement. The fine grained TiC(p)/SiC composite increased from 497 to 586 MPa as the TiC content increased from 13 to 25 vol%, while the coarse grained composite decreased from 497 to 418 MPa for the same increase in TiC content. This divergence in behavior was due to thermal expansion mismatch stresses. These stresses were much more severe in the coarse TiC composite. This caused microcracking that was detrimental to the strength of the coarse grain composite. Extensive microcracking did not occur in the fine grain composite, and the strength of this composite continued to increase with further addition of TiC. At 25 vol% reinforcement, the fracture toughness of the fine TiC composite was more than 1 MPa $m^{1/2}$ greater than the fracture toughness of the coarse TiC composite. This difference in toughness influenced the flexural strength of the two 25 vol% TiC(p)/SiC composites.

Figure 49: 3-pt. Flexural Strength of TiC(p)/SiC at 1500 °C

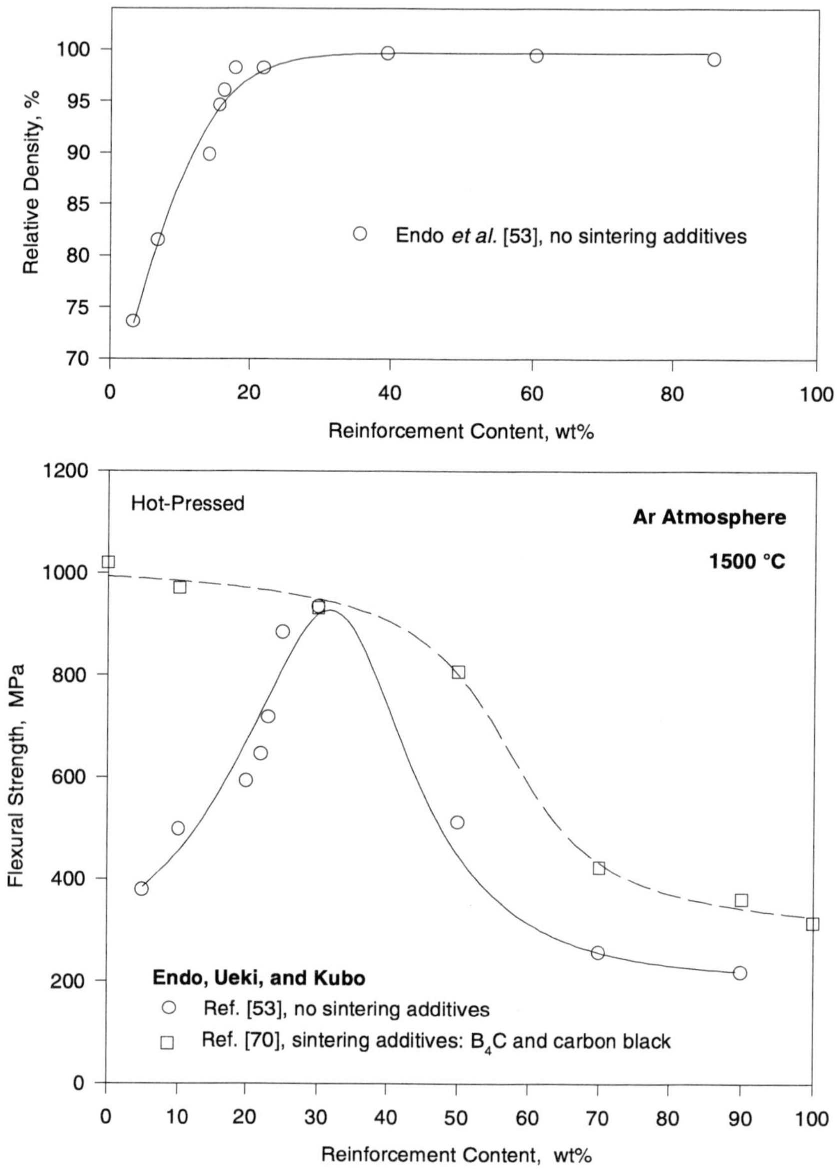

Discussion: Endo, Ueki, and Kubo [53 and 70] tested the flexural strength of TiC(p)/SiC composites at 1500 °C in an argon atmosphere. Reference [70] found that the cause of the decrease in strength with the addition of TiC at 1500 °C was due to the relative strengths of TiC and SiC, and due to the plasticity of TiC. Since SiC was stronger than TiC at elevated temperatures, the flexural strength decreased with the addition of TiC to the composite. The same was true for the composite tested by ref. [53] with greater than 30 wt% TiC. Below 30 wt% TiC, the reason for the low strength in the ref. [53] composite was the low relative density. This low density was the result of no sintering aids used. Contrary to this, the high SiC content composites tested by ref. [70] had high strength due to the B_4C and carbon black sintering aids that promoted densification. The composites were tested in Ar, so oxidation was not a factor in the these tests.

Figure 50a: Effect of Sintering Additives on Elevated Temperature Flexural Strength of 10, 18.5, and 30 wt% TiC(p)/SiC

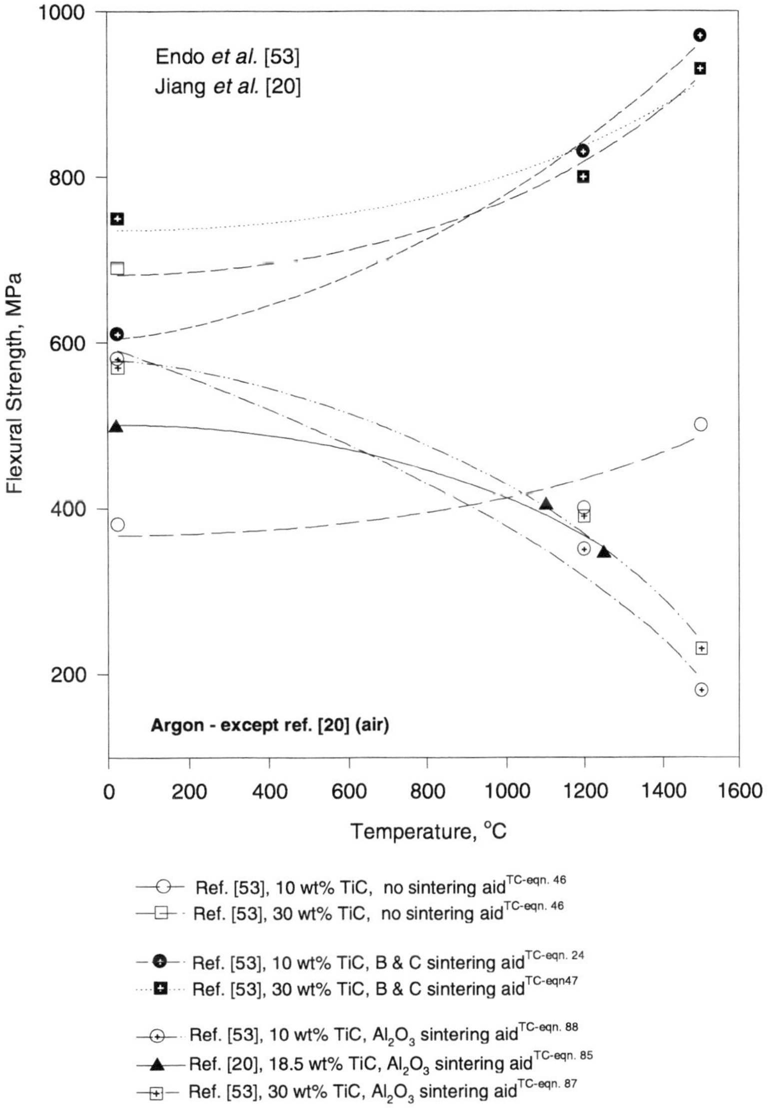

Discussion: The TiC(p)/SiC composite showed opposite trends with respect to the sintering additives. The general trends of all composite strengths increased except for the composites with Al_2O_3 additives. This was due to glassy phase that formed at the grain boundaries in the alumina containing composites. At about 1200 °C the effect of this phase was evident as the samples with Al_2O_3 had a flexural strength of ~200 MPa less than at room temperature because of the glass phase becoming less viscous at higher temperatures. This enabled the grain boundaries to slide past one another resulting in a lower flexural strength. Intergranular fracture was given as evidence for the grain boundary sliding fracture mechanism. Boron and carbon were effective densification aids for the TiC-SiC system. Carbon was especially effective in removing oxide phases at grain boundaries in the sintered body by reducing SiO_2 and TiO_2 on the surface of the starting powders [53]. Also noteworthy is the nearly parallel lines between the 10 wt% and 30 wt% TiC(p)/SiC composites with no sintering additives. The increase in TiC content from 10 to 30 wt% allowed for higher densification. This resulted in an increase in strength of approximately 300 MPa at all test temperatures.

Figure 50b: 3-pt. Flexural Strength vs. Temperature of 50 wt% TiC(p)/SiC With Various Sintering Additives

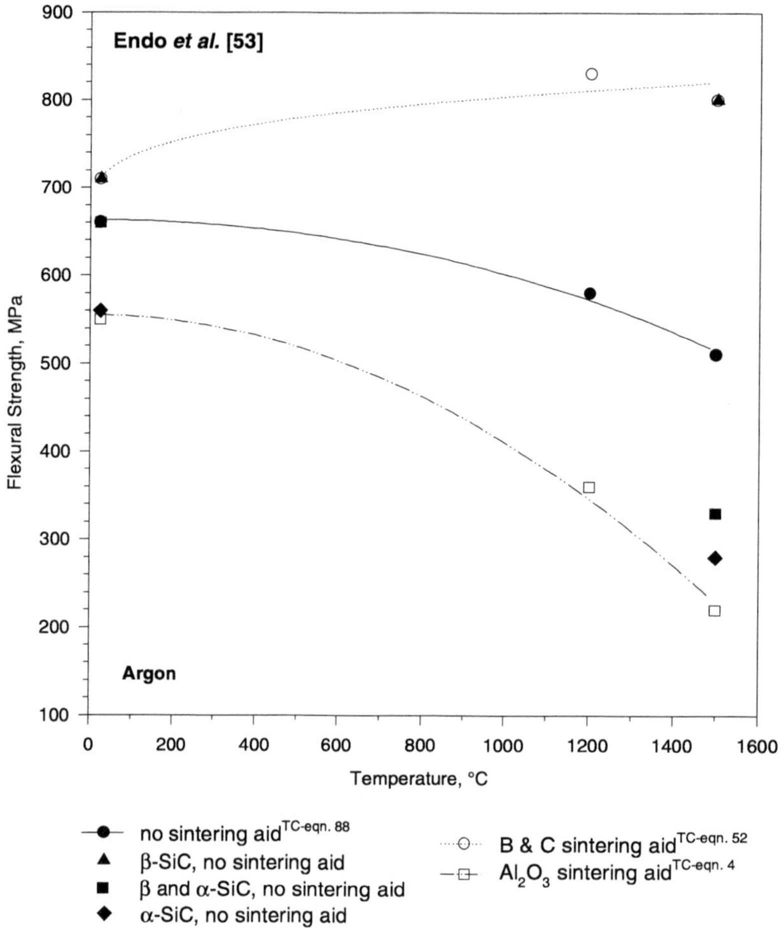

Discussion: The strength of the TiC(p)/SiC composites with alumina sintering additives decreased with increasing temperature while the composites with boron and carbon sintering additives increased with increasing temperature. The rapid decline of the strength of the composite with the Al_2O_3 sintering additives was due to the formation of a glassy phase at the grain boundaries. This weakened the samples because the glassy phase became more viscous as the temperature increased and enabled the grains to move past one another. The TiC softened at high temperatures; this may also have had a deleterious effect on the strength of the composite. The downward slope did not hold for the sample with the boron and carbon sintering aids. This trend was due to boron and carbon having had a higher resistance to thermal effects than Al_2O_3, as well as due to the ability of carbon to reduce oxide grain boundary phases.

The composites containing α and β forms of SiC were equivalent from the densification point of view, so the difference in strengthening was due to the phase of the SiC matrix. The β-SiC composites had the greater strength mainly due to the larger thermal mismatch between itself and the TiC matrix. β-SiC also had a lower amount of metallic impurities than α-SiC. These impurities had the similar deleterious effects as described above, and were particularly detrimental at high temperatures. Because the thermal mismatch was greater for the β–SiC than the α–SiC, and because β-SiC had less metallic impurities, flexural strength was greater for the β–SiC than the α–SiC or the α, β–SiC mixture.

Figure 51: 3-pt. Flexural Strength vs. Hot-Press Temperature of TiC(p)/SiC

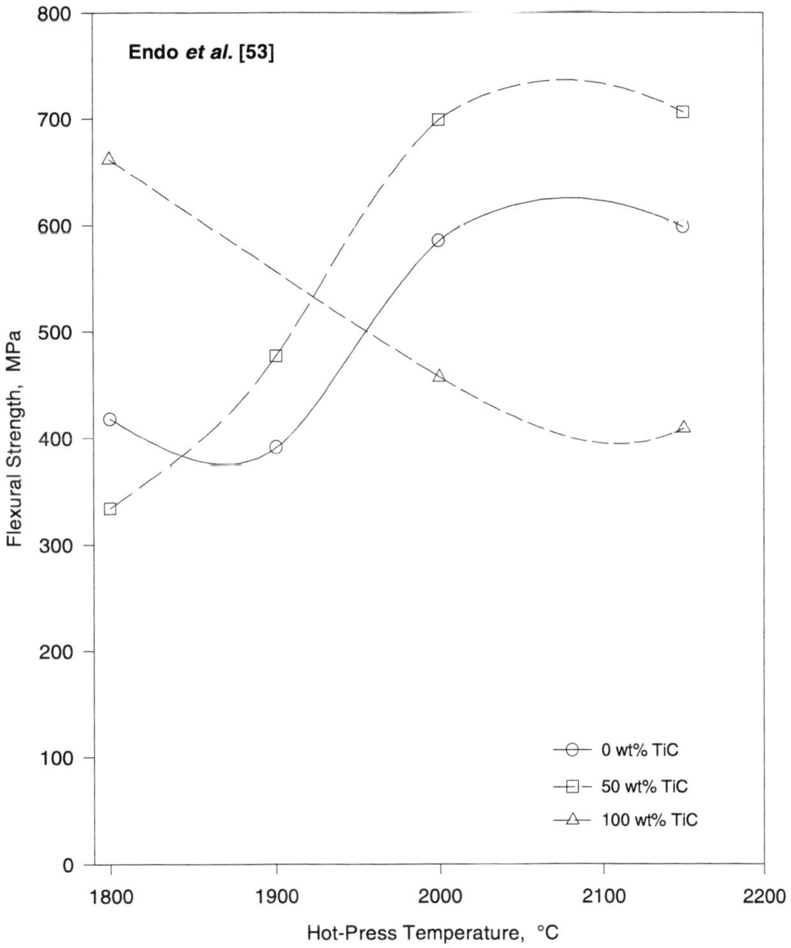

Discussion: For monolithic SiC, the upward shape of the curve was due to the increase in the density of the samples. Complete densification for silicon carbide did not occur until well above 2000 °C. In the case of monolithic TiC, complete densification happened at 1800 °C, so the decrease in the flexural strength at high hot-pressing temperatures was due to the growth of the TiC grains. For the 50-50 wt% composite the densification followed that of the SiC samples. Although bulk diffusion of the TiC may have initially occurred in the composite at temperatures as low as 1300 °C, SiC particles prevented substantial densification until HP temperatures above 1900 °C [53].

TiC(p)/SiC

Reference: 11

Composition
0 - 25 vol% TiC

Processing Method
Hot-Pressed

Detailed Description
The powders were mixed in an acetone solution of a phenolic resin, dried by stirring, and calcined in Ar to convert the resin to carbon. Al powder was added to the mixtures during grinding in a B_4C mortar and pestle. Specimens 38 mm in diameter by 25 mm high were hot-pressed in graphite dies lined with graphite foil at 2000 °C and 70 MPa in a vacuum (~10 MPa).

Sintering Additives
1 wt% Al
1 wt% C

Reinforcement Description
TiC particulates

Reinforcement Supplier
H. C. Starck GmbH, Berlin, Germany

Reinforcement Dimensions
dia. = 2-5 μm

Matrix Description
SiC powder

Matrix Supplier
H. C. Starck GmbH, Berlin, Germany

Matrix Designation
A-10

Matrix Properties
particle size < 1 μm

Test Type
4-pt. Flexural

Specimen Preparation
Specimens were ground to shape with 140/170- grit resin-bound diamond wheel. The surface was polished on the tensile face to a 9 μm finish and the edges were beveled.

Specimen Dimensions
2.3 x 22.5 x 25 mm

Test Description
crosshead speed = 0.51 mm/min

Test Temperature
RT, RT-1200 °C

Test Environment
Air

Room Temperature:

Reinforcement Content (volume percent)	Flexural Strength (MPa)	Range (MPa)
0	472	+45/-38
8	542	+86/-46
16	679	+171/-147
25	653	+30/-24

Elevated Temperature:

Temperature (°C)	Flexural Strength (MPa)	Range (MPa)	Reinforcement Content (volume percent)
25	464	+44/-36	0
600	383	+13/-14	0
1,000	227	+32/-25	0
25	681	±154	16
600	632	+34/-31	16
1,000	597	+60/-61	16
1,200	428	+36/-26	16

Remarks/Comments
The flexural strength increased with increasing reinforcement content. The strengthening was believed to be due to the thermal expansion mismatch between the TiC and the SiC. This produced a stress field around the particle, causing a propagating crack to be deflected. The fracture strength of the composite was maintained up to 1000 °C (90% of RT value), whereas the monolithic SiC retained 45% of its RT strength. The TiC particles were believed to inhibit viscous deformation of the grain boundaries.

TiC(p)/SiC

Reference: 90

Composition
0 and 33 vol% TiC

Processing Method
Hot-Pressed

Detailed Description
After the additives were mixed with the powders by wet ball-milling, the samples were prepared by hot-pressing at 2000 °C and 30 MPa for 1 hour in argon.

Sintering Additives
2 wt% AlN, Ti, or C

Reinforcement Description
TiC powder

Reinforcement Supplier
Nippon Shinkinzota Co., Osaka, Japan

Matrix Description
β-SiC powder

Matrix Supplier
Ibiden Co., Gifu, Japan

Test Type
4-pt. Flexural

Specimen Dimensions
2.5 x 3.5 x 20 mm
inner span = 7 mm
outer span = 14 mm

Test Description
crosshead speed = 0.5 mm/min

Test Temperature
RT-1400 °C

Test Environment
Air (RT)
5.3e-3 Pa vacuum (>RT)

Temperature (°C)	Flexural Strength (MPa)	Reinforcement Content (volume percent)	Density (percent)	Comments
25	452	0	99.6	*2 wt% AlN, C/Ti – 0.86*
1,000	443	0	99.6	
1,200	465	0	99.6	
1,400	202	0	99.6	
25	349	33	98.2	*2 wt% AlN, C/Ti = 0.86*
1,000	243	33	98.2	
1,200	194	33	98.2	
25	396	33	98.5	*2 wt% Ti, C/Ti = 0.86*
1,000	440	33	98.5	
1,200	419	33	98.5	
1,400	183	33	98.5	
25	299	33	82.8	*2 wt% C, C/Ti = 0.86*
800	246	33	82.8	
1,000	260	33	82.8	
1,200	166	33	82.8	
25	456	33	98.7	*2 wt% AlN, C/Ti = 0.97*
1,000	302	33	98.7	
1,200	278	33	98.7	

Remarks/Comments
All composites whose toughening mechanism was crack deflection degraded as temperature increased, while the Ti-containing composite, as well as the monolithic SiC, did not degrade until 1400 °C. In composites not containing the Ti additive, the toughening mechanism of crack deflection due to thermal expansion mismatch was no longer effective at high temperatures.

TiC(p)/SiC

Reference: 105

Composition
0 and 22 vol% TiC

Processing Method
Hot-pressed

Detailed Description
The SiC and TiC were mixed and/or milled with 1 wt% Al powder and 1 wt% carbon black or 2 wt% phenolic resin. The C-series were made by slurrying the carbon black, SiC, and TiC powders in isopropanol with 1 wt% polyvinylpyrollidone (PVPK15, GAF Corp., New York, NY) added as a dispersant. The slurry was agitated for 3 minutes with a 300-watt ultrasonic wand (Model 300, Fisher Scientific Co., Pittsburgh, PA). The P-series were made by slurrying the SiC, TiC, and Al powders in an acetone slurry of phenolic resin, sonicating the slurry for 2-3 minutes, and drying. For either series, milling was performed for 4 hours in a 500 mL polyethylene bottle with sintered α-SiC rods (supplied by Carborundum Corp., Niagara Falls, NY; 6.3 mm diameter by 100 mm long) mounted on a vibratory mill (Megapact, Pilamec, Ltd., Clearwell, Gloucestershire, UK). Al powder was added to the slurry after any milling step. The slurries were dried, calcinated at 500 °C (for C-series) or 600 °C (for P-series) in Ar, ground, and screened to 100- mesh. The powders were then hot-pressed in carbon dies lined with graphite foil at 2000 °C and 35 MPa for 1 hour to a final dimension of 38 mm diameter by 25 mm high.

Sintering Additives
1 wt% C or 2 wt% phenolic resin and 1 wt% Al

Reinforcement Description
TiC powders

Reinforcement Supplier
H.C. Starck GmbH, Berlin, Germany

Matrix Description
α-SiC powders

Matrix Supplier
H.C. Starck GmbH, Berlin, Germany

Matrix Designation
A-10

Test Type
4-pt. Flexural

Specimen Preparation
Specimens were cut from each hot-pressed billet with the tensile face perpendicular to the hot-pressing direction. The tensile face was polished to a 9-μm finish, and the tensile edges were beveled to eliminate edge flaws. Surfaces were ground parallel to the long axis with a 140/170-grit resin-bonded diamond wheel.

Specimen Dimensions
2.3 x 2.5 x 25 mm
inner span = 6.4 mm
outer span = 19 mm

Test Description
crosshead speed = 0.51 mm/min

Test Temperature
RT

Reinforcement Content (volume percent)	Flexural Strength (MPa)	Relative Density (percent)	Weibull's Modulus (dimensionless)	Std. Dev. (MPa)	Comments
0	280	98.5	10.0	33.9	*unmilled*
22	430	99.3	9.1	55.5	*carbon*
0	429	98.5	15.1	33.9	*milled*
22	550	99.4	15.6	41.8	*carbon*
0	341	98.9	5.8	71.6	*unmilled*
22	468	99.4	10.5	51.5	*phenolic*
0	438	99.0	9.0	56.9	*milled*
22	546	99.3	6.9	92.8	*phenolic*

Remarks/Comments
All samples were hot-pressed to densities between 98.5 % and 99.4 % of the theoretical density. Milling the starting powders prior to hot-pressing increased the mean strength. For the 0 vol% TiC samples, the mean strength increased from 280 to 429 MPa. For the 22 vol% TiC samples, the mean strength increased from 430 to 550 MPa. Adding TiC increased the mean strength by 54 % in the unmilled samples and by 25 % in the milled samples. The average strengths of the P-series samples were higher than those of the C-series (1 wt% carbon black added instead of phenolic resin). The milled samples were stronger than the corresponding unmilled samples, in agreement with the C-series samples. Adding TiC to SiC increased the mean strength of the P-series samples by 37 % and 28 %, respectively, for the unmilled and the milled samples.

Figure 52: 4-pt. Flexural Strength of Hot-Pressed TiC(p)/SiC

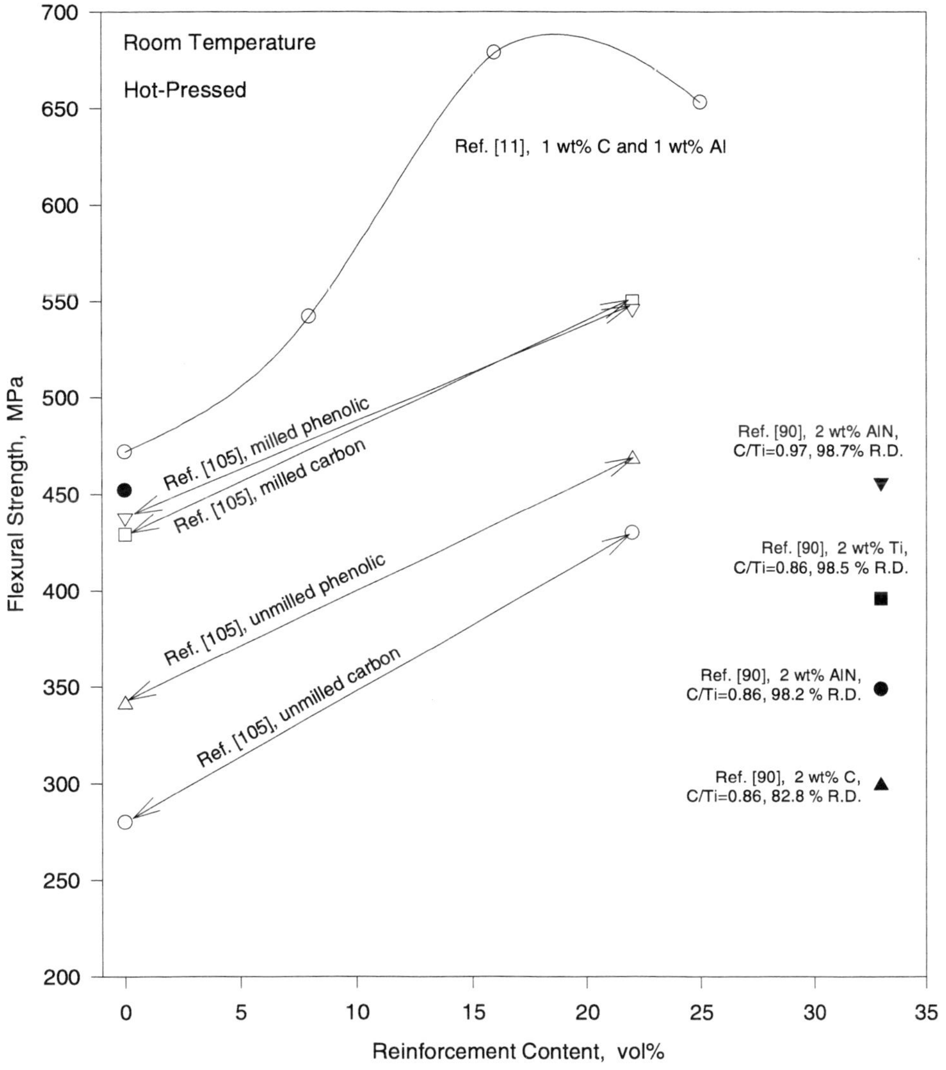

Discussion: The rationale for the toughening of SiC with TiC is that the thermal expansion mismatch between the particulate and the matrix phases creates a stress field around the TiC particles. This stress field caused a propagating crack to deflect around the particle. The longer path around the particles caused an increase in its distance of travel and altered the stress intensity at the crack tip [11]. This graph shows the relationship between the reinforcement with the sintering additive and the flexural strength. From Janney [105], the additives caused a denser material in the range of 99.3 to 99.4 % relative density, whereas the monolithic materials only had a relative density of 98.5 to 99.0 %. The increase in density along with reinforcement resulted in higher strength. Milling of the SiC and TiC powders with the sintering additives increased the strength of the hot-pressed composite [105]. Lin *et al.* [90] tested different TiC(p)/SiC composites with different sintering additives. The composite with 2 wt% C had a relatively low strength of 300 MPa due to a low relative density of 82.8 %. All other samples from ref. [90] possessed a strength in the range of 350 to 450 MPa. Strength depended on the type of sintering additive used and the C/Ti ratio. The composite with 2 wt% AlN and with a C/Ti ratio of 0.97 produced the strongest samples. Janney [11] found that TiC(p)/SiC composites with 1 wt% C and 1 wt% Al sintering additives achieved fairly high strengths, greater than 650 MPa at higher TiC contents.

Figure 53: 4-pt. Flexural Strength vs. Temperature of Hot-Pressed TiC(p)/SiC

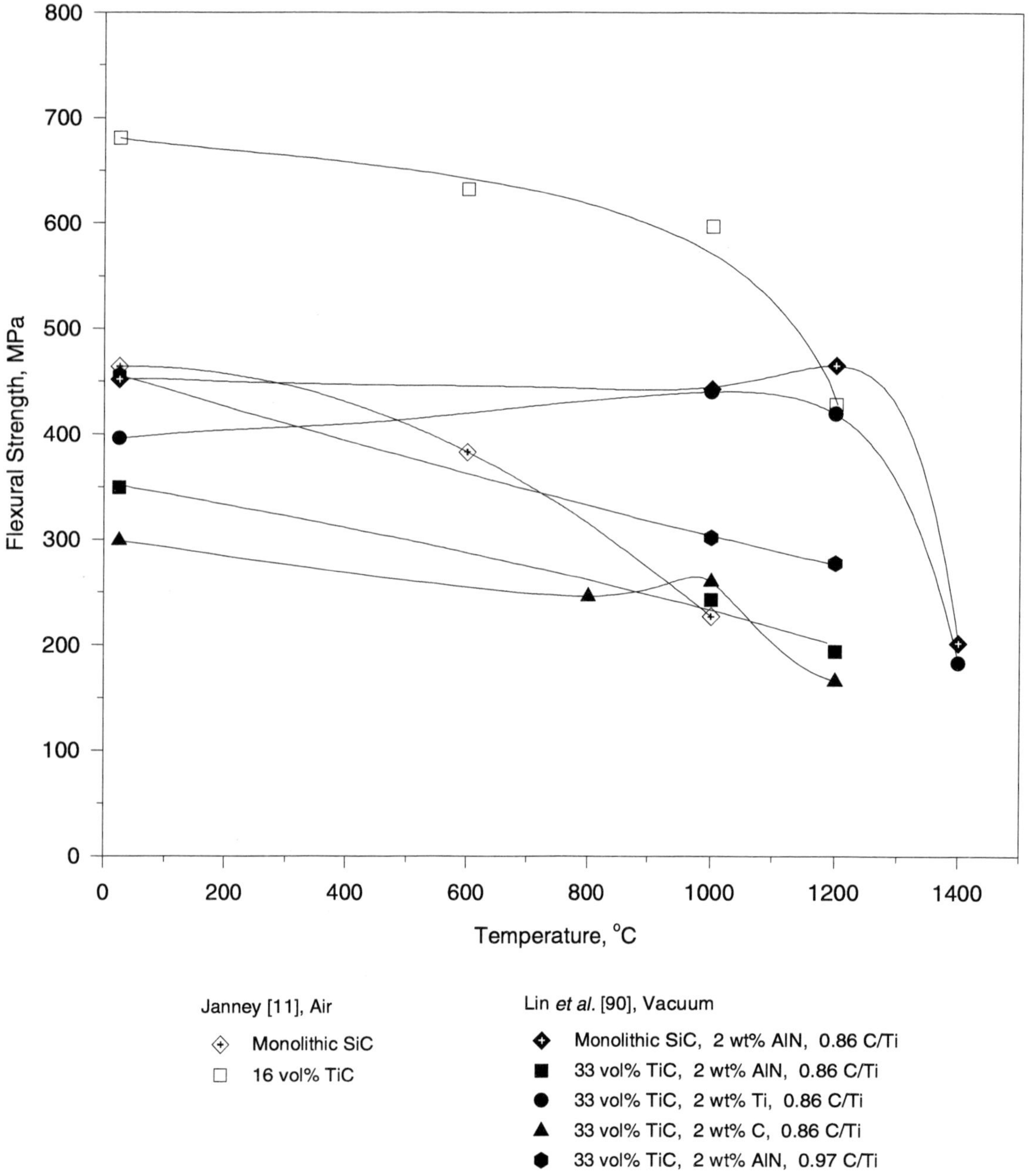

Discussion: Janney [11] showed that the strength of SiC increased by over 200 MPa with the addition of TiC reinforcement. The reason for the significant strength increase was the thermal expansion mismatch between the particulate and the matrix phase that created a stress field around the TiC particles. The stress field caused propagating cracks to deflect around the TiC particles, resulting in a more torturous crack path. Janney [11] also showed that the composite had improved elevated temperature strength. The mechanism of strength degradation was a weakening, or viscous sliding of the Al- containing grain boundaries under stress at high temperatures. In contrast, the 16 vol% TiC composite retained ~90% of its ambient-temperature strength at 1000 °C, and at 1200 °C still retained ~65% of its strength. The mechanical interactions of TiC with SiC caused relative strengthening of SiC at high temperatures. The TiC particles may have formed a rigid backbone that would tend to hinder the viscous deformation of the grain boundaries [11].

From Lin *et al.* [90], all composites that relied on crack deflection from coefficient of thermal expansion mismatch stresses as the method of strengthening degraded as temperature increased due to the relaxation of these stresses. The 2 wt% Ti containing composite and the monolithic SiC did not degrade until after 1200 °C. This was because the toughening mechanism relied on crack bridging that resulted from the chemical residual stresses that were less dependent on temperature than thermal residual stresses. The sharp drop in strength of the Ti and monolithic samples at 1400 °C was related to an amorphous phase at the grain boundaries that formed from intrinsic impurities and/or process contamination.

TiB_2(p)/SiC

Reference: 69

Composition
0 - 30 vol% TiB_2

Processing Method
Reaction Sintered/HIPed

Detailed Description
Fine TiB_2 particles were prepared by milling the TiB_2 agglomerates produced by the following carbothermal reaction at 2000 °C:

$$TiO_2 + 0.5\ B_4C + 1.5\ C \rightarrow TiB_2 + 2\ CO$$

The removal of the CO gas from the system was important to enhance the reaction at lower temperatures. The TiO_2 particle size was 0.4 μm, the B_4C powder size was 1.5 μm, and the carbon black powder size was 0.018 μm. The amounts of B_4C and carbon that were added were more than the stoichiometrically required amount. The mixing ratio of SiC and TiO_2 was designed so that the volume ratio of SiC matrix to TiB_2 particulates formed by the carbothermal reaction should be 8:2. The starting materials were mixed with water by ball milling for 20 hours, using a polyethylene pot and nylon-coated iron balls. The slurries were filtrated by suction, and dried at 100 °C. Screened powders under 32-mesh were molded in dies at a pressure of 20 MPa and isostatically pressed at 300 MPa. The green compacts in a graphite container were heated in a vacuum at 1500 °C for 4 hours. After the vacuum soaking treatment, argon gas was introduced into the furnace and the compacts were sintered in an argon atmosphere at 2150 °C for 1 hour. Some composites were also HIPed in a graphite container at 1900 °C for 1 hour in argon at 200 MPa.

Sintering Additives
B_4C, Denki Kagaku Kogyo
Carbon, Mitsubishi Chemical

Reinforcement Description
TiB_2 particulates

Reinforcement Supplier
Fuji Titan Kogyo

Supplier Designation
TM-1

Matrix Description
SiC powder

Matrix Supplier
Showa Denko Co.

Matrix Properties
ave. particle size = 0.48 μm

Test Type
4-pt. Flexural

Specimen Dimensions
4 x 3 x 35 mm
inner span = 10 mm
outer span = 30 mm

Test Description
crosshead speed = 0.5 mm/min

Test Temperature
RT, 1200 °C and 1400 °C

Test Environment
Air (RT) and vacuum (>RT)

Reinforcement Content (volume percent)	Flexural Strength (MPa)	Temperature (°C)	Range (MPa)
0	550	20	±50
10	350	20	±25
20	510	20	±15
30	500	20	±30
0	590	1,400	±30
10	380	1,400	±20
20	500	1,400	±35
30	600	1,400	±25

Temperature (°C)	Flexural Strength (MPa)	Range (MPa)	Comments
20	726	±60	*20 vol% TiB_2(p)/SiC Reaction Sintered/HIPed*
1,200	700	±45	
1,400	715	±90	
20	500		*20 vol% TiB_2(p)/SiC Reaction-Sintered*
1,200	500		
1,400	500		

Remarks/Comments

The flexural strength of the specimen HIPed at 1900 °C after sintering was constant from room temperature to 1400 °C. The average four point strengths were 726 MPa at 20 °C and 715 MPa at 1400 °C. The four-point flexural strengths were measured at 20 °C in air and at 1400 °C in a vacuum. The flexural strength of the sintered and HIPed specimen were about 40 % higher than those of the pressureless-sintered specimens.

TiB_2(p)/SiC

Reference: 83

Composition

0 and 16 vol% TiB_2

Processing Method

Colloidal Processed / Pressureless Sintered

Detailed Description

The commercially available SiC-TiB_2 (Hexoloy ST) CMC was fabricated by mixing presized TiB_2 and submicrometer-sized SiC powders using a nonaqueous blending and drying procedure. The homogeneous mixture, combined with a binder, was formed into green compacts that were then pressureless-sintered in an inert atmosphere at temperatures over 2000 °C.

Reinforcement Description

TiB_2 particulates

Reinforcement Typical Properties

melting point = 2790 °C
E = 531 GPa
Poisson's ratio = 0.28
CTE = 4.6×10^{-6} °C^{-1}

Reinforcement Dimensions

ave. particle size = 2 μm

Matrix Description

α–SiC

Matrix Supplier

Sohio Engineered Materials Co., Niagara Falls, NY

Matrix Designation

Hexoloy ST

Matrix Properties

ave. grain size = 4 μm
melting point = 2540 °C
E = 410 GPa
Poisson ratio = 0.19
CTE = 4.02×10^{-6} °C^{-1}

Test Type

Flexural Test

Test Temperature

RT

Reinforcement Content (volume percent)	Flexural Strength (MPa)	Density (g cm^{-3})
0	460	3.10
16	448	3.30

Remarks/Comments

The flexural strength decreased slightly with addition of TiB_2 particles.

TiB_2(p)/SiC

Reference: 106

Composition
0 and 16 vol% TiB_2

Processing Method
Pressureless-Sintered

Detailed Description
The mixture was combined with binders. The green compacts were pressureless-sintered in an inert atmosphere at temperatures > 2000 °C.

Reinforcement Description
h-TiB_2 particulates

Reinforcement Typical Properties
density = 4.52 g cm^{-3}
melting point = 2790 °C
CTE = 4.6×10^{-6} $°C^{-1}$
thermal conductivity = 66 W m^{-1} K^{-1}
flexural strength = 365 MPa
E = 531 GPa

Reinforcement Dimensions
particle size = 1-5 μm

Matrix Description
α-SiC

Matrix Properties
grain size = 4-8 μm
density = 3.21 g cm^{-3}
melting point = 2540 °C
CTE = 4.02×10^{-6} $°C^{-1}$
thermal conductivity = 126 W $m^{-1}K^{-1}$
flexural strength = 461 MPa
E = 410 GPa

Test Type
4-pt. Flexural

Specimen Dimensions
3.18 x 6.35 x 51 mm
inner span = 19 mm
outer span = 38 mm

Test Description
Bars were cut from sintered plates and ground to size with edges chamfered prior to testing. crosshead speed = 0.51 mm/min

Test Temperature
RT-1200 °C

Test Environment
Air

Temperature (°C)	Flexural Strength (MPa)	Reinforcement Content (volume percent)	Range (MPa)
30	356	0	±50
600	318	0	±55
800	325	0	±45
1,000	353	0	±39
1,200	377	0	±34
30	478	16	±67
600	507	16	±61
800	465	16	±49
1,000	448	16	±56
1,200	451	16	±46

Remarks/Comments
The Weibull modulus for the composite was in the range of 12-16. Pure SiC had a Weibull modulus between 8 and 10. Failure originated from flaws formed after processing. The composite showed a higher flexural strength than the monolithic SiC.

$TiB_2(p)/SiC$

Reference: 107

Composition
0 and 15 vol% TiB_2

Processing Method
Hot-Pressed

Detailed Description
The combined powders were vibromilled in isopropanol with 1 wt% polyvinylpyrollidone as a dispersant, using 6.4 mm diameter by 100 mm long SiC rods. The polyvinylpyrollidone was burned out in argon at 500 °C prior to hot-pressing. Samples were hot-pressed in carbon dies lined with graphitc foil at 2000 °C for 45 minutes at 35 MPa to final dimensions of 38 mm diameter by 25 mm high. Measurements were made on bars cut from the specimens.

Sintering Additives
1 wt% B
0.5 wt% C

Reinforcement Description
TiB_2 particulates

Reinforcement Supplier
KBI Division, Cabot Corp., Reading, PA

Reinforcement Dimensions
ave. particle size = 4-8 μm

Matrix Description
SiC powder

Matrix Supplier
H. C. Starck Co., Berlin, Germany

Matrix Designation
A-10

Matrix Properties
ave. particle size = 3-5 μm

Test Type
4-pt. Flexural

Specimen Preparation
Samples were ground parallel to the long axis with a 140/170-grit resin-bonded diamond wheel. The tensile face was ground to a 30 μm finish, and the tensile edges were beveled. Some samples were oxidized prior to testing.

Specimen Dimensions
2.3 x 2.5 x 25 mm
inner span = 0.64 cm
outer span = 1.9 cm

Test Description
crosshead speed = 0.51 mm/min

Test Temperature
RT

Reinforcement Content (volume percent)	Flexural Strength (MPa)	St. Dev. (MPa)	Comments
0	379	±38.7	
15	485	±44.1	
15	480	±46.6	*oxidized at 1000 °C for 24 hours*
15	498	±49.3	*oxidized at 1200 °C for 24 hours*
15	192	±27.1	*oxidized at 1400 °C for 24 hours*

Remarks/Comments
Data presented is an average of 9 specimens for non-oxidized tests, and 5 specimens for oxidized tests. Addition of TiB_2 to the SiC base increased the flexural strength. Ambient temperature strength determined after oxidation showed no reduction for samples oxidized at 1000 or 1200 °C. However, the average strength fell dramatically for samples oxidized at 1400 °C.

Figure 54: 4-pt. Flexural Strength vs. Temperature of TiB_2(p)/SiC

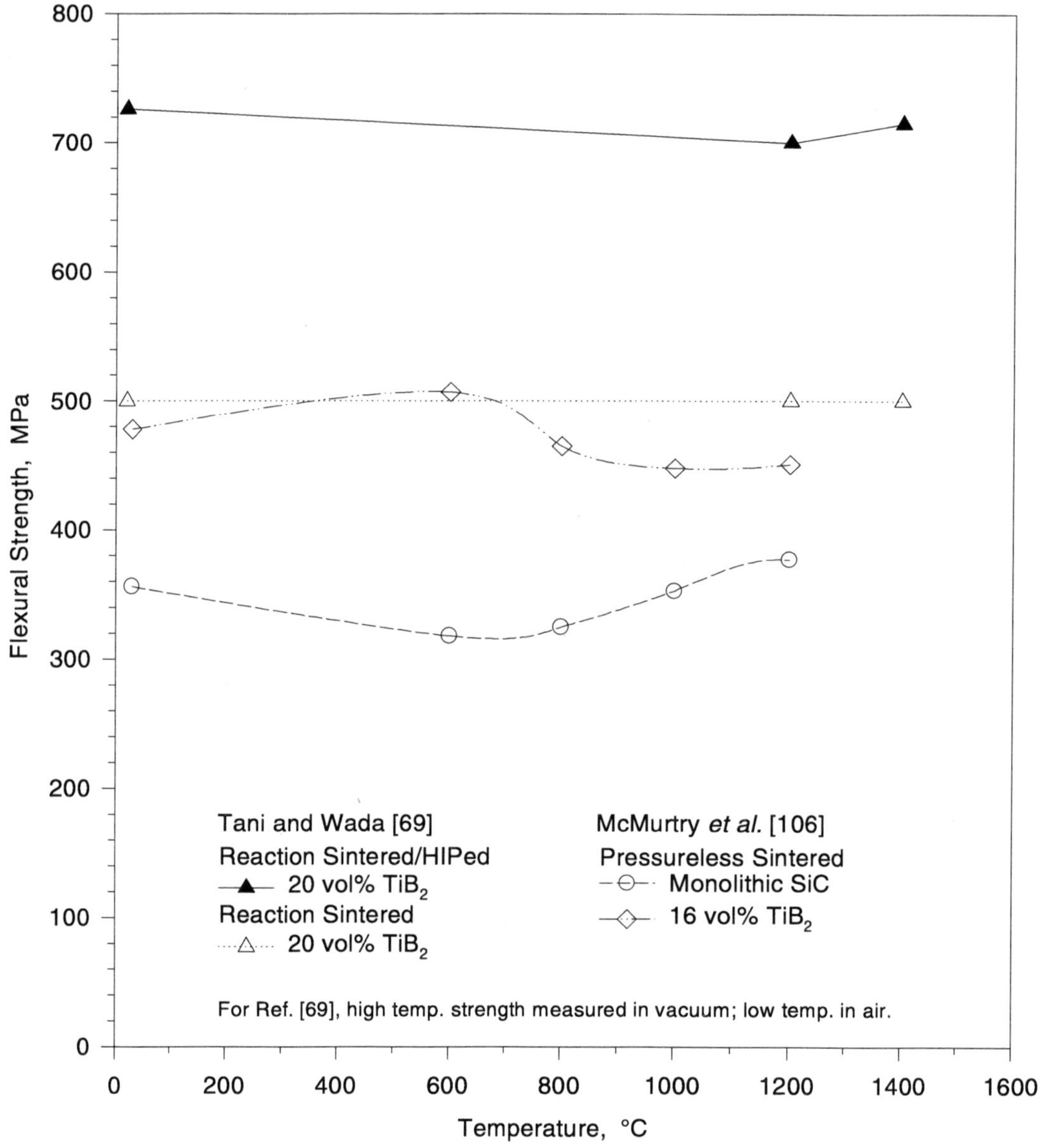

Discussion: McMurtry *et al.* [106] tested pressureless sintered TiB_2(p)/SiC. The flexural strength of the 16 vol% TiB_2(p)/SiC was approximately 30% greater (~480 MPa) than the monolithic SiC (360 MPa) at room temperature. The strength of the 16 vol% TiB_2(p)/SiC composite increased slightly at mid-range temperatures, and decreased slightly at high temperatures. The strength of the monolithic SiC showed the opposite behavior, decreasing at mid-range temperatures and increasing at high temperatures. The Weibull modulus for the 16 vol% TiB_2(p)/SiC composite was in the range of 12 to 16, whereas the Weibull modulus for monolithic SiC was between 8 and 10.

Tani and Wada [69] tested reaction sintered and HIPed 20 vol% TiB_2(p)/SiC composites and found an average four-point flexural strength of more than 700 MPa at both 20 °C and 1400 °C. This was about 40% greater than those of the reaction sintered specimen. Neither the as-sintered nor HIPed composite showed variation with temperature. The high temperature flexural strength tests were performed in a vacuum, so that oxidation effects were not a factor in the high temperature tests.

Figure 55: 4-pt. Flexural Strength of TiB_2(p)SiC

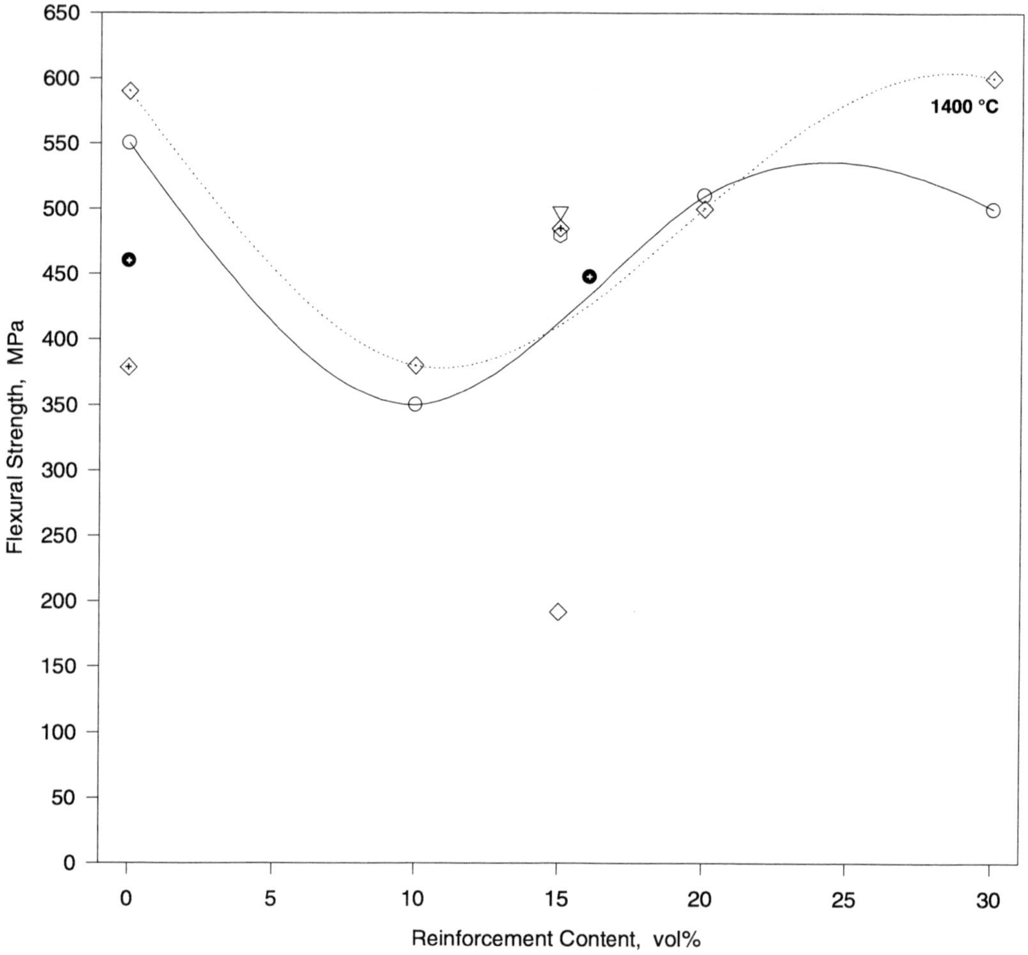

Room Temperature

- Ajayi *et al.* [83], Colloidal Pressed/Pressureless Sintered (Hexalloy ST)
- Tani and Wada [69], Reaction Sintered
- Tani and Wada [69], Reaction Sintered, 1400 °C

Janney [107]

- as-Hot-Pressed
- Hot-Pressed, oxidized at 1000 °C
- Hot-Pressed, oxidized at 1200 °C
- Hot-Pressed, oxidized at 1400 °C

Discussion: The flexural strength of the reaction sintered composite tested by Tani and Wada [69] decreased approximately 200 MPa with the addition of 10 vol% TiB_2 to SiC. This decrease in strength occurred despite an increase in toughness from 2 to 3 MPa $m^{1/2}$. The flexural strength of the composite increased about 150 MPa when the TiB_2 content was further increased to 20 or 30 vol%. The same samples tested at 1400 °C showed much the same behavior as they did at room temperature. Fracture analysis revealed spherical pores 20 to 50 μm in diameter at most of the fracture origins.

Ajayi *et al.* [83] tested Hexoloy ST (16 vol% TiB_2/SiC). The microstructure of the Hexoloy ST composite consisted of two distinct phases. The TiB_2 particles had an average grain size of approximately 2 μm, while the SiC matrix had an average grain size of approximately 4 μm. Compared to monolithic α-SiC, the grain size of the SiC matrix was more uniform, which was attributed to matrix grain boundary pinning by the TiB_2 particles during sintering. The flexural strength of the Hexoloy ST composite was 12 MPa less than the sintered α-SiC ceramic. On the other hand, the fracture toughness of the Hexoloy ST was 8 MPa $m^{1/2}$, compared to 4.6 MPa $m^{1/2}$ for monolithic SiC. Results from Tani and Wada [69] and Ajayi *et al.* [83] suggest that there can be significantly increased toughness without increased strength with the addition of TiB_2 to a SiC matrix. One possible explanation of this behavior is that the critical flaw size increases with the addition of TiB_2 as well as various toughening mechanisms.

The TiB_2(p)/SiC samples tested by Janney [107] were both stronger and tougher than the monolithic SiC. At 15 vol% reinforcement content, the flexural strength for the TiB_2(p)/SiC composites that were hot-pressed and oxidized at both 1000 °C and 1200 °C was approximately equal to the flexural strength of the as-hot-pressed TiB_2(p)/SiC composite. However, the TiB_2(p)/SiC composite that was oxidized at 1400 °C exhibited a substantial decrease in flexural strength of approximately 190 MPa after oxidation. The composite that was oxidized at 1400 °C exhibited the greatest rate of attack by oxygen. The oxidation attack occurred with both the SiC matrix phase and the TiB_2 particulate phase. Interconnected porosity and a glass phase extended deeply into the composite at the highest exposure temperature of 1400 °C [107]. In addition, all the composites that were oxidized had large bubbles containing glass and TiO_2 form on the surface. Boron oxide acted as a fluxing agent that accelerated the attack of the SiC matrix by forming a nonprotective oxide glass layer on the surface of the composite [107].

B_4C(p)/Graphite

Reference: 99

Composition
0-30 at% B

Processing Method
CIPed/HIPed

Detailed Description
The powders were mixed in alcohol, dried, and then separated. The mixture was CIPed prior to insertion in a metal capsule which was sealed under a vacuum and placed in a HIP press. Densification was performed in an argon atmosphere at 190 MPa and 1600 °C for 2 hours. After densification, the metal clad was removed by machining, and the test piece was characterized.

Reinforcement Description
B_4C particles

Reinforcement Dimensions
ave. grain size = 1 μm

Matrix Description
Graphite powder

Matrix Properties
ave. grain size = 4.1 μm

Test Type
3-pt. Flexural

Specimen Preparation
Two large, opposite faces of the test pieces were polished, and the edges were beveled.

Specimen Dimensions
24 x 4 x 4 mm

span = 20 mm

Test Description
crosshead speed = 0.2 mm/min

Test Temperature
RT

Reinforcement Content (atomic percent)	Flexural Strength (MPa)	Relative Density (percent)	Range (MPa)
0	61.6	98.5	±7.7
10	62.1	98.8	±8.3
20	63.2	98.8	±5.8
30	86.8	96.3	±2.9

Remarks/Comments
Excellent densification was observed for materials with 0-20 atomic% B. The flexural strength was constant up to 20 atomic% B, and then increased by about 40 %.

Figure 56: 3-pt. Flexural Strength of $B_4C(p)$/Graphite

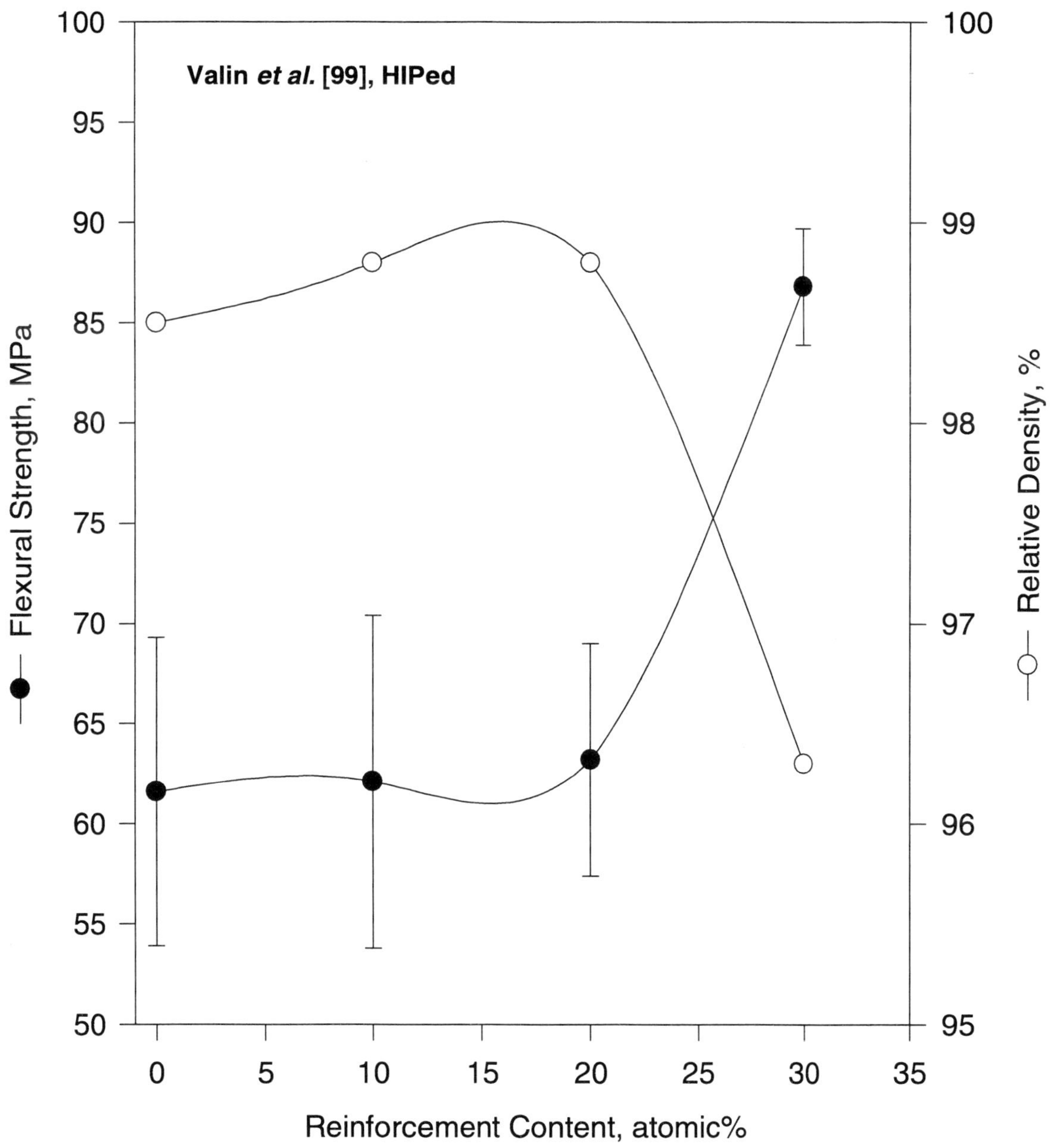

Discussion: Flexural strength was constant up to a boron content of 20%, then increased by about 40% with the further addition of 10 atomic % boron despite the significant decrease in relative density.

11. Fracture Toughness
(Section II)

SiC(p)/MgO

Reference: 77

Composition
0-50 vol% SiC

Processing Method
Hot-Pressed

Detailed Description
The composites containing various volume fraction of SiC particles were mixed by the ball milling technique in the highly pure toluene with Al_2O_3 balls for 24 hours. The dried mixtures were hot-pressed in N_2 atmosphere at 1500-1900 °C at 28 MPa.

Reinforcement Description
β-SiC particulates

Reinforcement Supplier
Ibiden Co., Ltd.

Reinforcement Dimensions
ave. particle size < 0.3 μm

Matrix Description
MgO particulates

Matrix Supplier
Ube Industries, Ltd., Japan

Matrix Properties
ave. particle size = 0.1 μm

Test Type
Vickers Indentation Microfracture (IM) Method

Test Description
indentor load = 0.49-18.6 N
Density was measured by an Archimedes method using toluene.

Test Temperature
RT

Reinforcement Content (volume percent)	Fracture Toughness, K_{IC} ($MPa\ m^{1/2}$)	Range ($MPa\ m^{1/2}$)
0.0	1.09	+0.15/-0.1
5.1	1.36	+0.17/-0.21
10	1.88	+0.21/-0.23
20	2.79	+0.56/-0.32
30	3.28	+0.26/-0.72
50	4.08	+0.22/-0.21

Remarks/Comments
The toughness and strength of MgO was significantly improved by incorporating the SiC particles within the matrix. Overall, both strength and toughness was greater for the SiC-Al_2O_3 composite than for the SiC-MgO composite.

Figure 57: Indentation Microfracture (IM) Fracture Toughness of SiC(p)/MgO

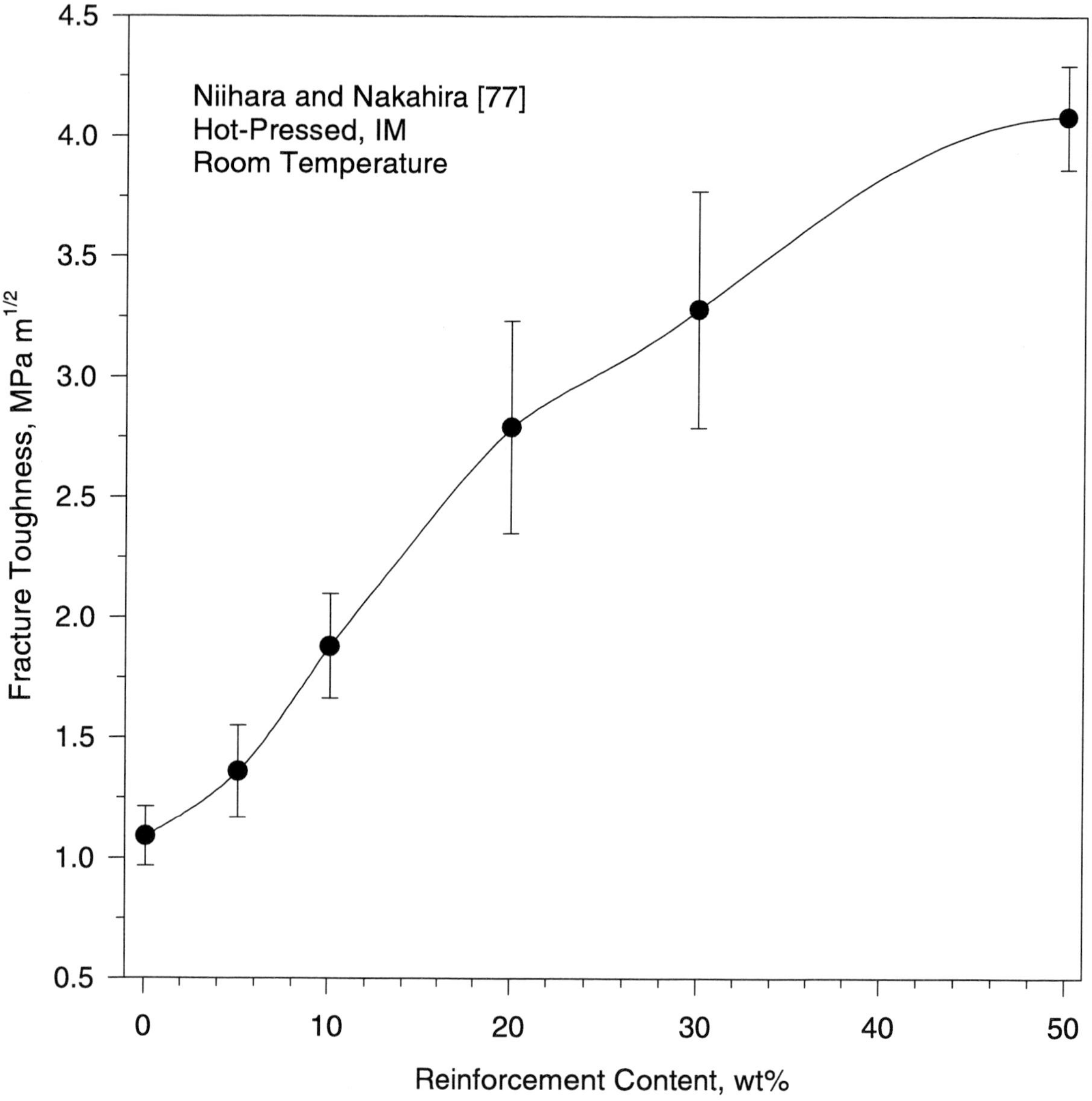

Discussion: Fracture toughness of the SiC(p)/MgO nanocomposite tested by Niihara and Nakahira [77] increased with increasing SiC content. The primary toughening mechanism was crack deflection. It was improbable that microcracking was a significant toughening mechanism, because of the very small size of the SiC particulates; however, microcrack toughening can not be ruled out. SEM analysis of the fracture surface revealed that the second phase dispersions within the matrix grains changed the fracture mode from intergranular to transgranular fracture. The change of propagation mode was due to the thermal mismatches between the SiC and the MgO [77].

ZrO_2(p)/Mullite

Reference: 10

Composition
0-20 vol% ZrO_2

Processing Method
Sintered

Detailed Description
Very fine pre-mullite powder was obtained by thermal and chemical treatment of kandite. Powder with the appropriate content of ZrO_2 was attrition milled with alumina balls in isopropyl alcohol for 1 hour. These mixtures were isostatically pressed at 200 MPa and sintered at 1570 °C for 2.5 hours.

Reinforcement Description
ZrO_2 particulates

Reinforcement Supplier
Magnesium Elektron Ltd., Manchester, UK.

Matrix Description
mullite powder

Test Type
Vickers Indentation Technique [I-21]

Applicable Test Standard
ASTM STP 678

Specimen Preparation
Surface polished

Test Description
Indentor load = 50 to 70 N for the monolith Indentor load = 100 to 120 N

Reinforcement Content (volume percent)	Fracture Toughness, K_{IC} (MPa $m^{1/2}$)	Relative t-ZrO_2 Content (percent)	Range (MPa $m^{1/2}$)
0.0	2.13	-	+0.14/-0.13
10	2.93	45	+0.17/-0.14
15	3.07	30	+0.15/-0.12
20	3.16	9	+0.14/-0.13

Remarks/Comments
The fractional tetragonal content of ZrO_2 does not influence the K_{IC} of the tested composites. An increase from 2-3 MPa $m^{1/2}$ in toughness was observed.

ZrO_2(p)/Mullite

Reference: 48

Composition
35.4 wt% ZrO_2

Processing Method
Hot-Pressed

Detailed Description
The dried mixture was vacuum hot-pressed in graphite dies at temperatures of 1300-1700 °C at a pressure of 34.5 MPa.

Reinforcement Description
0Y-, 0.41Y-, 0.83Y-, and 1.65Y-ZrO_2 particulates

Matrix Description
mullite powder

Test Type
Vickers Indentation Microfracture Technique [I-23]

Specimen Preparation
Ground on SiC papers through 600 grit, ran on 15-, 9-, and 6-μm diamond paste on silk cloth, and finally polished with 0.05 γ-Al_2O_3 on microcloth with the edges rounded throughout the process.

Specimen Dimensions
25 x 6.4 x 3.2 mm

Type of Specimen
Bar

Test Description
A diamond pyramid indentor was used with a 3 kg load

Test Temperature
RT

Stabilizer Content (weight percent)	Fracture Toughness, K_{IC} (MPa $m^{1/2}$)	Range (MPa $m^{1/2}$)	Comments
0.0	1.87	±0.25	*mullite*
0.0	4.18	+0.66/-0.62	*zirconia-mullite*
0.75	4.36	+1.35/-1.95	
1.5	4.24	+0.48/-0.34	
3.0	2.14	+0.27/-0.19	

Remarks/Comments
Each data point includes five sample tests and the spread is shown for all fracture toughness data. The data shows the effect of the addition of yttria stabilizer to the zirconia phase.

ZrO_2(p)/Cordierite

Reference: 73

Composition
0-10 vol% ZrO_2

Processing Method
CIPed/Pressureless Sintered

Detailed Description
Ingredients were mixed in a high density alumina ball mill with water and dried. The batches were isostatically pressed under 140 MPa and sintered at different temperatures from 1200-1350 °C for 4 hours. The heating and cooling rates were ±10 °C/min.

Reinforcement Description
m-ZrO_2 particulates

Reinforcement Dimensions
ave. surface area = 12 m^2g^{-1}

Matrix Description
cordierite $2MgO.2Al_2O_3.5SiO_2$

Matrix Supplier
commercial variety

Matrix Properties
ave. particle size = 18 μm

Test Type
Vickers Indentation Microfracture Technique (IM) [I-29]

Specimen Dimensions
5 x 7 x 30 mm

span = 20 mm

Type of Specimen
Bar

Test Temperature
RT

Reinforcement Content (volume percent)	Fracture Toughness, K_{IC} (MPa $m^{1/2}$)	Range (MPa $m^{1/2}$)
0.0	1.75	±0.2
1.0	1.85	±0.15
2.4	2.00	±0.1
4.5	2.20	±0.25
7.0	2.30	±0.6
10	2.40	±0.5

Remarks/Comments
The fracture toughness increased from 1.75-2.4 MPa $m^{1/2}$ with addition of m-ZrO_2 up to 10 vol%. This toughening was partly attributed, especially at higher ZrO_2 contents, to the residual compressive stresses imposed on the cordierite due to thermal expansion mismatch between cordierite and $ZrSiO_4$ second phase and partly to crack deflection around $ZrSiO_4$ interfaces.

ZrO_2(p)/Mullite

Reference: 50

Composition
30 wt% ZrO_2

Processing Method
CIPed/Sintered

Detailed Description
The sols in the solution (pH~3.0 and 5 wt% in concentration), mixed for one hour, were coprecipitated with diluted aqueous ammonia (pH 9.0). All gels were dispersed and filtered in methanol and later in acetone and subsequently dried in acetone with a rotary evaporator. After vacuum-drying at 60 °C and in oven at 120 °C the powders were milled in methanol for 6 hours so as to break down hard agglomerates before calcination. The powders were calcined for 30 minutes at 1150 °C for the composites and at 1250 °C for the mullite powders. The calcined powders were then milled in methanol for 48 hours in a nylon vibration mill with Si_3N_4-cylinders, dried and screened (150 mesh) for granulation. All powders were calcined at 800 °C for 1 hour in order to burn out inclusions. Green bodies were formed by die-pressing at 30 MPa and organic subsequent cold isostatically pressing at 200 MPa which were then sintered in air between 1550-1650 °C for 2 or 4 hours.

Reinforcement Description
0Y-ZrO_2 and 3Y-ZrO_2 sol powder

Reinforcement Supplier
powder (A) zirconia and yttria sols: Nyacol Products Inc., USA.

Reinforcement Typical Properties
powder (A) zirconia (particle size 75 Å), and yttria (particle size 100 Å)
powder (B) produced from aqueous solution of $ZrO(NO_3)_3$ and $Y(NO_3)_3$

Reinforcement Dimensions
ave. particle size (0 % Y_2O_3) = 1.02 μm
ave. particle size (3 % Y_2O_3) = 0.99 μm

Matrix Description
mullite sol powder

Matrix Supplier
powder (A) silica sol: Eka Nobel AB, Sweden; Al sol: Nyacol Products Inc., USA
powder (B) silica sol: Eka Nobel AB, Sweden; aluminum chlorohydrate sol: Albright and Wilson, UK.

Supplier Designation
powder (A) zirconia and yttria sols: Nyacol Products Inc., USA.

Matrix Properties
powder (A) produced from pure inorganic sols of silica (particle size 50 Å), alumina (particle size 200 Å) composition of mullite: 72.1 wt% Al_2O_3
powder (B) alumina (particle size 10 Å), composition of mullite: 72.9 wt% Al_2O_3 powder,

Test Type
Indentation Microfracture Technique [I-14]

Specimen Preparation
The sintered plates were cut and ground with #400 diamond wheel.

Test Description
load = 5 kg

Test Temperature
RT

Relative Density (percent)	Fracture Toughness, K_{IC} (MPa $m^{1/2}$)	Sintering Temperature (°C)	Sintering Time (hours)	Comments
				<u>Powder A</u>
98.5	1.9	1,650	4.0	*monolithic mullite*
96.2	2.10	1,650	2.0	*(sol-gel precursors)*
98.5	3.70	1,650	2.0	*30 % ZrO_2, 0 % Y_2O_3*
98.2	3.70	1,600	4.0	
96.2	4.00	1,600	2.0	
97.5	3.50	1,600	4.0	*30 % ZrO_2, 3 % Y_2O_3*
97.8	3.50	1,600	2.0	
				<u>Powder B</u>
99.2	1.80	1,650	4.00	*monolithic mullite*
98.0	1.90	1,650	2.00	*(sol-gel precursors)*
97.6	2.10	1,630	4.00	
99.0	3.40	1,650	2.00	*30 % ZrO_2, 0 % Y_2O_3*
97.7	3.30	1,600	4.00	
94.2	3.70	1,600	2.00	
97.4	4.10	1,600	4.00	*30 % ZrO_2, 3 % Y_2O_3*
97.9	4.00	1,600	2.00	
99.0	3.40	1,550	4.00	

Remarks/Comments

The composite demonstrated nearly double the fracture toughness as the monolithic mullite.

Figure 58: Fracture Toughness (IM) of ZrO_2(p)/Mullite and Cordierite

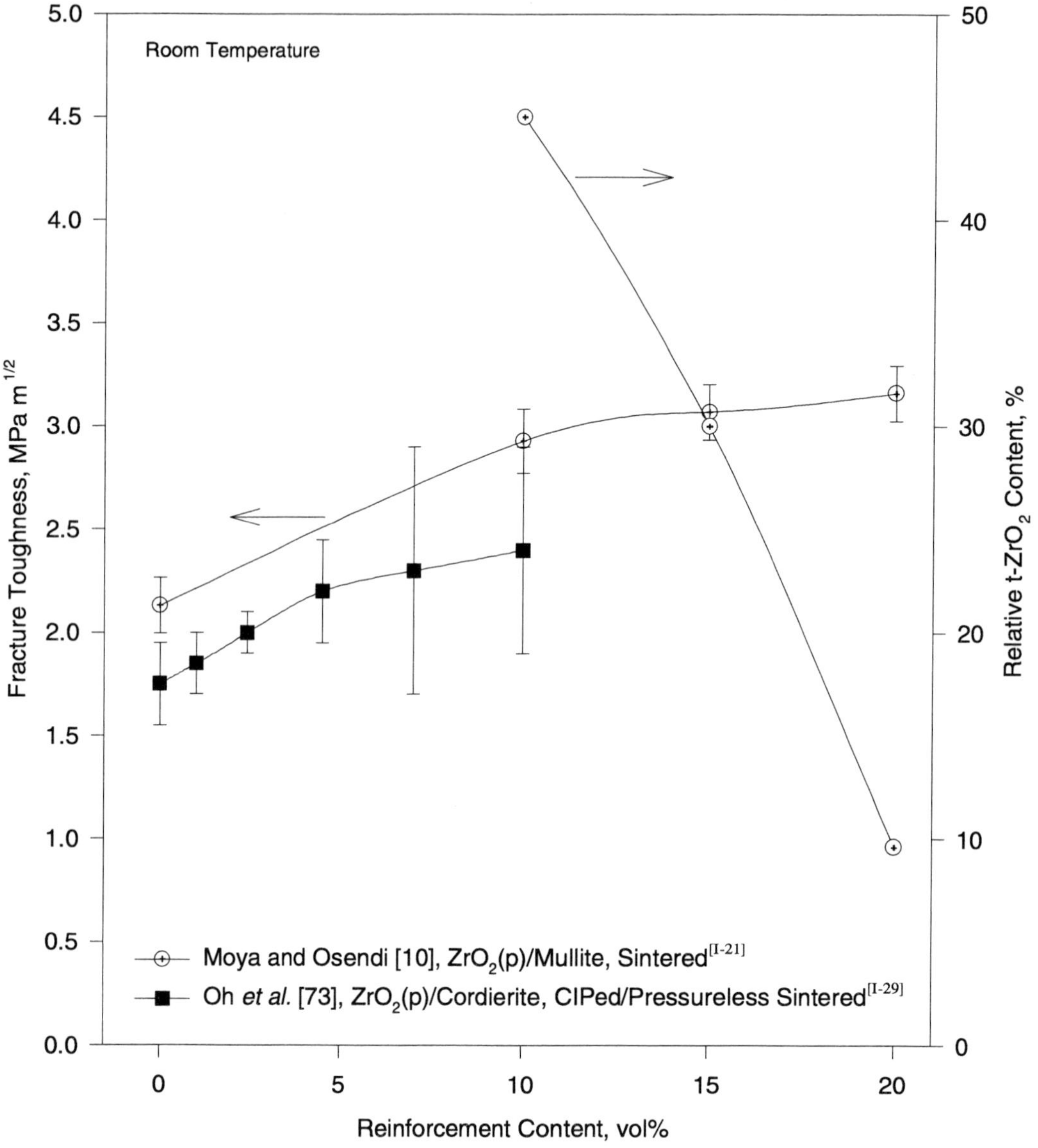

Discussion: Figure 58 describes the effect of ZrO_2 content on the toughness of zirconia·mullite and zirconia·cordierite composites. Normally, some transformation toughening is expected in zirconia toughened ceramics. Oh *et al.* [73] found that below 10 vol% reinforcement content all the ZrO_2 reacted with the cordierite to form $ZrSiO_4$. Thus, the expected transformation toughening effect did not occur. The toughening effect that did occur was due to the compressive stresses that were a result of the formation of $ZrSiO_4$ second phase. These stresses were large enough to deflect cracks. A solid solution formed at the boundaries between these two phases of the composite and caused a non-distinct boundary that had excellent bonding. This may also have contributed to the improved mechanical properties of the composite. These factors caused the increasing toughness observed in the ZrO_2 ($ZrSiO_4$) (p)/cordierite composite.

Moya and Osendi [10] examined the effects of transformation toughening on the ZrO_2(p)/mullite composite system. They showed that the operative toughening mechanism was not the transformation of ZrO_2 as thought because there was not an increasing relationship between the fracture toughness and the amount of t-ZrO_2. In fact, the relationship was an inverse one. Microcracking also was not the mode of toughening because the same relationship that affected the transformation change would have affected microcracking. The attributed reason for the ZrO_2(p)/mullite to have an increased toughness as reinforcement content increased was the increase in strength of the grain boundaries. Reaction between the matrix and the particulates produced a continuous metastable solid solution. This interaction between the two parts of the composite caused a transgranular fracture behavior that increased the toughness of the composite.

Figure 59: Fracture Toughness [I-14] (IM) vs. Relative Density of Zirconia·Mullite

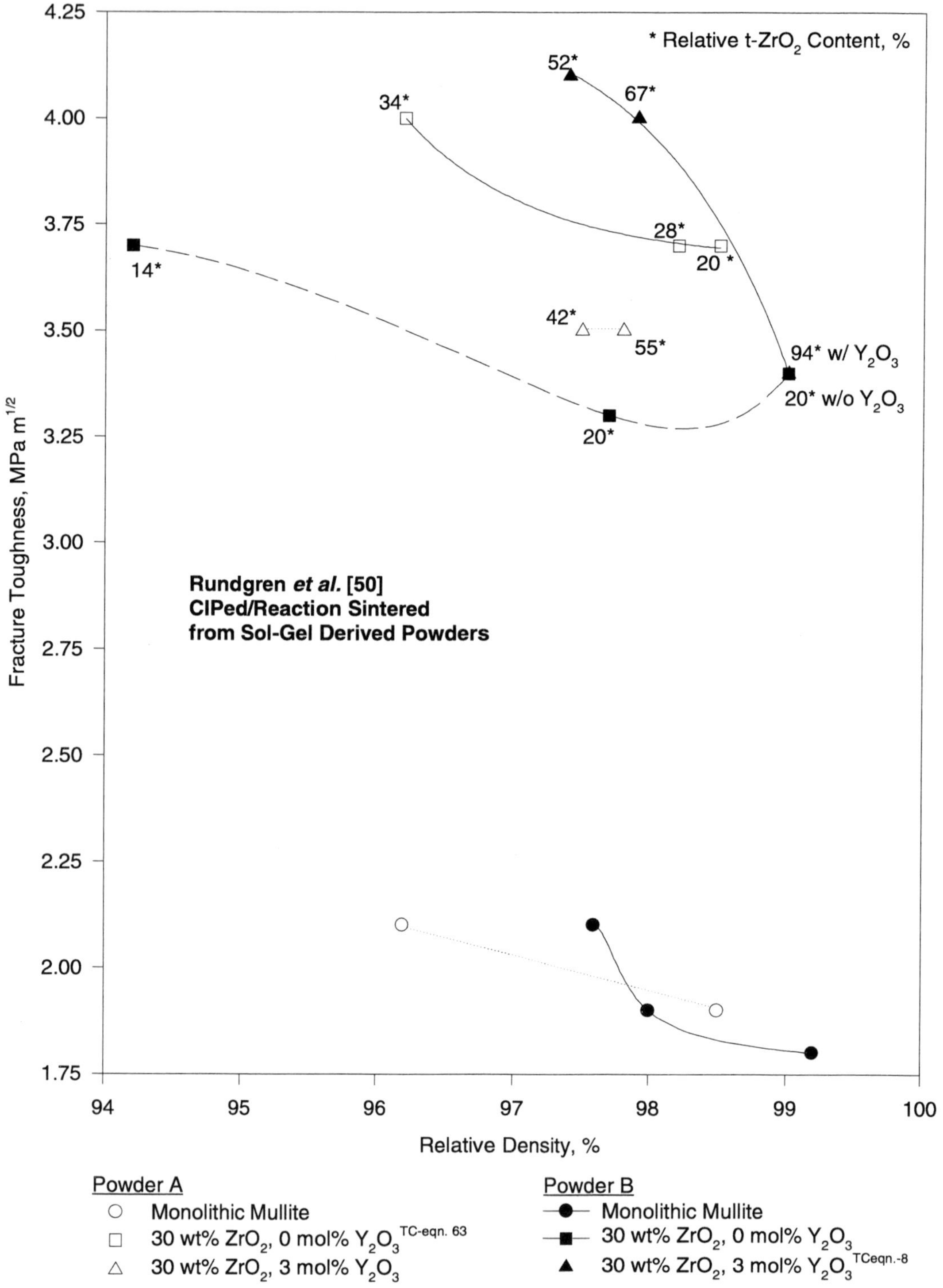

Discussion: Rundgren *et al.* [50] examined the use of commercially available sols and salts to make mullite and zirconia·mullite composites. Powders derived from two different precursors were used. Powder A (colloidal powder) was made from pure inorganic sols of silica, alumina, and yttria. Powder B (polymeric powder) was derived from very fine silica and alumina with aqueous solution of $ZrO(NO_3)_3$ and $Y(NO_3)_3$.

Microstructural analysis found that the matrix grain size of the powder B composites was much finer than that of the powder A composites. Significant silica-glass was found at the triple junctions of the powder A composite, which accounts for the larger grain growth. The mullite grains in both the A and B composites were slightly elongated, which enhanced toughness of both composites. The zirconia reinforcement in the powder B composite was sharp-edged and irregular in size. The morphology of the zirconia grains enhanced the toughness of the powder B composites. The ZrO_2 particulate phase in the powder A composite was large and equiaxed.

Figure 59 shows the fracture toughness of zirconia·alumina composites (3.25-4.25 MPa $m^{1/2}$) was almost double that of the monolithic mullite processed from either powders A or B (~ 2 MPa $m^{1/2}$). The powder A zirconia·mullite decreased in toughness with decreasing t-ZrO_2 content and increasing relative density. The toughness of the yttria containing powder A zirconia·mullite composite was lower despite higher t-ZrO_2 content, possibly due to greater grain growth. The yttria containing powder B zirconia·mullite composite dropped despite increasing t-ZrO_2 content and increasing relative density. This was due to the decrease in grain size as densification was enhanced, i.e., relative density was increased. At 99% R.D., the grain size was smaller than the critical grain size to allow transformation toughening, and despite being greater than 94 % t-ZrO_2, the primary toughening mechanism was crack deflection, not transformation toughening.

Figure 60: Fracture Toughness [I-23] vs. Stabilizer Content of ZrO_2(p)/Mullite

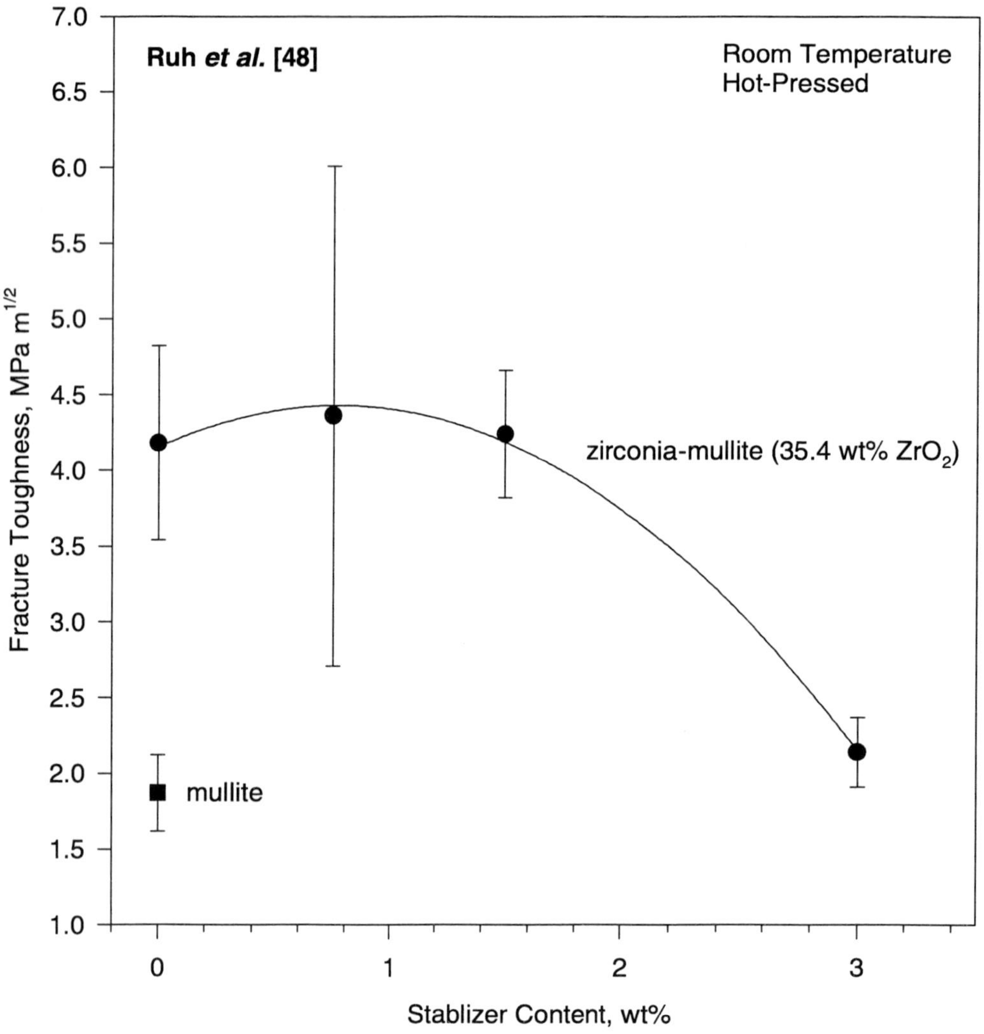

Discussion: Figure 60 shows the effect of yttria content in a 35.4 vol% YSZ(p)/mullite composite measured by Ruh *et al.* [48]. At low yttria contents, the yttria acts to increase and stabilize the tetragonal ZrO_2 phase. Since more t-ZrO_2 is available to transform in the matrix, allowing for greater transformation toughening, fracture toughness generally goes up. The results show that fracture toughness remained fairly constant from 0 to 1.5 wt% yttria, suggesting that the zirconia was primarily in the tetragonal phase, even without the yttria stabilizer. The toughness of the composite in the 0-1.5 wt% yttria range was more that double that of monolithic mullite. However, as the yttria content continues to increase, the YSZ was stabilized as cubic zirconia at the expense of the tetragonal phase, and fracture toughness may have decreased due to decreased transformation toughening.

SiAlON(p)/Mullite

Reference: 110

Composition
0 and 15 wt% SiAlON

Processing Method
CIPed/Pressureless Sintered

Detailed Description
X-phase SiAlON was produced by simultaneous carbothermal reduction and nitriding of kaolinitic clays. Powders were comminuted for 70 hours by ball-milling in a 2-proponal medium and then dried. The dried powders were then uniaxially compacted with a pressure of 14 MPa and isostatically cold-pressed at a pressure of 345 MPa into pellets of 14 mm diameter. Sintering was carried out in a nitrogen atmosphere at 1660-1740 °C for 2 or 6 hours.

Reinforcement Description
SiAlON particulates

Matrix Description
Mullite

Matrix Supplier
Baikalox Company, France

Matrix Designation
SA 193 CR

Test Type
Vickers Indentation Method

Type of Specimen
Beam

Test Temperature
RT

Reinforcement Content (weight percent)	Fracture Toughness, K_{IC} (MPa $m^{1/2}$)	Relative Density (percent)	Sintering Time (hours)	Sintering Temperature (°C)	Range (MPa $m^{1/2}$)
unmilled					
15	3.20	95.0	6	1660	±0.1
milled					
15	3.70	99.4	2	1700	±0.2
15	3.50	98.8	2	1720	±0.1
0.0	3.40	99.7	2	1740	±0.1
0.0	3.50	98.4	2	1720	±0.2

Remarks/Comments
Fracture toughness did not show a significant change in toughness with the SiAlON phase..

Figure 61: Fracture Toughness of SiAlON(p)/Mullite

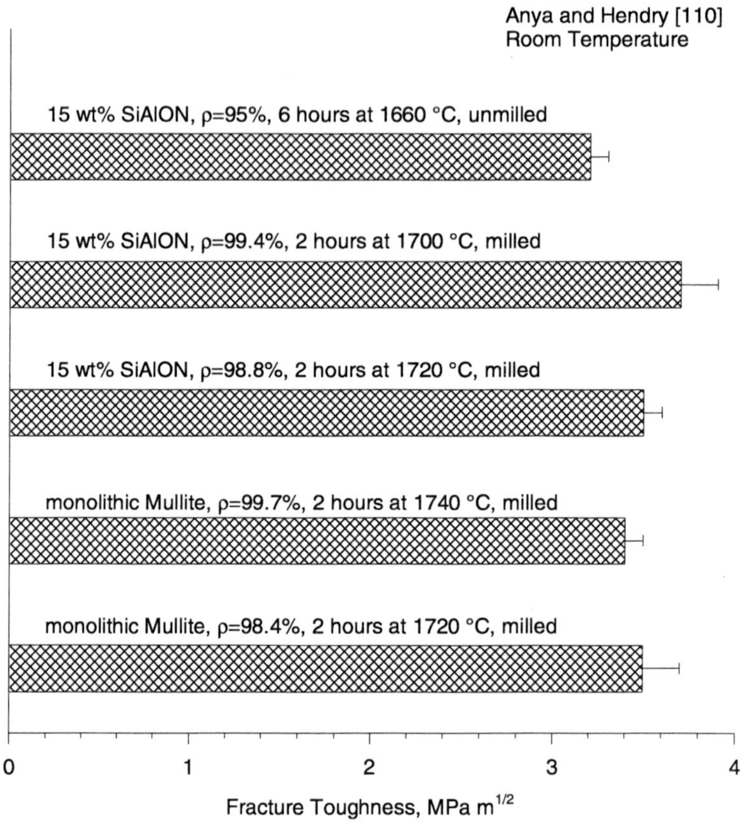

Discussion: Figure 61 examined fracture toughness of a SiAlON(p)/mullite composite in relation to the sintering time and temperature. The indentation microfracture method tends to overestimate values, so the Shetty *et al.* [I-6] equation was employed to correct this problem. For the monoliths, the sample that was sintered at the highest temperature was the most dense, but had the lowest fracture toughness. This was due to the grain size dependence of fracture toughness. The high sintering temperature encouraged grain growth that decreased toughness. The same held true with the composites. The composite with the smallest grain size and greatest toughness was sintered at 1700 °C. The composite sintered at 1660 °C did not reach as high a density, mainly because the powders were not milled like the other samples. The milling step produced powders that had a uniform particle size distribution which favored improved densification, which, in turn, resulted in higher toughness.

ZrO_2(p)/ZrO_2.Mullite

Reference: 71

Composition
15 vol% ZrO_2

Processing Method
Colloidal Processed/Hot-Pressed

Detailed Description
The powders were first produced by a sol-gel coprecipitation process. The powders were then mixed together with whiskers in a liquid medium, dried, and then hot-pressed. WRTZP sintered at 1620 °C and WRTZM sintered at 1500 °C.

Reinforcement Description
0Y-, 1.4Y-, and 2.2Y-ZrO_2 particulates

Matrix Description
ZrO_2.mullite

Test Type
3-pt Single Edge Notched Beam (SENB) Method

Test Description
Chevron notches were machined to 0.25 mm and three point bend test employed with a crosshead speed of 0.05 mm/min.

Test Temperature
RT

Stabilizer Content (weight percent)	Fracture Toughness, K_{IC} (MPa $m^{1/2}$)	Range (MPa $m^{1/2}$)
0.0	3.70	0.4
2.5	3.90	0.4
4.0	3.40	0.3

Remarks/Comments
Fracture toughness enhanced by microcracking. Whisker bridging and/ or crack deflection may also contribute to toughening.

Al_2O_3(p)/ZrO_2

Reference: 72

Composition
20 wt% Al_2O_3

Processing Method
Sinter/HIPed

Detailed Description
Specimens were sintered at 1400 °C and HIPed at 1500 °C. The composites were purchased as-fabricated from TOSOH Corporation of Tokyo. Some samples were heat-treated in air for 100 hours at 1000 °C prior to testing. Formed into billets 6 x 6 x 0.5 in.

Matrix Description
3Y-Al_2O_3.ZrO_2 powder

Matrix Supplier
TOSOH Corp. of Tokyo

Matrix Designation
TOSOH TZ-3Y20A ("Super Z")

Test Type
Indentation Fracture Technique (IF)[I-15]

Specimen Dimensions
3 x 4 x 50 mm

Type of Specimen
Bar

Test Description
load = 10 kg
Half of the bars were heat treated for 100 h at 1000 °C.

Test Temperature
RT

Fracture Toughness, K_{IC} (MPa $m^{1/2}$)	Density (g cm^{-3})	Weibull's Modulus (dimensionless)	Comments
5.0	5.50	6.00	*as received*
5.0	5.50	24.0	*after heat treatment*

Remarks/Comments
No noticeable change in fracture toughness occurred.

Figure 62: Fracture Toughness vs. Stabilizer Content of ZrO_2(p)/ZrO_2.Mullite and Al_2O_3(p)/ZrO_2

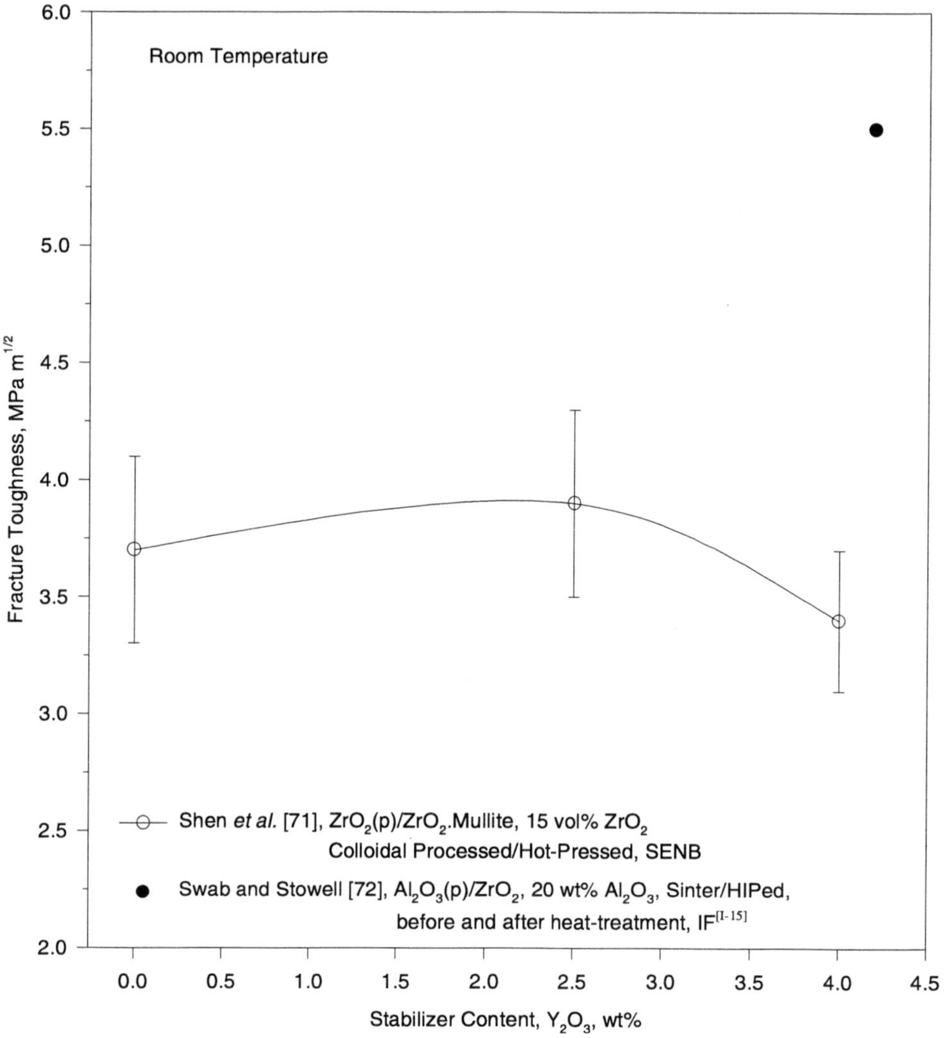

Discussion: Swab and Stowell [72] tested the fracture toughness of the zirconia·alumina composite as-received from the supplier. Then they heat-treated the composite for 100 hours at 1000 °C and tested it again. No noticeable change occurred in the fracture toughness or density. However, the Weibull modulus did change, increasing from 6 to 24.

Shen *et al.* [71] tested zirconia(p)/zirconia-mullite composites to compare whisker and particulate reinforcement. They discovered that with the addition of stabilizer content, fracture toughness increased for up to 2.2 wt% Y_2O_3. Above 2.2 wt%, any increase in stabilizer content resulted in a decrease in toughness.

TiC(p)/Al_2O_3

Reference: 8

Composition
0-40 wt% TiC

Processing Method
Hot-Pressed

Detailed Description
Purchased as-fabricated

Reinforcement Description
TiC particulates

Matrix Description
Al_2O_3 powder

Matrix Supplier
Feldmuhl AG, Plochingen, Germany (composite manufacturer)

Test Type
3-pt. SENB

Specimen Preparation
Triangulary notched on test face, relative notch depth was maintained in all cases at 0.75 ± 0.02 mm, or machine-cut straight edge notch (0.30 ± 0.01 mm deep) cut with a 0.01 mm thick diamond wheel.

Specimen Dimensions
span = 16 mm

Type of Specimen
Notched Bar

Test Description
For triangular notched: crosshead speed = 500 µm/min.
For machine notched: span = 30 mm and crosshead speed = 100 µm/min.
The relation $K_{IC} = \sigma_b c^{1/2} \cdot Yx$, where σ_b = flexural strength, c = notch depth and Yx is a correction factor that is a function of relative crack depth (depth/width).

Test Equipment
Instron universal testing machine

Reinforcement Content (volume percent)	Fracture Toughness, K_{IC} (MPa $m^{1/2}$)	Range (MPa $m^{1/2}$)	Comments
0.0	4.50	+0.13/-0.06	*natural crack SENB*
5.0	4.31	+0.32/-0.29	
10	4.24	+0.06/-0.10	
20	4.60	+0.12/-0.10	
30	4.75	+0.09/-0.11	
40	5.04	+0.11/-0.14	
0.0	4.08	+0.10/-0.11	*machined crack SENB*
5.0	6.22	+0.14/-0.14	
10	6.12	+0.10/-0.11	
20	6.63	+0.16/-0.20	
30	6.19	+0.18/-0.14	
40	6.31	+0.39/-0.34	

Remarks/Comments
The natural (triangular notch) crack showed a general increasing trend, while the machine-cut (straight notch) specimen was independent of the TiC content.

$TiC(p)/Al_2O_3$

Reference: 23

Composition
25 vol% TiC

Processing Method
CIPed/Pressureless Sintered

Detailed Description
The powders were mixed by ball milling with ethanol for 24 hours. After drying, the mixtures were uniaxially pressed and finally isostatically pressed at 200 MPa. Sintering of the compact was done in a radio-frequency induction heating furnace. Specimens were set in an SiC crucible to prevent reaction with the graphite susceptor, and were fired at 1600 to 1800 °C for 1 hour under an argon gas flow at atmospheric pressure. Hot pressing was also carried out at 1600 °C under a pressure of 30 MPa for one hour in an argon gas atmosphere.

Sintering Additives
MgO 0.5 wt% of matrix

Reinforcement Description
TiC particulates

Reinforcement Supplier
H.C. Stark

Supplier Designation
c.a.s. graded TiC

Reinforcement Dimensions
ave. grain size = 1.5 μm

Matrix Description
Al_2O_3

Matrix Supplier
Shinkosha

Matrix Properties
ave. grain size = 0.2 μm

Test Type
SENB

Type of Specimen
Bar

Test Temperature
RT

Fracture Toughness, K_{IC} (MPa $m^{1/2}$)	Comments
6.60	*Pressureless sintering*
6.16	*HP -from reference*

Remarks/Comments
The measured value was comparable to the reference value.

$TiB_2(p)/Al_2O_3$

Reference: 43

Composition
0-20 vol% TiB_2

Processing Method
Hot-Pressed

Detailed Description
The powders were mixed in a ball mill in ethanol for 2 hours using alumina balls in a plastic jar. The resulting slurries were oven dried. The Al_2O_3-TiB_2 powders were hot-pressed in BN coated graphite dies at 1470 °C for 20 minutes.

Reinforcement Description
TiB_2 powder

Reinforcement Dimensions
ave. particle size = 3 µm
range: 1-15 µm

Matrix Description
Al_2O_3 powder

Matrix Supplier
ALCOA

Matrix Designation
A-16SG

Test Type
Single Edge Notched Beam (SENB)

Specimen Preparation
diamond cut

Specimen Dimensions
5 x 5 x 25.4 mm
notch depth = 1.5 mm
span = 19 mm

Type of Specimen
Bar

Test Description
crosshead speed = 0.5 mm/min
applied pressure = 32 MPa

Test Temperature
RT

Reinforcement Content (volume percent)	Fracture Toughness, K_{IC} ($MPa\ m^{1/2}$)
0.0	4.67
5.0	6.10
10	5.62
15	6.12
20	6.10

Remarks/Comments
Each point is the average of 5-7 specimens.

$SiC(p)/Al_2O_3$

Reference: 80

Composition
~55 vol% SiC

Processing Method
Slip-Cast/DIMOX™

Detailed Description
Commercially available SiC particle/Al_2O_3 (Du Pont Lanxide composite Inc., USA) was obtained. This material was prepared using DIMOX™ Method. An SiC preform was slip-cast and then an Al alloy was infiltrated and oxidized in the furnace, resulting in an SiC-particle/alumina composite. In testing, the heat treatment condition was air-annealed at 1300 °C for 8 hours.

Reinforcement Description
SiC particulates

Matrix Description
aluminum alloy

Test Type
Single Edge Precracked Beam Method (SEPB)

Specimen Dimensions
4 x 3.2 x 40 mm
inner span = 10 mm
outer span = 30 mm

Type of Specimen
Bar

Test Description
crosshead speed = 0.5 mm/min

Test Temperature
RT

Reinforcement Content (volume percent)	Fracture Toughness, K_{IC} (MPa $m^{1/2}$)
22	5.20

SiC(p)/Al_2O_3

Reference: 88

Composition
0-30 vol% SiC

Processing Method
Hot-Pressed

Detailed Description
First, the alumina was ball milled for 24 hours to break up the agglomerates in a deflocculated, water-based slurry (Darvan 821-A). Next, the slurry was passed through a 325 mesh sieve to remove any large agglomerates. Afterwards, SiC was added and the mixture was milled for one hour. The material was then dried, dry pressed at 12 MPa, and then pre-fired to 600 °C for one hour to remove any residual volatile species. The pre-fired discs were then inductively hot-pressed in a nitrogen atmosphere for one hour at 1600 °C. Pressure was applied in a gradual rate beginning at 400 °C and held at 15 MPa.

Reinforcement Description
SiC particulates

Supplier Designation
Lonza LS-13

Reinforcement Typical Properties
ave. size = 13 μm
density = 3.2 g cm^{-3}
CTE = 4.3 x10^{-6} C^{-1}
E = 470 GPa
fracture toughness = 3-5 MPa. $m^{1/2}$

Matrix Description
Al_2O_3

Matrix Designation
Ferro A-310

Matrix Properties
ave. grain size = 0.35 μm
density = 3.96 g cm^{-3}
CTE = 7.7-8.3 x10^{-6} C^{-1}
E = 380 GPa
fracture toughness = 2-3 MPa. $m^{1/2}$

Test Type
Single Edge Notched Beam (SENB) Technique

Specimen Preparation
polished to 1 μm depth of notches was one-half the thickness of the sample and had width of 0.127 mm tensile edges were chamfered (tensile side was perpendicular to the hot-pressing direction)

Specimen Dimensions
3 x 4 x 50 mm

Type of Specimen
Single Edge Notched Bar (SENB)

Test Temperature
RT

Reinforcement Content (volume percent)	Fracture Toughness, K_{IC} (MPa $m^{1/2}$)	Range (MPa $m^{1/2}$)
0.0	4.35	±0.13
10	4.32	±0.13
20	5.05	±0.45
30	5.42	±1.17

Remarks/Comments
Fracture toughness increased with additions of particulates. Particulate SiC reinforcement did not result in quite the same increase in the fracture toughness as the platelet reinforced composite. Respective peak values of 5.42 and 5.68 MPa $m^{1/2}$ were determined.

Figure 63: Fracture Toughness of $TiC(p)/Al_2O_3$, $TiB_2(p)/Al_2O_3$, and $SiC(p)/Al_2O_3$

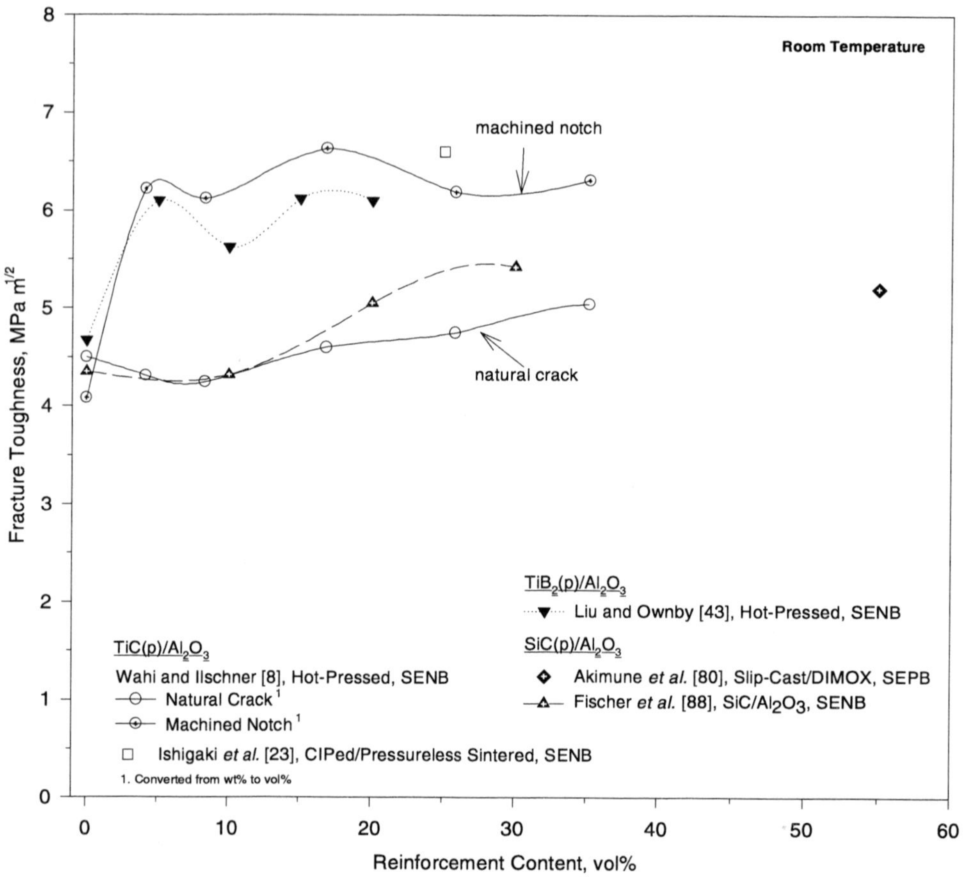

Discussion: Wahi and Ilschner [8] tested the fracture toughness of two $TiC(p)/Al_2O_3$ composites by SENB, one with a machined notch with a notch root radius of 0.05 mm, and the other with a sharp natural crack. In the machined notch sample, it appears that the critical notch root radius was smaller than 0.05 mm, which resulted in measurements that overestimated the toughness of the composite. The sharp natural crack SENB showed that toughness increased slightly with increasing TiC content from 10 to 40 wt% (8.3 to 35.1 vol%). The fracture toughness of the $TiC(p)/Al_2O_3$ composite tested by Ishigaki *et al.* [23] was very close to the toughness of the machined notched sample tested by Wahi and Ilschner [8]. Despite this fact, there was no indication that the notch root radius was too large. Therefore, it was likely that the fracture toughness from this source was not significantly overestimated.

The DIMOX™ 55 vol% $SiC(p)/Al_2O_3$ showed residual Al alloy in the grain boundaries, which resulted in some ductile blunting of the crack tip and a room temperature fracture toughness of 5.2 MPa $m^{1/2}$ [80]. Fischer *et al.* [88] found that the fracture toughness increased with increasing SiC content in the hot-pressed composite, reaching a maximum toughness of 5.42 MPa $m^{1/2}$ at 30 vol% reinforcement. The observed rough fracture surface was evidence for crack deflection and crack branching as the toughening mechanisms. Crack branching was enhanced in this system due to the tensile matrix stress developed.

In the TiB_2/Al_2O_3 composites tested by Liu and Ownby [43], crack interaction with the particulates was the toughening mechanism. Also, jet-milling of the TiB_2 formed irregular particles that had an increased aspect ratio. This irregular shape led to an increase in the deflection angle and an increase in fracture toughness.

$TiB_2(p)/Al_2O_3$

Reference: 43

Composition
0-20 vol% TiB_2

Processing Method
Hot-Pressed

Detailed Description
The powders were mixed in a ball mill in ethanol for 2 hours using alumina balls in a plastic jar. The resulting slurries were oven dried. The Al_2O_3-TiB_2 powders were hot-pressed in BN coated graphite dies at 1470 °C for 20 minutes.

Reinforcement Description
TiB_2 powder

Reinforcement Dimensions
ave. particle size = 3 μm
range: 1-15 μm

Matrix Description
Al_2O_3 powder

Matrix Supplier
ALCOA

Matrix Designation
A-16SG

Test Type
Vickers Indentation Microfracture Technique [I-16]

Specimen Preparation
Specimens were diamond cut and ground.

Specimen Dimensions
5 x 5 x 25.4 mm
notch depth = 5.06 mm
span = 19 mm

Type of Specimen
Bar

Test Description
load = 30 kg

Test Temperature
RT

Test Equipment
Fractometer, Terra Tek Systems

Reinforcement Content (volume percent)	Fracture Toughness, K_{IC} (MPa $m^{1/2}$)
0.0	3.78
5.0	5.01
10	4.92
15	4.41
20	5.33

Remarks/Comments
Each point is the average of 5-7 specimens.

$TiC(p)/Al_2O_3$

Reference: 47

Composition
0-80 wt% TiC

Processing Method
Hot-Pressed

Detailed Description
The composites were wet-mixed for 20 hours in a ball mill. Specimens were hot-pressed in a graphite die at 1500-1750 °C and 20 MPa for 1 hour. Applied pressure for the temperature range from room temperature to the hold temperature was one-half the final pressure. Hot-pressed specimens were removed from the die after cooling to room temperature.

Sintering Additives
TiO_2 and MgO powder

Reinforcement Description
TiC powder

Reinforcement Typical Properties
S < 400 ppm
Fe < 200 ppm
Ni < 20 ppm
Cr < 10 ppm

Reinforcement Dimensions
ave. particle size = 1.0 μm

Matrix Description
α-Al_2O_3 powder

Matrix Properties
Si < 50 ppm
Na < 10 ppm
Mg < 10 ppm
Cu < 10 ppm
Fe < 20 ppm
ave. particle size = 0.8 μm

Test Type
Vickers Indentation Microfracture (IM) Technique [I-14]

Specimen Preparation
Specimens were polished on the surface with 6- and 1- μm diamond paste.

Specimen Dimensions
10 x 10 x 3 mm

Type of Specimen
Bar

Test Description
indentor load = 10 kg

Test Temperature
RT

Reinforcement Content (weight percent)	Fracture Toughness, K_{IC} ($MPa\ m^{1/2}$)	Hot Pressing Temperature (°C)
0.0	3.00	1500
10	3.80	1600
20	4.30	1650
30	5.70	1680
40	6.10	1700
60	6.90	1700
80	5.80	1750

Remarks/Comments
Addition of TiC resulted in a small but distinct increase in fracture toughness. Fracture toughness reached a maximum at 60 wt% TiC.

TiC(p)/Al_2O_3

Reference: 58

Composition
26 wt% TiC

Processing Method
Cold Pressed / Sintered (some HIPed)

Detailed Description
Powders were dispersed and milled in a 15 cm diameter stainless steel mill with hexane for 24 hours with a ball to charge ratio of 10:1. Larger batches were attrition milled in deionized water and a binder was added prior to spray drying. Powders were dry screened to -80 mesh and wet screened to -325 mesh. Powders were uniaxially pressed and sintered in N_2 at 1840 °C for 10 minutes. HIPing was performed on some specimens by stacking the parts on a graphite setter plate and heating for 1 hour while increasing the argon pressure to 100 or 200 MPa and holding at 1300-1700 °C for 15-30 minutes.

Sintering Additives
3.7 wt% TiH_2

Reinforcement Description
TiC powder

Reinforcement Supplier
Hermann C. Starck Inc.,
New York, NY

Supplier Designation
Grade CA

Matrix Description
Al_2O_3 powder

Matrix Supplier
ALCOA Aluminum Co.,
Pittsburgh, PA

Matrix Designation
A-16 Superground

Test Type
Vickers Indentation Microfracture (IM) Technique [I-16]

Specimen Preparation
Specimens were ground to a 30 μm surface finish.

Type of Specimen
Bar

Test Description
Indentation load = 50-75 N

Test Equipment
Model 1123 Universal Testing Machine Instron Corporation. Canton, MA

Relative Density (percent)	Fracture Toughness, K_{IC} (MPa m$^{1/2}$)	Range (MPa m$^{1/2}$)	HIP Temp. (°C)	HIP Pressure (MPa)	Comments
97.5	4.30	±0.4	----	----	*as sintered*
97.9	4.20	±0.4	1300	100	*sintered/HIPed*
100	4.10	±0.2	1600	100	
99.5	4.20	±0.3	1700	200	

Remarks/Comments
Each data point was the average of 5-6 indentations.

$TiC(p)/Al_2O_3$

Reference: 58

Composition
26 wt% TiC

Processing Method
Cold Pressed/Sintered (some HIPed) or Hot-Pressed

Detailed Description
Powders were dispersed and milled in a 15 cm diameter stainless steel mill with hexane for 24 hours with a ball to charge ratio of 10:1. Larger batches were attrition milled in deionized water and a binder was added prior to spray drying. Powders were dry screened to -80 mesh and wet screened to -325 mesh. Powders were uniaxially pressed and sintered in Ar at 1875 °C for 10 minutes. HIPing was performed on some specimens by stacking the parts on a graphite setter plate and heating for 1 hour while increasing the argon pressure to 100 MPa and holding at 1600 °C for 15-30 minutes. Other specimens were hot-pressed at 1650 °C for 10 minutes in argon at 35 MPa.

Sintering Additives
3.7 wt% TiH_2

Reinforcement Description
TiC powder

Reinforcement Supplier
Hermann C. Starck Inc., New York, NY

Supplier Designation
Grade CA

Matrix Description
Al_2O_3 powder

Matrix Supplier
ALCOA Aluminum Co., Pittsburgh, PA

Matrix Designation
A-16 Superground

Test Type
Vickers Indentation Microfracture (IM) Technique [I-16]

Specimen Preparation
Specimens were ground to a 30 μm surface finish.

Type of Specimen
Bar

Test Description
Indentation load = 50-75 N

Test Equipment
Model 1123 Universal Testing Machine Instron Corporation. Canton, MA

Relative Density (percent)	Fracture Toughness, K_{IC} (MPa $m^{1/2}$)	Comments	Comments
94.9	4.35	±0.24	*sintered*
99.5	4.42	±0.52	*sintered and HIPed*
98.6	4.53	±0.24	*hot -pressed*

Remarks/Comments
Fracture toughness was highest at 4.53 MPa $m^{1/2}$ when the specimen was hot-pressed. Each data point was the average of 5-6 indentations.

$TiC(p)/Al_2O_3$

Reference: 68

Composition
30 wt% TiC

Processing Method
CIPed/Sintered (some HIPed) or Hot-Pressed

Detailed Description
The powders were milled in acetone. The milled slurry was dried, uniaxially pressed at 50 MPa, and isostatically pressed at 200 MPa. Some compacts were sintered using a heating rate of 5 °C/min. under helium with a 10 minute hold at 1870 °C. Some samples were also embedded with C powder to prevent chemical reactions and vaporization from the specimen. Final densification for some of the sintered specimens was done by HIPing at 1550 °C for 30 minutes under 150 MPa pressure of argon. For comparison, other samples were prepared by hot-pressing at 1650 °C and 40 MPa in a vacuum for 30 minutes.

Sintering Additives
0.01 wt% MgO

Reinforcement Description
TiC powder

Reinforcement Supplier
H.C. Starck, West Berlin, Germany

Reinforcement Dimensions
particle size = 1-3 μm

Matrix Description
Al_2O_3 powder

Matrix Supplier
Sumitomo Chemical Co., Osaka, Japan

Matrix Properties
particle size = 0.4 μm

Test Type
Vickers Indentation Microfracture (IM) technique

Specimen Preparation
Surfaces and edges were polished with an 800 grit diamond wheel and 1 μm diamond paste.

Specimen Dimensions
3 x 4 x 25 mm

Type of Specimen
Bar

Test Description
Indentor load = 100 N

Test Temperature
RT

Relative Density (volume percent)	Fracture Toughness, K_{IC} (MPa $m^{1/2}$)	Range (MPa $m^{1/2}$)	Comments
96.7	4.20	±0.5	*sintered with embedding*
99.8	4.30	±0.3	*sintered/HIPed*
99.9	4.30	±0.3	*hot-pressed*

Remarks/Comments
Fracture toughness was taken as average of 12 samples. Typical fracture toughness of Al_2O_3-30 wt% TiC composites which were sintered with embedding powder and hot isostatically-pressed was 4.3 MPa $m^{1/2}$. Fracture toughness did not change much with the increase of sintered density.

$TiB_2(p)/Al_2O_3$

Reference: 75

Composition
0-25 vol% TiB_2

Processing Method
Cold Pressed/Sintered

Detailed Description
The TiB_2 powder was first classified into five different size fractions: 1-3, 3-6, 6-10,10-15 and 1-15 μm. This was performed by air elutriation using an Aminco-Roller Particle Size Analyzer 5-445. The powders were ultrasonically dispersed and mixed in methanol, oven dried, uniaxially cold-pressed into pellets and rectangular bars, and subsequently isostatically pressed. The pressed samples were then sintered in a tube furnace at 1600 °C in a hydrogen atmosphere and held for 1 hour.

Reinforcement Description
TiB_2 powder

Reinforcement Supplier
H.C. Starck

Supplier Designation
Grade A

Reinforcement Dimensions
ave. particle size = 5.2 μm
range = 1-15 μm

Matrix Description
Al_2O_3 powder

Matrix Supplier
Sumitomo

Matrix Designation
AKP-50

Matrix Properties
ave. particle size = 0.2 μm
>99.995 % pure

Test Type
Vickers Indentation Microfracture (IM)[I-16] Technique and Multiple Indentation Flexural Strength (MIFS)[I-10]

Test Description
30 kg load

Test Temperature
RT

Test Equipment
LECO V-100-A2 hardness tester

Reinforcement Content (volume percent)	Fracture Toughness, K_{IC} (MPa $m^{1/2}$)	Relative Density (percent)	Comments
0.0	3.60	97.4	*1-3 μm particles*
5.0	5.18	97.0	
10	5.58	96.4	
15	5.71	95.7	
20	5.71	94.7	
25	5.68	94.2	
0.0	3.57	97.4	*3-6 μm particles*
5.0	5.62	96.0	
10	6.13	96.5	
20	6.07	95.3	
25	6.40	94.7	
0.0	3.59	97.4	*6-10 μm particles*
5.0	4.72	97.2	
10	4.97	96.7	
15	4.95	96.7	
20	4.97	95.6	
25	5.03	95.2	
0.0	3.58	97.4	*10-15 μm particles*
5.0	4.47	97.2	
10	4.54	97.2	
15	4.66	97.1	
20	4.63	96.3	
25	4.74	95.9	
0.0	3.59	97.4	*1-15 μm particles*
5.0	4.94	97.0	
10	5.34	96.0	
15	5.52	94.9	
20	5.54	93.5	
25	5.57	93.3	

Remarks/Comments

The 3-6 μm particulates demonstrated the highest fracture toughness values. The composite reinforced with unseparated particulates had intermediate fracture toughness. The DCM measurements gave similar results as the CNSB method.

SiC(p)/Al_2O_3

Reference: 77

Composition

0-32 vol% SiC

Processing Method

Hot-Pressed

Detailed Description

The composites containing various volume fraction of SiC particles were mixed by the ball milling technique in the highly pure toluene with Al_2O_3 balls for 24 hours. The dried mixtures were hot-pressed in N_2 atmosphere at 1500-1900 °C at 28 MPa.

Reinforcement Description

β-SiC particulates

Reinforcement Supplier

Ibiden Co., Ltd.

Reinforcement Dimensions

ave. particle size < 0.3 μm

Matrix Description

α-Al_2O_3 particulates

Matrix Supplier

Asahi Chemical Co.

Matrix Properties

ave. particle size = 0.3 μm

Test Type

Vickers Indentation Microfracture (IM) Technique[I-8]

Test Description

indentor loads of 0.49-18.6 N

Test Temperature

RT

Reinforcement Content (volume percent)	Fracture Toughness, K_{IC} (MPa $m^{1/2}$)	Range (MPa $m^{1/2}$)
0.0	3.12	+0.21/-0.12
5.1	4.70	+0.20/-0.18
11	4.32	+0.30/-0.27
17	4.32	+0.33/-0.26
32	4.22	+0.26/-0.23

Remarks/Comments

The toughness and strength of Al_2O_3 was significantly improved by incorporating the SiC particles within the matrix. The maximum toughness of over 4.7 MPa $m^{1/2}$ was observed for 5 vol% SiC reinforcement. Overall, strength and toughness was greater for the SiC-Al_2O_3 composite than for the SiC-MgO composite.

TiN(p)/Al_2O_3

Reference: 84

Composition

0 and 30 vol% TiN

Processing Method

Hot-Pressed

Detailed Description

Al_2O_3 powders were homogenized with 30 vol% TiN in plastic jars with alumina balls and isobuthyl alcohol. The composites were hot-pressed in a vacuum at 1600-1650 °C and 30 MPa.

Reinforcement Description

TiN particulates

Matrix Description

Al_2O_3

Test Type

Vickers Indentation Microfracture (IM) technique

Test Description

load = 98 N

Test Temperature

RT

Reinforcement Content (volume percent)	Fracture Toughness, K_{IC} (MPa $m^{1/2}$)	Relative Density (percent)	Standard Deviation (MPa $m^{1/2}$)
0.0	3.20	100	±0.2
30	5.20	99.3	±0.4

Remarks/Comments

Fracture toughness increased noticeably for the composites compared to the base Al_2O_3 matrix.

TiC(p)/Al_2O_3

Reference: 84

Composition
0 and 30 vol% TiC

Processing Method
Hot-Pressed

Detailed Description
Al_2O_3 powders were homogenized with 30 vol% TiC in plastic jars with alumina balls and isobuthyl alcohol. The composites were hot-pressed in a vacuum at 1600-1650 °C and 30 MPa.

Reinforcement Description
TiC particulates

Matrix Description
Al_2O_3

Test Type
Vickers Indentation Microfracture (IM) technique

Test Description
load = 98 N

Test Temperature
RT

Reinforcement Content (volume percent)	Fracture Toughness, K_{IC} (MPa $m^{1/2}$)	Relative Density (percent)	Standard Deviation (MPa $m^{1/2}$)
0.0	3.20	100	±0.2
30	4.30	99.3	±0.5

Remarks/Comments
Fracture toughness increased noticeably for the composite compared to the base Al_2O_3 matrix.

TiB_2(p)/Al_2O_3

Reference: 84

Composition
0 and 30 vol% TiB_2

Processing Method
Hot-Pressed

Detailed Description
Al_2O_3 powders were homogenized with 30 vol% TiB_2 in plastic jars with alumina balls and isobuthyl alcohol. The composites were hot-pressed in a vacuum at 1600-1650 °C and 30 MPa.

Reinforcement Description
TiB_2 particulates

Matrix Description
Al_2O_3

Test Type
Vickers Indentation Microfracture (IM) Technique

Test Description
load = 98 N

Test Temperature
RT

Reinforcement Content (volume percent)	Fracture Toughness, K_{IC} (MPa $m^{1/2}$)	Relative Density (percent)	Standard Deviation (MPa $m^{1/2}$)
0.0	3.20	100	±0.2
30	5.70	99.3	±0.6

Remarks/Comments
Fracture toughness increased noticeably for the composites compared to the base Al_2O_3 matrix.

SiC(p)/Al_2O_3

Reference: 112

Composition
~55 vol% SiC

Processing Method
Slip-Cast/DIMOX™

Detailed Description
Commercially available SiC(p)/Al_2O_3 (Du Pont Lanxide composite Inc., USA) was obtained. This material was prepared using a DIMOX™ method. An SiC preform was slip cast and then an Al alloy was infiltrated and oxidized in the furnace, resulting in the composite.

Reinforcement Description
SiC particulates

Reinforcement Properties
ave. particle diameter = 5-20 μm

Matrix Description
Al_2O_3 powder

Test Type
Chevron-Notched Beam (CNB)

Test Temperature
RT to 1400 °C

Test Environment
Air

Temperature (°C)	Fracture Toughness, K_{IC} (MPa $m^{1/2}$)	Comments
20	7.6	*7 μm filler*
391	4.8	
648	3.7	
1000	3.8	
1400	3.8	
20	4.5	*64 μm filler*
649	4.0	
1000	4.0	
1400	4.2	

Figure 64: Indentation Microfracture (IM) Fracture Toughness of TiB_2(p)/Al_2O_3 and SiC(p)/Al_2O_3

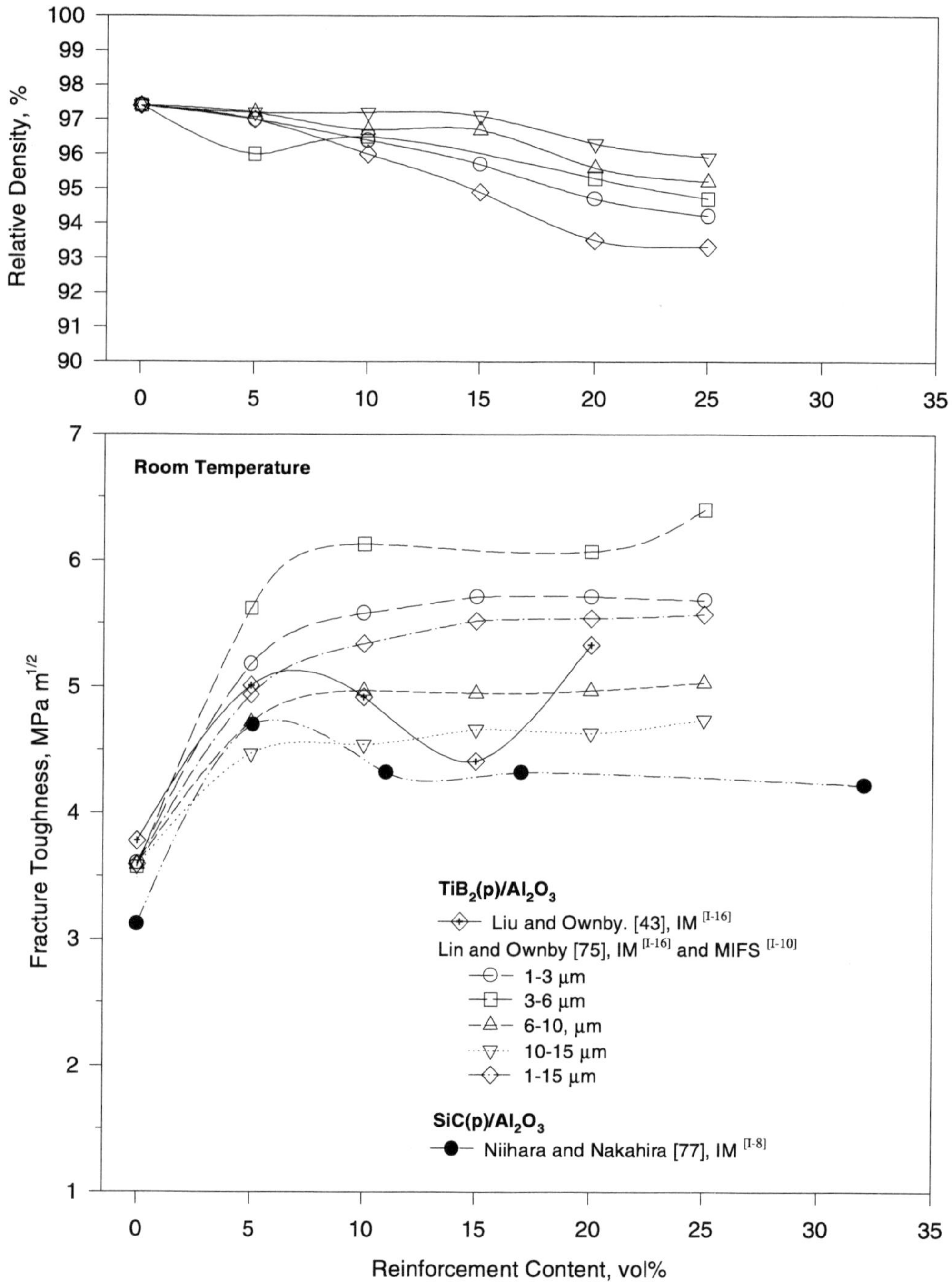

Discussion: Figure 64 presents indentation fracture toughness data on TiB_2 and SiC particulate reinforced alumina. All monolithic alumina samples had a toughness between 3 and 4 MPa $m^{1/2}$ and increased by 1 MPa $m^{1/2}$ with the addition of 5 vol% particulate reinforcement. For the most part, the fracture toughness was not affected by changing the reinforcement content in the range of 5 to 30 vol%.

Liu, Lin, and Ownby [43 and 75] measured the IM and MIFS fracture toughness of $TiB_2(p)/Al_2O_3$ composites. Chevron Notched Short Rod (CNSR) data for these same composites are given in figure 68. The CNSR data corresponded well with the IM and MIFS tests. The primary toughening mechanism for these composite types was crack deflection, which was highly dependent on the reinforcement grain size. Lin and Ownby [75] examined the effect of various TiB_2 particle sizes by first separating the TiB_2 powder into 5 size catagories. The 3-6 μm particulates produced the toughest composites. In general, the smaller particles increase the likelihood of an encounter between an inclusion and an on-coming crack tip. However, if the size of the particulate reinforcement drops below a critical size, residual stress fields due to thermal expansion mismatches disappear, and toughness may drop. The lower toughness of the 1-3 μm composite compared to the 3-6 μm demonstrated this effect. Relative density decreased with an increase in reinforcement content. This seemed to have little effect on the toughness of the $TiB_2(p)/Al_2O_3$ composites. In summary, the two competing effects result in an optimum particle size for toughening [75].

Niihara and Nakahira *et al.* [77] tested $SiC(p)/Al_2O_3$ nanocomposites with extremely small SiC particulates. Fracture toughness of these nanocomposites was lower than the composites tested by Liu, Lin, and Ownby [43 and 75]. This suggested that the SiC reinforcement may be below the critical size for sufficient CTE mismatch stresses to result in optimum crack deflection and microcrack toughening, although several other variables, such as test type and reinforcement type, must be accounted for to make such a comparison.

Figure 65: Fracture Toughness of TiB_2, TiC, and $TiN(p)/Al_2O_3$

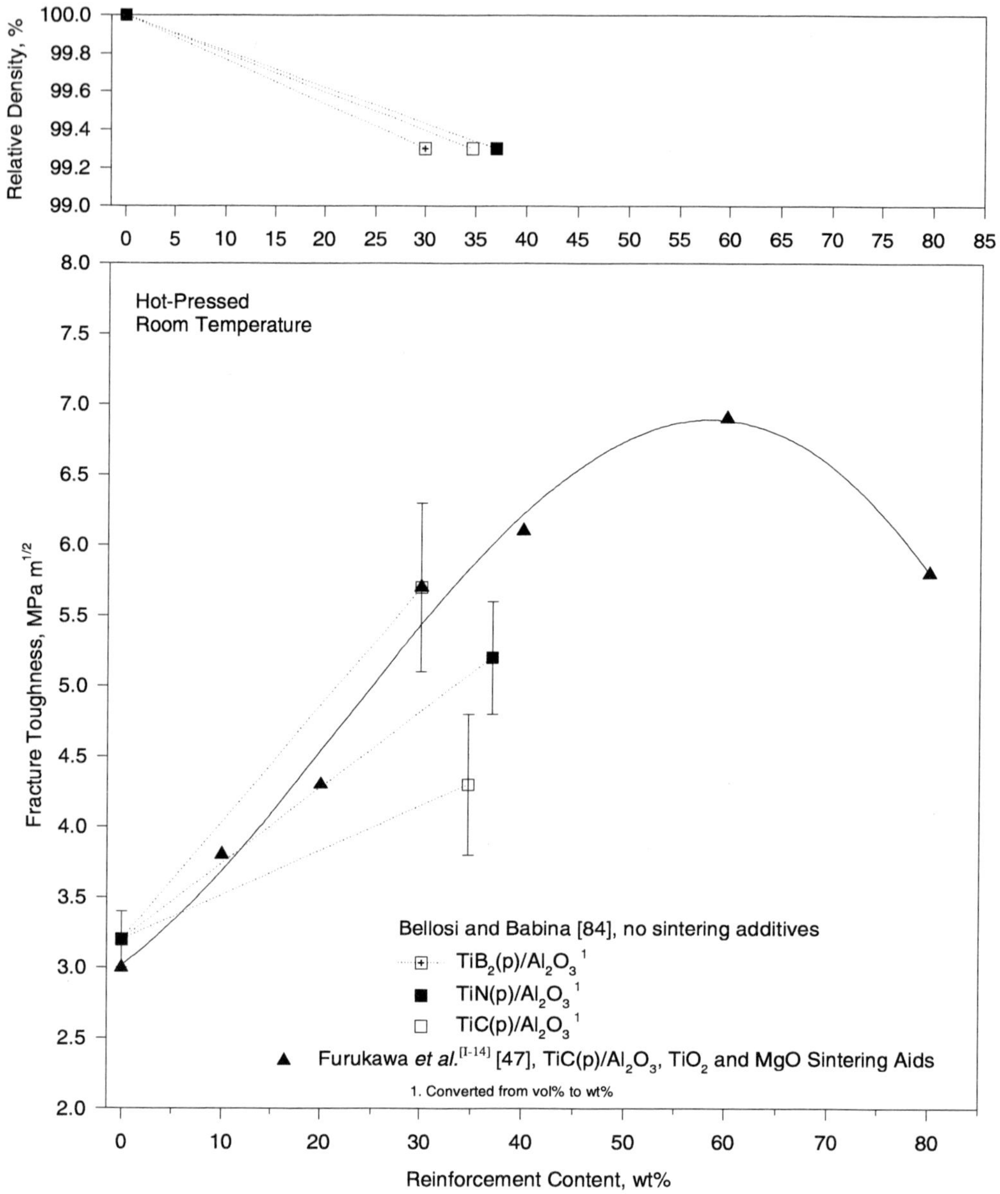

Discussion: Furukawa *et al.* [47] found that toughness of $TiC(p)/Al_2O_3$ composites increased with increasing TiC content until a maximum of 6.9 MPa $m^{1/2}$ was reached at 60 wt% TiC. The reason for the increase in toughness at low reinforcement contents was the increase in fracture surface energy, which was dependent on the difference in grain size. As TiC content increased up to 60 vol%, the increase in toughness was due to transgranular fracture of the TiC grains, crack pinning, and microcrack toughening [47].

Bellosi and Babina [84] showed that with a reinforcement content increase, relative density decreased, but fracture toughness increased. The relative density decrease was due to the limited sinterability of TiN, TiB_2 and TiC. At the same time, all the composites on this graph were tougher because of crack interactions with the hard particulates resulting in crack deflection and microcrack toughening.

Figure 66: Effect of Processing Conditions on Fracture Toughness of $TiC(p)/Al_2O_3$

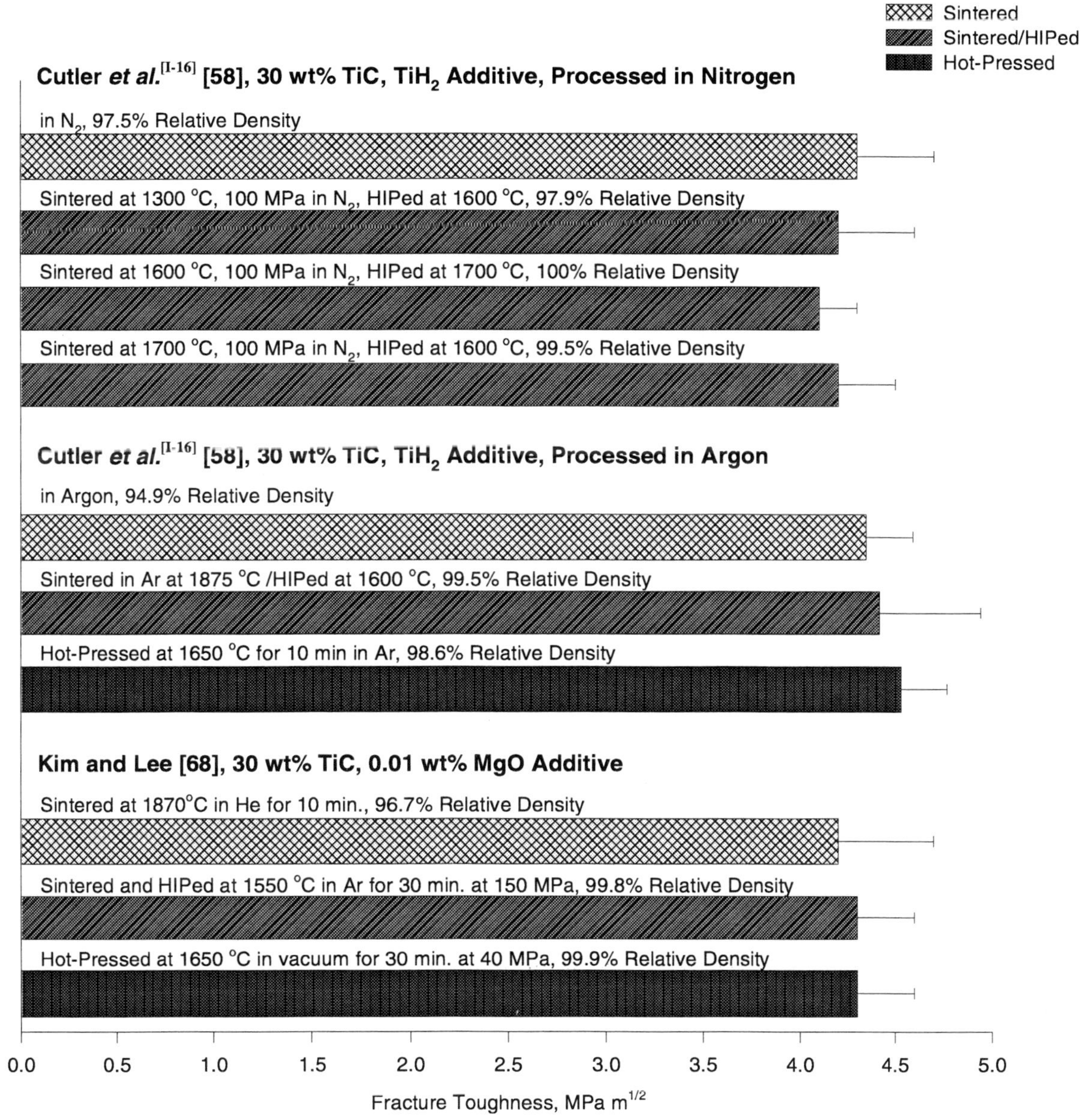

Discussion: Cutler *et al.* [58] and Kim and Lee [68] looked at processing methods for $TiC(p)/Al_2O_3$ composites. They discovered that among the sintered samples it made little difference which sintering atmosphere was used. Samples were between 95 and 98% dense, and had fracture toughness values within 0.5 MPa $m^{1/2}$ of any other sample. The sinter/HIPed samples showed dependence on the temperature of sintering, HIPing temperatures and atmosphere used. Samples that were HIPed the same but pre-sintered differently showed a change in density, but showed the same toughness. Relative to this, samples that were HIPed at higher temperatures but sintered the same showed a decrease in toughness. Samples hot-pressed in argon or a vacuum showed comparable or better results compared to the samples fabricated by the other processing techniques.

Figure 67: Fracture Toughness vs. Temperature of DIMOX™ Produced $SiC(p)/Al_2O_3$

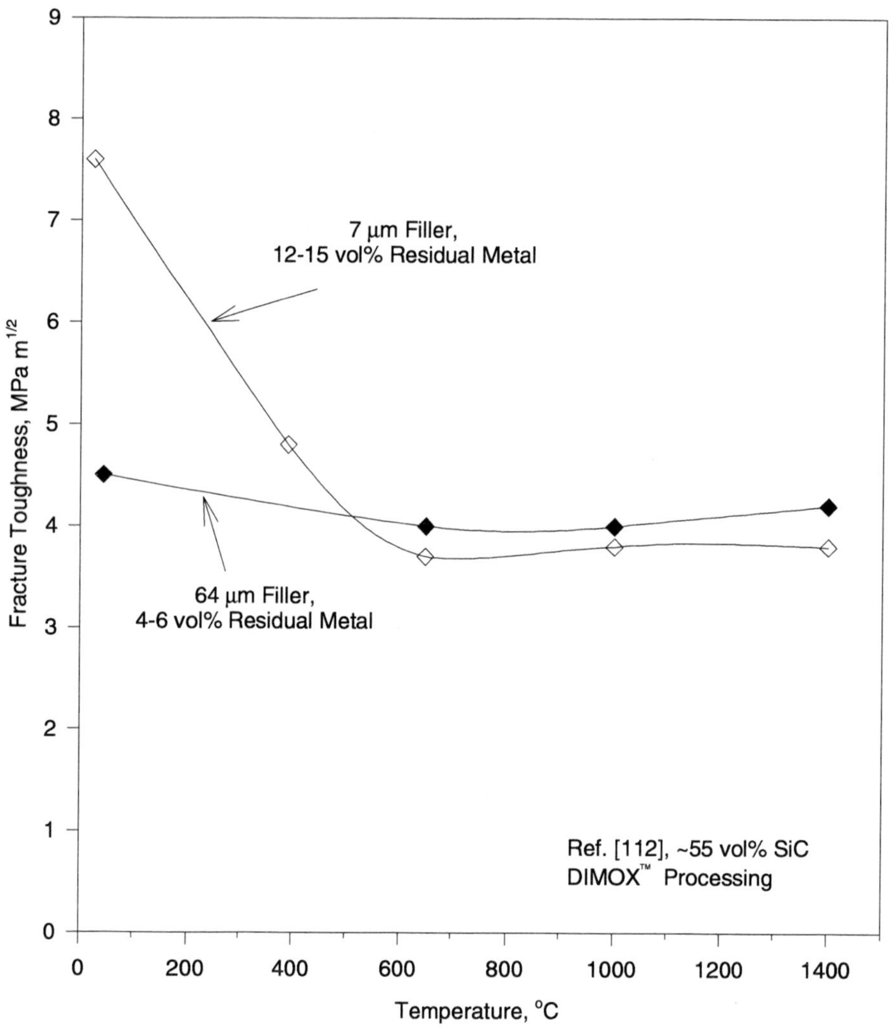

Discussion: The fracture toughness of the DIMOX™ $SiC(p)/Al_2O_3$ was slightly below that of the other composites. The toughness of the composite with 7 µm filler was much greater than the toughness of the 64 µm filler composite, due to higher residual metal content. As the temperature increased, toughness declined until it was comparable to the 64 µm filler composite. The residual metal softened and melted in the range of 400 to 600 °C, which essentially equalized the toughness of the two composites. The 64 µm filler composite stayed relatively constant because of low contents of residual metal. The data appears to show that the fracture toughness was not dependent on which SiC particle size used. However, different toughening mechanisms may have been active in the two composites, and the similar fracture toughness of the two composites may have been coincidental.

Diamond(p)/Al_2O_3

Reference: 59

Composition
0-15 vol% diamond

Processing Method
Hot-Pressed

Detailed Description
The diamond powder was mixed with α-alumina and ball milled in methanol for 2 hours using alumina balls in a plastic jar. The resulting slurries were oven dried. The granulated powders were hot-pressed in a boron-nitride coated graphite die at 32 MPa and 1250 °C under an argon atmosphere.

Reinforcement Description
diamond powder

Reinforcement Supplier
Smith Mega Diamond, Inc., Provo, UT

Supplier Designation
(1) S-diamond
(2) M-diamond

Reinforcement Dimensions
(1) 0.5-1.0 μm
ave. particle size = 0.48 μm
(2) 0.1-5.0 μm
ave. particle size = 1.56 μm

Matrix Description
α-Al_2O_3 powder

Matrix Supplier
(1) ALCOA Industrial Chemical, Bauxite, AR
(2) Sumitomo Chemical America, New York

Matrix Designation
(1) A16SG
(2) AKP 50

Matrix Properties
(1) ave. particle size = 0.83 μm
(2) ave. particle size = 0.33 μm

Test Type
Chevron Notched Short Rod (CNSR)

Test Temperature
RT

Reinforcement Content (volume percent)	Fracture Toughness, K_{IC} (MPa $m^{1/2}$)	Relative Density (percent)	Comments
0.0	4.09	99.3	*A16SG Alumina S-Diamond*
5.0	5.25	96.1	
10	5.30	94.3	
15	4.94	93.9	
0.0	3.89	97.3	*AKP50 Alumina S-Diamond*
5.0	4.26	97.3	
10	3.83	95.2	
15	4.75	92.5	
5.0	4.28	98.6	*AKP50 Alumina M-Diamond*
10	4.51	97.1	
15	5.75	95.4	

Remarks/Comments
The AKP50 matrix composite with 15 vol% m-diamond reinforcement was the toughest composite tested. The composites with the A16SG matrix had high density at low reinforcement contents.

$B_4C(p)/Al_2O_3$

Reference: 60

Composition
0-20 vol% B_4C

Processing Method
Hot-Pressed

Detailed Description
Powders were jet milled in methanol for 2 hours using alumina balls. Slurries were oven dried at 1500 °C and 1600 °C for 3 hours. Hot-pressed at 1520 °C for 20 minutes in a BN coated graphite die in an argon atmosphere.

Reinforcement Description
B_4C particulates

Reinforcement Supplier
Eagle-Picher Industries Inc.

Reinforcement Dimensions
ave. particle size = 0.2-7 μm

Matrix Description
Al_2O_3

Matrix Supplier
ALCOA industrial Chemicals

Matrix Designation
A16 SG

Test Type
Chevron Notched Short Rod (CNSR) [I-19]

Specimen Dimensions
dia. = 0.95 cm
length = 1.43 cm

Type of Specimen
Chevron Notched Rod

Test Description
Fracture toughness tests were made parallel to the hot pressing direction.

Test Temperature
RT

Reinforcement Content (volume percent)	Fracture Toughness, K_{IC} ($MPa\ m^{1/2}$)
0.0	3.98
2.5	4.81
5.0	5.30
10	4.74
15	4.64
20	4.61

Remarks/Comments
Maximum fracture toughness occurred at 5 vol% particulates. Resistance to crack propagation appeared to be associated with crack interactions with the B_4C inclusions.

$TiB_2(p)/Al_2O_3$

Reference: 75

Composition
0-25 vol% TiB_2

Composite Architecture
particulate reinforcement

Processing Method
Cold Pressed/Sintered

Detailed Description
The TiB_2 powder was first classified into five different size fractions: 1-3, 3-6, 6-10, 10-15 and 1-15 μm. This was performed by air elutriation using an Aminco-Roller Particle Size Analyzer 5-445. The powders were ultrasonically dispersed and mixed in methanol, oven dried, uniaxially cold-pressed into pellets and rectangular bars, and subsequently isostatically pressed. The pressed samples were then sintered in a tube furnace at 1600 °C in a hydrogen atmosphere and held for 1 hour.

Reinforcement Description
TiB_2 powder

Reinforcement Supplier
H.C. Starck

Supplier Designation
Grade A

Reinforcement Dimensions
ave. particle size = 5.2 μm
range = 1-15 μm

Matrix Description
Al_2O_3 powder

Matrix Supplier
Sumitomo

Matrix Designation
AKP-50

Matrix Properties
ave. particle size = 0.2 μm
>99.995 % pure

Test Type
Chevron Notched Short Bar Method (CNSB) [I-7], [I-19]

Test Description
Notches were cut with a diamond saw of thickness 0.75 mm.

Test Equipment
Terra Tek Systems Model 4200 Fractometer I

Test Temperature
RT

Reinforcement Content (volume percent)	Fracture Toughness, K_{IC} (MPa $m^{1/2}$)	Relative Density (percent)	Comments
0.0	3.64	97.4	*1-3 μm particles*
5.0	5.30	97.0	
10	5.92	96.4	
15	6.02	95.7	
20	6.13	94.7	
25	6.14	94.2	
0.0	3.67	97.4	*3-6 μm particles*
5.0	5.81	96.0	
10	6.15	96.5	
20	6.47	95.3	
25	6.38	94.7	
0.0	3.64	97.4	*6-10 μm particles*
5.0	4.53	97.2	
10	5.13	96.7	
15	5.49	96.7	
20	5.62	95.6	
25	5.73	95.2	
0.0	3.64	97.4	*1-15 μm particles*
5.0	4.82	97.0	
10	5.41	96.0	
15	5.71	94.9	
20	5.83	93.5	
25	5.95	93.3	

Remarks/Comments

The 3-6 μm particulates demonstrate the highest fracture toughness values. The composite reinforced with unseparated particulates have intermediate fracture toughness. The DCM measurements gave similar results as the CNSB method.

$TiB_2(p)/Al_2O_3$

Reference: 76

Composition
0-20 vol% TiB_2

Processing Method
Hot-Pressed

Detailed Description
TiB_2 powders were dispersed with Al_2O_3 powders by slurry mixing. The oven dried mixture was then hot pressed at 1470 °C for α-Al_2O_3 composites and 1350 °C for β-Al_2O_3 composites.

Reinforcement Description
TiB_2 particulates

Matrix Description
α– or β-Al_2O_3

Test Type
Chevron Notched Short Rod Technique (CNSR)

Type of Specimen
Notched Rod

Test Temperature
RT

Reinforcement Content (volume percent)	Fracture Toughness, K_{IC} (MPa $m^{1/2}$)	Range (MPa $m^{1/2}$)	Comments
0.0	3.98		*α-Al_2O_3*
2.5	5.96		
5.0	6.55		
7.5	5.47	+0.27 /-0.30	
10	5.66	+0.21 /-0.23	
15	5.54		
20	5.14	+0.36 /-0.45 [2]	
0.0	2.09		*β-Al_2O_3*
5.0	3.19		
10	3.40		
15	4.35		
20	3.42		

Remarks/Comments
The α-Al_2O_3 composites were much tougher than the β-Al_2O_3 composites. The α-Al_2O_3 composites reached a maximum toughness at 5 vol% TiB_2 composites. All samples were greater than 98.0 %.

Figure 68: Chevron Notched Short Rod Fracture Toughness [I-7], [I-9] of $TiB_2(p)/Al_2O_3$

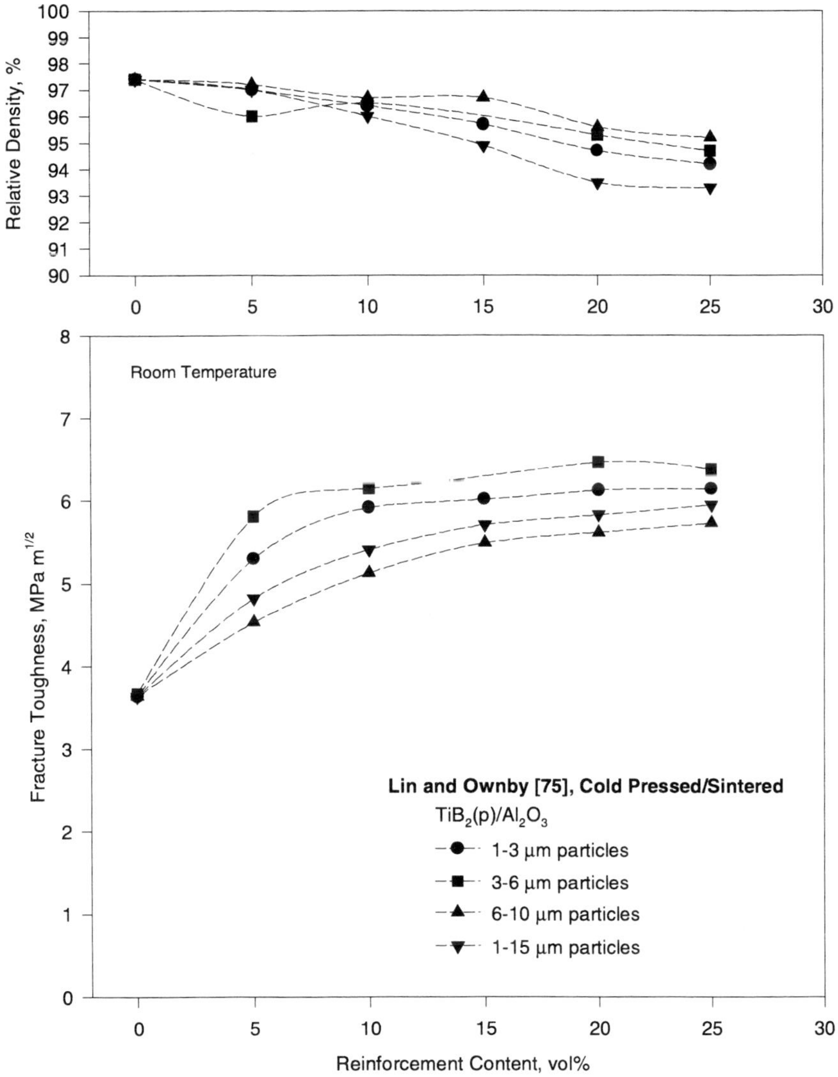

Discussion: Lin, Liu, and Ownby *et al.* [43 and 75] measured the fracture toughness of $TiB_2(p)/Al_2O_3$ composites by the Chevron Notched Short Rod (CNSR) technique. Fracture toughness data for the same materials measured by Vickers indentation and the MIFS technique is given in figure 64. The primary toughening mechanism for these composite types was crack deflection, which was highly dependent on the reinforcement grain size. Lin and Ownby [75] examined the effect of various TiB_2 particle sizes by first separating the TiB_2 powder into 5 size categories. The 3-6 µm particulates produced the toughest composites. In general, smaller particles increase the likelihood of an encounter between an inclusion and an on-coming crack tip. However, if the size of the particulate reinforcement drops below a critical size, residual stress fields due to thermal expansion mismatches disappear, and toughness may drop. The lower toughness of the 1-3 µm composite compared to the 3-6 µm composite demonstrated this effect. Relative density decreased with increasing reinforcement content, which seemed to have little effect on the toughness of the $TiB_2(p)/Al_2O_3$ composites. In summary, the two competing effects resulted in an optimum particle size for toughening [75].

Figure 69: Chevron Notched Short Rod Fracture Toughness of Diamond, B_4C, and TiB_2 Particulate Reinforced Alumina

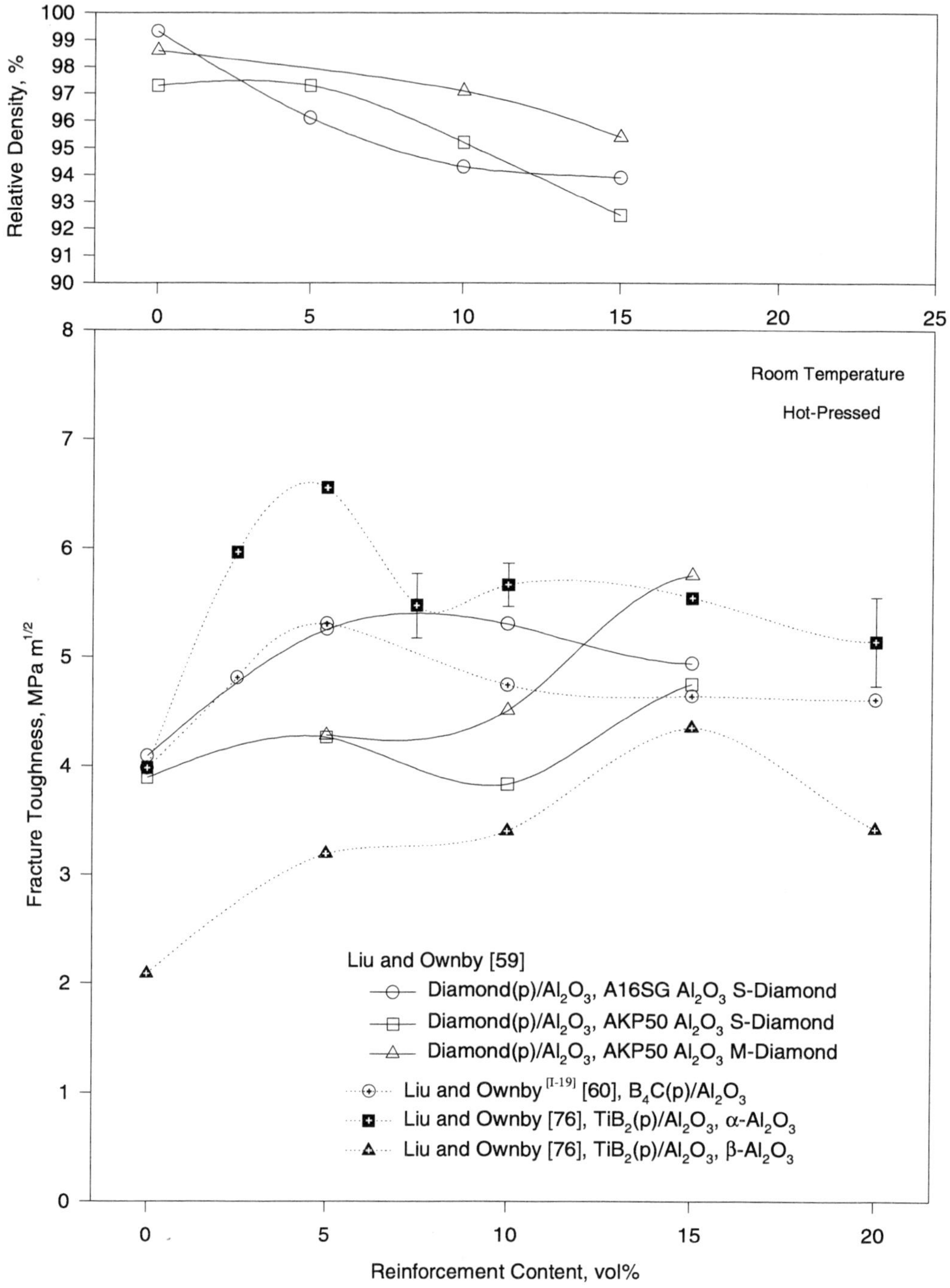

Discussion: Liu and Ownby [59, 60, and 76] tested diamond, B_4C, and TiB_2 as potential reinforcements for alumina. These were chosen as a potential reinforcement for alumina because of their extremely high hardness and elastic properties. The addition of any one of these particulates to an alumina matrix increases toughness by crack deflection and by the interaction of an approaching crack tip with the induced stress field around the particulate reinforcement. Ref. [59] tested diamond reinforced alumina composites fabricated from two types of alumina powders and two types of diamond powders. The toughness of the two monolithic aluminas were the same. However, the addition of 5 to 10 vol% S-diamond increased the toughness of the A16SG matrix more than the AKP50 matrix, while the addition of 15 vol% increased the toughness of the AKP50 matrix above that of the 15 vol% diamond reinforced A16SG alumina. The larger diamond particle size, M-diamond, improved the toughness of the composite more than the addition of the smaller particle size, S-diamond. The relative density of the material had little discernible effect on the overall toughness of the composite.

Results from the fracture toughness tests by Liu and Ownby [60 and 76] showed that the CNSR fracture toughness increased with a small addition (5 vol%) of B_4C and TiB_2 particulate reinforcement. The B_4C reached a maximum at 5 vol% and dropped slightly with increasing reinforcement content. Ref. [76] also investigated α-and β-Al_2O_3 in $TiB_2(p)/Al_2O_3$ materials. The α-Al_2O_3 composites were much tougher than the β-Al_2O_3 composites. Both reached a maximum toughness, α-Al_2O_3 at ~6.5 MPa $m^{1/2}$ with 5 vol% reinforcement content and β-Al_2O_3 at ~4.5 MPa $m^{1/2}$ with 15 vol% reinforcement content. Since α-Al_2O_3 is twice as tough as β-Al_2O_3, it was expected the α-Al_2O_3 composites would be tougher. However, this difference in toughness narrowed as reinforcement content increased. These references cited crack interactions with B_4C particulates for the improvement in toughness, much as in the case with diamond particulates.

ZrO_2(p)/Al_2O_3

Reference: 7

Composition
0-100 vol% ZrO_2

Processing Method
Hot-Pressed

Detailed Description
Sub-micron composite powders, including yttrium nitrate (a Y_2O_3 precursor), were mixed by ball-milling with methanol and Al_2O_3 balls in a plastic container. All powders were dried, those with yttrium nitrate were calcined at 400 °C for 4 hours. Most Y_2O_3 containing composites were hot-pressed at 1600 °C for 2 hours. Composites containing 80 and 100 vol% ZrO_2 were hot-pressed at 1400 °C in order to achieve a smaller grain size which allowed the retention of tetragonal ZrO_2. Non-yttria composites were hot-pressed at 1500 °C, again to achieve smaller grain size. The pure Al_2O_3 end-member was hot-pressed at 1400 °C to achieve a grain size comparable to the two-phase materials.

Reinforcement Description
0Y-, 2Y-, and 7.5Y-ZrO_2 particulates

Reinforcement Supplier
Zircas Corp.

Matrix Description
Al_2O_3 powder

Matrix Supplier
Lindy B, Union Carbide Corp.

Test Type
Vickers Indentation Microfracture (IM) Technique [I-23]

Test Description
The equation $K_{IC} = (H*d^{1/2})/(3(H/3E)^{0.4})*f(c/d)$;
where:
H = Vickers hardness,
d = half diagonal of indent,
c = half-crack length,
E = Young's modulus and
f(c/d) is an experimentally derived function of the relative crack length.

Reinforcement Content (volume percent)	Fracture Toughness, K_{IC} (MPa m$^{1/2}$)	Range (MPa m$^{1/2}$)	Comments
0.0	4.89	+0.18/-0.06	*0 mol% Y_2O_3*
7.5	5.88	+0.19/-0.27	
10	6.73	+0.50/-0.48	
13	6.21	+0.54/-0.56	
15	5.71	+0.48/-0.47	
20	4.90		
0.0	4.89	+0.19/-0.06	*2.0 mol% Y_2O_3*
6.0	5.97	+0.07/-0.09	
12	6.22	+0.55/-0.58	
18	6.58	+0.29/-0.30	
24	6.38	+0.58/-0.40	
30	7.43	+0.30/-0.35	
45	8.12	+0.32/-0.28	
60	7.45	+0.26/-0.44	
80	6.79	+0.38/-0.42	
100	6.62	+0.14/-0.44	
20	4.54	+0.18/-0.39	*7.5 mol% Y_2O_3*
40	3.75	+0.15/-0.27	
60	3.50	+0.14/-0.32	
80	3.14	+0.14/-0.19	
100	3.90	+0.16/-0.26	

Remarks/Comments
The composite with 2 mol% Y_2O_3 yielded a maximum at 40 vol%. This was believed to be due to the retention of the tetragonal phase. Beyond 40 vol% the tetragonal content decreased, causing a similar decrease in fracture toughness. However the addition of 7.5 mol% Y_2O_3 caused a significant reduction in the fracture toughness. A minimum occurred at 80 vol%, but at 100 vol% the toughness increased slightly.

Al_2O_3(p)/ZrO_2

Reference: 12

Composition
0-7.5 wt% Al_2O_3

Processing Method
Cold-Pressed/Sintered

Detailed Description
The YSZ was washed with distilled water to remove chlorine. Fifty gram batches were prepared by adding the appropriate amount of Al_2O_3 to the washed YSZ and ball milling for 12 hours. Milling was done in polypropylene jars using zirconia balls in a fish oil dispersing agent with distilled water and isopropanol. A binder mixture was added and milled for an additional 1.5 hours. Next the mixtures were spray dried then cold pressed into pellets, 1.6 cm in diameter and ~1.2 mm thick, and bars 3.1 x 0.4 x 0.4 cm at 220 MPa and 280 MPa, respectively. The specimens were fired in a $MoSi_2$ resistance furnace on ZrO_2 setters in air. Sintering was for 8 hours at 1300 °C.

Reinforcement Description
Al_2O_3 particulates

Reinforcement Supplier
Apache Chemicals Inc., Seward, IL

Reinforcement Dimensions
ave. particle size = 0.03 μm

Matrix Description
4.5Y-ZrO_2

Matrix Supplier
Zircar Corp., Florida, NY

Matrix Properties
ave. particle size = 0.03 μm

Test Type
Vickers and Knoop Indentation Technique [I-16]

Test Description
Loads of 2.0, 2.5 and 3.0 kg were used. The equation $K_{IC} = 0.016(E/H)^{1/2} * P/C_0^{3/2}$ where E/H is the elasticity to hardness ratio form the Knoop indentations, P is the indenter load, and C_0 is the crack size.

Test Equipment
Tukon microhardness tester

Reinforcement Content (weight percent)	Fracture Toughness, K_{IC} (MPa m$^{1/2}$)	Standard Deviation (MPa m$^{1/2}$)
0.0	4.70	±0.5
0.3	5.20	±0.6
0.7	5.30	±0.6
1.3	4.20	±0.5
3.3	9.00	±0.8
5.0	8.40	±0.8
7.5	6.60	±0.6

Remarks/Comments
All samples were > 99 % T.D. The fracture toughness of the alumina-zirconia showed a sharp maxima at 3.3 wt% Al_2O_3.

$ZrO_2(p)/Al_2O_3$

Reference: 15

Composition
0-100 vol% ZrO_2

Processing Method
Sintered

Detailed Description
The starting powders were mixed by ball milling using plastic containers and balls with acetone. After drying, the composite powders were uniaxially pressed at 300 MPa to form pellets 5 mm in diameter and 5 mm thick. These pellets were sintered at 1600 °C for 10 hours in an argon atmosphere.

Reinforcement Description
$ZrO_2.Y_2O_3$ particulates

Reinforcement Typical Properties
0, 3, and 6 mol% Y_2O_3 stabilizer

Matrix Description
Al_2O_3 powder

Matrix Designation
AKP-20

Test Type
Vickers Indentation Technique

Test Description
Load = 50 and 100 N

Reinforcement Content (volume percent)	Fracture Toughness, K_{IC} (MPa $m^{1/2}$)	ZrO_2 Phase
0 mol% Y_2O_3		
0.0	4.92	48.5 % t, 51.5 % m
5.0	5.54	37.7 % t, 65.3 % m
10	6.72	21.2 % t, 78.8 % m
15	6.00	8.7 % t, 91.3 % m
20	5.99	3.9 % t, 96.1 % m
30	4.94	fully monoclinic
40	4.62	fully monoclinic
60	4.37	fully monoclinic
80	4.06	fully monoclinic
3 mol% Y_2O_3		
20	6.58	fully tetragonal
40	8.30	
60	8.34	
80	8.65	
100	6.58	
6 mol% Y_2O_3		
5.0	4.61	fully cubic
20	5.27	
40	4.45	
60	4.76	
80	4.80	

Remarks/Comments
All samples were approximately 99 % dense. The addition of ZrO_2 had varying effects on the fracture toughness of the composite. Zero Y_2O_3 content caused an increase for low volume fractions, but a decrease for more than ~10 vol% was noted. The addition of ZrO_2 with 3 mol% Y_2O_3 caused an increase up to 100 vol%, where the toughness dropped, still above that of the monolith. The addition of the ZrO_2 with 6 mol% Y_2O_3 to the Al_2O_3 caused little change in the fracture toughness.

$ZrO_2(p)/Al_2O_3$

Reference: 22

Composition
0-100 wt% ZrO_2

Processing Method
Slip-Cast/Sintered

Detailed Description
Powders were milled for 3 days with 1 % dispersant (Celuna D-305) and formed by slip casting in a rectangular mould 4x10x80 mm. Cast samples were sintered at 1550 °C for 3 hours.

Reinforcement Description
(1) stabilized 3Y-PSZ or
(2) unstabilized monoclinic ZrO_2

Reinforcement Supplier
(1) TSK Co. 3Y-PSZ
(2) Daiichi Kigenso Chemicals Co. EP-ZrO_2

Supplier Designation
(1) TZ-3Y

Matrix Description
Al_2O_3

Matrix Supplier
ALCOA

Matrix Designation
A-16

Test Type
Indentation Microfracture (IM) Technique [I-14]

Specimen Preparation
Surface polished

Type of Specimen
Beam

Test Description
Indentation under load of 1 kg. Developed by Marshall and Evans.
$K_{IC} = 0.036*E0.4*p0.6*a-0.7*(c/a)-1.5$

Test Temperature
RT

Reinforcement Content (weight percent)	Fracture Toughness, K_{IC} (MPa $m^{1/2}$)	Comments
0.0	2.60	*unstabilized ZrO_2*
10	4.00	
20	6.50	
30	5.10	
40	5.80	
50	6.50	
60	6.30	
70	6.30	
80	5.20	
90	3.20	
0.0	3.18	*stabilized 3Y-PSZ*
20	4.18	
30	5.55	
40	5.68	
50	5.91	
60	6.73	
70	7.63	
80	7.55	
90	7.04	
100	6.66	

Remarks/Comments
Maximum toughening observed for 70 wt% ZrO_2 for the 3Y-PSZ and maximum toughening observed for 50 wt% ZrO_2 for unstabilized ZrO_2 system.

$ZrO_2(p)/Al_2O_3$

Reference: 26

Composition
15 wt% ZrO_2

Processing Method
CIPed/Pressureless Sintered

Detailed Description
A 0.333 wt% water solution of polyvinyl alcohol was filtered through a 10-20 μm glass frit. Some of the filtered solution, 400 mL, was poured into a polypropylene and 40 g of an appropriate mixture of alumina, zirconia, and zirconia-yttria powders were then added. The bottle's contents were adjusted to pH 2-3 with 0.3 ml concentrated nitric acid, and ultrasonically agitated at high power for 8 minutes. This agitation significantly warmed the slurry. The bottle containing the slurry was resealed, and the contents were allowed to gravitationally settle for 24 hours. The middle 300 ml of the supernatant was removed using a cannulation procedure at a rate of about 5 ml/sec. Filtered air was used as the cover gas during cannulation. The cannulated supernatant was homogenized before further use. It was then flocculated with 375 ml filtered reagent-grade acetone. The flocculated slurry was allowed to stand for 15 minutes, and was then poured onto a 100-mesh stainless steel screen. The retained flocs were scooped into a colloid press, where they were consolidated at a pressure of 35 MPa over a period of 15-30 minutes. The sample was held in the press for 30 minutes after the press came to equilibrium, then was ejected from the press, dried at 60-80 °C in laboratory air, and CIPed at 280 MPa. Binder burnout and bisque firing were accomplished by ramping the specimen at 2 °C/min from room temperature to 600 °C and at 5 °C/min from 600 to 1200 °C.

Reinforcement Description
ZrO_2 and Y-ZrO_2 particulates

Reinforcement Supplier
Toyo Soda, Atlanta, Georgia

Supplier Designation
ZrO_2: TZ-O
Y-ZrO_2: TZ-3Y

Matrix Description
Al_2O_3 powder

Matrix Supplier
Sumitomo Chemical America, New York

Matrix Designation
AKP-HP

Test Type
Vickers Microindentation Technique [II-23]

Specimen Preparation
surfaces polished

Test Description
load = 10 kg
indentation rate = 0.05 mm/min
hold time after loading = 1 min

Test Environment
Air

Sintering Temperature (°C)	Fracture Toughness, K_{IC} (MPa m$^{1/2}$)	Range (MPa m$^{1/2}$)
1,350	5.02	±0.09
1,400	4.98	±0.13
1,450	5.13	±0.20
1,500	4.69	±0.16
1,550	5.19	±0.25
1,600	4.92	±0.15

Anneal Time (hours)	Fracture Toughness, K_{IC} (MPa m$^{1/2}$)	Range (MPa m$^{1/2}$)	Relative t-ZrO_2 Content (percent)
0.0	4.98	±0.20	-
1.0	4.98	±0.09	35
3.0	5.05	±0.16	39
9.0	5.04	±0.08	57
27	5.55	±0.25	85

Remarks/Comments
Five indents made on each specimen. Five bars were tested for each experiment. Fracture toughness remained relatively constant for different sintering temperatures and increasing annealing times. Only after 27 hours of annealing time did fracture toughness increase.

$ZrO_2(p)/Al_2O_3$

Reference: 64

Composition
0-25 vol% ZrO_2

Processing Method
Hot-Pressed

Detailed Description
Alumina powder was mixed with 10, 15, 20, and 25 vol% ZrO_2 by ball milling for 24 hours in anhydrous alcohol. Mixtures were green compacted at 300 MPa, then hot-pressed at 1650 °C and 25 MPa.

Reinforcement Description
2Y-ZrO_2 particulates

Reinforcement Supplier
Toso Co. Ltd.

Reinforcement Typical Properties
54 % tetragonal ZrO_2 & 46 % monoclinic ZrO_2

Reinforcement Dimensions
particle size = 0.06 μm

Matrix Description
Al_2O_3 powder

Matrix Properties
particle size = 0.1 μm

Test Type
Indentation Microfracture (IM) Technique [I-23]

Specimen Dimensions
thickness = 5 mm
dia. = 20 mm

Type of Specimen
Discs

Test Description
load = 500 N

Test Temperature
RT

Reinforcement Content (volume percent)	Fracture Toughness, K_{IC} (MPa m$^{1/2}$)	Alumina Grain Size (μm)	Zirconia Grain Size (μm)	Relative t-ZrO_2 Content (percent)	Relative Density (percent)	Range (MPa m$^{1/2}$)
0.0	4.70	35.9	--	--	95.3	+0.45/-0.40
10	6.78	31.5	3.09	79.9	96.1	+0.42/-0.40
15	7.21	29.8	2.97	86.2	99.6	+0.32/-0.41
20	8.88	27.8	3.74	92.1	99.8	+0.47/-0.55
25	8.75	24.8	4.78	88.7	99.9	+0.54/-0.49

Remarks/Comments
Maximum fracture toughness was achieved at 20 vol% ZrO_2.

Figure 70a: Indentation Fracture Toughness [I-23] of 2Y-ZrO_2(p)/Al_2O_3

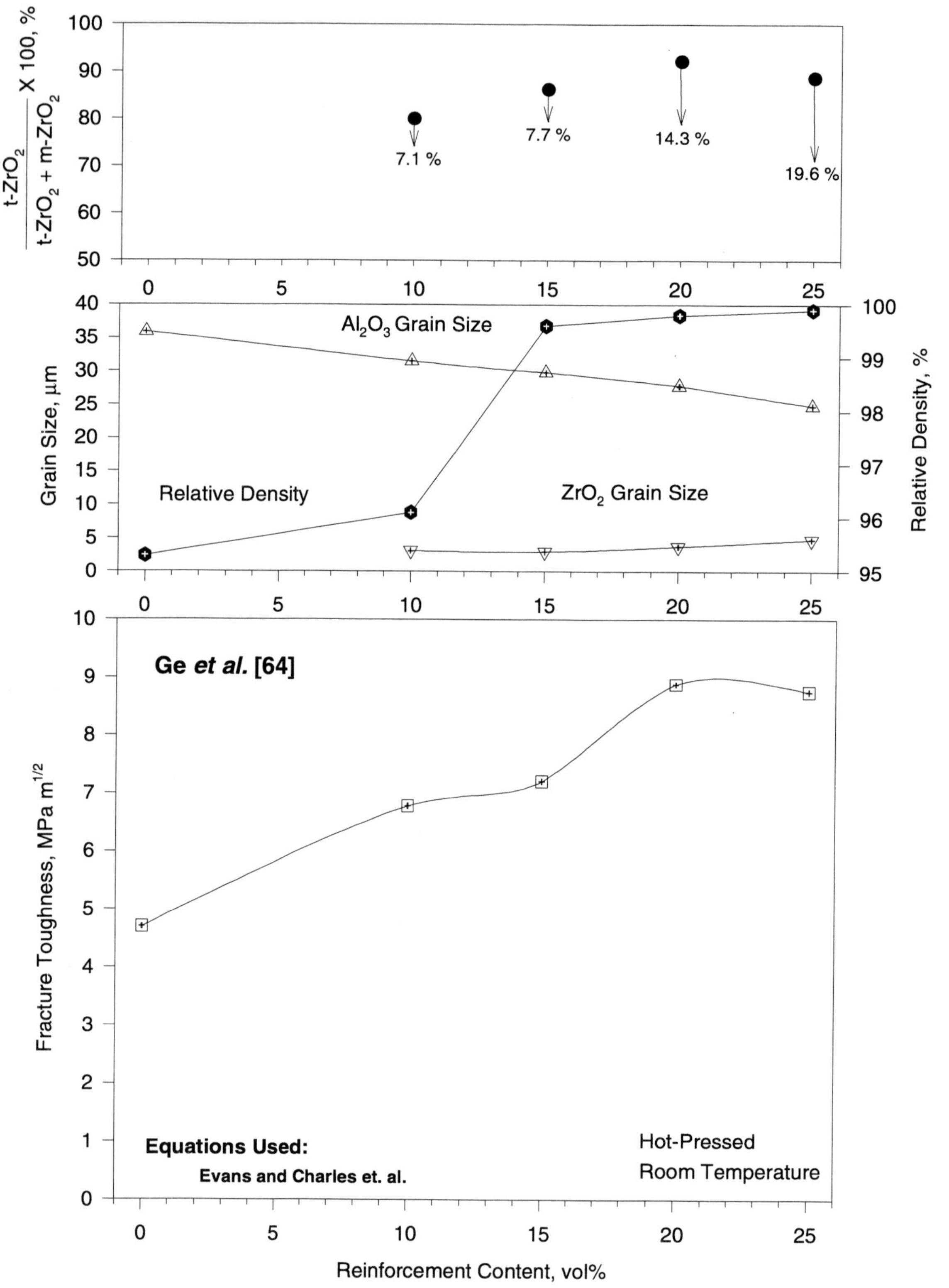

Discussion: Figure 70a shows the indentation fracture toughness and corresponding physical characteristics of the zirconia-alumina ceramics [64]. In particular, figure 70a examines the effect of grain size, relative density, and drop in t-ZrO_2 content during fracture. The relative density was low, around 96 %, with 0 and 10 vol% YSZ reinforcement. However, relative density jumped from 96 % to nearly full density when the YSZ content was increased to above 15 vol%. This shows that the YSZ was an effective sintering aid when added in quantities of 15 vol% or more. The YSZ particulates inhibited grain growth of the alumina matrix, which partially accounts for this improved densification. At the same time, the grain size of the YSZ particulates grew. This increased grain growth of the YSZ was the result of decreasing grain size of the alumina matrix and increased matrix grain boundaries which were mass transport paths for YSZ grain coalescence.

There were two toughening mechanisms active during fracture of the zirconia-alumina composites. The dominant toughening mechanism was t→m transformation toughening. Figure 70a also shows the percentage of the zirconia present in the tetragonal phase in the as-fabricated sample. As the reinforcement content increased, more YSZ was available to transform. Examination of the fracture surface showed that there was a drop in tetragonal content, which indicates that a t→m transformation, thus transformation toughening, occurred. As the YSZ content increased and the YSZ grains became larger, t→m transformation became more favorable, and the percent drop increased from 7.7 to 14.3 to 19.6 percent for the 15, 20 and 25 vol% YSZ composites, respectively. The result was that there was substantially greater volume of t-ZrO_2 transforming at higher reinforcement contents. The secondary toughening mechanism was matrix microcracking in the crack process zone caused by the t→m transformation toughening and corresponding volume expansion. The effectiveness of the two toughening mechanisms reached a maximum at 20 vol% YSZ. For the 25 vol% YSZ reinforced composite, the effect of transformation toughening continued to increase, however, it was likely that extensive microcracking caused the slight decrease in toughness.

Figure 70b: Fracture Toughness of $ZrO_2(p)/Al_2O_3$

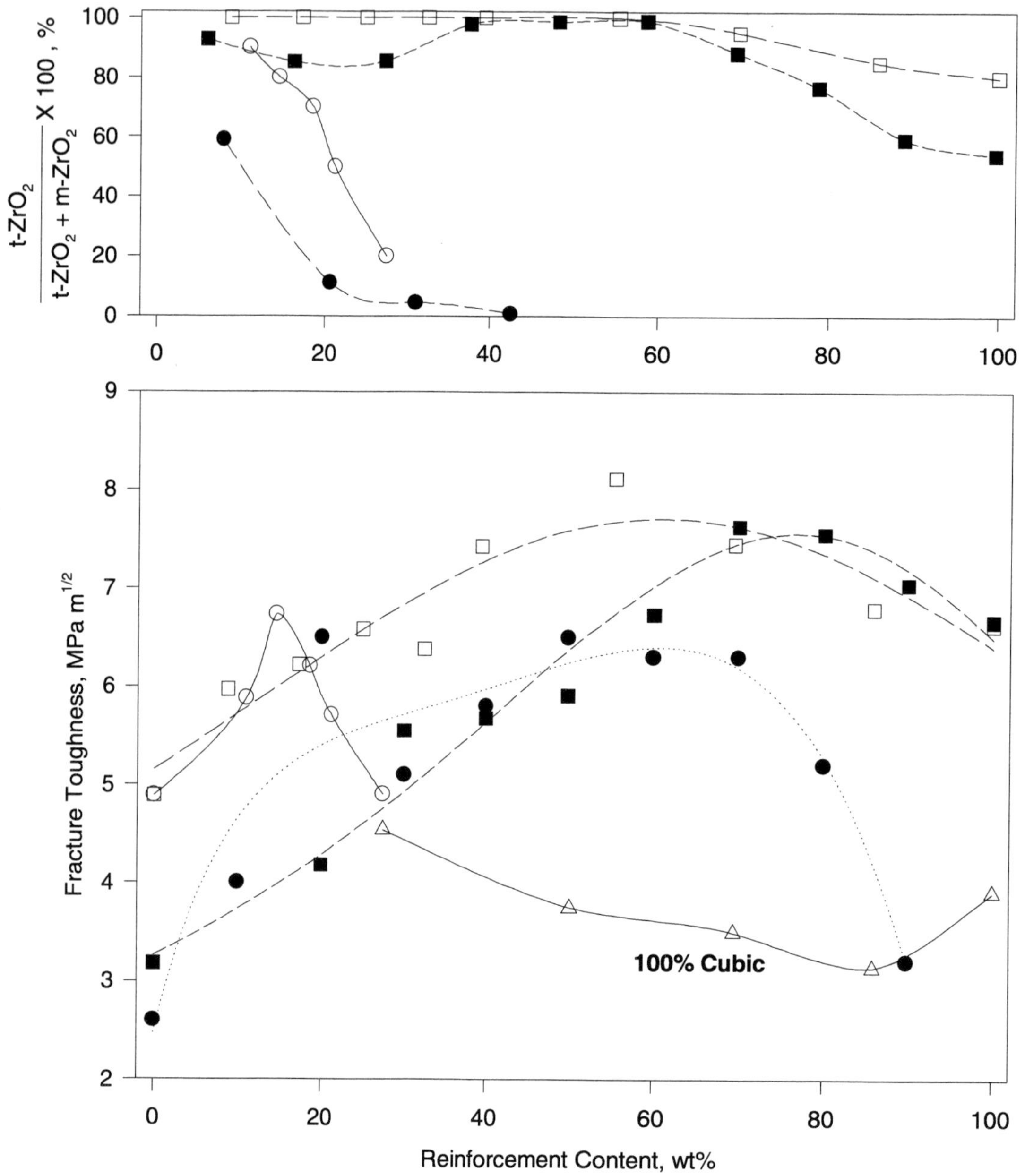

Lange [7] [I-23]
Hot-Pressed [1]

- ○ 0 mol% Y_2O_3
- □ 2 mol% Y_2O_3 TC-eqn. 8072
- △ 7.5 mol% Y_2O_3

1. Converted from vol% to wt%

Ono *et al.* [22] [I-14]
Slip Cast/Sintered

- ● 0 mol% Y_2O_3 TC-eqn. 4341
- ■ 3 mol% Y_2O_3 TC-eqn. 7002

Discussion: Figure 70b presents indentation fracture toughness data of zirconia-alumina composites reinforced with unstabilized (0 mol% Y_2O_3), stabilized (2 or 3 mol% Y_2O_3), or cubic zirconia (7.5 mol% Y_2O_3) [7, 22]. Equations from Evans *et al.* [I-23, I-14] were used to calculate indention fracture toughness. The type of zirconia powder used and the volume fraction of the zirconia controlled the phase content of the zirconia reinforcement. The zirconia phase content, in turn, controlled the toughening mechanisms active in the composite. The composites fabricated from stabilized zirconia contained significant volume fractions of t-ZrO_2 at all volume fractions of ZrO_2. The tetragonal phase was stable between 40 and 60 vol% zirconia, in this range the zirconia phase was 100 % tetragonal. Two toughening mechanisms were present in these composites. The dominant mechanism was transformation toughening and the secondary mechanism was microcrack toughening. Because the t-ZrO_2 content remained high, the addition of zirconia reinforcement increased the transformation toughening effect and fracture toughness increased. The fracture toughness of the composite reached a maximum in the range of 60 to 80 vol% zirconia. Increasing zirconia content decreased the t-phase because there was little alumina matrix left to constrain the zirconia and prevent an undesirable t→m phase change during processing. The fracture toughness decreased slightly with this decrease in t-ZrO_2 ratio.

The addition of a small amount of unstabilized zirconia produced a composite with some t-ZrO_2 phase. The reason for the presence of t-ZrO_2 was that the high modulus alumina matrix constrained the zirconia in the tetragonal phase, and no yttria stabilization was necessary. However, the t-ZrO_2 disappeared quickly as the zirconia content increased. This occurred mainly because the modulus of the material surrounding the zirconia grains decreased and could no longer constrain the zirconia in the tetragonal phase. The fracture toughness of the unstabilized zirconia composite tested by Lange [7] increased and reached a maximum at 10 vol%, and declined with further addition of ZrO_2 reinforcement as the tetragonal phase disappeared. The unstabilized zirconia-alumina composite tested by Ono *et al.* [22] indicated another toughening mechanism besides transformation toughening. The t-ZrO_2 content drops to near 0 percent above 30 vol% zirconia reinforcement; nevertheless, the fracture toughness continued to increase. The high fracture toughness between 30 and 70 wt% zirconia was attributed to microcracking [22]. Above 70 wt% zirconia, the fracture toughness declined rapidly.

Lange [7] also tested a zirconia-alumina composite reinforced with cubic zirconia. The fracture toughness of these composites was not as high as the composites fabricated with either the stabilized or unstabilized zirconia. No analysis of the fracture mechanisms was performed on the cubic zirconia-alumina composites.

Figure 70c: Fracture Toughness of ZrO_2(p)/Al_2O_3

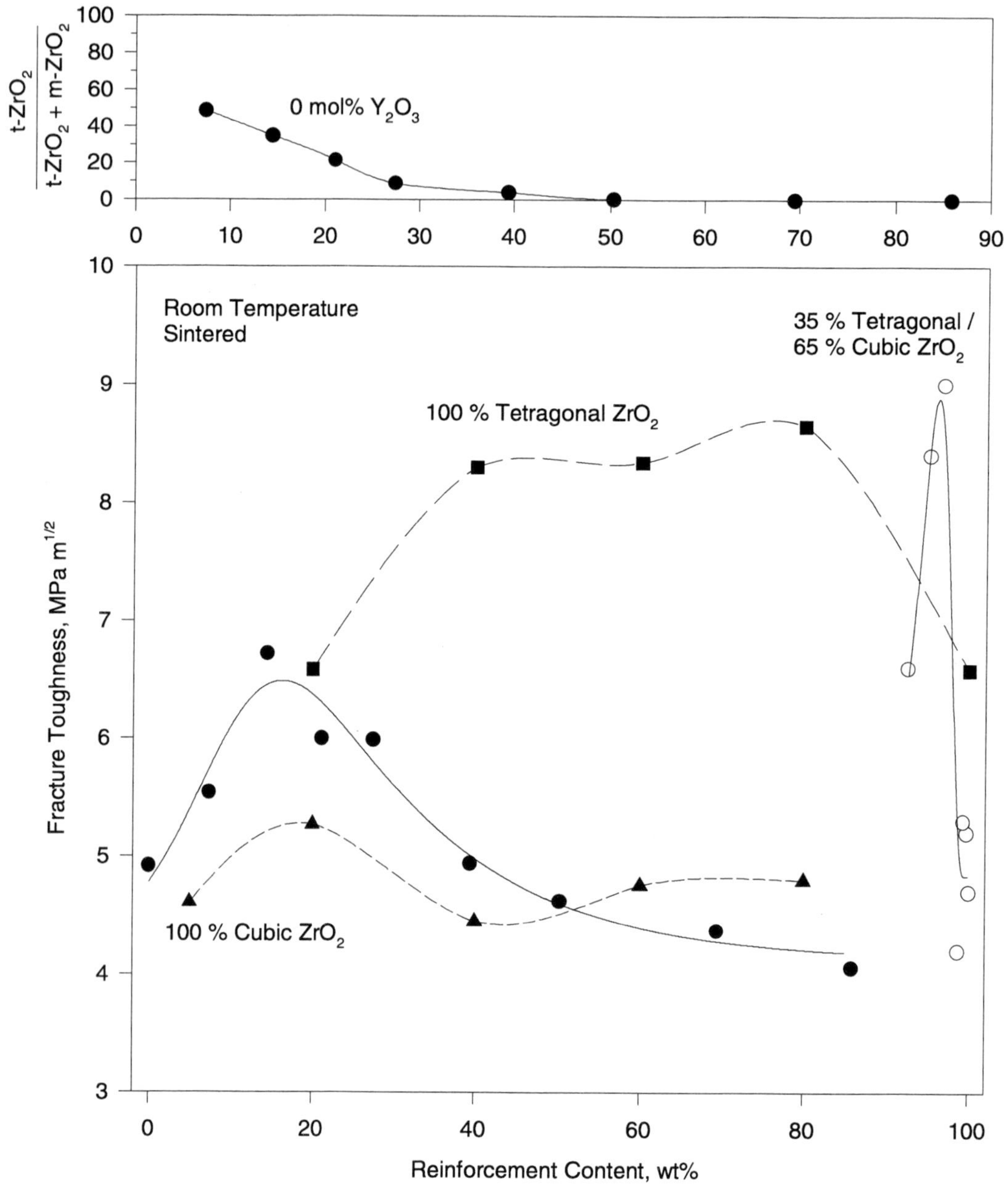

Equations Used:

Niihara *et al.* [I-12]	**Anstis *et al.*** [I-16]
Sato *et al.* [15] [1]	○ Buchanan and Davison [12]
● 0 mol% Y_2O_3 TC-eqn. 8063	4.5 mol% Y_2O_3
■ 3 mol% Y_2O_3	
▲ 6 mol% Y_2O_3	

1. Converted from vol% to wt%

Discussion: Figure 70c presents indentation fracture toughness data of zirconia-alumina composites calculated by equations proposed by Anstis *et al.* [I-16] or Niihara *et al.* [I-12]. The behavior of these composites was similar to the composites presented in figure 70b. Sato *et al.* [15] tested alumina matrix composites reinforced by unstabilized, stabilized, and cubic zirconia particulates. Phase analysis of the pure zirconia reinforced alumina showed that the zirconia was partially stabilized at low reinforcement contents, but the t-ZrO_2 content decreased rapidly with increasing reinforcement content to nearly 0 % at 40 vol% ZrO_2. The fracture toughness increased by 2 MPa $m^{1/2}$ with the addition of 10 vol% zirconia reinforcement. It was believed that transformation toughening alone could not explain this increase, since only 35 - 50 % of the zirconia was in the tetragonal phase. This increase in toughness was partially attributed to the formation of microcracks in the crack-tip region by the expansion of ZrO_2 during the tetragonal to monoclinic phase transformation [15]. The fully stabilized (3 mol% Y_2O_3) reinforced composite had a toughness of greater than 8 MPa $m^{1/2}$ with 40 to 80 vol% 3Y-ZrO_2 reinforcement. The high toughness of this composite was primarily attributed to transformation toughening. The fracture toughness dropped below 40 vol% reinforcement and above 80 vol% reinforcement. The cubic zirconia reinforced alumina did not show a strong dependence on reinforcement content [15].

The alumina reinforced zirconia tested by Buchanan and Davison [12] showed a very strong dependence on alumina content. The fracture toughness of the zirconia matrix increased from around 5 MPa $m^{1/2}$ to 9 MPa $m^{1/2}$ with the addition of only 3.3 wt% Al_2O_3, and fell to 6.6 MPa $m^{1/2}$ at 7.5 wt%. The microstructure of these alumina-zirconia composites consisted of well dispersed Al_2O_3 inclusions in a fine grain size (0.2 - 0.4 μm) ZrO_2 matrix. The matrix consisted of ~ 35 % tetragonal phase and the rest comprised of cubic zirconia. The zirconia existed as a continuum between the two phases rather than a mixture of discrete cubic and tetragonal grains. No monoclinic phase was found in the matrix either before or after fracture, which eliminated transformation toughening as a potential toughening mechanism [12]. In contrast, the alumina and zirconia showed no solubility and existed as discrete grains. The primary toughening mechanism appeared to be crack deflection. Between 3 and 5 wt% alumina, the dispersed phase was optimally situated and presented a barrier to crack propagation by crack deflection. Also, some deformation and crack arrest mechanisms associated with this deformation process appeared within the alumina grains [12].

Figure 71: Effect of Processing Conditions on Indentation Fracture Toughness[I-23] of ZrO_2(p)/Al_2O_3

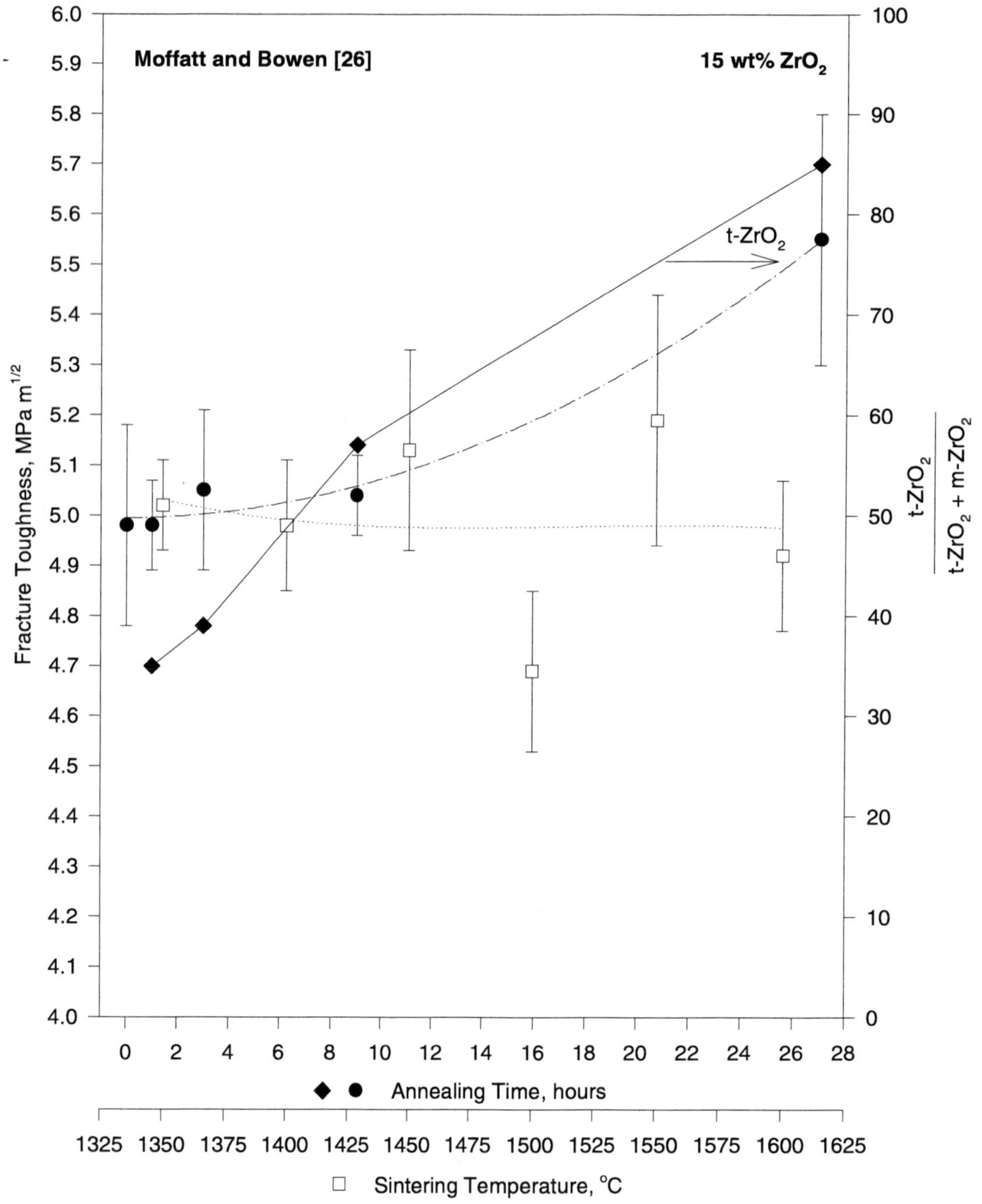

Discussion: Figure 71 shows the affect of sintering temperature on a zirconia-alumina composite sintered for 2 hours, and the affect of a post-sintering annealing step in an attempt to increase the tetragonal ZrO_2 content. Results showed that the fracture toughness was not highly dependent on sintering temperature. However, a post-sintering annealing step was found to increase the t-ZrO_2 content and thus increase fracture toughness. The tetragonal phase increased linearly with annealing time and the increase in fracture toughness corresponded with the increase in t-phase [26].

$SiC(p).ZrO_2(p)/Al_2O_3$

Reference: 15

Composition
10 vol% ZrO_2
0-10 vol% SiC

Processing Method
Sintered (some HIPed)

Detailed Description
The starting powders were mixed by ball milling using plastic containers and balls with acetone. After drying, the composite powders were uniaxially pressed at 300 MPa to form pellets 5 mm in diameter and 5 mm thick. These pellets were sintered at 1600 °C for 10 hours in an argon atmosphere. Some specimens were HIPed at 1500 °C and 200 MPa for 30 min.

Reinforcement Description
0Y-ZrO_2 or 3Y-ZrO_2, and SiC particulates

Matrix Description
Al_2O_3 powder

Matrix Designation
AKP-20

Test Type
Vickers Indentation Technique[I-12]

Test Description
Load = 50 and 100 N

Test Temperature
RT

SiC Reinforcement Content (volume percent)	Fracture Toughness, K_{IC} (MPa $m^{1/2}$)	Relative Density (percent)	Stabilizer Content (atomic percent)	Comments
0.0	6.60	98.8	0.0	*as-sintered*
1.0	6.00	98.7	0.0	
3.0	5.80	96.0	0.0	
5.0	5.60	95.4	0.0	
10	5.20	93.4	0.0	
0.0	8.10	98.3	3.0	*as-sintered*
1.0	5.20	96.8	3.0	
3.0	5.50	96.0	3.0	
5.0	5.10	96.6	3.0	
10	4.80	95.6	3.0	
1.0	6.90	97.6	3.0	*HIPed*
5.0	7.10	96.7	3.0	
10	6.60	96.7	3.0	

Remarks/Comments
The addition of the nonoxides to the composite reduced the fracture toughness slightly. The addition of SiC particulates did not improve the strength of the zirconia-alumina composites.

$TiC(p).ZrO_2(p)/Al_2O_3$

Reference: 15

Composition
10 vol% ZrO_2
1-10 vol% TiC

Processing Method
Sintered

Detailed Description
The starting powders were mixed by ball milling using plastic containers and balls with acetone. After drying, the composite powders were uniaxially pressed at 300 MPa to form pellets 5 mm in diameter and 5 mm thick. These pellets were sintered at 1600 °C for 10 hours in an argon atmosphere.

Reinforcement Description
0Y-ZrO_2 or 3Y-ZrO_2 and TiC particulates

Matrix Description
Al_2O_3 powder

Matrix Designation
AKP-20

Test Type
Vickers Indentation Technique[I-12]

Test Description
Load = 50 and 100 N

Test Temperature
RT

TiC Reinforcement Content (volume percent)	Fracture Toughness, K_{IC} (MPa $m^{1/2}$)	Relative Density (percent)	Stabilizer Content (mol percent)
1.0	6.00	96.8	0.0
3.0	5.40	95.6	0.0
5.0	5.90	94.7	0.0
10	6.10	92.5	0.0
1.0	4.90	97.2	3.0
3.0	4.80	96.3	3.0
5.0	5.80	96.9	3.0
10	6.30	95.9	3.0

Remarks/Comments
The addition of the nonoxides to the composite reduced the fracture toughness slightly. Fracture toughness did not increase with TiC content without yttria stabilizers, but increased by more than 1 MPa $m^{1/2}$ with the addition of 10 vol% TiC when 3 mol% yttria was added.

$SiC(p).ZrO_2(p)$/Mullite

Reference: 49

Composition
20 and 40 vol% SiC
5-20 vol% ZrO_2

Processing Method
CIPed/Pressureless Sintered

Detailed Description
The powders were mixed and ground with mullite balls and ion-exchanged water in Al_2O_3 pot mill for 3 hours. The powder mixtures were pressed under 31 MPa and formed with the cold isostatic press under 196 MPa. The pressureless sintering of the compacts were done in an elastic furnace under an air atmosphere at 1700 °C for 2 hours. The test pieces were annealed in air at 1300 °C for 2 hours.

Reinforcement Description
SiC powder
ZrO_2 powder

Reinforcement Supplier
Ibiden (SiC)
Dow Chemical (ZrO_2)

Reinforcement Typical Properties
purity of SiC: 98.0 %
impurity composition: SiO_2 0.3 %, Fe 0.04 %, C 0.4 %, Al 0.03 %
purity of ZrO_2: 99.9 %
impurity composition: SiO_2 0.01 %, Al_2O_3 0.026 %, Na_2O 0.0012 %

Reinforcement Dimensions
ave. particle size of all powders = 0.2 µm

Matrix Description
Mullite powder

Matrix Supplier
Nihon Cement

Matrix Properties
composition: 72 wt% Al_2O_3
purity: 99.9 %
impurity composition: Fe_2O_3 0.01 %, TiO_2 0.006 %, K_2O 0.006 %, Na_2O 0.01 %

Test Type
Vickers Indentation Microfracture (IM) or Vickers Indentation Strength in Bending (ISB)

Specimen Preparation
Specimens were sliced and ground by #600 diamond wheel polished at finish with 0.04 µm γ-Al_2O_3.

Specimen Dimensions
3 x 4 x 40 mm

Type of Specimen
Bar

Test Description
crosshead speed = 0.5 mm/min
IM: load = 49 N for 15 sec
ISB: load = 98 N for 15 sec at center of test piece

Test Temperature
RT

ZrO_2 Reinforcement Content (volume percent)	Fracture Toughness, K_{IC} (MPa m$^{1/2}$)	Comments
4.9	2.82	*IM, 20 vol% SiC*
10	2.90	
15	4.38	
20	5.16	
4.9	3.56	*IM, 40 vol% SiC*
10	3.79	
15	4.66	
20	4.86	
9.7	2.50	*ISB, 20 vol% SiC*
15	3.25	
20	3.80	
10	2.43	*ISB, 40 vol% SiC*
15	3.23	
20	3.84	

Remarks/Comments
The fracture toughness values measured by IM were higher than the ones measured by ISB. Overall, the fracture toughness measured by both methods was increased with addition of ZrO_2.

$SiC(p).ZrO_2(p)$/Mullite

Reference: 86

Composition
0-35 vol% SiC
20 vol% 2.8Y-ZrO_2

Processing Method
Hot-Pressed

Detailed Description
The mullite, SiC, and Y-TZP powders were mixed with the desired volume ratio by wet milling with distilled water using Al_2O_3 balls in a polyethylene container for 6 hours, and dried. The obtained mixed powders were sieved to under 120 mesh, and were molded in a BN-coated graphite die for sintering. Hot-pressing was carried out at 1600 °C for 40 minutes at nitrogen atmosphere under 25 MPa for all composites.

Reinforcement Description
α-SiC and 2.8Y-ZrO_2 particulates

Reinforcement Typical Properties
SiC composition: total C = 27.0 wt%; free C = 0.41 wt%; total Si = 68.5 wt%; free Si = 1.08 wt%; O = 2.15 wt%; Al_2O_3 = 0.13 wt%; Fe_2O_3 = 0.18 wt%; CaO = 0.02 wt%

Reinforcement Dimensions
SiC ave. dia. = 0.93 μm
SiC specific surface = 7.53 $m^2 g^{-1}$

Matrix Description
Mullite powder

Matrix Properties
ave. grain size = 2 μm
composition: Al_2O_3 = 75.92 wt%
SiO_2 = 23.34 wt%
other oxides = 0.74 wt%

Test Type
3-pt Single Edge Notched Beam (SENB) Technique

Specimen Preparation
Specimens were cut and ground by a diamond wheel.

Specimen Dimensions
5 x 2.5 x 28 mm
notch width = 0.22-0.25 mm
notch depth = 2.5 mm
span = 20 mm

Type of Specimen
Notched Beam

Test Description
crosshead speed = 0.005 mm/min

Test Temperature
RT

SiC Reinforcement Content (volume percent)	Fracture Toughness, K_{IC} (MPa $m^{1/2}$)	Standard Deviation (MPa $m^{1/2}$)
0.1	4.86	±0.3
9.8	4.99	±0.3
20	5.59	±0.3
25	6.07	±0.2
30	6.46	±0.2
35	6.58	±0.2

Remarks/Comments
Fracture toughness was increased with additions of SiC particles. The maximum value of fracture toughness was 6.7 MPa $m^{1/2}$ at 35 vol% SiC particles. The increasing of toughness with SiC particle content was smaller than that of the SiC(p)/mullite composite at lower amount of SiC particles. At high amount of SiC particles, the SiC(p).Y-TZP/mullite composite was greater than the SiC(p)/mullite composite. The effect combined by Y-TZP and SiC particles on toughness appears to be additive or synergetic. The toughness of the mullite matrix composites reinforced by SiC particles and Y-TZP was dramatically improved.

Figure 72a: Fracture Toughness of SiC(p).ZrO_2(p)/Mullite

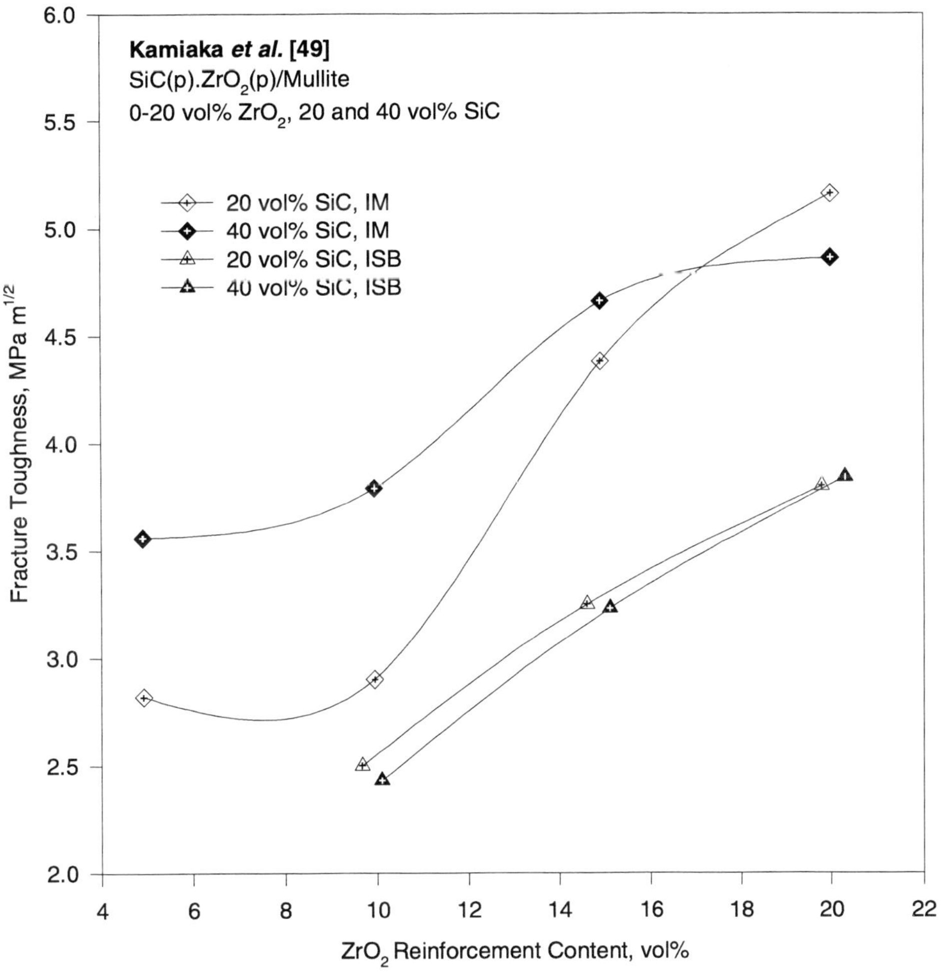

Discussion: Kamiaka *et al.* [49] tested SiC(p)/mullite composites with varying amounts of unstabilized zirconia. It was found that the zirconia and silicon carbide particulates had opposite affects on mullite grain growth kinetics. The addition of SiC particulates pinned mullite grain boundaries and inhibited grain growth and densification, whereas the addition of zirconia particulates enhanced grain growth and acted as an effective sintering additive, especially above 5 vol% ZrO_2.

The size of the composite grains dominated the materials fracture behavior. The composites with 20 vol% SiC reinforcement had relatively large grain structures. Fracture appeared to be transgranular, and the crack routinely branched within grains at small SiC and ZrO_2 inclusions. Some zirconia particulates may have been sufficiently large for phase transformation, providing an additional toughening mechanism. The composites with 40 vol% SiC reinforcement had a much smaller grain size, and fracture was mainly inter-granular. Crack branching along the grain boundaries of the fine mullite grains was the primary toughening mechanism in the high SiC content composites. No transformation toughening was possible in the 40 vol% SiC composite. The toughness of the 20 and 40 vol% SiC composites at a particular zirconia content was similar. The addition of zirconia, which increased density and grain size, increased the toughness of all composites.

Figure 72b: Fracture Toughness of $SiC(p).ZrO_2(p)/Al_2O_3$, $TiC(p).ZrO_2(p)/Al_2O_3$, and $SiC(p).ZrO_2(p)$/Mullite

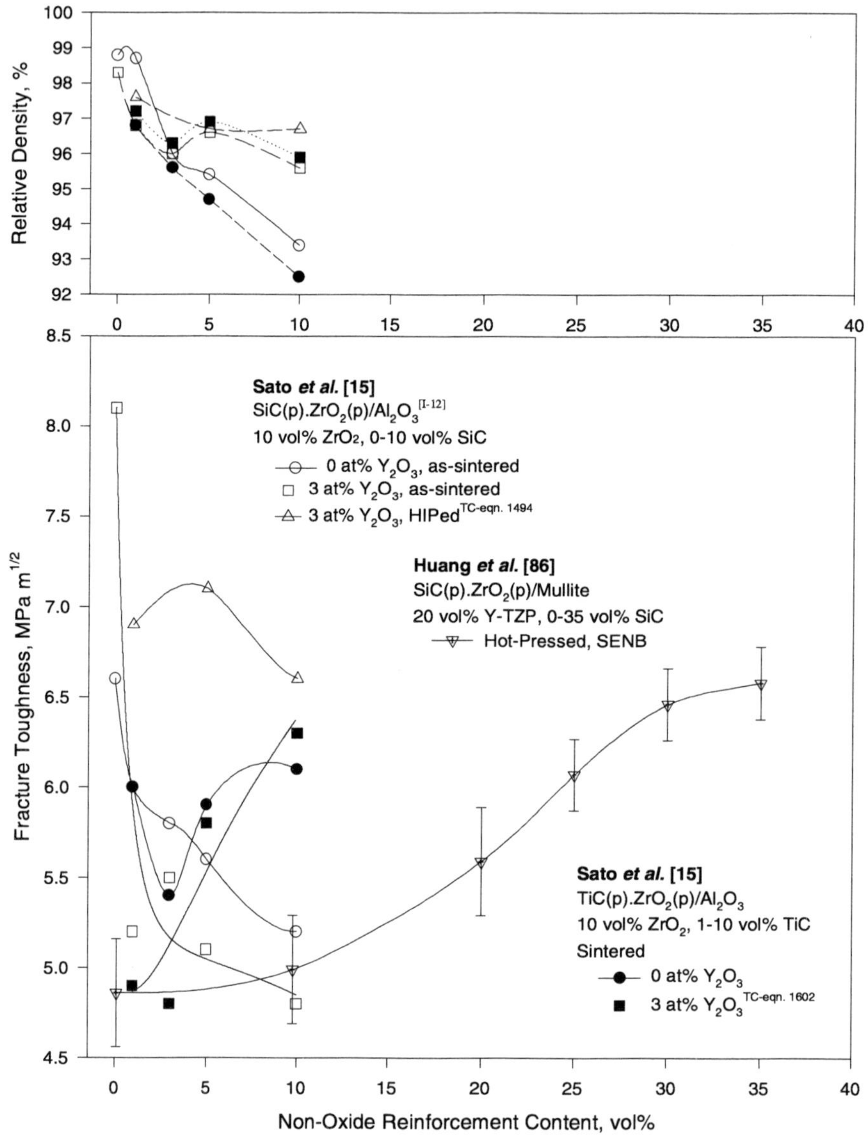

Discussion (cont.): Huang *et al.* [86] showed that the toughening mechanisms provided by both the silicon carbide and zirconia reinforcement were additive, and that the toughness reached a maximum of 6.7 MPa $m^{1/2}$ at 35 vol% SiC and 20 vol% Y-ZrO_2. This composite was successfully tested as an adiabatic engine pistol cap for over 100 hours and remained undamaged.

Sato *et al.* [15] tested both SiC and TiC particulate reinforced zirconia-alumina, and compared different amounts of stabilizer content. They found that the SiC-ZrO_2 with 3 wt% Y_2O_3 in the as-sintered condition were the toughest. Adding Y_2O_3 to the samples with TiC and ZrO_2 increased fracture toughness after they added 10 vol% reinforcement content. Inflections in the fracture toughness tended to be due to the influence of the relative density of the composites.

$ZrO_2.HfO_2(p)/Al_2O_3.Cr_2O_3$

Reference: 13

Composition
10 and 30 mol% HfO_2
0-15 vol% ZrO_2

Processing Method
CIPed/Sintered

Detailed Description
Particulate and matrix powders were attrition milled separately in proponal. These powders were then mixed by ball-milling with water. The mixture was dried, CIPed at 28 MPa, and sintered at 1550 °C for 2 h in argon

Reinforcement Description
$ZrO_2.HfO_2$ particulates (HfO_2--10 and 30 mol%)

Matrix Description
$Al_2O_3.Cr_2O_3$ (Cr_2O_3-5 and 20 mol%)

Test Type
4-pt Indentation Fracture Technique (IF) [I-15]

Test Temperature
RT

$ZrO_2 \cdot HfO_2$ Reinforcement Content (volume percent)	Fracture Toughness, K_c (MPa $m^{1/2}$)	Comments
0.0	3.59	*20 mol% Cr_2O_3, 10 mol% HfO_2*
5.0	4.55	
10	5.24	
15	4.58	
0.0	3.37	*20 mol% Cr_2O_3, 30 mol% HfO_2*
5.0	5.05	
10	4.52	
15	3.88	
0.0	4.10	*5 mol% Cr_2O_3, 10 mol% HfO_2*
7.5	5.12	
12.5	5.55	
15	5.65	
0.1	4.21	*5 mol% Cr_2O_3, 30 mol% HfO_2*
5.1	5.41	
10	5.51	
12.5	5.30	
15	5.20	

Remarks/Comments
Fracture toughness was greatest for the composite with 5 mol% Cr_2O_3 and 10 mol% HfO_2.

$ZrO_2.HfO_2$(p)/$Al_2O_3.Cr_2O_3$

Reference: 13

Composition
5-15 vol% ZrO_2

Processing Method
CIPed/Sintered

Detailed Description
Particulate and matrix powders were attrition milled separately in proponal. These powders were then mixed by ball-milling with water. The mixture was dried, CIPed at 28 MPa, and sintered at 1550 °C for 2 h in argon.

Reinforcement Description
$ZrO_2.HfO_2$ particulates (10 mol% HfO_2)

Matrix Description
$Al_2O_3.Cr_2O_3$ (5 mol% Cr_2O_3)

Test Type
4-pt Indentation Fracture Technique (IF)[I-15]

Test Temperature
RT - 1000 °C

Test Environment
Air

Temperature (°C)	Fracture Toughness, K_c (MPa $m^{1/2}$)	$ZrO_2 \cdot HfO_2$ Reinforcement Content (volume percent)
20	4.10	5.0
300	5.80	5.0
800	6.70	5.0
1,000	5.50	5.0
20	4.40	7.5
300	6.40	7.5
500	7.10	7.5
800	6.30	7.5
1,000	4.60	7.5
20	4.60	12.5
300	6.90	12.5
500	6.90	12.5
800	6.20	12.5
1,000	4.70	12.5
20	5.50	15
300	7.70	15
500	6.60	15
800	5.70	15

Remarks/Comments
Fracture toughness was greatest for the composites with 15 vol% ZrO_2.

Figure 73: Fracture Toughness[I-15] of $ZrO_2.HfO_2$(p)/$Al_2O_3.Cr_2O_3$ Ceramics

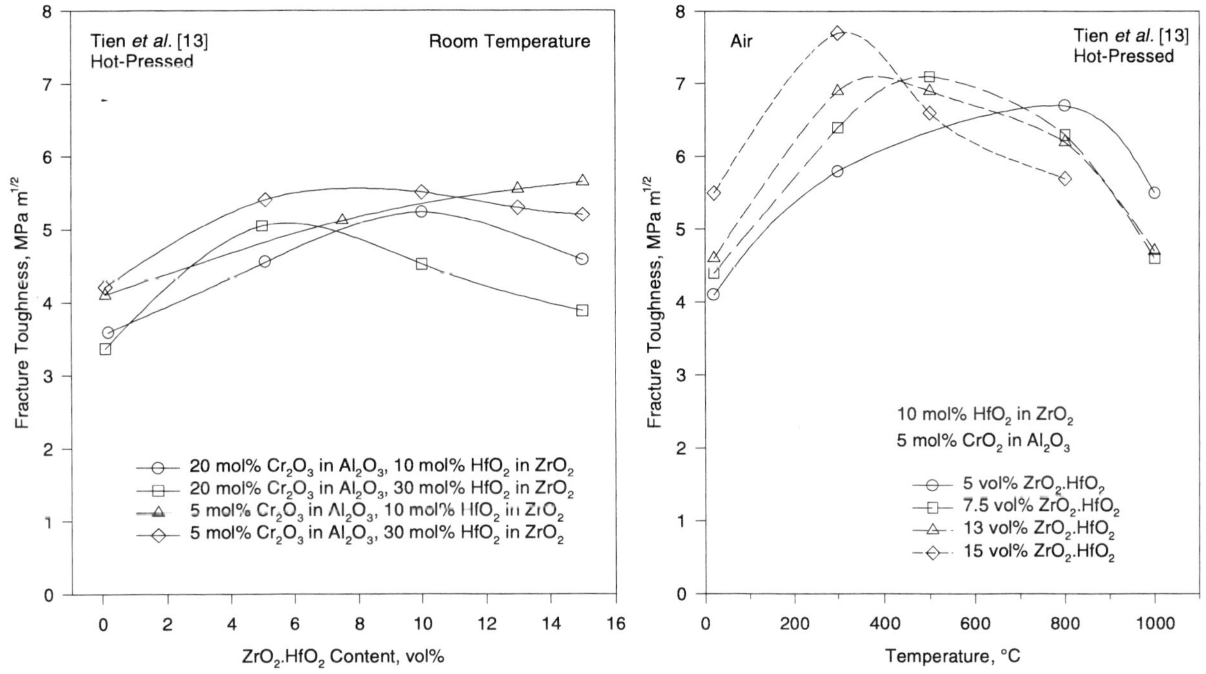

Discussion: Tien *et al.* [13] considered a zirconia-alumina composite for application in an advanced heat engine. In order to meet the requirement of a heat engine application, chromia was added to alumina to decrease thermal conductivity, and hafnia was added to the zirconia to increase t-m transformation temperature. Increasing the transformation temperature increased the maximum usable temperature for the composite. The relationship between fracture toughness and $ZrO_2.HfO_2$ content was complex. The 30 mol% HfO_2 composites had a lower t-ZrO_2 content than the 10 mol% HfO_2; and the higher additions of Cr_2O_3 decreased t-ZrO_2 content as well. However, all the composites had varying amounts of microcracking, and it was difficult to discern whether the microcracking had a positive or negative impact on toughness. The effect of temperature on composite toughness was more evident. The composites showed an increase in toughness in the range of 200 to 800 °C. No explanation for this increase in toughness was offered by ref. [13].

SiC(p)/ZrO_2

Reference: 85

Composition
0 - 30 wt% SiC

Composite Architecture
particulate reinforcement

Processing Method
Sol-Gel/Hot-Pressed

Detailed Description
ZrO_2Cl_2-$8H_2O$, $Y(NO_3)_3$-$6H_2O$, and α–SiC were used as starting materials. Y-TZP and SiC(p)/Y-TZP were prepared by the coprecipitation method from the starting solution. The mixture was hot-pressed in a boron-coated graphite mold at 1600 °C and 25 MPa for 1 hour.

Reinforcement Description
α–SiC particulates

Matrix Description
2.8Y-TZP

Test Type
3-pt. Single Edge Notched Beam (SENB)

Specimen Dimensions
5 x 2.5 x 26 mm
span = 20 mm

Test Description
crosshead speed = 0.05 mm/min.

Test Temperature
RT

Test Equipment
Instron 1195 testing machine

Reinforcement Content (weight percent)	Fracture Toughness, K_{IC} (MPa $m^{1/2}$)	Relative Density (percent)
0	14.6	100
5	10.7	99.2
10	10.8	98.8
15	10.8	98.4
20	8.7	97.2
30	3.5	88.4

Remarks/Comments
Fracture toughness decreased with increasing reinforcement content, but was fairly constant from 5 wt% to 15 wt% SiC. The large decrease in toughness above 20 wt% SiC was due mainly to poor sinterability of the composite.

Figure 74: SENB Fracture Toughness of Hot-Pressed SiC(p)/ZrO_2

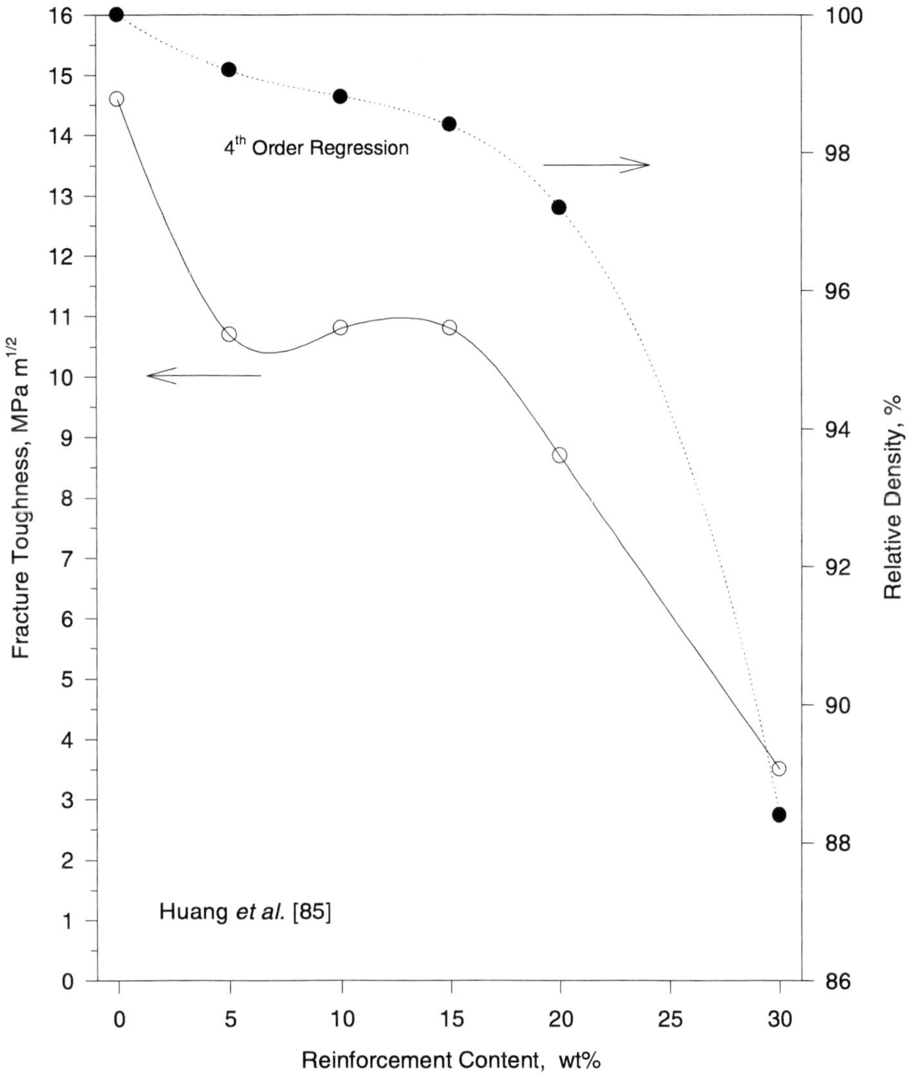

Discussion: Huang *et al.* [85] found that fracture toughness of YSZ decreased with the addition of SiC particulates to a YSZ matrix. Toughness decreased with the addition of a small quantity of SiC reinforcement (below 6 wt%) because the SiC reinforcement decreased the amount of transformable t-zirconia in the matrix. With further addition of SiC particulates, from 6 to 15 wt%, the t-zirconia content continued to decrease. Strengthening and toughening mechanisms created by the SiC reinforcement, on the other hand, compensated for the decrease in tetragonal phase, and resulted in no change in toughness. Further addition of SiC reinforcement dramatically decreased relative density, resulting in a sharp decrease in toughness.

ZrO_2(p)/ZrC

Reference: 18

Composition
0-35 wt% ZrO_2

Processing Method
Slip Cast/Isostatic Pressed/Sintered

Detailed Description
Type A zirconia was a pre-reacted stabilized material. Particles below 1 μm were extracted from the powders by sedimentation. ZrC was dispersed in the aqueous solution. Test blocks of each material were cast and allowed to go to completion, the side molds used were removed and the specimens were air dried for 12 hours. The specimens were then dried at 50 °C for 48 hours and cut to size by hand sanding. A graphite tube furnace with an Ar atmosphere was used for sintering. Samples were set in a graphite crucible and packed in loose powder of the same ZrC-ZrO_2 composition. The samples were heated at a rate of 30 °C/min. to the desired temperature and held there for one hour. Type B zirconia was a physical mixture of ZrO_2 and Y_2O_3. ZrC and Type B ZrO_2 mixtures were ball-milled in polyethylene mills using ZrO_2 balls for up to 16 hours. Specimens from this were slip cast, isostatically pressed, and sintered the same as the Type A sample.

Reinforcement Description
Type A: 3Y-ZrO_2 (5.24 wt% Y_2O_3)
Type B: Y-ZrO_2 (4.76 wt% Y_2O_3)

Reinforcement Supplier
Toyo Soda Manufacturing Co.,Tokyo, Japan (Type A)
Wah Chang Corp., Albany, Oregon, USA (Type B)

Matrix Description
ZrC powder

Matrix Supplier
Furrchi Chemical Co., Japan

Test Type
3-pt. Single Edge Notched Beam (SENB) [I-20]

Specimen Preparation
Notched with a 0.15 mm thick diamond saw with a a/w ratio of 0.3

Specimen Dimensions
3.0 x 3.5 mm cross-section
span = 12 mm

Type of Specimen
Notched Bar

Test Description
Crosshead speed = 0.05 mm/min.
The equation
$K_{IC} = 3YPsa^{1/2}bw^2$
was used to calculate the fracture toughness of the samples where P is the breaking load, a the notch depth, b the specimen width, w the specimen depth, s the outer span, and Y a dimensionless geometric parameter

Reinforcement Content (weight percent)	Fracture Toughness, K_{IC} (MPa $m^{1/2}$)	Sintering Temperature (°C)	Range (MPa $m^{1/2}$)	Comments
20	4.32	2,000	+0.48/-0.56	*Type A ZrO_2*
30	4.20	2,000		
40	5.83	2,000		
5.0	4.49	2,200	+0.79/-0.69	*Type B ZrO_2*
6.0	4.91	2,200	+0.80/-0.79	
7.0	4.38	2,200	+0.0.45/-0.57	
0.0	4.07	2,400	+0.24/-0.28	*Type B ZrO_2*
35	5.39	2,400		

Remarks/Comments
A marginal increase in fracture toughness was noted. This increase in toughness can be attributed to the presence of tetragonal ZrO_2 and/or the occurrence of residual stress in the two-phase microstructure.

Figure 75: Fracture Toughness[I-20] of ZrO_2(p)/ZrC

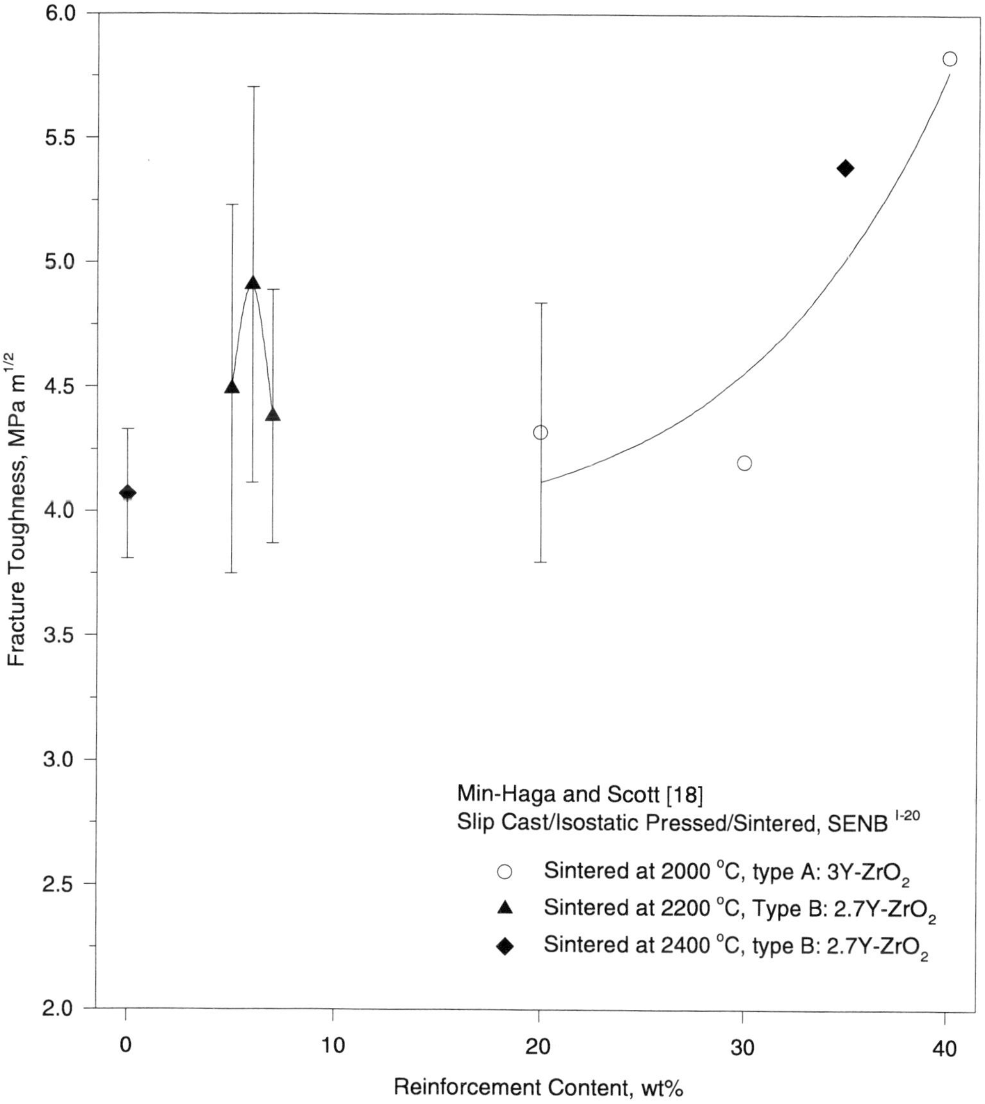

Discussion: Figure 75 shows the effect of zirconia addition to zirconium carbide. The data shows a slight increasing trend with increasing ZrO_2; however, the effect of the reinforcement was unclear because of the large scatter of the data.

$SiC(p)/Si_3N_4$

Reference: 35

Composition

12 wt% Si-rich SiC
10 and 20 wt% C-rich SiC

Processing Method

Colloidal Pressed / Reaction Bonded

Detailed Description

Four sample groups were fabricated. The first group, the monolithic RBSN, were fabricated either in methanol with CIP or in octanol without CIP. The remaining RBSN composite samples were fabricated in octanol without CIP. Octanol was substituted for methanol to reduce the probability of differential sedimentation between the dissimilar Si and SiC powders before samples were pressed. A high-intensity ultrasonic probe was used to break up agglomerates in the dispersed powder mixture rather than the low-intensity ultrasonic bath used for the methanol processed baseline group. Si and SiC powders were ultrasonically dispersed separately, mixed together by hand to visual uniformity, and again sonicated. Total sonication time was 5-8 minutes at up to 40 W. Oxygen exposure was avoided except during pressing. Samples were colloidally pressed at 17 MPa to yield 1.27 diameter x 0.2 cm thick samples. A 0.4 °C/min. ramp followed by a 24 hour, 200 °C soak in N_2 was used to dry the samples. Nitridation was accomplished in ultra-high purity N_2 at atmospheric pressure. The firing schedule for the samples are:
(A) RBSN (methanol): 1 °C/min. to 1200 °C; 0.5 °C/min. to 1400 °C; 1 hour hold
(B) RBSN (octanol): 1 °C/min. to 1250 °C; 4 hour hold; 0.5 °C/min. to 1400 °C; 7 hour hold
(C) RBSN + 12 % Si-rich SiC: 2 hours at 1250-1400 °C; 4 hour hold; 10 hours at > 1400 °C
(D) RBSN + 10 % C-rich SiC: 3 °C/min. to 1250 °C; 4 hour hold; 0.5 °C/min. to 1400 °C; 10 hour hold
(E) RBSN + 20 % C-rich SiC: 3 °C/min. to 1250 °C; 4 hour hold; 0.5 °C/min. to 1400 °C; 10 hour hold

Reinforcement Description

SiC particulates

Reinforcement Dimensions

ave. Si dia. = 0.23 μm
ave. C-rich SiC dia. = 0.04 μm
ave. Si-rich SiC dia. = 0.05 μm

Matrix Description

Si powder

Matrix Designation

F93 S

Test Type

Vickers Indentation Microfracture (IM) Technique[I-16]

Specimen Preparation

Samples were ground and polished to 1 μm on the tensile surface.

Specimen Dimensions

dia. = 12.7 mm
thickness = 1 mm

Test Temperature

RT

Relative Density (percent)	Fracture Toughness, K_{IC} ($MPa\ m^{1/2}$)	Range ($MPa\ m^{1/2}$)	Comments
77	2.4	±0.3	*A*
77	2.3	±0.1	*A*
76	2.5	±0.1	*B*
73	2.4	±0.1	*C*
76	2.4	±0.2	*D*
73	2.2	±0.1	*E*

Remarks/Comments

Toughness measured by the indentation technique were similar to one another for all sample sets. No significant toughening was achieved by adding SiC particles to the RBSN.

SiC(p)/Si_3N_4

Reference: 46

Composition
0-52 vol% SiC

Processing Method
Hot-Pressed

Detailed Description
An amorphous Si-C-N powder was prepared by introducing vaporized $((CH_3)_3Si)_2NH$ and NH_3 gas into an Al_2O_3 tube. Various powders with different C:N ratios were prepared by controlling the volume of NH_3 gas. These composite powders were stabilized by heat treatment in N_2 gas. 6 wt% Y_2O_3 and 2 wt% Al_2O_3 were added to the composite powder. This powder was hot-pressed for 2 hours under 34 MPa at 1800 °C.

Sintering Additives
6 wt% Y_2O_3
2 wt% Al_2O_3

Reinforcement Description
SiC particulates

Matrix Description
Si_3N_4 powder

Test Type
Indentation Microfracture (IM) Technique

Test Temperature
RT

Reinforcement Content (volume percent)	Fracture Toughness, K_{IC} (MPa $m^{1/2}$)	Standard Deviation (MPa $m^{1/2}$)
0.0	5.07	±0.25
2.2	5.58	±0.2
9.4	6.99	±0.6
17	6.22	±0.25
23	5.71	±0.2
34	5.44	±0.2
52	5.32	±0.15

Remarks/Comments
The fracture toughness increased up to 10 vol% SiC. With further addition of SiC reinforcement, the toughness decreased. Fracture toughness dropped by 1.3 MPa $m^{1/2}$ from 10 to 23 vol% SiC; from 23 to 52 vol% SiC the decrease in toughness was much more gradual.

SiC(p)/SiAlON

Reference: 78

Composition
0-12 wt% SiC

Processing Method
CIPed / Pressureless Sintered

Detailed Description
β-SiAlON powder was prepared by carbo-thermal reduction at 1450 °C for 5 hours in N_2 atmosphere. Raw materials were high purity SiO_2 powder, carbon powder and aluminum hydroxide which was prepared by hydrolysis of aluminum isopropoxide. Ingredients were mixed. 100 g of above mentioned powder was milled with 250 ml of ethanol for 24 hours using a pot mill with silicon nitride beads then drying under diminished pressure. Each composite powder was formed into green bodies 50 mm diameter x 5 mm thick under 33 MPa and cold-isostatic-pressing at 98 MPa. These green bodies were sintered in a carbon crucible at 1780 °C for 3 hours under N_2 atmosphere by pressureless sintering..

Sintering Additives
5 wt% Y_2O_3

Reinforcement Description
β-SiC particulates

Reinforcement Dimensions
grain size = 0.2-0.3 μm

Matrix Description
β-SiAlON

Matrix Supplier
$Si_{5.5}Al_{0.5}O_{0.5}N_{7.5}$

Test Type
Vickers Indentation Microfracture (IM) Technique[I-22]

Specimen Preparation
Specimens were machined from sintered bodies to test bars.

Test Description
Indenter load = 20 kg
loading time = 30 s

Test Temperature
RT

Reinforcement Content (weight percent)	Fracture Toughness, K_{IC} (MPa m$^{1/2}$)
0.0	4.65
3.0	5.04
6.0	5.40
9.0	5.15
12	4.85

Remarks/Comments
Fracture toughness increased for composites containing up to 6 wt% SiC, and then decreased as the SiC content increased. Fracture toughness for composites containing up to 12 wt% β-SiC was higher than that for β-SiAlON ceramics. The maximum value was increased by 15 % to 5.4 MPa m$^{1/2}$ at 6 wt% β-SiC content in comparison with β-SiAlON ceramics (at 4.7 MPam$^{1/2}$). Fracture toughness was at a maximum for composites containing 6 wt% β-SiC, when relative densities were about 100 %. Fracture toughness decreased with decreasing relative density. The fracture toughness for composites up to 12 wt% β-SiC were higher than for β-SiAlON ceramics.

SiC(p)/Si_3N_4

Reference: 93

Composition
0-20 vol% SiC

Processing Method
Hot-Pressed

Detailed Description
These powders were mixed by ball-milling with Al_2O_3 ball using ethanol solvent for 77.8 h in a vessel made by polyethylene. Hot-pressing was carried out at 1800 °C for 1 hour in nitrogen atmosphere under a pressure of 30 MPa.

Sintering Additives
8 wt% Y_2O_3

Reinforcement Description
β-SiC powders

Reinforcement Supplier
Ibiden Co., Ltd.

Supplier Designation
UF grade

Reinforcement Dimensions
ave. particle size = 0.3 μm

Matrix Description
α-Si_3N_4 powder

Matrix Supplier
Ube Industries, Ltd., Japan

Matrix Designation
E-10 grade

Matrix Properties
ave. particle size = 0.2 μm

Test Type
Indentation Microfracture (IM) Technique

Test Description
Density measured by Archimedes' method.

Test Temperature
RT

Reinforcement Content (volume percent)	Fracture Toughness, K_{IC} (MPa $m^{1/2}$)	Relative Density (percent)	Range (MPa $m^{1/2}$)
0.0	4.18	99.9	+0.33/-0.34
5.0	6.24	99.9	+1.06/-1.02
10	5.17	98.4	+0.88/-0.91
20	4.14	98.7	+0.65/-0.60

Remarks/Comments
Two types of dispersion configurations were evident: nano-meter scale particles uniformly, and sub-micron particles forming aggregates at grain boundaries of the matrix. Grains changed to rod-like structure as SiC content increased. Fracture toughness reached a maximum at 5 vol% SiC-Si_3N_4.

Figure 76: Fracture Toughness of SiC(p)/Si_3N_4 and SiAlON

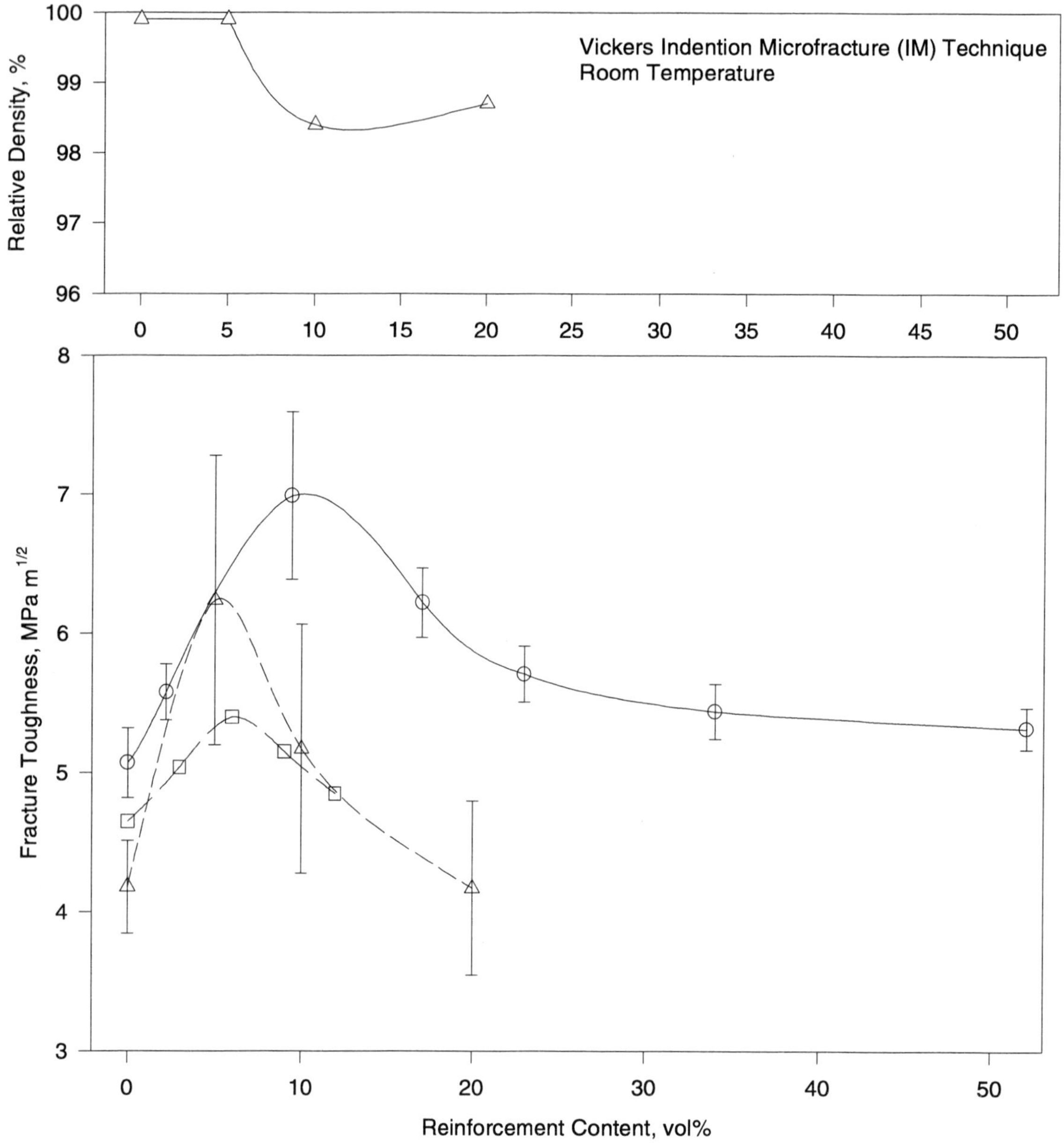

Reference	Sintering Additives	Equation Used	Processing
—○— Sawaguchi *et al.* [46]	6 wt% Y_2O_3, 2 wt% Al_2O_3	Niihara *et al.* [I-12]	Hot-Pressed
—△— Sasaki *et al.* [93]	8 wt% Y_2O_3	Faber and Evans [I-31]	Hot-Pressed
—□– Yamagishi *et al.* [78]	5 wt% Y_2O_3	Lawn and Evans [I-22]	SiC(p)/SiAlON * CIPed/Pressureless Sintered

*SiC(p)/SiAlON wt% to vol% conversion using a 1:1 ratio

Discussion: Sawaguchi *et al.* [46] and Sasaki *et al.* [93] performed indentation fracture toughness tests on silicon carbide particulate reinforced Si_3N_4 nanocomposites. They mechanically mixed and hot-pressed fine Si_3N_4 and SiC powders. The hot-pressed nanocomposites showed a 2 MPa $m^{1/2}$ increase in toughness with the addition of 5-10 vol% reinforcement. The fracture toughness decreased with an increase in SiC particulate content above 10 vol%. The toughness reached a maximum at 5-10 vol% for two reasons. Firstly, the Si_3N_4 grains changed from an equiaxed structure to a rod-like structure. Secondly, crack deflection from the particulate reinforcement was observed. The decrease in toughness at 20 vol% SiC reinforcement occurred because the α- to β-phase transformation was suppressed, indicating inhibition of sintering process by the SiC particulates. This caused the Si_3N_4 matrix to revert to an equiaxed grain structure and the porosity to increase. The grain boundary phase was changed from $Y_5N(SiO_4)_3$ to Y_2SiO_5 with the increase of SiC content [93].

Yamagishi *et al.* [78] tested a pressureless sintered SiC(p)/SiAlON composite. The β-SiAlON powders were obtained by carbo-thermal reduction, which produced a highly homogeneous matrix. All samples reached a density of greater than 98.5%, and only β-SiAlON and β-SiC crystalline phases were observed in XRD patterns, showing no intermediate phases. The toughness reached a maximum of 5.4 MPa $m^{1/2}$ at 6 wt% SiC content.

Figure 77: Fracture Toughness[I-16] of $SiC(p)/Si_3N_4$

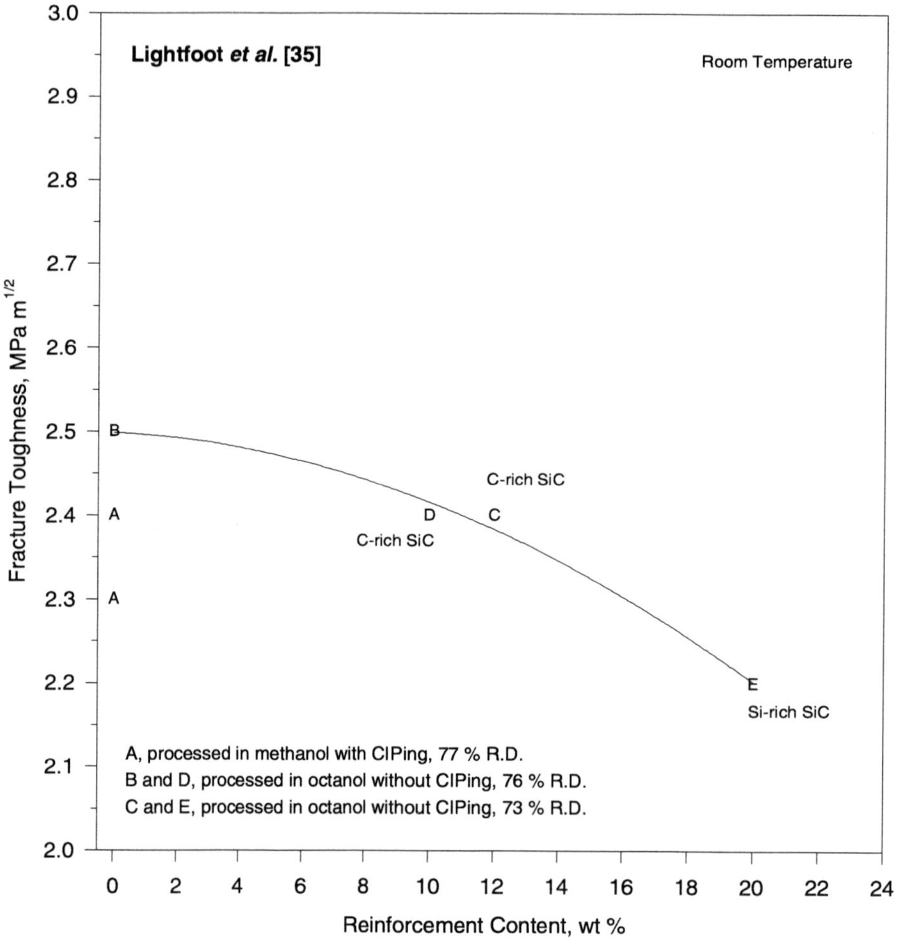

Discussion: Lightfoot *et al.* [35] shows the effect of adding different amounts and types of SiC to a $SiC(p)/Si_3N_4$ composite. Densities of RBSN tend to be lower than conventional Si_3N_4 materials. In this case the densities of samples A and B are about the same as those of the composites with the Si or C rich SiC. This shows that the voids between the Si particles were not filled effectively by the SiC particles. Even with these low densities, samples were free of large processing flaws, and average flaw sizes were approximately the same for all sample types. RBSN without SiC reinforcement was the toughest (sample B). Because of this fact, it was decided that the addition of SiC did not contribute to the toughening of the samples. Monolithic RBSN processed in octanol (sample B) achieved slightly better results than the RBSN processed in methanol (sample A). At these low densities the indentation toughness is not very reliable due to crushing under the indenter load instead of plastic deformation; this resulted in a relaxation of the crack-propagating stresses.

$SiC(p)/Si_3N_4$

Reference: 31

Composition
0, 10, and 40 wt% SiC

Processing Method
CIPed/Pressureless Sintered (Some HIPed)

Detailed Description
All powders were mixed together in an attrition mill. This mixture was then cold isostatically pressed at 630 MPa. Next, sintering was performed in a graphite element furnace at 0.1 MPa in nitrogen. Some specimens were then HIPed at 2000 °C, 100 MPa in nitrogen. The HIPed specimens were presintered at 1850 °C without encapsulation.

Sintering Additives
10.5 wt% Al_2O_3 (Alcoa, A16)
4.5 wt% Y_2O_3 (Ventron, 4N)

Reinforcement Description
β-SiC powder

Reinforcement Supplier
H. C. Starck

Supplier Designation
B 10

Reinforcement Dimensions
ave. grain size = 1.0 μm

Matrix Description
α-Si_3N_4 powder

Matrix Supplier
H. C. Starck

Matrix Designation
LC 10

Matrix Properties
ave. grain size = 0.5 μm

Test Type
3-pt Indentation Strength in Bending (ISB)[I-15]

Specimen Dimensions
span = 12 mm

Test Description
indentation forces were 29.4-196.2 N

Test Temperature
RT

Reinforcement Content (weight percent)	Fracture Toughness, K_{IC} (MPa $m^{1/2}$)	Relative Density (percent)	Range (MPa $m^{1/2}$)	Comments
0.0	3.95	96.0	±0.13	*Normally sintered*
10	3.94	94.9	±0.11	
40	3.17	91.7	±0.15	
0.0	3.85	96.9	±0.20	*After HIP*
10	3.55	96.6	±0.35	
40	3.30	96.6	±0.08	

Remarks/Comments
The fracture toughness increased with the addition of SiC particulates.

$SiC(p)/Si_3N_4$

Reference: 37

Composition
0 and 10 vol% SiC

Processing Method
CIPed/Pressureless Sintered (some HIPed)

Detailed Description
The starting powders were ball-milled, dried and sieved. This mixture was cold isostatically pressed at 400 MPa and pressureless sintered at 1750 °C under 0.1 MPa of N_2 for 3 hours. Half of the samples were HIPed for 1 hour at 1850 °C under 100 MPa of N_2.

Sintering Additives
10 wt% Y_2O_3
5 wt% Al_2O_3

Reinforcement Description
SiC particulates

Reinforcement Supplier
Ibiden Co., Ohgaki, Japan

Reinforcement Typical Properties
specific density = 3.21 g cm^{-3}

Reinforcement Dimensions
ave. particle dia. = 0.26 μm

Matrix Description
Si_3N_4

Matrix Supplier
Ube Kosan Co., Japan

Matrix Designation
E-10

Matrix Properties
specific density = 3.18 g cm^{-3}
ave. particle dia. = 0.5 μm

Test Type
Vickers Indentation Microfracture (IM) Technique[I-24]

Specimen Preparation
Specimens were polished using 3 and 6 μm paste.

Specimen Dimensions
6 x 3 x 45 mm

Type of Specimen
Bar

Test Temperature
RT

Reinforcement Content (volume percent)	Fracture Toughness, K_{IC} (MPa $m^{1/2}$)	Relative Density (percent)	Standard Deviation (MPa $m^{1/2}$)	Comments
0	5.70	98	±0.2	*Pressureless Sintered*
10	4.60	97	±0.1	
0	5.40	99	±0.2	*Sintered/HIPed*
10	5.40	99	±0.2	

Remarks/Comments
Fifteen specimens were tested for each composition.

$SiC(p)/Si_3N_4$

Reference: 40

Composition
0 and 10 vol% SiC

Processing Method
Hot-Pressed

Detailed Description
The starting powders were mixed, dried and sieved. The powders were hot-pressed at 30 MPa under 0.1 MPa nitrogen at 1700 °C and 1780 °C.

Sintering Additives
5 wt% Y_2O_3 (in addition to the Y_2O_3 and Al_2O_3 premixed in the matrix powder)

Reinforcement Description
β-SiC particulates

Reinforcement Supplier
Ibiden Co. Ltd., Oogaki-city, Gifu-prefecture, Japan

Reinforcement Typical Properties
specific density = 3.21 g cm^{-3}

Reinforcement Dimensions
ave. particle dia. = 0.26 µm
surface area = 15.1 $m^2.g^{-1}$

Matrix Description
Si_3N_4

Matrix Supplier
Ube Industries., Ltd., Japan

Matrix Designation
COA

Matrix Properties
specific density = 3.18 g cm^{-3}
ave. particle dia. = 0.5 µm
contains 5 wt% Y_2O_3 and 5 wt% Al_2O_3

Test Type
Vickers Indentation Microfracture (IM) Technique [I-24]

Specimen Preparation
Specimens were polished with 6 and 3 µm diamond paste.

Specimen Dimensions
6 x 3 x 50 mm

Test Temperature
RT

Reinforcement Content (volume percent)	Fracture Toughness, K_{IC} (MPa $m^{1/2}$)
0	5.70
10	5.40

Remarks/Comments
The fracture toughness decreased with the addition of particulates.

TiC(p)/Si_3N_4

Reference: 42

Composition
0-30 vol% TiC

Processing Method
Hot-Pressed

Detailed Description
The particulate powder, matrix powder, and densification aids were hot pressed for 90 minutes at constant pressure and temperature.

Sintering Additives
6.0 wt% Y_2O_3
1.5 wt% Al_2O_3

Reinforcement Description
TiC particulates

Reinforcement Dimensions
ave. particle dia. = 2 μm

Matrix Description
Si_3N_4 particulates

Test Type
4-pt. Indentation Strength in Bending (ISB) [I-3]

Type of Specimen
Bar

Test Temperature
RT

Reinforcement Content (weight percent)	Fracture Toughness, K_{IC} (MPa $m^{1/2}$)	Range (MPa $m^{1/2}$)
0.0	4.13	+0.1/-0.08
10	4.13	+0.13/-0.09
20	4.16	+0.29/-0.34
30	4.34	+0.24/-0.18

Remarks/Comments
Fracture toughness at room temperature did not change with increasing TiC content as predicted.

TiC(p)/Si_3N_4

Reference: 42

Composition
0 and 30 vol% TiC

Processing Method
Hot-Pressed

Detailed Description
The particulate powder, matrix powder, and densification aids were hot pressed for 90 minutes at constant pressure and temperature.

Sintering Additives
6.0 wt% Y_2O_3
1.5 wt% Al_2O_3

Reinforcement Description
TiC particulates

Reinforcement Dimensions
ave. particle dia. = 2 μm

Matrix Description
Si_3N_4 particulates

Test Type
4-pt. Indentation Strength in Bending (ISB) [I-3]

Type of Specimen
Bar

Test Temperature
RT-1200 °C

Test Environment
Air

Temperature (°C)	Fracture Toughness, K_{IC} (MPa $m^{1/2}$)	Reinforcement Content (volume percent)	Range (MPa $m^{1/2}$)
22	4.48	0	+0.2/-0.17
900	3.99	0	+0.16/-0.2
950	4.39	0	+0.48/-0.17
1,000	4.78	0	+0.58/-0.49
1,100	5.74	0	+0.65/-0.11
1,200	6.39	0	+0.62/-0.35
22	4.58	30	+0.11/-0.14
900	3.97	30	+0.40/-0.10
950	4.31	30	+0.24/-0.15
1,100	5.75	30	+0.31/-0.18
1,200	6.27	30	+0.44/-0/40

Remarks/Comments

At high temperatures the TiC reduced Si_3N_4 grain growth, preventing toughening of the composite.

SiC(p)/Si_3N_4

Reference: 42

Composition
0-30 vol% SiC

Processing Method
Hot-Pressed

Detailed Description
The particulate powder, matrix powder, and densification aids were hot pressed for 110 minutes at constant pressure and temperature.

Sintering Additives
6.0 wt% Y_2O_3
1.5 wt% Al_2O_3

Reinforcement Description
SiC particulates

Reinforcement Dimensions
ave. particle dia. = 0.5 and 8 μm

Matrix Description
Si_3N_4 particulates

Test Type
4-pt. Indentation Strength in Bending (ISB)[I-3]

Type of Specimen
Bar

Test Temperature
RT

Reinforcement Content (volume percent)	Fracture Toughness, K_{IC} (MPa $m^{1/2}$)	Comments
0.0	4.59	*submicron SiC*
10	4.69	
10	5.01	
20	4.02	
20	4.38	
30	3.77	
30	4.14	
0.0	4.61	*8 μm SiC*
10	4.99	
20	4.84	
30	4.89	

Remarks/Comments

This data shows the average of two trials for the submicron SiC reinforcement. The size of the SiC dispersoids affect toughening mechanisms.

SiC(p)/SiAlON

Reference: 44

Composition
10 vol% SiC

Processing Method
Hot-Pressed (some HIPed)

Detailed Description
The mixed powders were ball-milled for 24 hours with a sufficient amount of ethanol. The slurry was dried in a rotary evaporator and sieved to a particle size of <210 μm. The mixed powders were then hot-pressed at 1720 °C for 30 minutes under 24.5 MPa in a 0.1 MPa nitrogen atmosphere. Some were also HIPed at 1850-1950 °C under 10-100 MPa nitrogen gas pressure.

Reinforcement Description
SiC particulates

Reinforcement Supplier
Ibiden Co., Ltd

Supplier Designation
UF30

Reinforcement Dimensions
ave. particle size = 0.28 μm

Matrix Description
α-SiAlON and α-Si_3N_4 powders

Matrix Supplier
Ube Kosan Co., Ltd.

Matrix Properties
60 wt% Si_3N_4
40 wt% SiAlON

Test Type
Single Edge Precracked Beam (SEPB) Technique[I-4]

Specimen Dimensions
4 x 3 x 40 mm

Type of Specimen
Bar

Test Temperature
RT

HIP Temperature (°C)	Fracture Toughness, K_{IC} (MPa m$^{1/2}$)	Weibull Modulus (dimensionless)	Density (g cm^{-3})	Range (MPa m$^{1/2}$)	Comments
-	4.60	8.2	3.20	±0.1	*as-hot-pressed*
1,850	5.00	10.9	3.27	±0.1	*HIPed at 10 MPa N_2 for 1 hour*
1,850.	4.70	14.3	3.72	±0.1	*HIPed at 100 MPa N_2 for 1 hour*
1,950	5.00	16.5	3.27	±0.1	*HIPed at 100 MPa N_2 for 1 hour*

Remarks/Comments
Fracture toughness results were not significantly affected by a post hot-pressing HIPing step.

$SiC(p)/Si_3N_4$

Reference: 62

Composition
0 and 30 vol% SiC

Processing Method
Hot-Pressed

Detailed Description
The Si_3N_4 (AY6) was milled with appropriate content of SiC particulates, then screened prior to hot-pressing. Hot-pressing was performed in BN-coated graphite dies with a flowing N_2 atmosphere. The hot-pressing pressure was increased in a stepwise fashion as the temperature increased. Maximum pressure was reached at 1500 °C and held there. The maximum temperature was <1800 °C.

Sintering Additives
6 wt% Y_2O_3

1.5 wt% Al_2O_3

Reinforcement Description
SiC particulates

Reinforcement Supplier
Milwaukee, WI,

Supplier Designation
Cerac, (α: 12H)

Reinforcement Dimensions
mean particle size = 8 μm

Matrix Description
Si_3N_4 powder

Matrix Supplier
GTE Chemical and Metallurgical Div., Towanda, PA

Matrix Designation
SN502

Matrix Properties
ave. particle size = 0.5 μm

Test Type
Indentation Microfracture (IM) Technique[I-25]

Specimen Preparation
The tensile face of each was wet polished to a 1 μm diamond paste finish. Randomly selected groups of test bars were precracked with a single Knoop indentation. A 100 N load was used. These samples were annealed in Ar at 1200 °C and slowly cooled to relieve any residual stress.

Specimen Dimensions
1.27 x 2.59 x 25 mm

Type of Specimen
Bar

Test Description
Knoop indenter load = 10 N for IM Vickers indenter load = 100 N for IM

Test Temperature
RT

Reinforcement Content (volume percent)	Fracture Toughness, K_{IC} (MPa m$^{1/2}$)	Range (MPa m$^{1/2}$)	Comments
0.0	3.75	±0.20	*in plane*
10	3.55	±0.25	*parallel to*
20	3.55	±0.15	*HP direction*
30	3.35	±0.20	
0.0	3.65	±0.20	*in plane*
10	3.80	±0.20	*perpendicular*
20	3.80	±0.20	*to HP direction*
30	4.00	±0.20	

Remarks/Comments
The toughness decreased with dispersoid content for the smaller dispersoid, while the coarser dispersoid slightly increased the toughness. For the 8 μm particulate, in the plane parallel to the HP direction, the fracture toughness decreased with increased content when measured in either parallel or perpendicular. However, in the plane perpendicular to the HP direction, the fracture toughness increased with increasing SiC content when measured in either the parallel or perpendicular direction. The IM in a plane perpendicular to the HP direction was lower in all cases than that measured parallel to the HP direction.

$SiC(p)/Si_3N_4$

Reference: 67

Composition
24 wt% SiC

Processing Method
CIPed/Polymer Pyrolyzed/Sintered

Detailed Description
The amorphous Si_3N_4 and C containing powder was prepared by the pyrolysis of a polymethylsilazane at 1000 °C in argon. Polymethylsilazane was dissolved in tetrahydrofuran (THF) in an argon atmosphere. 10.5 wt% Al_2O_3 and 4.5 wt% Y_2O_3 were added. The THF was evaporated in vacuum at 0.1 Pa and room temperature. Dried residue was heated at 550 °C for 1 hour, and then at 1000 °C for 1 hour in a quartz tube. The pyrolyzed powder was attrition milled for 3 hours at 1000 rpm, and cold pressed at 630 MPa into green compacts of 35x17x15 mm. Reference samples were made from α–Si_3N_4, β–SiC, 10.5 wt% Al_2O_3 and 4.5 wt% Y_2O_3. They were attrition milled, dried, sintered and hot-pressed in same manner as pyrolyzed sample. Both the as-pyrolyzed sample and the reference samples were sintered in a graphite resistivity heated furnace at 1750 °C and 1850 °C under 0.1 MPa N_2.

Sintering Additives
10.5 wt% Al_2O_3 (A16, ALCOA, Pittsburgh, PA)
4.5 wt% Y_2O_3 (Ventron Alta Produckte, Germany)

Reinforcement Description
β-SiC particulates

Reinforcement Supplier
H.C. Stark, Berlin, Germany

Supplier Designation
B10

Matrix Description
α-Si_3N_4 powder

Matrix Supplier
H.C. Stark, Berlin, Germany

Matrix Designation
LC10

Matrix Properties
Polymethylsilazane powder A (7.1 wt% °C, 22.5 wt% N, 10.4 wt% O)
Reference powder B (9 wt% C, 20.6 wt% N, 9.2 wt% O)

Test Type
Indentation Strength in Bending (ISB)[I-15]

Type of Specimen
Bar

Test Temperature
RT

Reinforcement Content (weight percent)	Fracture Toughness, K_{IC} (MPa $m^{1/2}$)	Relative Density (percent)	Comments
24	6.30	95.0	*sample A, polymer derived*
24	6.10	93.0	*sample B, reference*

Remarks/Comments
The polymer derived sample had better homogeneity and reproducibility than the reference sample. The as-pyrolyzed polymer derived powder was amorphous silicon carbonitride, while the reference sample was fabricated from crystalline powders. The amorphous polymer derived powders sintered much faster than the crystalline powders and resulted in smaller grain size and higher relative densities.

$SiC(p)/Si_3N_4$

Reference: 74

Composition
0 and 22 vol% SiC

Processing Method
Colloidal Pressed/Reaction Bonded

Detailed Description
Si and SiC powders were colloidally pressed from an octanol dispersion. The bodies were then reaction bonded to form the composite.

Reinforcement Description
SiC particulates

Reinforcement Dimensions
dia. = 15 µm

Matrix Description
Si powder

Matrix Properties
dia. = 0.2-0.3 µm

Test Type
3-pt Indentation Strength in Bending (ISB)[I-15]

Specimen Preparation
Specimens were polished to a 9 µm surface finish.

Specimen Dimensions
10.4 x 2 x 1.5 mm

Type of Specimen
Bar

Test Description
load = 10 - 100 N

Test Temperature
RT

Reinforcement Content (volume percent)	Fracture Toughness, K_{IC} (MPa $m^{1/2}$)	Range (MPa $m^{1/2}$)
0.0	1.63	±0.08
22	2.46	±0.14

Remarks/Comments

Fracture toughness of the composite increased slightly.

SiC(p)/Si_3N_4

Reference: 81

Composition

0 - 26 vol% SiC

Processing Method

Hot-Pressed

Detailed Description

Materials were mixed together by ball milling in a polypropylene jar with cyclohexane and Si_3N_4 media for 3 hours. The composite was mixed for another hour after pre-wetted SiC was added. After ball-milling, the slurry was dried in a thin film rotating evaporator, granulated and pre-fired to burn off residual organics. Monolithic silicon nitride and all composites were hot-pressed in a nitrogen atmosphere to 1725 °C and at a pressure of 26 MPa.

Sintering Additives

5 wt% Y_2O_3
2 wt% Al_2O_3

Reinforcement Description

SiC particulates

Reinforcement Supplier

Lonza Ltd.

Supplier Designation

LS-13

Reinforcement Dimensions

ave. size = 13 μm

Matrix Description

Si_3N_4 powder

Matrix Supplier

Ube Industries, Ltd., Japan

Matrix Designation

SN E-10

Test Type

Indentation Strength in Bending (ISB)

Specimen Preparation

Test bars were sectioned from 1 μm finish polished samples. The edges on the tensile side of the bars were chamfered. A flaw was induced into the center of the samples by using a Vickers indenter. Bars were fractured within one hour of indenting.

Specimen Dimensions

3 x 4 x 50 mm

Type of Specimen

Bar

Test Description

load = 10 kg

Test Temperature

RT

Reinforcement Content (volume percent)	Fracture Toughness, K_{IC} (MPa $m^{1/2}$)	Range (MPa $m^{1/2}$)
0.0	6.83	+0.23/-0.20
5.4	7.15	+0.26/-0.22
15	7.18	+0.10/-0.09
26	7.22	+0.21/-0.21

Remarks/Comments

Fracture toughness increased slightly with the addition of SiC particulates.

$TiN(p)/Si_3N_4$

Reference: 115

Composition
10 vol% TiN

Processing Method
CIPed/Pressureless Sintered

Detailed Description
The powders were mixed in a vibratory mill using methanol, and followed by drying. Then the powders were dispersed in water at pH ~10 by means of ultrasonification. The mixed slurry was washed with acetone and filtered off on the funnel. Then the obtained powder cake was dispersed again in toluene, and residual water was removed by utilizing azeotropy between water and toluene. After drying in N_2-5% H_2 atmosphere, the powders were CIPed at 490 MPa and was sintered at 1850 °C for 4 hours in a 0.9 MPa N_2 atmosphere.

Sintering Additives
5 wt% Y_2O_3
2 wt% Al_2O_3
2 wt% MgO

Reinforcement Description
TiN particulates

Reinforcement Properties
specific surface area = 11.2 $mm^2\ g^{-1}$
CTE = 9.4 x $10^{-6}\ C^{-1}$

Matrix Description
α-Si_3N_4

Matrix Supplier
Denki Kaguku Co.

Matrix Designation
SNP21 C3 grade

Matrix Properties
specific surface area = 11.2 $mm^2\ g^{-1}$
92.2 % α– phase
oxygen content = 1.57 wt%
CTE = 3 x $10^{-6}\ C^{-1}$

Test Type
Vickers Indentation Microfracture (IM) Technique

Test Description
Density was measured by Archimedes' method.
indentation load = 20 kg

Test Temperature
RT

Grain Size (μm)	Fracture Toughness, K_{IC} (MPa $m^{1/2}$)	Relative Density (percent)	Standard Deviation (MPa $m^{1/2}$)
0.27	7.34	98.0	±0.20
0.98	7.34	98.5	±0.25
1.8	7.61	98.3	±0.15
3.8	7.80	98.2	±0.25
7.3	7.55	98.0	±0.20
13.7	7.31	98.1	±0.30

Remarks/Comments
Fracture toughness was measured against the grain size of the TiN particulates. The fracture toughness reached a peak at around 4 μm; otherwise, fracture toughness did not change significantly.

Figure 78a: Fracture Toughness of SiC(p) and TiC(p)/Si_3N_4

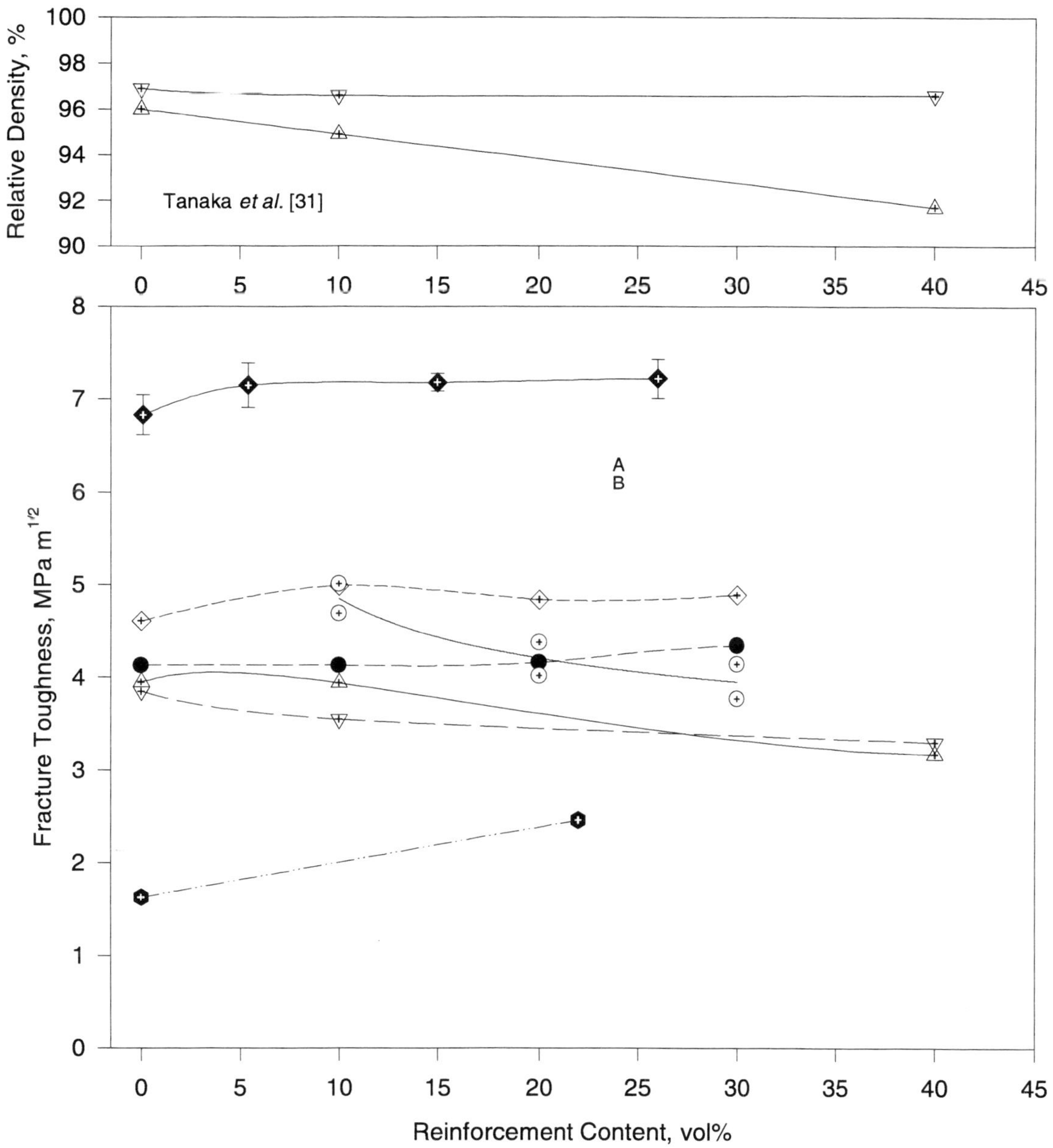

SiC Particulates

Tanaka *et al.* [31], ISB[I-15]
- Sintered
- Sintered then HIPed

Buljan *et al.* [42], Hot-Pressed, ISB[I-3]
- Hot-Pressed, submicron SiC[TC-eqn. 14]
- Hot-Pressed, 8 μm SiC

Nair *et al.* [74], RBSN, ISB[I-15]

Hänninen *et al.* [81], Hot-Pressed, ISB

Riedel *et al.* [67], CIPed/Sintered, polymer derived, ISB[I-15]
- A sample A
- B sample B

TiC Particulates

Buljan *et al.* [42], Hot-Pressed, ISB[I-3]

Figure 78b: Fracture Toughness[I-25] of $SiC(p)/Si_3N_4$

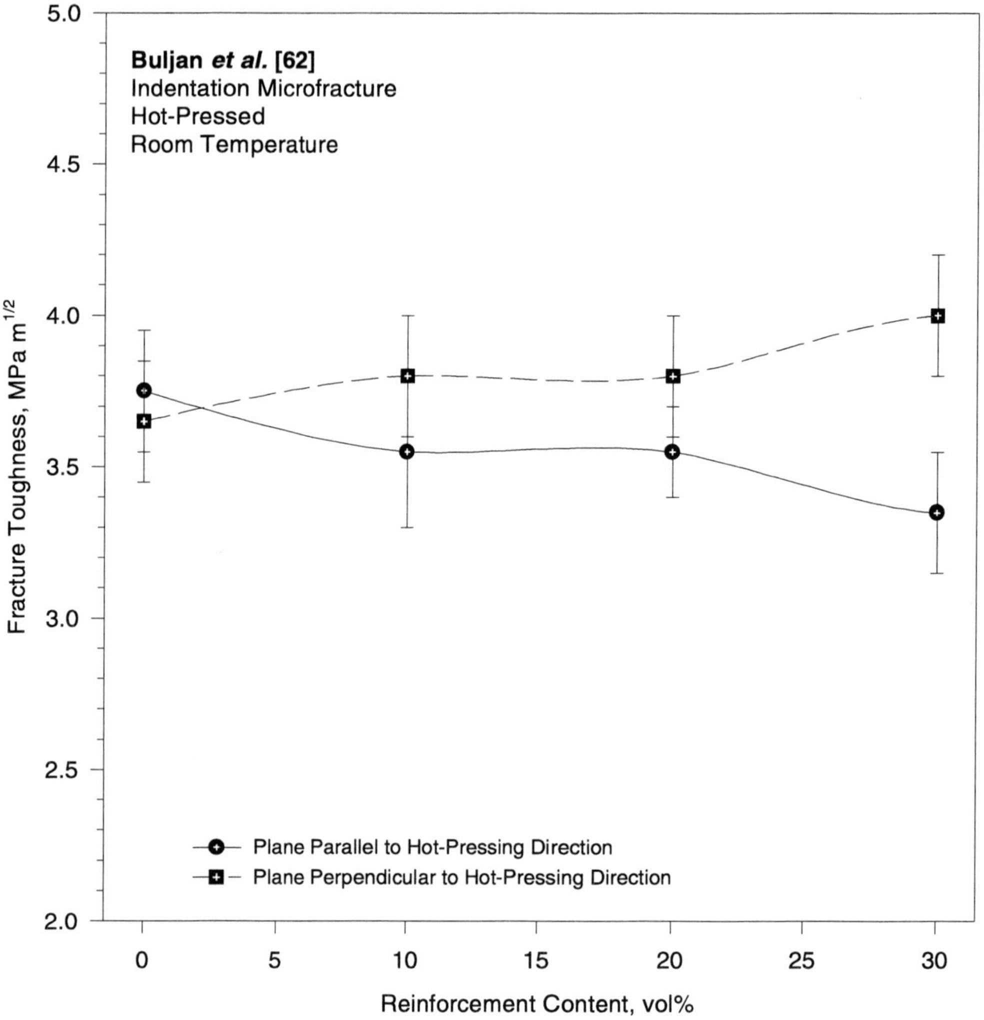

Discussion: Figure 78a shows that the fracture toughness of a SiC or TiC particulate reinforced silicon nitride typically falls in the range of 3.5 to 5 MPa $m^{1/2}$. However, the toughness was as high as 7 MPa $m^{1/2}$ in the hot-pressed composite tested by Hänninen *et al.* [81]. All composites were fabricated with 7-15 wt% yttria/alumina sintering additives, except for the reaction bonded composite tested by Nair *et al.* [74]. Because the porosity of the reaction bonded materials was very high, the toughness of the reaction bonded materials was low, ~ 2 MPa $m^{1/2}$.

It is also apparent from figure 78a that fracture toughness neither increases nor declines with the addition of carbide reinforcement. This is the result of competing fracture mechanisms that cancel each other and result in no toughness change. On the positive side, the addition particulate reinforcement increases the toughness of a silicon nitride matrix by mechanisms such as crack deflection and branching. On the other hand, monolithic silicon nitride ceramics with yttria/alumina sintering additives are typically tough due to the formation of elongated β-Si_3N_4 grains that act as 'whisker' reinforcement. The addition of particulate reinforcement inhibits the formation of *in situ* reinforcement As a result, there is typically no net gain in toughness.

There was another set opposing mechanisms related to CTE mismatch stresses that effectively caused no change in fracture toughness. Slight increases in fracture toughness caused by dispersed particles produced a deflection of cracks due to residual stresses. This stress occurred from a mismatch in elastic and thermal properties. Counteracting this benefit is the fact that the addition of particulate reinforcement introduced flaws and increased the critical flaw size. Thus the addition of the particulates had two effects on the toughness of the composite, one was an increase due to residual stresses, and the other was a decrease in toughness due to the introduction of flaws. Taking into account these two effects canceling each other, a somewhat stable curve for fracture toughness developed.

The 25 vol% SiC(p)/Si_3N_4 composites fabricated by Riedel *et al.* [67] from polymer derived powders had a respectable toughness of 6 MPa $m^{1/2}$. These composites had high density despite being sintered rather than hot-pressed. The primary advantage of polymer derived powders was that they were amorphous rather that crystalline. This attribute significantly increased sintering rates and decreased the grain size in the as-processed composite.

One way to decrease the negative effect of the critical flaw size was to reduce the grain size of the samples. Many of the sintering processes used in the cited references concentrated on this aspect. For example, Buljan *et al.* [62] showed that a smaller grain size directly related to the energy needed to propagate a crack in a material with intergranular fracture by the equation of:

$$\Delta K_{IC}=C\ K^{\circ}_{IC}(D/D^{\circ}-1)$$

where: C is a proportionality constant,
K°_{IC} is the fracture toughness of the material,
D° is the average grain size in the absence of other toughening effects and other modes of fracture,
D/D° is the change in grain size proportionality

Thus tailoring the grain size for the smallest flaw size greatly affected fracture toughness. There was a limit of how small grains should be in order not to cause transgranular fractures, but all references had grains sizes larger than this value.

Figure 78b shows that the addition of SiC particulate reinforcement caused a slight anisotropy to appear, Buljan *et al.* [62]. At higher reinforcement contents, the toughness on the plane perpendicular to the HP direction became increasingly tougher than the toughness on the plane parallel to the HP direction.

Figure 79: Effect of Process Conditions on Fracture Toughness of SiC(p)/SiAlON

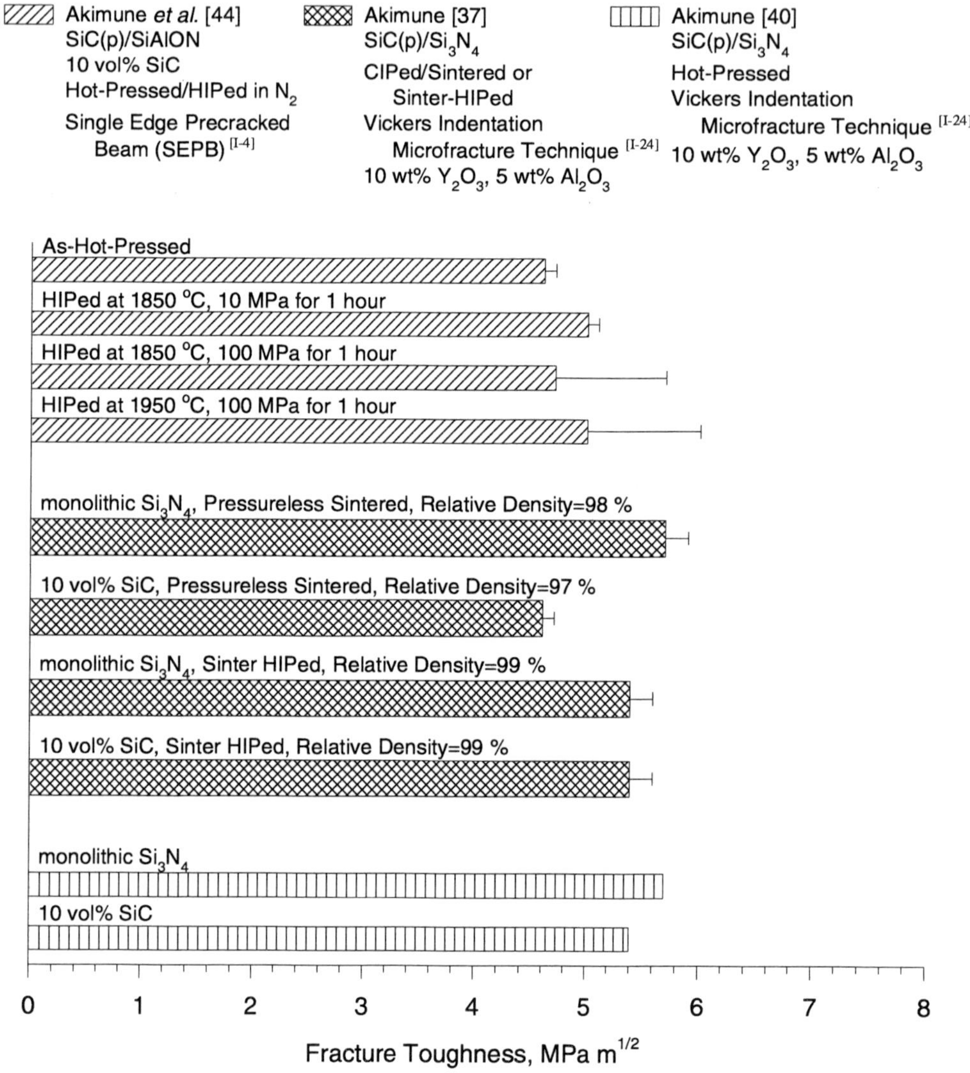

Discussion: Akimune [37 and 40] tested monolithic Si_3N_4 and 10 wt% SiC(p)/Si_3N_4 composites for use in gas turbine engines and internal combustion engines at the Nissan Motor Co. These materials were fabricated by conventional powder techniques, either pressureless sintered, hot-pressed, or sintered then HIPed. The materials contained significant quantity of sintering additives (15 wt% alumina-yttria) and all were at least 97% dense. The toughness of the sintered silicon nitride fell ~ 1 MPa $m^{1/2}$ with the addition of 10 vol% SiC, while the fracture toughness of both the hot-pressed and HIPed silicon nitride did not significantly change with the addition of 10 vol% reinforcement. All materials except the pressureless sintered composite had a toughness around 5.5 MPa $m^{1/2}$.

Akimune *et al.* [44] showed that sintering conditions affected SiC particles inhibiting sialon grain growth. The pressureless sintered 10 vol% SiC(p)/Si_3N_4 composites had a slightly lower toughness than the hot-pressed or HIPed composites. The pressureless sintered monolithic silicon nitride, on the other hand, was as tough as any of the hot-pressed or HIPed composites due to the development of elongated grain structure.

Figure 80: Fracture Toughness vs. Grain Size of TiN(p)/Si_3N_4

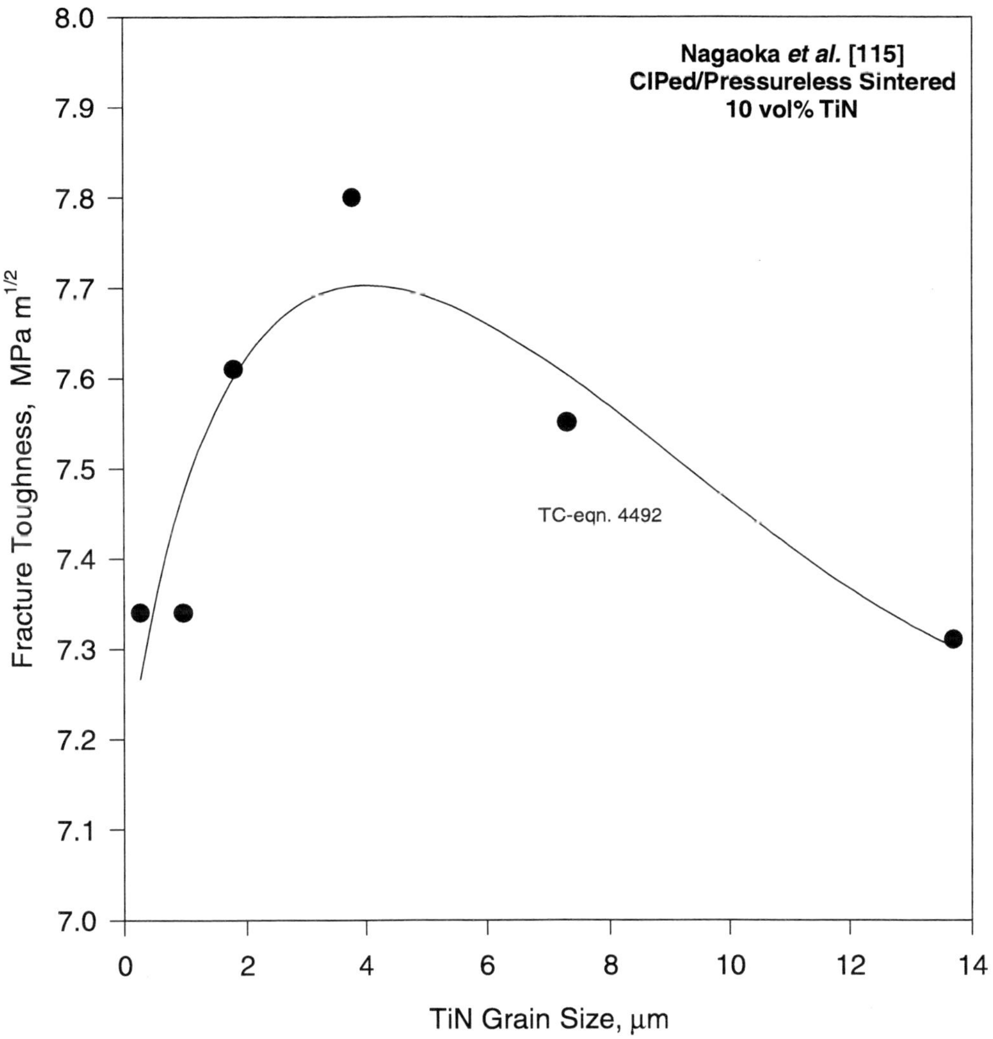

Discussion: Nagaoka *et al.* [115] found that the fracture toughness of TiN(p)/Si_3N_4 composites varied as a function of TiN grain size. The primary toughening mechanism of particulate reinforced CMCs in which there is a large difference in thermal expansion between the two phases, as there is in the TiN-Si_3N_4 composite, is likely to be microcracking. In the 10 vol% TiN(p)/Si_3N_4 composite, the effectiveness of microcracking as a toughening mechanism is highly dependent on the density of microcracks, which, in turn, is dependent on the size of the TiN particulates. Even at small TiN grain sizes, some microcracks were observed. Microcracking was enabled at small TiN grain size by the presence of glassy or semi-crystalline grain boundaries phases which have a lower surface energy than the silicon nitride matrix. As the TiN grain size increased, thermal expansion mismatch stresses increased causing an increase in microcrack density. Optimum toughening occurred at TiN grain size of 4 µm, and fracture toughness reached a maximum of 7.8 MPa $m^{1/2}$. Fracture toughness decreased with a continued increase in TiN grain size. This was attributed to the high density of microcracks, which made it easier for primary crack propagation through microcrack coalescence mechanisms and through microcrack-primary crack interactions [115].

$SiC(p)/Si_3N_4$

Reference: 27

Composition
0-20 vol% SiC

Processing Method
Isostatically Pressed / Reaction Bonded

Detailed Description
SiC and Si_3N_4 powders were isostatically pressed then nitrided in a dry N_2 atmosphere in a temperature range of 1350-1380 °C.

Reinforcement Description
α-SiC particulates

Reinforcement Supplier
Grindwell Norton, India

Matrix Description
Si powder

Matrix Supplier
Indian Metals and Ferro Alloys, India

Test Type
4-pt Single Edge Notched Beam (SENB)

Specimen Dimensions
45 x 4.5 x 3.5 mm
notch tip radius = 130 μm
normalized notch length = 0.3-0.6
inner span = 10.04 mm
outer span = 40.1 mm

Type of Specimen
Notched Beam

Test Description
loading rate = 1.25 N/min

Test Temperature
RT-1400 °C

Test Environment
Air

Test Equipment
High Temperature Bending Strength Tester (Model 422S, Netzsch, Germany)

Reinforcement Content (volume percent)	Fracture Toughness, K_{IC} ($MPa\ m^{1/2}$)	Range ($MPa\ m^{1/2}$)	Relative Density (percent)	Density ($g\ cm^{-3}$)
0.0	1.90		72	2.22
5.0	2.70	±0.5	77	2.24
10	2.50	±0.4	81	2.26
15	1.70	±0.3	73	2.16
20	2.30	±0.4	76	2.22

Temperature (°C)	Fracture Toughness, K_{IC} (MPa $m^{1/2}$)	Reinforcement Content (volume percent)	Density (g cm^{-3})
30	1.9	0.0	2.22
1,000	2.5	0.0	2.22
1,200	2.90	0.0	2.22
1,400	2.10	0.0	2.22
30	2.70	5.0	2.24
1,000	3.00	5.0	2.24
1,200	3.30	5.0	2.24
1,400	2.10	5.0	2.24
30	2.50	10	2.26
1,000	2.50	10	2.26
1,200	2.70	10	2.26
1,400	2.30	10	2.26
30	1.80	15	2.16
1,000	2.60	15	2.16
1,200	2.00	15	2.16
1,400	2.30	15	2.16
30	2.30	20	2.22
1,000	1.70	20	2.22
1,200	1.90	20	2.22
1,400	1.90	20	2.22

Remarks/Comments

The maximum fracture toughness occurred at 5 vol% SiC particulates. Fracture toughness tended to be greater for the composites with the higher density values.

BN(p)/Si_3N_4

Reference: 27

Composition
0 and 10 vol% BN

Processing Method
Isostatically Pressed / Reaction Bonded

Detailed Description
SiC and Si_3N_4 powders were isostatically pressed then nitrided in a dry N_2 atmosphere in a temperature range of 1350-1380 °C.

Reinforcement Description
BN particulates

Reinforcement Supplier
H.C. Starck

Matrix Description
Si powder

Matrix Supplier
Indian Metals and Ferro Alloys, India

Test Type
4-pt Single Edge Notched Beam (SENB)

Specimen Dimensions
45 x 4.5 x 3.5 mm
notch tip radius = 130 μm
normalized notch length = 0.3-0.6
inner span = 10.04 mm
outer span = 40.1 mm

Type of Specimen
Notched Beam

Test Description
loading rate = 1.25 N/min

Test Temperature
RT

Test Equipment
High Temperature Bending Strength Tester (Model 422S, Netzsch, Germany)

Reinforcement Content (volume percent)	Fracture Toughness, K_{IC} (MPa $m^{1/2}$)	Range (MPa $m^{1/2}$)	Relative Density (percent)	Density (g cm^{-3})
0.0	1.90		72	2.22
10	1.90	±0.1	80	2.21

TiC(p)/Si_3N_4

Reference: 27

Composition
0 and 5 vol% TiC

Processing Method
Isostatically Pressed / Reaction Bonded

Detailed Description
SiC and Si_3N_4 powders were isostatically pressed then nitrided in a dry N_2 atmosphere in a temperature range of 1350-1380 °C.

Reinforcement Description
TiC particulates

Reinforcement Supplier
H.C. Starck

Matrix Description
Si powder

Matrix Supplier
Indian Metals and Ferro Alloys, India

Test Type
4-pt Single Edge Notched beam (SENB)

Specimen Dimensions
45 x 4.5 x 3.5 mm
notch tip radius = 130 μm
normalized notch length = 0.3-0.6
inner span = 10.04 mm
outer span = 40.1 mm

Type of Specimen
Notched Beam

Test Description
bending arm length = 10.04 mm

Test Temperature
RT

Test Equipment
High Temperature Bending Strength Tester (Model 422S, Netzsch, Germany)

Reinforcement Content (volume percent)	Fracture Toughness, K_{IC} (MPa $m^{1/2}$)	Relative Density (percent)	Density (g cm^{-3})
0.0	1.90	72	2.22
5.0	1.30	80	2.29

Remarks/Comments

Fracture toughness decreased with the addition of 5 vol% TiC particulates to the Si_3N_4 matrix.

BN(p)/Si_3N_4

Reference: 27

Composition

10 vol% BN

Processing Method

Isostatically Pressed / Reaction Bonded

Detailed Description

SiC and Si_3N_4 powders were isostatically pressed then nitrided in a dry N_2 atmosphere in a temperature range of 1350-1380 °C.

Reinforcement Description

BN particulates

Reinforcement Supplier

H.C. Starck

Matrix Description

Si powder

Matrix Supplier

Indian Metals and Ferro Alloys, India

Test Type

4-pt Single Edged Notched beam (SENB)

Specimen Dimensions

45 x 4.5 x 3.5 mm
notch tip radius = 130 μm
normalized notch length = 0.3-0.6
inner span = 10.04 mm
outer span = 40.1 mm

Type of Specimen

Notched Beam

Test Description

Loading rate = 1.25 N/min

Test Temperature

RT-1400 °C

Test Environment

Air

Test Equipment

High Temperature Bending Strength Tester (Model 422S, Netzsch, Germany)

Temperature (°C)	Fracture Toughness, K_{IC} (MPa $m^{1/2}$)
30	1.90
1,000	2.00
1,200	1.90
1,400	2.90

Remarks/Comments

Density of the composites was 2.21 g cm^{-3}. Fracture toughness at 1400 °C was substantially greater than it was at room temperature.

TiC(p)/Si_3N_4

Reference: 27

Composition
5 vol% TiC

Processing Method
Isostatically Pressed / Reaction Bonded

Detailed Description
SiC and Si_3N_4 powders were isostatically pressed then nitrided in a dry N_2 atmosphere in a temperature range of 1350-1380 °C.

Reinforcement Description
TiC particulates

Reinforcement Supplier
H.C. Starck

Matrix Description
Si powder

Matrix Supplier
Indian Metals and Ferro Alloys, India

Test Type
4-pt Single Edge Notched Beam (SENB)

Specimen Dimensions
45 x 4.5 x 3.5 mm
notch tip radius = 130 μm
normalized notch length = 0.3-0.6
inner span = 10.04 mm
outer span = 40.1 mm

Type of Specimen
Notched Beam

Test Description
loading rate = 1.25 N/min

Test Temperature
RT-1400 °C

Test Environment
Air

Test Equipment
High Temperature Bending Strength Tester (Model 422S, Netzsch, Germany)

Temperature (°C)	Fracture Toughness, K_{IC} (MPa $m^{1/2}$)
30	1.40
1,000	1.50
1,200	1.90
1,400	2.60

Remarks/Comments
Density of the composites was 2.29 g cm^{-3}. Fracture toughness increased as temperature increased.

TiC(p)/Si_3N_4

Reference: 38

Composition
0 and 15 vol% TiC

Processing Method
Hot-Pressed

Detailed Description
The starting powders were homogenized and mixed together. The mixtures were filter pressed at 2.07 MPa for ~45 minutes over a 0.02 μm polymer filter. The mixtures were dried for 2 weeks. The resulting green bodies were hot-pressed under 41 MPa at 1800 °C for 2 hours.

Sintering Additives
6 wt% CeO_2 (Cerac Inc., Milwaukee, WI)

Reinforcement Description
TiC particulates

Reinforcement Supplier
Hermann C. Starck, Inc., New York, NY

Supplier Designation
c.a.s. Grade

Reinforcement Dimensions
ave. particle size = 1.25 μm

Matrix Description
α–Si_3N_4

Matrix Supplier
Hermann C. Starck, Inc., New York, NY

Matrix Designation
Grade LC12

Matrix Properties
ave. particle size = 0.55 μm

Test Type
4-pt Chevron Notched Beam (CNB)

Specimen Dimensions
50.8 x 4.47 x 5.56 mm
inner span = 22.2 mm
outer span = 44.4 mm

Type of Specimen
Chevron Notched Beam

Test Description
crosshead speed = 0.51 mm/s

Test Temperature
RT

Reinforcement Content (volume percent)	Fracture Toughness, K_{IC} (MPa m$^{1/2}$)
0	5.40
15	7.20

Remarks/Comments

Fracture toughness increased significantly with the addition of TiC particulates to the Si_3N_4 matrix.

TiC(p)/Si_3N_4

Reference: 95

Composition

0 and 20 vol% TiC

Processing Method

Gas Pressure Sintered

Detailed Description

The powders were mixed wet or attrition-milled under isoproponal or hexane. Sintering was done in a graphite resistance furnace under positive pressure of N_2 (10 mm Hg) and at a temperature of 1750 °C. The pressureless sintered materials were in the form of billets initially isostatically-pressed in the green state (67 x 17 x 17 mm size).

Sintering Additives

Y_2O_3
AlN
quartz powder

Reinforcement Description

TiC powder

Reinforcement Supplier

H.C. Starck, Goslar, Germany

Reinforcement Dimensions

surface area = 4.5 $m^2\ g^{-1}$

Matrix Description

Si_3N_4 powder

Matrix Properties

surface area = 8.78 $m^2\ g^{-1}$

Test Type

Single Edge Notched Beam (SENB)

Specimen Dimensions

dia. = 25 mm
thickness = 8-10 mm
notch width = 0.2-0.3 mm

Type of Specimen

Notched Beam

Test Temperature

RT-1250 °C

Temperature (°C)	Fracture Toughness, K_{IC} (MPa m$^{1/2}$)	Reinforcement Content (volume percent)
22	6.81	0.0
800	6.85	0.0
1,000	5.56	0.0
1,250	5.26	0.0
22	5.80	20
1,000	4.22	20
1,250	1.52	20

Remarks/Comments

TiC converted to TiN during sintering in presence of N_2. This conversion was less in the case of hot-pressed products. Higher amounts of sintering aid with Si_3N_4 had hardly any effect on fracture toughness. TiC can improve fracture toughness by increasing the crack propagation path. All percentages of TiC in composites decreased fracture toughness of base Si_3N_4 because of its conversion to TiN.

Figure 81: Fracture Toughness vs. Temperature of $SiC(p)/Si_3N_4$, $BN(p)/Si_3N_4$, and $TiC(p)/Si_3N_4$

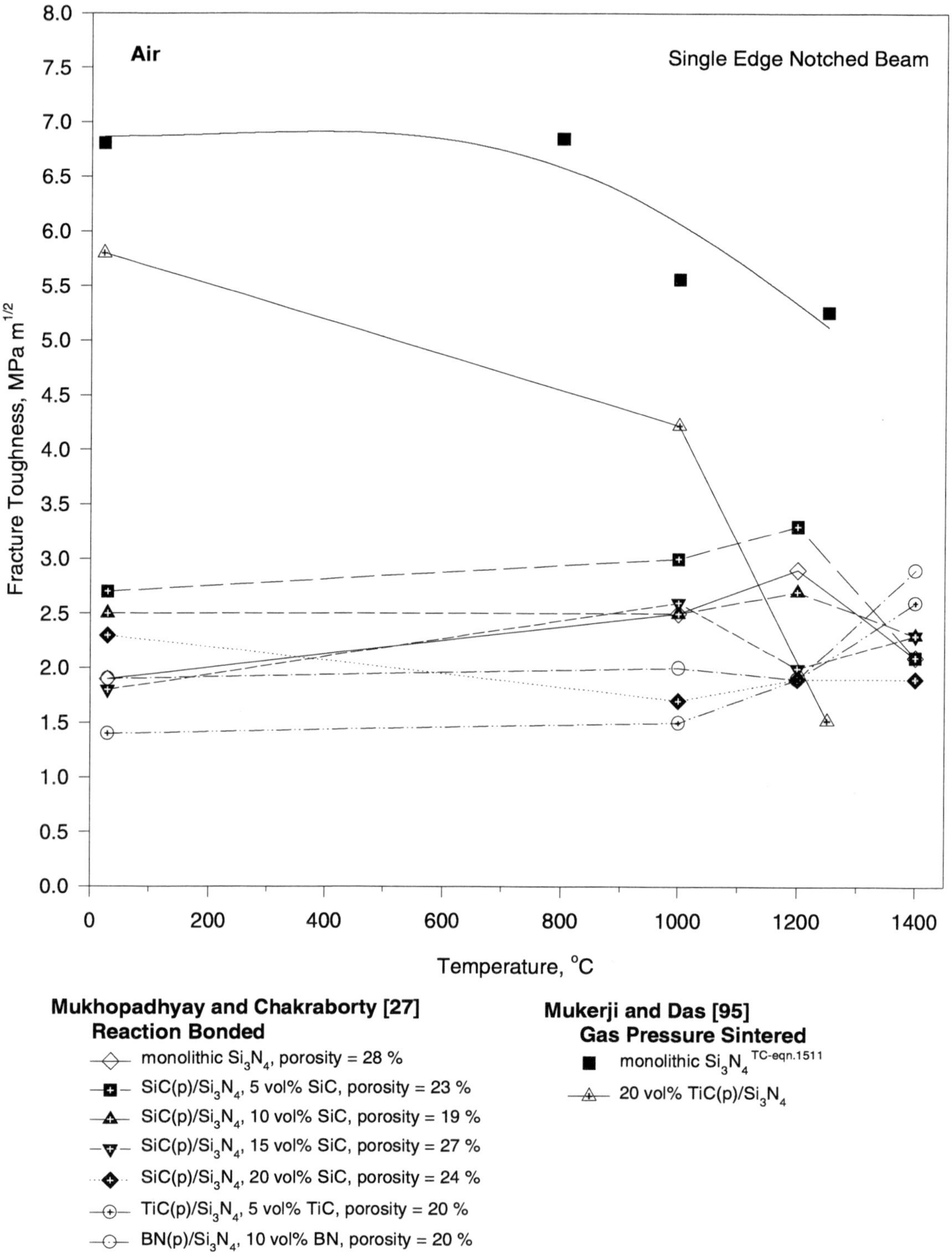

Discussion: Mukhopadhyay and Chakraborty [27] mixed BN, TiC, or varying quantities of SiC filler powders to the silicon powder prior to nitridation to form particulate reinforced silicon nitride composites. All the composites and the monolithic RBSN contained significant porosity, between 20 to 30 %, after nitridation. The room temperature fracture toughness was controlled by residual thermal expansion mismatch stresses which caused crack deflection and branching as well as microcracking. At room temperature, the 5 vol% SiC(p)/RBSN composite showed the highest toughness because the reinforcement created residual stresses, yet did not cause extensive microcracking. Adding further SiC content decreased composite toughness because overlapping stress fields caused microcracks to link up and cause flaw initiation sites and weak points within the material. BN or TiC particulates were not as effective toughening agents as SiC particulates at room temperature.

All reaction bonded composites showed an increase in toughness in the temperature range of 1000 to 1400 °C. This increase in toughness was not due to crack deflection or microcracking caused by residual stresses, since residual stresses disappear at temperatures approaching the processing temperature of the composite. The increase in toughness was likely caused by oxidation of the RBSN matrix and particulate reinforcement. Since the composites were highly porous, air could easily infiltrate deep into the composite. Oxidation of silicon nitride shows self healing properties, and the SiO_2 reaction product likely healed some of the microcracks and filled some open porosity, creating a composite with improved toughness.

The gas pressure sintered composites tested by Mukerji and Das [95] showed 2-3 times higher toughness than the RBSN composites, primarily because of the greater densification. The TiC acted to inhibit densification, therefore lowering the relative density of composite and resulted in a material with lower RT toughness. The higher porosity contained in the TiC(p)/Si_3N_4 composites left the structure open to oxidation at elevated temperatures, and the TiC particulates easily oxidized. This resulted in a rapid decrease in toughness above 1000 °C. The fracture toughness of the sintered monolithic ceramic remained fairly high up to 1200 °C.

Figure 82: Fracture Toughness of SiC(p)/Si_3N_4, BN(p)/Si_3N_4, and TiC(p)/Si_3N_4

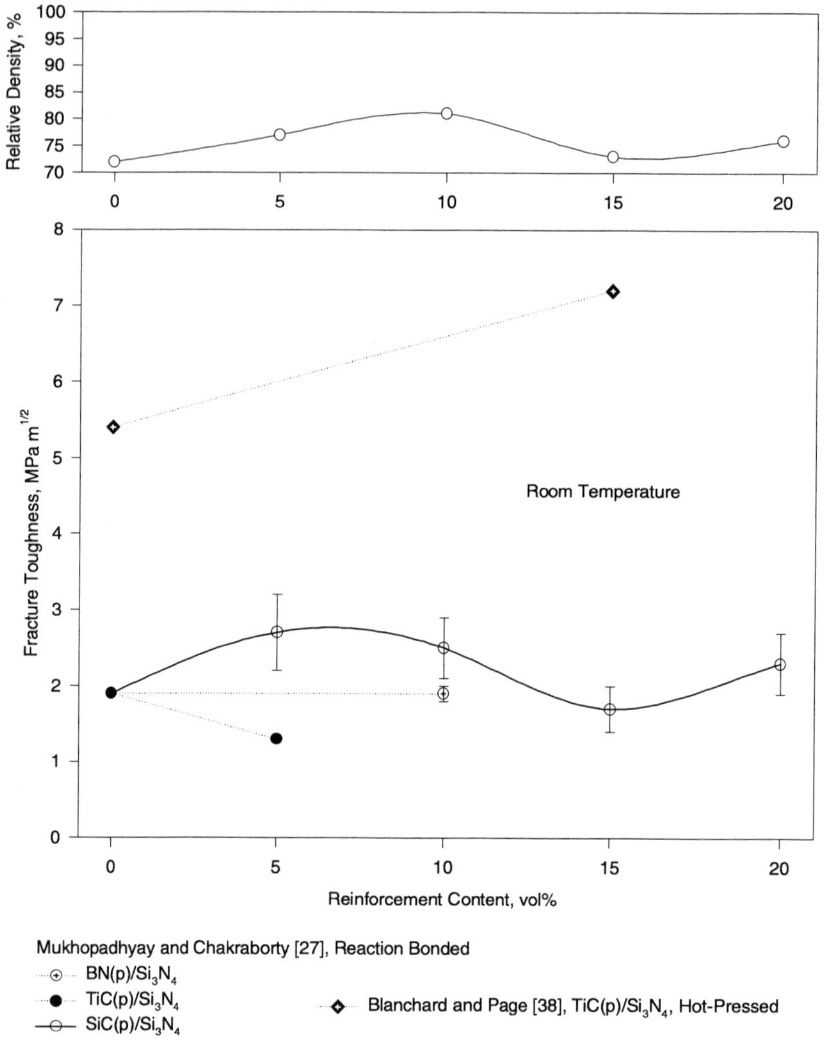

Discussion: This graph shows the relationships between reinforcement content and fracture toughness. All reaction bonded monolithic Si_3N_4 had a K_{IC} of 1.9 MPa $m^{1/2}$, whereas the hot-pressed monolithic Si_3N_4 started at 5.7 MPa $m^{1/2}$. The addition of 5 vol% TiC particulates to the RBSN caused a decrease in the fracture toughness, while the addition of 15 vol% to the hot-pressed samples caused an increase in the toughness. This seemingly low toughness of the RBSN samples in comparison to the hot-pressed samples was due to the large amount of porosity. RBSN samples had a significantly higher porosity than the other samples because of the nature of the processing. The addition of 10 vol% BN seemed not to change the toughness, while the addition of SiC caused fluctuations in the relative density that caused the change in toughness [27].

TiC(p)/Si_3N_4

Reference: 21

Composition
0 and 15 vol% TiC

Processing Method
Colloidal-Processed / Hot-Pressed

Detailed Description
The starting powders were homogenized and mixed together. The mixtures were filter pressed at 2.07 MPa for approximately 45 minutes over a 0.02 μm polymer filter. The mixtures were dried for 2 weeks. The resulting green bodies were hot-pressed under 41 MPa at 1800 °C for 2 hours.

Sintering Additives
6 wt% CeO_2
(Cerac Inc., Milwaukee, WI)

Reinforcement Description
TiC particulates

Reinforcement Supplier
Hermann C. Starck, Inc., New York, NY

Supplier Designation
c.a.s. Grade

Reinforcement Dimensions
ave. particle size = 1.25 μm

Matrix Description
α-Si_3N_4

Matrix Supplier
Hermann C. Starck, Inc., New York, NY

Matrix Designation
Grade LC12

Matrix Properties
ave. particle size = 0.55 μm

Test Type
Vickers Indentation Microfracture (IM) Technique[I-23]

Specimen Preparation
Specimens' surfaces were polished prior to testing.

Test Description
Indentations were made both parallel and perpendicular to the hot-pressing direction.
Indenter load = 10 kg
K_{IC} was calculated using the equation: $K_{IC} = 0.129(H*a^{1/2}) / ((c/a)^{3/2}*(H/E*\phi)^{2/5})$ where H is the hardness, a is the half-diagonal of the Vickers indent, c is the radius of the surface crack, E is the Young's modulus, and ϕ is the constraint factor. H, a, c, and E are determined experimentally from the sample.

Reinforcement Content (volume percent)	Fracture Toughness, K_{IC} (MPa $m^{1/2}$)	Comments
0.0	5.05	*perpendicular to the hot-pressed direction*
15	8.26	*perpendicular to the hot-pressed direction*
0.0	7.85	*parallel to the hot-pressed direction*
15	10.6	*parallel to the hot-pressed direction*

Remarks/Comments
Fracture toughness increased by about 63 % perpendicular and 35 % parallel to the hot-pressed direction for the TiC composites in both cases.

TiN(p)/Si_3N_4

Reference: 54

Composition
0-40 vol% TiN

Processing Method
Hot-Pressed

Detailed Description
The Si_3N_4 powder and sintering aids were homogenized in water for 48 hours and then freeze dried. The second phase (TiN) was added to the matrix mix and mixed in isobutyl alcohol for 48 hours. The materials were then hot-pressed at 1800 °C and 30 MPa for 10 minutes to nearly full density.

Sintering Additives
3 wt% Al_2O_3
8 wt% Y_2O_3

Reinforcement Description
TiN particulates

Reinforcement Supplier
Starck, Germany

Supplier Designation
Grade C

Reinforcement Dimensions
ave. size = 0.3-3 μm

Matrix Description
Si_3N_4 powder

Matrix Supplier
Starck, Germany

Matrix Designation
LC12

Test Type
Indentation Microfracture (IM) Technique[I-16]

Test Temperature
RT

Reinforcement Content (volume percent)	Fracture Toughness, K_{IC} (MPa $m^{1/2}$)	Relative Density (percent)
0.0	4.80	100
20	6.40	99.4
30	9.20	99.6
40	6.90	99.4

Remarks/Comments
The fracture toughness reached its maximum value when 30 vol% TiN was added (9.2 MPa $m^{1/2}$). All additions of TiN increased the fracture toughness over the monolithic Si_3N_4.

TiN(p)/Si_3N_4

Reference: 65

Composition
12 - 40 wt% TiN

Processing Method
Hot pressed

Detailed Description
Specimens were hot-pressed in graphite dies. Pressure was applied uniaxially. Heating was done by an inductor from a high-frequency generator. Some were heat treated for 2 hours at 1450 °C in a reducing atmosphere.

Sintering Additives
Y_2O_3

Reinforcement Description
TiN particulates

Matrix Description
Si_3N_4 matrix

Test Type
Vickers Indentation Microfracture Technique (IM)

Specimen Dimensions
12.7 x 12.7 x 3.5 mm

Test Description
Fracture toughness taken from crack length made by Vickers indenter.

Test Temperature
RT

Reinforcement Content (weight percent)	Fracture Toughness, K_{IC} (MPa $m^{1/2}$)	Comments
12	5.30	*before heat treatment*
30	5.80	
40	6.00	
12	7.00	*after heat treatment*
30	8.20	
40	8.30	

Remarks/Comments

Fracture toughness increased after heat treatment in a reducing atmosphere due to the crystallization of the grain boundary phases.

TiC(p)/Si_3N_4

Reference: 95

Composition
0-20 vol% TiC

Processing Method
Hot-Pressed

Detailed Description
The powders were mixed wet or attrition-milled under isoproponal or hexane. Hot-pressing was done in a vacuum hot press in BN coated graphite die and punched under a pressure of around 30 MPa at 1650 °C. The hot-pressed billet size was around 25 mm in dia. and 8 to 10 mm in thickness.

Sintering Additives
Y_2O_3
AlN
quartz powder

Reinforcement Description
TiC particulates

Reinforcement Supplier
H.C. Starck, Goslar, Germany

Reinforcement Dimensions
surface area = 4.5 m^2/g

Matrix Description
Si_3N_4 powder

Matrix Properties
surface area = 8.78 m^2/g

Test Type
Indentation Microfracture (IM) Technique[I-16]

Specimen Dimensions
dia. = 25 mm
thickness = 8-10 mm
notch width = 0.2-0.3 mm

Test Temperature
RT

Reinforcement Content (volume percent)	Fracture Toughness, K_{IC} (MPa $m^{1/2}$)	Density (g cm^{-3})
0.0	8.52	3.11
10	6.67	3.38
20	6.59	3.62

Remarks/Comments

TiC converted to TiN during sintering in the presence of N_2. This conversion was less in the case of hot-pressed products. Increasing the amount of sintering additives with Si_3N_4 had hardly any effect on fracture toughness. TiC can improve fracture toughness by increasing the crack propagation path. All percentages of TiC in composites decreased fracture toughness of base Si_3N_4 because of its conversion to TiN.

Figure 83: Fracture Toughness (IM) of TiC(p) and TiN(p)/Si_3N_4

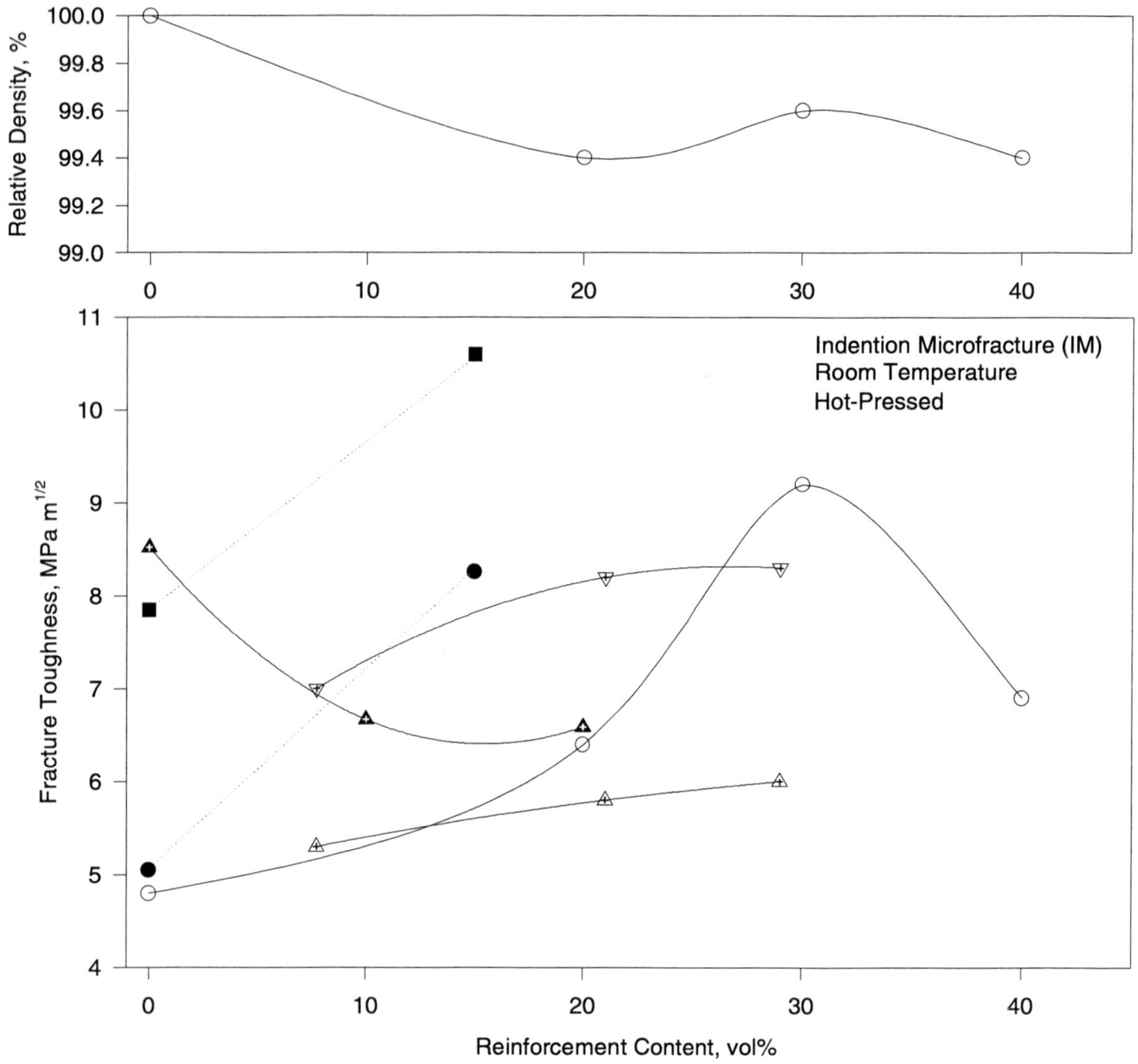

		Material	Sintering Additive	Equation Used	Comments
●	Blanchard and Page [21]	TiC	6 wt% CeO_2	[I-23]	⊥ to Hot-Pressing Direction
■	Blanchard and Page [21]	TiC	6 wt% CeO_2	[I-23]	ll to Hot-Pressing Direction
○	Bellosi *et al.* [54]	TiN	3 wt% Al_2O_3, 8 wt% Y_2O_3	[I-16]	
△	Gnesin *et al.* [65]	TiN	Y_2O_3	unknown	Before Heat Treatment
▽	Gnesin *et al.* [65]	TiN	Y_2O_3	unknown	After Heat Treatment
▲	Mukerji and Das [95]	TiC	Y_2O_3, AlN, quartz	[I-16]	

Discussion: Figure 83 presents indentation fracture toughness data for TiC and TiN particulate reinforced silicon nitride composites. TiC and TiN particulate reinforced Si_3N_4 composites have been examined as potential materials for cutting tool and wear applications. To increase the lifetime of cutting tools and wear parts, the hardness and fracture toughness must be simultaneously maximized. Gnesin *et al.* [65] and Mukerji and Das [95] attempted to maximize both hardness and fracture toughness of silicon nitride cutting tools and wear parts by the addition of particulate reinforcement. Fracture toughness of the as-hot-pressed TiN(p)/Si_3N_4 composite tested Gnesin *et al.* [65] increased slightly with increasing TiC content; however, after annealing the fracture toughness greatly increased by 2 MPa $m^{1/2}$ or more. The fracture toughness, and consequently wear resistance, increased due to the crystallization of yttria containing grain boundary phases to Y_2SiO_5.

Mukerji and Das [95] also examined TiC as a particulate reinforcement of materials used for cutting tools. The hot-pressed samples had a higher toughness when indented with a load parallel to the hot-pressing direction. The addition of TiC lowers the fracture toughness by 23 percent. The flexural strength also decreased with the addition of TiC. The conversion of the TiC to TiN during processing was detrimental to the mechanical properties of this composite [95].

Because structural ceramics typically come in contact with other parts or are hit by debris, improved contact damage resistance is considered critical for successful application of ceramics in real applications. For this reason, titanium carbide-silicon nitride composites are being examined as structural components. Studies have shown that contact damage resistance is a function of fracture toughness, hardness and coefficient of friction [21]. Wear tests have shown that TiC(p)/Si_3N_4 composites have high hardness and toughness. One study found that the coefficient of friction (μ_f) of TiC(p)/Si_3N_4 was 0.38. This is much lower than similar materials, such as Si_3N_4, SiC(p)/ Si_3N_4 and BN(p)/Si_3N_4 materials, for which $\mu_f = 0.62$. The low coefficient of friction is achieved by the formation of a lubricious oxide film containing Si and Ti [21].

Results from the fracture toughness tests performed by Blanchard and Page [21] show that the addition of TiC reinforcement was effective in toughening the material, (K_{IC} increased by ~ 3 MPa $m^{1/2}$). The fracture toughness also showed significant anisotropy due to preferred orientation of the grains during hot-pressing. The fracture toughness of the composites tested parallel to the hot-pressing direction was much greater than the fracture toughness tested perpendicular to the hot-pressing direction.

TiN and TiC particulate reinforced ceramics are also being examined as electrically conductive ceramics. One advantage of electrical conductivity is the ability to be machined by electrical discharge machining (EDM). Bellosi *et al.* [54] tested an electrically conductive TiN(p)/Si_3N_4 composite. The composite reached a maximum toughness of 9.2 MPa $m^{1/2}$ at 30 vol% TiN, which is almost double that of the monolithic silicon nitride which had a toughness of 4.8 MPa $m^{1/2}$.

TiC(p)/SiC

Reference: 11

Composition
0-24 vol% TiC

Processing Method
Hot-Pressed

Detailed Description
The powders were mixed in an acetone solution of a phenolic resin, dried by stirring, and calcined in Ar to convert the resin to carbon. Al powder was added to the mixtures during grinding in a B_4C mortar and pestle. Specimens 38 mm in diameter by 25 mm high were hot-pressed in graphite dies lined with graphite foil in vacuum (~10 MPa) at 2000 °C and 70 MPa.

Sintering Additives
1 wt% Al
1 wt% C

Reinforcement Description
TiC particulates

Reinforcement Supplier
H. C. Starck GmbH, Berlin, Germany

Reinforcement Dimensions
dia. = 2-5 μm

Matrix Description
SiC powder

Matrix Supplier
H. C. Starck GmbH, Berlin, Germany

Matrix Designation
A-10

Matrix Properties
particle size < 1 μm

Test Type
Applied Moment Double Cantilever Beam Method (DCB)

Specimen Preparation
Flat face opposite the groove was ground to a 37 μm finish.

Specimen Dimensions
10.2 x 2.5 x 31.8 mm

Type of Specimen
Bar

Test Temperature
RT

Reinforcement Content (volume percent)	Fracture Toughness, K_{IC} (MPa $m^{1/2}$)	Range (MPa $m^{1/2}$)
0.0	3.90	
8.0	5.29	+0.64/-0.42
16	5.27	+0.46/-0.48
24	6.19	

Remarks/Comments
The fracture toughness increased with increasing volume fraction of TiC. The toughening was believed to be due to the thermal expansion mismatch between the TiC and the SiC. This produced a stress field around the particle, which caused a propagating crack to be deflected.

TiC(p)/SiC

Reference: 20

Composition
0-25 vol% TiC

Processing Method
Hot-Pressed

Detailed Description
Powders and sintering aids were mixed together. Hot pressed at 2000-2100 °C and 40 MPa.

Sintering Additives
1.5 wt% Al_2O_3
1 wt% C

Reinforcement Description
TiC particulates

Reinforcement Dimensions
coarse TiC (ave. particle size = 7.2 μm)
fine TiC (ave. particle size = 0.82 μm)

Matrix Description
α-SiC powder

Matrix Properties
ave. particle size = 2.0 μm

Test Type
Single Edge Notched Beam (SENB), Indentation Strength in Bending (ISB)

Specimen Dimensions
2.5 x 30 mm cross-section (SENB)
notch depth = 2.0 mm (SENB)
3 x 4 x 26 mm (ISB)

Type of Specimen
Bar

Test Description
indentation load = 10 kg

Test Temperature
RT

Reinforcement Content (volume percent)	Fracture Toughness, K_{IC} ($MPa\ m^{1/2}$)	Density ($g\ cm^{-3}$)	Comments
0.0	4.50	3.17	*SENB (coarse)*
7.0	5.70	3.22	
13	6.60	3.37	
25	6.20	3.65	
7.0	6.57	3.18	*ISB (fine)*
13	6.48	3.38	
25	7.15	3.63	

Remarks/Comments
Fine particles showed a higher fracture toughness but not a significant difference in densities.

TiC(p)/SiC

Reference: 53

Composition
50 wt% TiC

Processing Method
Hot-Pressed

Detailed Description
Powders of either α– or β– types of SiC were mixed with TiC together with the sintering additives. Well-mixed powders were obtained by ball-milling in n-hexane for 24 hours using polyamide media. After drying, the powders were loaded into a graphite die with an inner diameter of 90 mm, and hot pressed at 2150 °C for 2 hours under 40 MPa in Ar gas atmosphere.

Sintering Additives
0.5 wt% B and 1 wt% C

Reinforcement Description
TiC particulates

Reinforcement Supplier
Kojundo Chem. Lab. Co. Ltd.

Reinforcement Typical Properties
average particle size = 1.2 μm
purity > 99 %
CTE = 7.9 x 10^{-6} C^{-1}

Matrix Description
SiC powder (β– or α– + β– (1:1 ratio) or α–)

Matrix Supplier
Ibiden Co., Ltd. (β–) Showa Denko K. K. (α–)

Matrix Designation
Ultrafine (β–) DU A-1 (α–)

Matrix Properties
ave. particle size = 0.3 μm (β–)
purity > 98 %
ave. particle size = 0.4 μm (α–)
purity > 98 %
α-SiC: CTE = 4.8 x 10^{-6} C^{-1}
β-SiC: CTE = 4.4 x 10^{-6} C^{-1}

Test Type
3-pt Single Edge Precracked Beam (SEPB)[I-4]

Applicable Test Standard
JIS R-1601 and JIS R-1604 (for RT and elevated temps, respectively)

Specimen Dimensions
3 x 4 x 38 mm

Type of Specimen
Bar

Test Temperature
RT

Relative Density (percent)	Fracture Toughness, K_{IC} (MPa $m^{1/2}$)	Comments
98.80	6.00	*β– only*
98.50	5.20	*α– only*
98.40	5.50	*α– + β–*

Remarks/Comments
The composite made with β-SiC has the highest fracture toughness (6.0 MPa $m^{1/2}$). The fracture toughness decreased with increasing α-SiC content and decreasing β-SiC content.

TiC(p)/SiC

Reference: 70

Composition
0-100 wt% TiC

Processing Method
Hot-Pressed

Detailed Description
Powders of β-SiC and TiC were mixed with B_4C powder and carbon black as sintering aids for SiC. Well-mixed powders were attained by ball-milling in n-hexane for 24 hours using a polyamide media. After drying, the powders were loaded into a graphite die with id = 90 mm and hot-pressed at 2150 °C for 2 hours under 40 MPa in an Ar gas atmosphere. Hot-pressed discs of SiC-TiC composites were machined into pieces.

Sintering Additives
B_4C: (Denki Kagaku Kogyo KK)
ave. particle size = 1.2 μm
purity > 99 %

C: (Mitsubishi Chem. Ind.)
ave. particle size = 0.02 μm
purity > 99 %

Reinforcement Description
TiC powders

Reinforcement Supplier
Kojundo Chem. Lab. Co. Ltd

Reinforcement Typical Properties
ave. particle size = 0.4 μm
purity > 99 %

Matrix Description
β-SiC powders

Matrix Supplier
Ibiden Co. Ltd

Matrix Designation
Ultrafine

Matrix Properties
ave. particle size = 0.3 μm
purity > 98 %

Test Type
Single Edge Precracked Beam (SEPB) Technique [I-4]

Specimen Dimensions
3 x 4 x 38 mm

Type of Specimen
Single-Edge Pre-Cracked Beam

Test Temperature
RT

Reinforcement Content (weight percent)	Fracture Toughness, K_{IC} (MPa $m^{1/2}$)	Range (MPa $m^{1/2}$)
0.0	1.69	
10	2.63	
30	5.00	
50	5.94	±0.16
70	5.40	
90	3.82	±0.40
100	3.09	

Remarks/Comments
Fracture toughness peaked at nearly 6 MPa $m^{1/2}$ at 50 wt% TiC.

TiC(p)/SiC

Reference: 90

Composition
0 or 33 vol% TiC

Processing Method
Hot-Pressed

Detailed Description
After the additives were mixed with the powders by wet ball milling, the samples were hot-pressed at 2000 °C and 30 MPa for 1 hour in argon atmosphere.

Sintering Additives
2 wt% AlN, Ti, or C

Reinforcement Description
TiC powder

Reinforcement Supplier
Nippon Shinkinzota Co., Osaka, Japan

Supplier Designation
TiC or TiC-M

Supplier Dimensions
TiC: ave. particle size = 1.36 μm
TiC-M: ave. particle size = 1.45 μm

Matrix Description
β-SiC powder

Matrix Supplier
Ibiden Co., Gifu, Japan

Test Type
4-pt Single Edge Precracked Beam (SEPB) [I-4]

Specimen Dimensions
2 x 3 x 20 mm
inner span = 2 mm
outer span = 14 mm

Type of Specimen
Precracked Beam

Test Description
crosshead speed = 0.5 mm/min

Test Temperature
RT-1600 °C

Test Environment
vacuum (5.3e-3 Pa)

Temperature (°C)	Fracture Toughness, K_{IC} ($MPa\ m^{1/2}$)	Relative Density (percent)	Comments
22	3.74	99.6	*monolithic (0 vol% TiC)*
337	3.54	99.6	*2 wt% AlN*
610	3.51	99.6	*C/Ti = 0.86*
1,010	3.59	99.6	
1,190	3.52	99.6	
1,370	2.81	99.6	
1,570	1.89	99.6	
22	6.88	98.2	*composite (33 vol% TiC)*
334	5.50	98.2	*2 wt% AlN*
623	5.35	98.2	*C/Ti = 0.86*
1,000	4.50	98.2	
1,180	3.83	98.2	
1,370	3.42	98.2	
1,570	2.59	98.2	
22	4.96	98.5	*composite (33 vol% TiC)*
335	4.51	98.5	*2 wt% Ti*
622	4.48	98.5	*C/Ti = 0.86*
1,000	4.20	98.5	
1,190	3.98	98.5	
1,370	3.73	98.5	
1,570	3.63	98.5	
22	3.21	82.8	*composite (33 vol% TiC)*
338	3.20	82.8	*2 wt% C*
606	2.52	82.8	*C/Ti = 0.86*
998	2.78	82.8	
1,180	2.53	82.8	
1,370	2.12	82.8	
1,560	2.29	82.8	
22	5.72	98.7	*composite (33 vol% TiC)*
334	5.07	98.7	*2 wt% AlN*
618	4.56	98.7	*C/Ti = 0.97*
1,000	3.99	98.7	
1,180	3.69	98.7	
1,360	3.31	98.7	
1,570	3.39	98.7	

Remarks/Comments

The fracture toughness of SiC matrix composite was improved in the temperature range from room temperature to 1570 °C by TiC particle dispersion. Between room temperature and the critical temperature, the toughening mechanism was either crack deflection or crack bridging depending on the additive. Above the critical temperature, the toughening mechanism was thought to be plastic deformation of TiC particles.

Figure 84: Fracture Toughness of TiC(p)/SiC

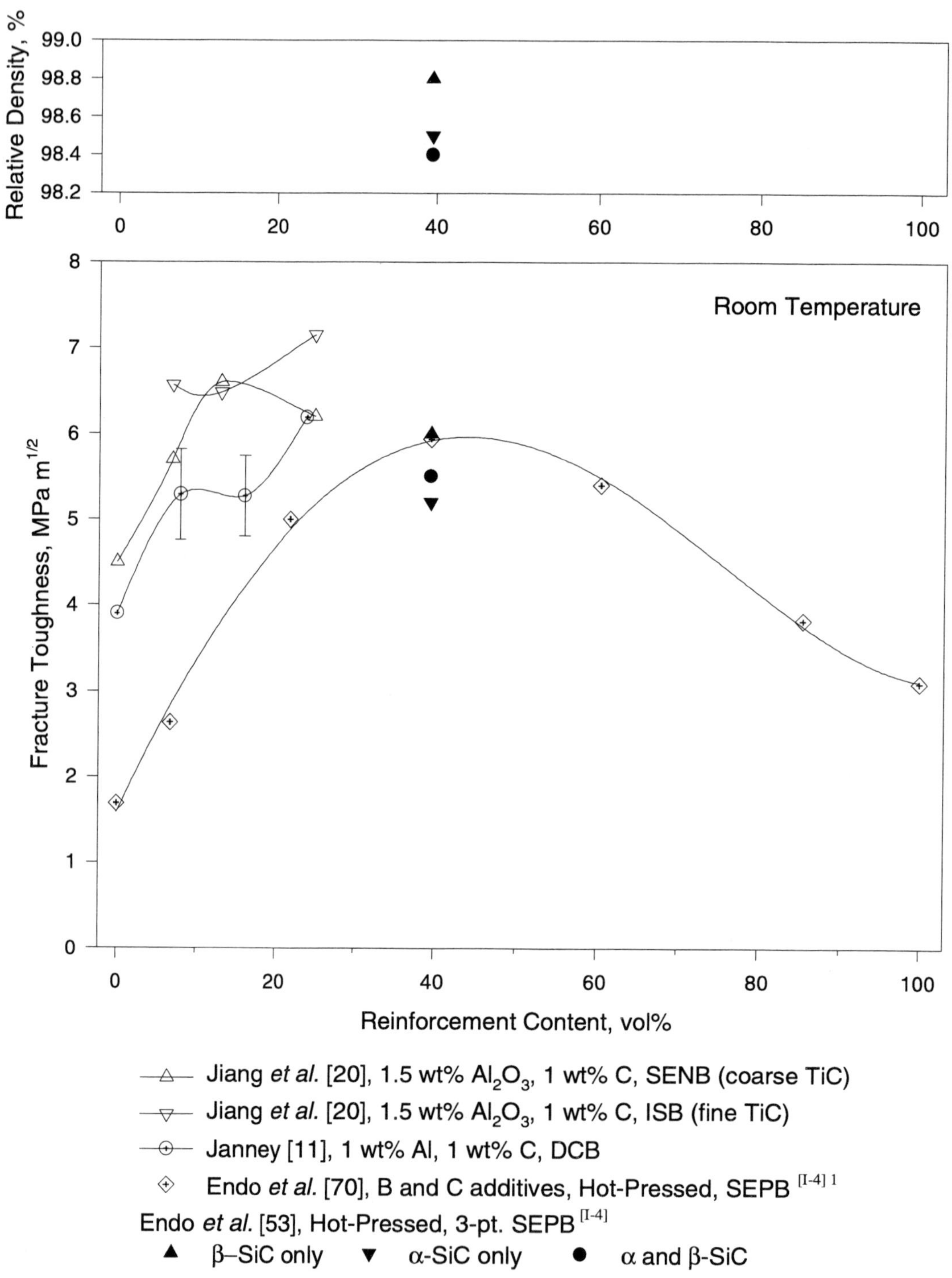

1. Converted from wt% to vol%

Discussion: Endo *et al.* [53 and 70] examined hot-pressed composites in the TiC-SiC system. First, composites were produced in which the TiC content was varied from 0 to 100 vol% in a β-SiC matrix by hot-pressing at 2150 °C for 2 h under 40 MPa pressure with B_4C and C sintering additives. All compositions were near fully dense. Crack deflection was rarely observed in the monolithic SiC, and fracture toughness of the SiC was less than 2 MPa $m^{1/2}$. The addition of TiC caused considerable crack deflection. Fracture toughness, consequently, crack deflection, reached a maximum at 40 vol% TiC where the fracture toughness reached 6 MPa $m^{1/2}$. Crack deflection was caused by the residual thermal stresses from CTE mismatches [70]. Microstructural analysis showed some evidence of solid-solution formation between SiC and TiC, but for the most part SiC and TiC existed as discrete phases. Endo *et al.* [53] also produced TiC-SiC composites at the optimum composition using α-SiC or β-SiC powders, or a mixture of the two. β-SiC produced composites with greater strength and toughness. The superior mechanical properties of the β-SiC composites was partly due to the greater CTE mismatch between TiC and β-SiC. It also may be due to the higher impurity (metallic elements) content found in α-SiC powders. Metallic impurities may migrate to the grain boundaries during sintering which would be detrimental to mechanical properties.

Jiang *et al.* [20] showed that at 25 vol% TiC reinforcement, the TiC(p)/SiC composite with fine TiC particulates was about 1 MPa $m^{1/2}$ tougher than the coarse TiC composite. The toughening mechanisms found in these composites were the same observed by Endo *et al.* [53 and 70]. At 13 vol% TiC, the fracture toughness was about the same for the composites reinforced by coarse TiC and fine TiC. At 25 vol%, the fracture toughness of the fine TiC composites was greater than the coarse TiC composites. Larger TiC grains may have increased microcrack density because of a greater thermal expansion mismatch between the matrix SiC and discrete TiC. Coarse TiC may have resulted in lower fracture toughness due to extensive microcracking. This difference may also have been due to the different test methods. Janney [11] found that the fracture toughness (measured by DCB) increased from 4 MPa $m^{1/2}$ for monolithic SiC to 6 MPa $m^{1/2}$ for the 24 vol% TiC(p)/SiC composite.

Figure 85: Fracture Toughness[I-4] vs. Temperature of TiC(p)/SiC

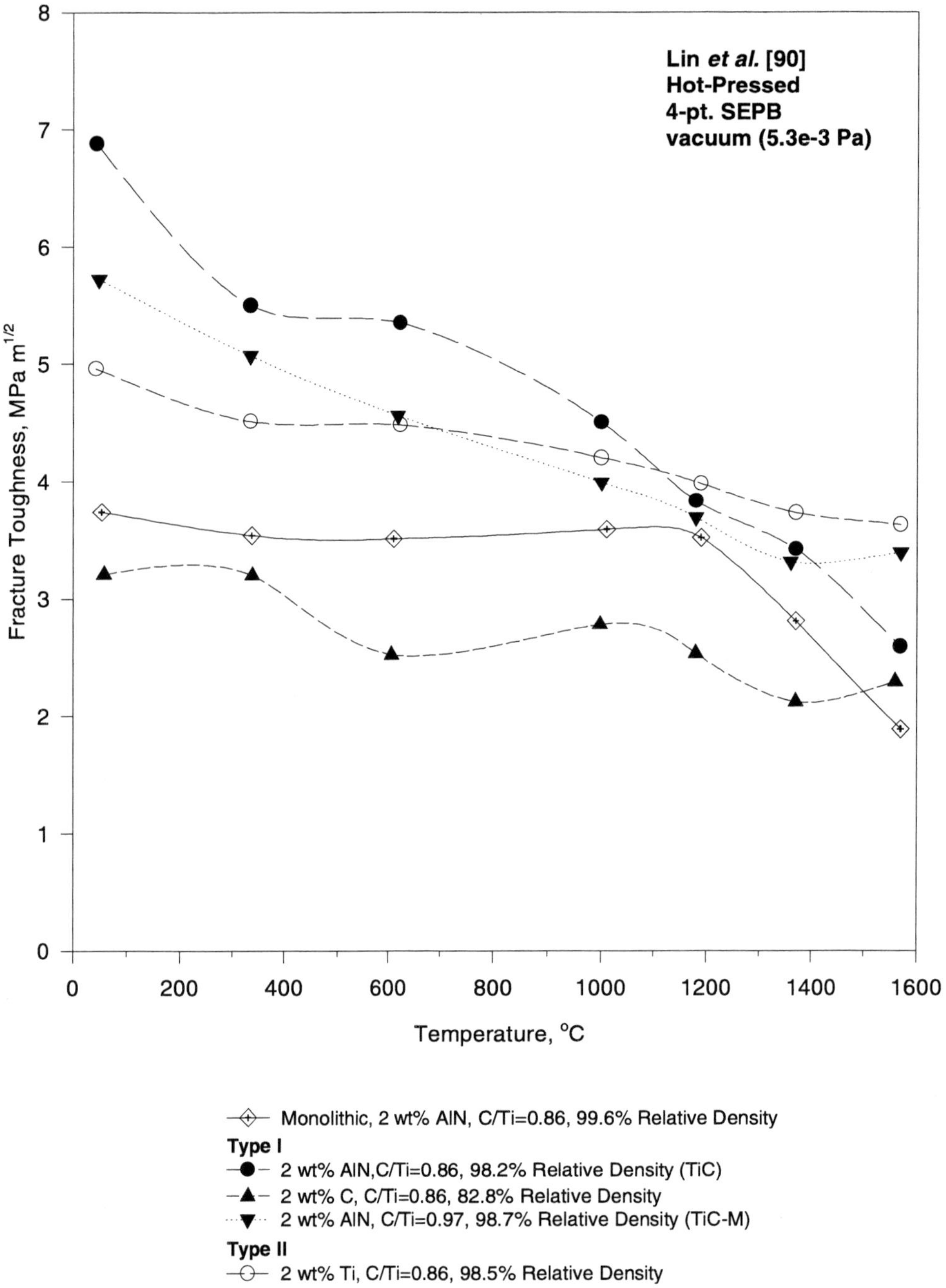

Discussion: Lin *et al.* [90] tested two types of TiC/SiC composites that were created by changing the dopants used in processing. The type I composites were fabricated with AlN or C dopants. The primary toughening mechanism of these composites was crack deflection caused by a thermal residual stress field surrounding TiC particles. The SiC matrix was under radial matrix tension and tangential matrix compression, while TiC particles were under hydrostatic tension. The type II composite was fabricated by the addition of 2 wt% Ti. In this composite, the SiC matrix was under radial matrix tension and tangential matrix compression generated by thermal expansion mismatch, and TiC particles were under internal compression due to Ti diffusion into the TiC particles creating chemical residual stress [90].

Figure 85 shows that fracture toughness decreased as temperature increased for all samples tested. This was due to the room temperature toughening mechanism of crack deflection no longer being effective at high temperatures. The type II composite with 2 wt% Ti decreased in toughness less than the others because the toughening mechanism of crack bridging due to the chemical residual stress was less dependent on temperature [90]. Above 1200 °C, plastic deformation of TiC particulates was an effective toughening mechanism. Plastic deformation was particularly effective above 1400 °C in the type II composites and in the type I composites fabricated with TiC-M powders. The monolithic SiC did not decrease in toughness until temperatures were more than 1100 °C, where the drop in toughness was related to an amorphous phase which existed in the grain boundaries and formed from intrinsic impurities and/or process conditions [90]. The overall low toughness of the composites with 2 wt% C was due to the low density.

TiB_2(p)/SiC

Reference: 51

Composition
16 vol% TiB_2

Processing Method
Hot-Pressed

Detailed Description
Composite was supplied by Standard Oil Engineered Materials Co., Niagara Falls, NY, and is designated Hexoloy ST.

Reinforcement Description
TiB_2 powder

Matrix Description
α-SiC

Test Type
Straight Notch Beam[I-18]
Chevron Notch Beam[I-26]

Specimen Preparation
The notches were machined with a rotating, diamond-tipped saw blade which produced a notch width of 0.30 mm.

Specimen Dimensions
6.36 x 6.36 x 50 mm
span = 40 mm
notch depth ratios(a_0/W) = 0.44 (Chevron) and 0.50 (Straight Notch)

Type of Specimen
Notched Bar

Test Description
crosshead speed = 0.01 mm/min; A laser interferometric strain gage (LISG) was used to monitor specimen crack mouth opening displacement (CMOD).

Test Temperature
RT-1400 °C

Test Environment
Air

Test Equipment
Instron tester, model TTDML

Temperature (°C)	Fracture Toughness, K_{IC} (MPa $m^{1/2}$)	Range (MPa $m^{1/2}$)	Comments
20	4.63	±0.23	*Straight Notched Beam Test*
200	4.21	±0.14	
400.	4.49	±0.26	
600	4.73	±0.22	
800	4.26	±0.19	
1,000	4.32	±0.35	
1,200	4.63	±0.64	
1,400	4.13	±0.35	
20	4.07	±0.17	*Chevron Notched Beam Test*
200	3.47	±0.04	
400	3.59	±0.11	
600	3.14	±0.09	
800	3.15	±0.18	
1,000	2.75	±0.39	
1,200	2.81	±0.11	
1,400	2.86	±0.27	

Remarks/Comments
The fracture toughness values were much higher for the straight notch test than the chevron notch test. The fracture toughness did not vary much with temperature for the straight notched beam test, but for the chevron test fracture toughness values decreased with increasing temperature.

TiB_2(p)/SiC

Reference: 69

Composition
0-30 vol% TiB_2

Processing Method
Reaction Sintered

Detailed Description
Fine TiB_2 particles were prepared by milling the TiB_2 agglomerates produced by the following carbothermal reaction at 2000 °C:

$$TiO_2 + 0.5\ B_4C + 1.5\ C \dashrightarrow TiB_2 + 2\ CO$$

The removal of the CO gas from the system was important to enhance the reaction at lower temperatures. The TiO_2 particle size was 0.4 μm, the B_4C powder size was 1.5 μm, and the carbon black powder size was 0.018 μm. The amounts of B_4C and carbon that were added were more than the stoichiometrically required amount. The mixing ratio of SiC and TiO_2 was designed so that the volume ratio of SiC matrix to TiB_2 particulates formed by the carbothermal reaction should be 7:3, 9:1, and 8:2. The starting materials were mixed with water by ball milling for 20 hours, using a polyethylene pot and nylon-coated iron balls. The slurries were filtrated by suction, and dried at 100 °C. Screened powders under 32-mesh were molded in dies at a pressure of 20 MPa and isostatically pressed at 300 MPa. The green compacts in a graphite container were heated in a vacuum at 1500 °C for 4 hours. After the vacuum soaking treatment, argon gas was introduced into the furnace and the compacts were sintered in an argon atmosphere at 2150 °C for 1 hour. Composites were also HIPed in a graphite container at 1750-1900 °C for 1 hour in argon at 200 MPa.

Sintering Additives
B_4C, Denki Kagaku Kogyo
C, Mitsubishi Chemical

Reinforcement Description
TiB_2 particulates

Reinforcement Supplier
Fuji Titan Kogyo

Supplier Designation
TM-1

Matrix Description
SiC powder

Matrix Supplier
Showa Denko Co.

Matrix Properties
ave. particle size = 0.48 μm

Test Type
Indentation Microfracture (IM) Technique [I-13] or Single Edge Precracked Beam (SEPB) Technique [I-4]

Specimen Dimensions
4 x 3 x 35 mm

Type of Specimen
Bar

Test Temperature
RT

Reinforcement Content (volume percent)	Fracture Toughness, K_{IC} (MPa m$^{1/2}$)	Range (MPa m$^{1/2}$)	Comments
0.0	3.00	±0.5	*IM*
10	4.40	±0.2	
20	5.80	±0.3	
30	6.90	±0.75	
0.0	2.00	±0.4	*SEPB*
10	3.00	±0.4	
20	3.80		
30	4.00	±0.3	

Remarks/Comments
Fracture toughness was averaged from 3 test results. Fracture toughness increased with increasing TiB_2 content. Fracture toughness values were greater for the IM method than for the SEPB method.

TiB_2(p)/SiC

Reference: 107

Composition
0 and 15 vol% TiB_2

Processing Method
Hot-Pressed

Detailed Description
The combined powders were vibromilled in isopropanol with 1 wt% polyvinylpyrollidone as dispersant, using 6.4 mm diameter by 100 mm long SiC rods. The polyvinylpyrollidone was burned out in argon at 500 °C prior to hot-pressing. Samples were hot-pressed in carbon dies lined with graphite foil at 2000 °C for 45 minutes at 35 MPa to final dimensions of 38 mm diameter by 25 mm high. Measurements were made on bars cut from the specimens.

Sintering Additives
1 wt% B
0.5 wt% C

Reinforcement Description
TiB_2 particulates

Reinforcement Supplier
KBI Division, Cabot Corp., Reading, PA

Reinforcement Dimensions
ave. particle size = 4-8 μm

Matrix Description
SiC powder

Matrix Supplier
H. C. Starck Co., Berlin, Germany

Matrix Designation
A-10

Matrix Properties
ave. particle size = 3-5 μm

Test Type
4-pt Multiple Indentation Flexural Strength (MIFS) [I-10]

Specimen Preparation
Samples were ground parallel to long axis with a 140/170 grit resin-bonded diamond wheel. The tensile face was ground to a 30 μm finish, and the tensile edges beveled.

Specimen Dimensions
2.3 x 2.5 x 25 mm

Type of Specimen
Bar

Test Description
crosshead speed = 0.51 mm/min

Test Temperature
RT

Reinforcement Content (volume percent)	Fracture Toughness, K_{IC} (MPa $m^{1/2}$)
0.0	3.10
15	4.50

Remarks/Comments
Data presented is an average of 9 specimens. Addition of TiB_2 to the SiC base increased the fracture toughness significantly.

TiB_2(p)/SiC

Reference: 32

Composition
15 vol% TiB_2

Processing Method
Hot-Pressed

Detailed Description
Composite was supplied by Carborundum Co., Niagara Falls, NY and is designated Hexoloy ST.

Reinforcement Description
TiB_2 particulates

Matrix Description
SiC

Test Type
Single Edge Notched Beam (SENB)

Specimen Dimensions
76.2 x 6.4 x 6.4 mm
notch depth = 4.55 mm
notch width = 0.25 mm

Type of Specimen
Notched Bar

Test Temperature
RT

Test Equipment
Laser Interferromagnetic Displacement Gauge

Fracture Toughness, K_{IC} (MPa $m^{1/2}$)	Comments
6.80	*static*

Figure 86: Fracture Toughness of TiB_2(p)/SiC

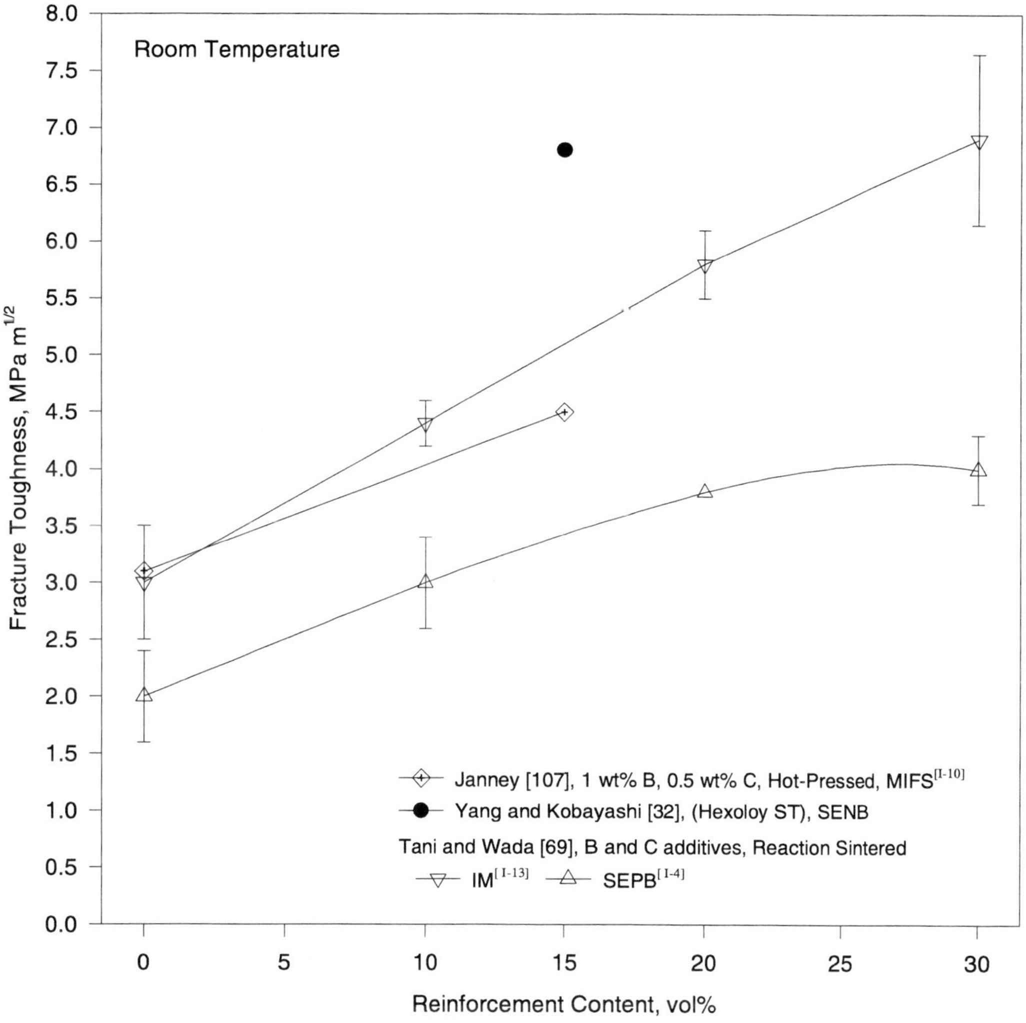

Discussion: Tani and Wada produced TiB_2(p)/SiC composites by reaction sintering in which TiO_2, B_4C and carbon powders reacted *in situ* to form TiB_2 reinforcement. The fracture toughness of these composites was measured by the indentation microfracture technique and SEPB method developed by Nose and Fuji [I-4]. The fracture toughness increased up to 20 vol% TiB_2 when measured by either method. Toughness increased again at 30 vol% when measured by IM, while the SEPB method found that the 30 vol% TiB_2 composite was only slightly higher than the 20 vol% TiB_2 composite. There was a wide difference between the toughness values measured by the two methods. The SEPB test method likely depicts the true toughness of the composite better than the IM method which is subject to errors in crack length measurement.

Janney [107] produced a 15 vol% TiB_2(p)/SiC composite by conventional hot-pressing of SiC and TiB_2 powders. The addition of 15 vol% TiB_2 to the SiC matrix increased the toughness from 3.0 to 4.0 MPa $m^{1/2}$. The fracture toughness of Hexoloy ST, which is a commercially available 15 vol% TiB_2(p)/SiC, was measured to be 6.8 MPa $m^{1/2}$ by SENB [32].

Figure 87: Fracture Toughness[I-18], [I-26] vs. Temperature of TiB_2(p)/SiC

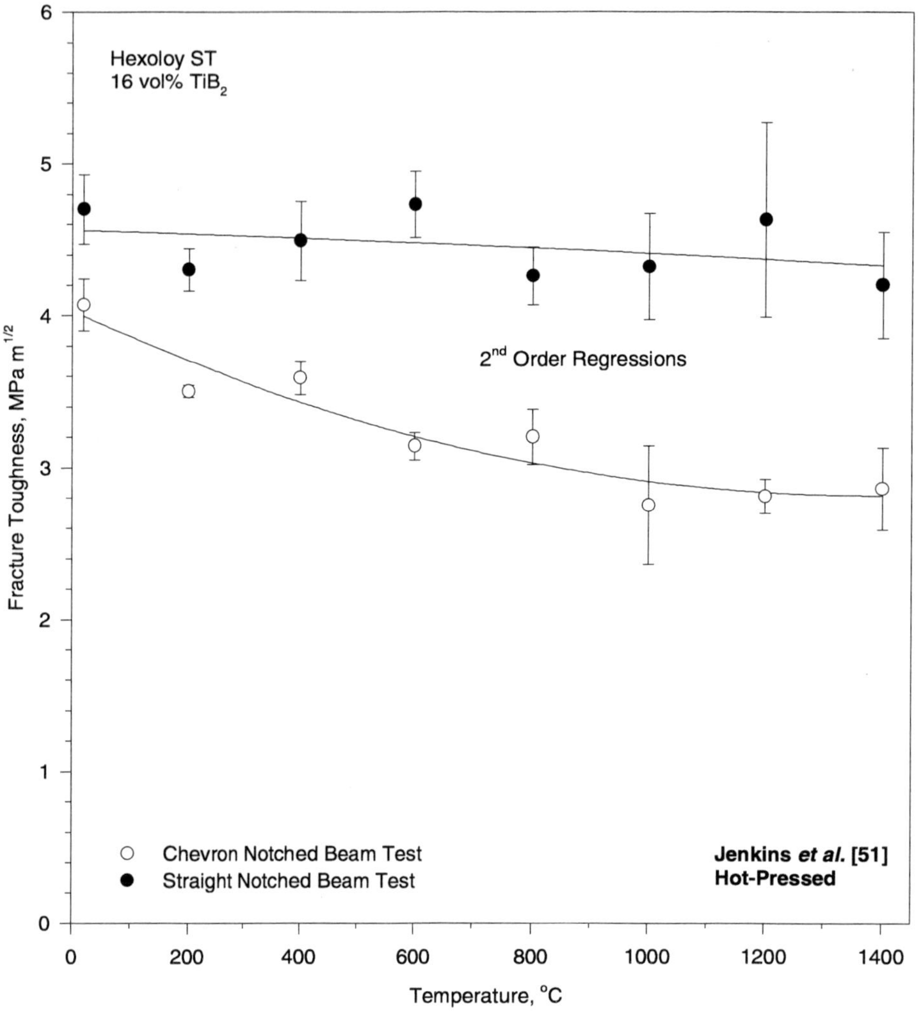

Discussion: Jenkins *et al.* [51] examined the differences in the straight edge notched bar test (SENB) and the chevron notched bar test (CNB) on a Hexoloy ST TiB_2(p)/SiC composite over a range of temperatures, shown in figure 87. Analysis of the fracture surfaces showed the TiB_2 was weakly bonded to the SiC matrix, which could cause some crack deflection and branching. At room temperature, the primary toughening mechanism was microcracking due to thermal expansion mismatch stresses. At elevated temperatures, the primary mechanisms were deformation mechanisms, such as TiB_2-SiC boundary separation.

It was determined that the fracture toughness data gathered from the chevron notched samples better represented the fracture behavior of a sharp crack than the relatively blunt notch tip of the straight notched samples [51]. The CNB showed the dependence of fracture toughness on temperature whereas the SENB showed little temperature dependence. The values from the CNB are lower than those for the SENB due to the geometry of the chevron notch not allowing a stress over-load. Using CNB over the SENB resulted in stable crack growth and led to reliable crack growth resistance curves (R-Curves).

Diamond(p)/SiC

Reference: 61

Composition
0-31 vol% diamond

Processing Method
Hot-Pressed

Detailed Description
The SiC powder and diamond particles were mixed in aqueous suspensions. The powder suspensions were evaporated during ultrasonic vibration to provide thorough dispersion of the of the individual diamond particles in the dry powder. This composite was pressed at 5.5 GPa and 1500 °C in a press capable of operation in the diamond stability range.

Reinforcement Description
Diamond particulates

Reinforcement Supplier
E. I. du Pont de Nemours & Co., Gibbstown, New Jersey

Supplier Designation
du Pont Mypolex

Reinforcement Dimensions
ave. particle size = 11 nm

Matrix Description
SiC powder

Matrix Supplier
Showa Denko K. K., Minato-Ku, Japan

Matrix Designation
DENSIC Type 2

Test Type
Vickers Indentation Microfracture Technique[I-16]

Specimen Preparation
Specimens were polished with 1 μm diamond paste, and ultrasonically cleaned to remove the polishing medium.

Test Description
load = 30 kg

Density measured by Archimedes method

Test Temperature
RT

Reinforcement Content (volume percent)	Fracture Toughness, K_{IC} (MPa $m^{1/2}$)	Relative Density (percent)	Range (MPa $m^{1/2}$)
0.0	3.55	97.5	+0.275/-0.328
10	3.51	95.2	+0.298/-0.301
18	7.12	95.8	+0.524/-0.503
25	7.10	93.6	+0.507/-0.480
31	3.47	91.5	+0.430/-0.378

Remarks/Comments
No increase in toughness was seen at <10 and >30 vol% diamond reinforcement. A remarkable increase (more than double) is produced between 18 and 25 vol% diamond. The decrease after 30 vol% may be due to the lower composite density.

TiB_2(p)/SiC

Reference: 83

Composition
0 and 16 vol% TiB_2

Processing Method
Colloidal Processed / Pressureless Sintered

Detailed Description
The commercially available SiC-TiB_2 (Hexoloy ST) CMC was fabricated by mixing presized TiB_2 and submicrometer-size SiC powders using a nonaqueous blending and drying procedure. The homogeneous mixture, combined with a binder, was formed into green compacts that were then pressureless-sintered in an inert atmosphere at temperatures of over 2000 °C.

Reinforcement Description
TiB_2 particulates

Reinforcement Typical Properties
melting point = 2790 °C
E = 531 GPa
Poisson ratio = 0.28
CTE = $4.6x10^{-6}$ °C^{-1}

Reinforcement Dimensions
ave. particle size = 2 μm

Matrix Description
α-SiC

Matrix Supplier
Sohio Engineered Materials Co., Niagara Falls, NY

Matrix Designation
Hexoloy ST

Matrix Properties
ave. grain size = 4 μm
melting point = 2540 °C
E = 410 GPa
Poisson ratio = 0.19
CTE = $4.02x10^{-6}$ °C^{-1}

Test Type
unknown

Test Temperature
RT

Reinforcement Content (volume percent)	Fracture Toughness, K_{IC} (MPa $m^{1/2}$)	Density (g cm^{-3})	Comments
0.0	4.60	3.10	***sintered α-SiC***
16	8.00	3.30	***composite***

Remarks/Comments
The fracture toughness increased significantly with the addition of TiB_2 particles.

Figure 88: Fracture Toughness of Diamond(p)/SiC and TiB_2(p)/SiC

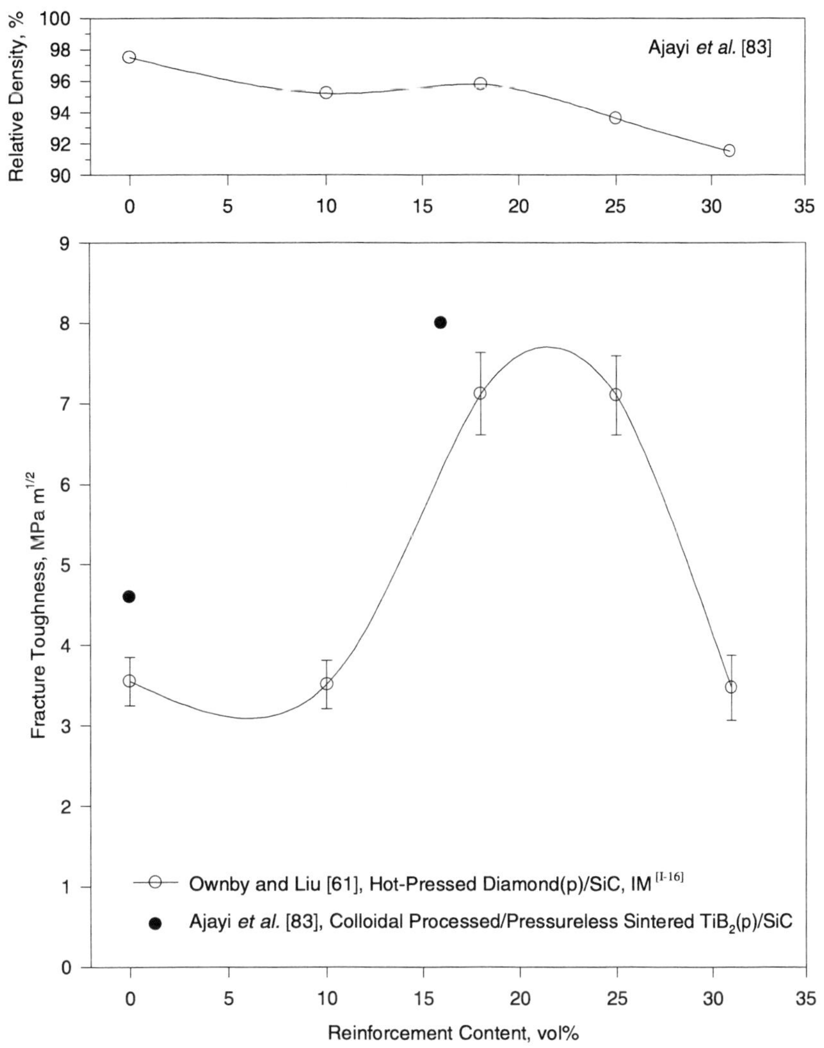

Discussion: Ownby and Liu [61] tested the fracture toughness of small samples of hot-pressed diamond particulate reinforced SiC by the Anstis *et al.* [I-16] technique. The diamond particles were extremely small, ~11 nm in diameter, and the composites were processed at a temperature and pressure within the diamond stability range (5.5 GPa). The composites did not show any increase in toughness at 10 vol%; however, there was a remarkable increase in toughness, approximately double that of monolithic SiC, at 18 and 25 vol% diamond reinforcement. This increase in toughness may be due to crack deflection by very hard and strong diamond particulates and due to overlapping stress fields caused by thermal expansion mismatch. Below 10 vol%, there were not enough of the diamond particles and the stress fields did not overlap. Above 30 vol% fracture toughness decreased, possibly due to poor densification and the formation of cracks due to large residual thermal stresses.

2Y-ZrO_2(p)/TiB_2

Reference: 34

Composition
20 wt% ZrO_2

Processing Method
CIPed/HIPed

Detailed Description
Starting powders were ball-milled in ethanol. The mixed powder was dried, uniaxially pressed at 9.8 MPa and CIPed at 300 MPa. The green bodies were HIPed at 1600 °C and 200 MPa for 2 hours in Ar gas by the glass-encapsulation method.

Sintering Additives
0-5 wt% SiC

Reinforcement Description
2Y-ZrO_2 particulates

Reinforcement Dimensions
particle dia. = 1-3 μm

Matrix Description
TiB_2

Matrix Properties
particle dia. = 2-4 μm

Test Type
Single Edge Pre-Cracked Beam (SEPB)

Test Temperature
RT

Specimen Dimensions
3 x 4 x 36 mm

Type of Specimen
Bar

SiC Content (weight percent)	Fracture Toughness, K_{IC} (MPa $m^{1/2}$)
0.0	5.1
3.0	5.1
5.0	5.2

Remarks/Comments
The SiC greatly improved the sinterability of the composite. However, at near full density, the fracture toughness was independent of the SiC content.

2Y-ZrO_2(p)/TiB_2

Reference: 34

Composition
0-40 wt% ZrO_2

Processing Method
CIPed/HIPed

Detailed Description
Starting powders were ball-milled in ethanol. The mixed powder was dried, uniaxially pressed at 9.8 MPa and CIPed at 300 MPa. The green bodies were HIPed at 1600 °C and 200 MPa for 2 hours in Ar gas by the glass-encapsulation method.

Reinforcement Description
2Y-ZrO_2 particulates

Reinforcement Dimensions
particle dia. = 1-3 μm

Matrix Description
TiB_2

Matrix Properties
particle dia. = 2-4 μm

Test Type
Single Edge Pre-Cracked Beam (SEPB)

Specimen Dimensions
3 x 4 x 36 mm

Type of Specimen
Bar

Test Temperature
RT

Reinforcement Content (weight percent)	Fracture Toughness, K_{IC} (MPa $m^{1/2}$)
0.0	3.4
5.0	3.7
10	4.6
20	5.5
40	7.5

Remarks/Comments
Fracture toughness increased with increasing reinforcement content due to crack deflection.

$Al_2O_3(p)/TiB_2$

Reference: 100

Composition
0 and 70 vol% Al_2O_3

Processing Method
Hot-Pressed

Detailed Description
Monolithic TiB_2 was hot-pressed at 1850 °C for 1 hour and at 30 MPa. TiB_2-Al_2O_3, 30 vol% TiB_2, was produced by hot-pressing at 1600 °C for 30 minutes and at 30 MPa.

Reinforcement Description
TiB_2 powder

Reinforcement Supplier
Starck

Supplier Designation
Grade F

Matrix Description
Al_2O_3 particulates

Matrix Supplier
ALCOA

Matrix Designation
A16 S. G.

Test Type
Indentation Microfracture (IM) Technique

Type of Specimen
Billet

Test Temperature
RT

Reinforcement Content (volume percent)	Fracture Toughness, K_{IC} ($MPa\ m^{1/2}$)	Range ($MPa\ m^{1/2}$)
0.0	5.40	±0.8
70	5.70	±0.6

Remarks/Comments
The $Al_2O_3(p)/TiB_2$ composite was slightly tougher than the monolithic TiB_2.

$B_4C(p)/TiB_2$

Reference: 101

Composition
0-30 vol% B_4C

Processing Method
Cold-Pressed/Pressureless Sintered

Detailed Description
Powder compacts formed by cold-pressing were fabricated by pressureless sintering at 1600 °C in an argon atmosphere.

Sintering Additives
8.6 mol% Ni
1.4 mol% C

Reinforcement Description
B_4C powder

Matrix Description
TiB_2 powder

Test Type
4-pt Indentation Fracture (IF) Technique

Specimen Dimensions
3 x 4 mm cross-section
inner span = 10 mm
outer span = 30 mm

Type of Specimen
Bar

Test Description
crosshead speed = 0.5 mm/min

Test Temperature
RT

Reinforcement Content (volume percent)	Fracture Toughness, K_{IC} ($MPa\ m^{1/2}$)	Relative Density (percent)	Range ($MPa\ m^{1/2}$)
0.0	5.10	94.3	-0.42/+0.28
10	6.07	94.7	-0.56/+0.29
20	5.95	93.7	-0.37/+0.34
30	4.99	88.3	-0.18/+0.15

Remarks/Comments
The fracture toughness and bending strength both increase, then decrease with the addition of B_4C, with a maximum of approximately 6 $MPa\ m^{1/2}$ and 533 MPa, respectively.

SiC(p)/ZrB_2

Reference: 109

Composition
10 wt% SiC

Processing Method
Hot-Pressed

Reinforcement Description
SiC particulates

Matrix Description
ZrB_2

Test Type
unknown

Specimen Preparation
Test specimens were produced by wire EDM.

Test Temperature
RT

Reinforcement Content (weight percent)	Fracture Toughness, K_{Ic} (MPa $m^{1/2}$)
10	4.20

Remarks/Comments

ZrB_2 ceramics have a high melting point and a high hardness.

Figure 89: Fracture Toughness of YSZ(p)/TiB_2, Al_2O_3(p)/TiB_2, B_4C(p)/TiB_2, SiC(p)/ZrB_2

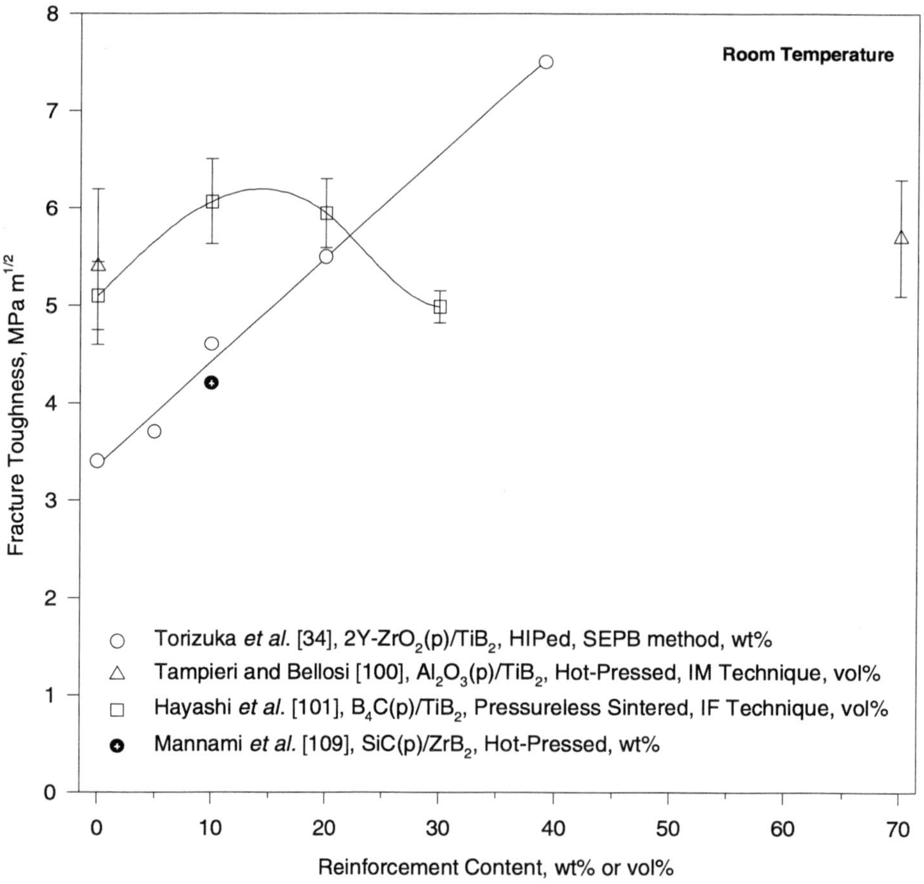

Discussion: The mechanical properties of the 20 vol% 2Y-ZrO_2/TiB_2 were excellent. The flexural strength was ~1200 MPa, hardness was >21 GPa, and fracture toughness was 5.5 MPa $m^{1/2}$ [34]. The 2Y-ZrO_2 was very effective in improving the toughness; the fracture toughness for the 2Y-ZrO_2(p)/TiB_2 composite increased linearly with the addition of the Y-ZrO_2 particulates.

The Al_2O_3(p)/TiB_2 composite tested by Tampieri and Bellosi [100] is being studied because of the versatility it has as electrodes, cutting tools, and wear parts. The toughness of the Al_2O_3-TiB_2 composite (w/ 30 vol% TiB_2) was slightly tougher than the monolithic TiB_2. The addition of Al_2O_3 greatly increased its hardness and strength.

The pressureless sintered B_4C(p)/TiB_2 composite showed a 1 MPa $m^{1/2}$ increase followed by a 1 MPa $m^{1/2}$ decrease with the addition of reinforcement content up to 30 vol%. In the B_4C(p)/TiB_2 composite, X-ray diffraction revealed the presence of TiB_2, B_4C, and NiB [101]. The maximum fracture toughness was 6.1 MPa $m^{1/2}$ in contrast to 5.1 MPa $m^{1/2}$ for a sintered body without B_4C [101]. The dispersion of B_4C particulates inhibited the grain growth of TiB_2 that helped improve the fracture toughness. However, crack deflection by B_4C particles was considered to be the main toughening mechanism for this composite [101].

The SiC(p)/ZrB_2 composite tested by Mannami *et al.* [109] has a high melting point and low electrical resistance. Mannami *et al.* [109] studied SiC(p)/ZrB_2 composite system because of its ability of being machined by EDM. Because of its adherence to the electrodes, wear of the electrodes decreases. Cutting rate for this composite is 50-70% of conventional steel, and 1.5-2 times that of cemented carbide [109].

Figure 90: Single Edge Pre-Cracked Beam Fracture Toughness vs. Sintering Additive of 20 wt% YSZ(p)/TiB_2

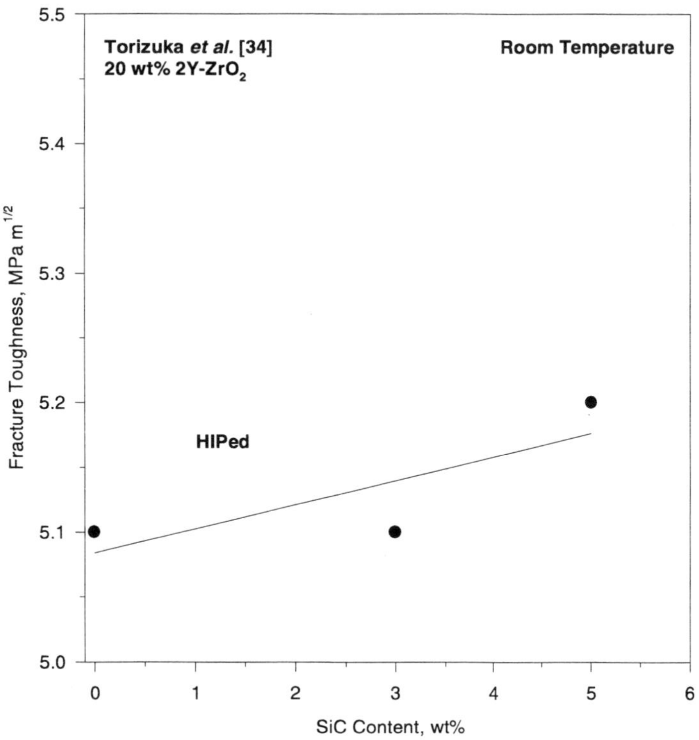

Discussion: The fracture toughness of the YSZ(p)/TiB_2 composite that was HIPed increased only slightly with the addition of SiC. The SiC was added to minimize the grain growth of TiB_2 and ZrO_2. The addition of SiC was effective in restraining the grain growth of TiB_2 and ZrO_2 in the HIPed composites. The grain sizes of TiB_2 and ZrO_2 in 2Y-ZrO_2 HIPed composites were 4-6 μm and 2-5 μm, respectively; those of TiB_2 and ZrO_2 in TiB_2-20 wt% (2Y-ZrO_2)-2.5 or 5 wt% SiC HIPed composites were 2-5 μm and 1-3 μm, respectively [34]. The SiC addition improved the sinterability and strength of the YSZ(p)/TiB_2 composite, but did not have much of an effect on the fracture toughness.

12. Elastic Properties
(Section III)

ZrO_2(p)/Al_2O_3

Reference: 7

Composition
0-100 vol% ZrO_2

Composite Architecture
particulate reinforcement

Processing Method
Hot-Pressed

Detailed Description
Sub-µm composite powders, including yttrium nitrate (a Y_2O_3 precursor), were mixed by ball-milling with methanol and Al_2O_3 balls in a plastic container. All powders were dried, and those with yttrium nitrate were calcined at 400 °C for 4 hours. Most Y_2O_3 containing composites were hot-pressed at 1600 °C for 2 hours. Composites containing 80 and 100 vol% ZrO_2 were hot-pressed at 1400 °C in order to achieve a smaller grain size which allowed the retention of tetragonal ZrO_2. Non-yttria composites were hot-pressed at 1500 °C, again to achieve smaller grain size. The pure Al_2O_3 end-member was hot-pressed at 1400 °C to achieve a grain size comparable to the two-phase materials.

Reinforcement Description
2Y-ZrO_2 particulates

Reinforcement Supplier
Zircas Corp.

Matrix Description
Al_2O_3 powder

Matrix Supplier
Lindy B, Union Carbide Corp.

Test Type
Resonance Frequency Method

Test Description
The test was conducted in two modes of vibration: flexural (9 kHz) and extensional (60 kHz).

Test Temperature
RT

Reinforcement Content (volume percent)	Young's Modulus (GPa)	Comments
0	440	*monolithic alumina*

18.2	402	*flexural (9 kHz)*
45	280	
80	232	
100	205	

18.2	405	*extension (60 kHz)*
45	290	
80	236	
100	210	

Remarks/Comments
The elastic (Young's) modulus was found to decrease with increasing ZrO_2 content through the complete range of compositions.

ZrO_2(p)/Al_2O_3

Reference: 15

Composition
0 - 100 vol% ZrO_2

Composite Architecture
particulate reinforcement

Processing Method
Sintered

Detailed Description
The starting powders were mixed by ball-milling using plastic containers and balls with acetone. After drying, the composite powders were uniaxially pressed at 300 MPa to form pellets 5 mm in diameter and 5 mm thick. These pellets were sintered at 1600 °C for 10 hours in an argon atmosphere.

Reinforcement Description
0, 3, and 6Y-ZrO_2 particulates

Matrix Description
Al_2O_3 powder

Matrix Designation
AKP-20

Test Type
Knoop Indentation

Test Description
Indenter load = 20 and 50 N

Test Temperature
RT

Reinforcement Content (volume percent)	Young's Modulus (GPa)	ZrO_2 Phase
0 mol% Y_2O_3		
0	348	48.5 % t, 51.5 % m
5	357	37.7 % t, 65.3 % m
10	345	21.2 % t, 78.8 % m
15	315	8.7 % t, 91.3 % m
20	260	3.9 % t, 96.1 % m
30	210	fully monoclinic
40	183	fully monoclinic
60	131	fully monoclinic
80	119	fully monoclinic
3 mol% Y_2O_3		
20	348	fully tetragonal
40	318	
60	291	
80	249	
100	237	
6 mol% Y_2O_3		
5	317	fully cubic
20	337	
40	281	
60	257	
80	241	

Remarks/Comments
Small amounts of unstabilized ZrO_2 increased the Young's modulus. The addition of more ZrO_2 to Al_2O_3 reduced the elastic modulus of the material. The Y_2O_3 stabilizer in the zirconia improved the Young's modulus of the composites, particularly at high zirconia contents, by decreasing the phase content of low modulus monoclinic phase. Three mol% Y_2O_3 gave the highest modulus over the largest range.

ZrO_2(p)/Mullite

Reference: 50

Composition
0 and 30 wt% ZrO_2

Processing Method
CIPed/Reaction Sintered

Detailed Description
The sols in the solution (pH ~ 3.0 and 5 wt% in concentration), mixed for one hour, were coprecipitated with diluted aqueous ammonia. All gels were dispersed and filtered in methanol and acetone then subsequently dried in acetone with a rotary evaporator. After vacuum-drying at 60 °C and in an oven at 120 °C, the powders were milled in methanol for 6 hours. The powders were calcined for 30 minutes at 1150 °C for the composites and 1250 °C for the mullite powders. The calcined powders were then milled in methanol for 48 hours in a nylon vibration mill with Si_3N_4-cylinders, dried, and screened (150-mesh) for granulation. All powders were calcined at 800 °C for 1 hour in order to burn out organic inclusions. Green bodies were formed by die-pressing at 30 MPa and subsequent cold isostatically pressing at 200 MPa which were then sintered in air between 1550 to 1650 °C for 2 or 4 hrs.

Reinforcement Description
0 or 3Y-zirconia-mullite composite powder

Reinforcement Supplier
silica sol: Eka Nobel AB
alumina sol(A), zirconia, and yttria sol: Nyacol Products Inc.
alumina sol(B): Albright & Wilson

Supplier Designation
silica sol: Bindzil 15-500
alumina sol(A): Nyacol AL20
alumina sol(B): Alumina Chlorohydrate

Reinforcement Typical Properties
(A) produced from pure inorganic sols of silica (50 Å), alumina (200 Å), zirconia (75 Å), and yttria (100 Å)
(B) produced from silica and alumina (10 Å) sols with aqueous solution of $ZrO(NO_3)_3$ and $Y(NO_3)_3$

Reinforcement Dimensions
ave. particle size (0 % Y_2O_3) = 1.02 μm
ave. particle size (3 % Y_2O_3) = 0.99 μm

Matrix Description
(see reinforcement description)

Matrix Properties
composition of mullite:
(A) 72.1 wt% Al_2O_3
(B) 72.9 wt% Al_2O_3

Test Type
4-pt. Flexural

Specimen Preparation
The sintered plates were cut and ground with a #400 diamond wheel.

Specimen Dimensions
3 x 4 x 40 mm
inner span = 10 mm
outer span = 30 mm

Test Temperature
RT

Relative Density (percent)	Young's Modulus (GPa)	Sintering Temperature (°C)	Sintering Time (hours)	Reinforcement Content (weight percent)	Comments
A					
98.5	216	1,650	4	0	*0 % Y_2O_3*
96.2	202	1,650	2	0	
98.5	218	1,650	2	30	*0 % Y_2O_3*
98.2	214	1,600	4	30	
96.2	205	1,600	2	30	
97.5	216	1,600	4	30	*3 % Y_2O_3*
97.8	214	1,600	2	30	
B					
99.2	223	1,650	4	0	*0 % Y_2O_3*
98.0	207	1,650	2	0	
97.6	204	1,630	4	0	
99.0	221	1,650	2	30	*0 % Y_2O_3*
97.7	214	1,600	4	30	
94.2	198	1,600	2	30	
97.4	213	1,600	4	30	*3 % Y_2O_3*
97.9	213	1,600	2	30	
99.0	180	1,550	4	30	

Remarks/Comments
Young's modulus was dependent on composite density and sintering schedule. The composites sintered at 1550 °C showed much lower modulus than the samples sintered at 1600 °C or higher despite high density.

Al_2O_3(p)/ZrO_2

Reference: 72

Composition
20 wt% Al_2O_3

Composite Architecture
particulate reinforcement

Processing Method
Sintered/HIPed

Detailed Description
Composite billets were formed by a sinter/ HIP process. Primary sintering was done at 1400 °C, followed by HIPing at 1500 °C.

Sintering Additives
4.2 wt% Y_2O_3

Reinforcement Description
Al_2O_3 particulates

Reinforcement Supplier
TOSOH Corp. of Tokyo

Supplier Designation
TOSOH TZ-3Y20A

Matrix Description
ZrO_2

Matrix Supplier
TOSOH Corp. of Tokyo

Matrix Designation
TOSOH TZ-3Y20A

Test Type
Pulse-Echo Technique

Specimen Dimensions
3 x 4 x 50 mm

Test Description
Half of the bars were heat treated for 100 hours at 1000 °C.

Test Temperature
RT

Reinforcement Content (weight percent)	Young's Modulus (GPa)	Comments
20	258	*as received*

20	259	*after heat treatment @ 1000 °C*

Remarks/Comments
Values agree with the manufacturer's values. Values are an average of five samples at each condition. Density was 5.50 g/cm^3.

Figure 91: Young's Modulus of ZrO_2(p)/Al_2O_3

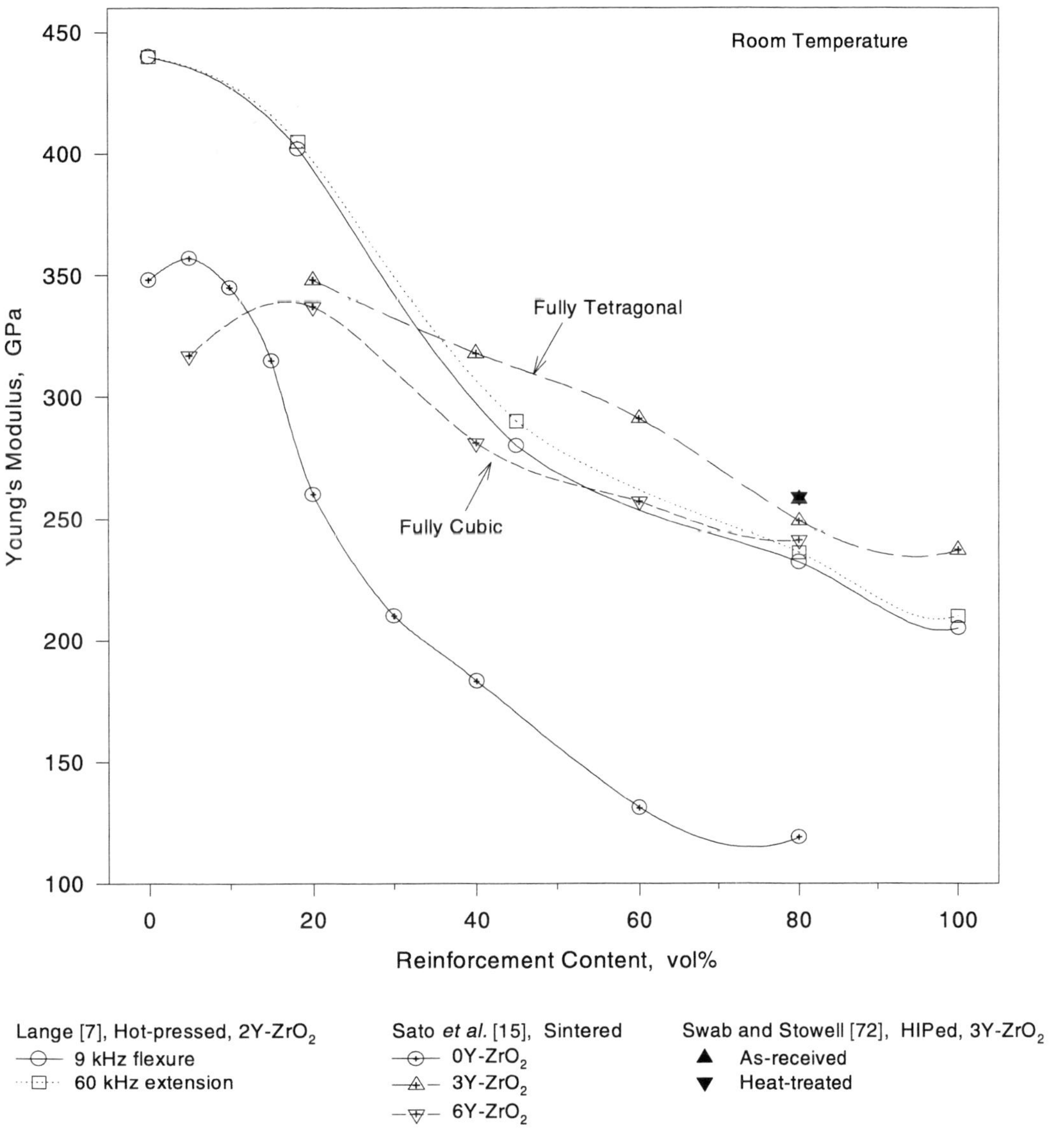

Discussion: The Young's modulus of alumina decreased with the addition of zirconia reinforcement due to the lower modulus of zirconia. Sato *et al.* [15] showed that the zirconia phase strongly influenced the composite modulus. The 3Y-ZrO_2 was fully tetragonal at all reinforcement contents, and the 6Y-ZrO_2 was fully cubic; both composite types followed the rule of mixtures and the modulus decreased linearly with increasing zirconia content. The data showed that the tetragonal phase had a higher modulus than the cubic phase. The unstabilized ZrO_2 contained some tetragonal phase at low reinforcement contents, but the tetragonal phase disappeared rapidly with increasing reinforcement content until the zirconia phase was fully monoclinic at 30 vol% and above. Since m-ZrO_2 has a much lower modulus than t-ZrO_2, the modulus of the unstabilized zirconia-alumina composites decreased rapidly with increasing reinforcement content. Lange [7] found that the modulus of the hot-pressed monolithic alumina was nearly 100 GPa greater than the sintered alumina; however, this difference gradually disappeared as zirconia was added. At 80 vol% t-ZrO_2 (2 or 3Y), there was little difference between the hot-pressed, sintered, and HIPed composites. Swab and Stowell [72] found that a 1000 °C heat treatment did not change the modulus of the HIPed 80 vol% 3Y-ZrO_2-Al_2O_3 composite.

Figure 92: Young's Modulus vs. Relative Density of Sintered ZrO_2(p)/Mullite

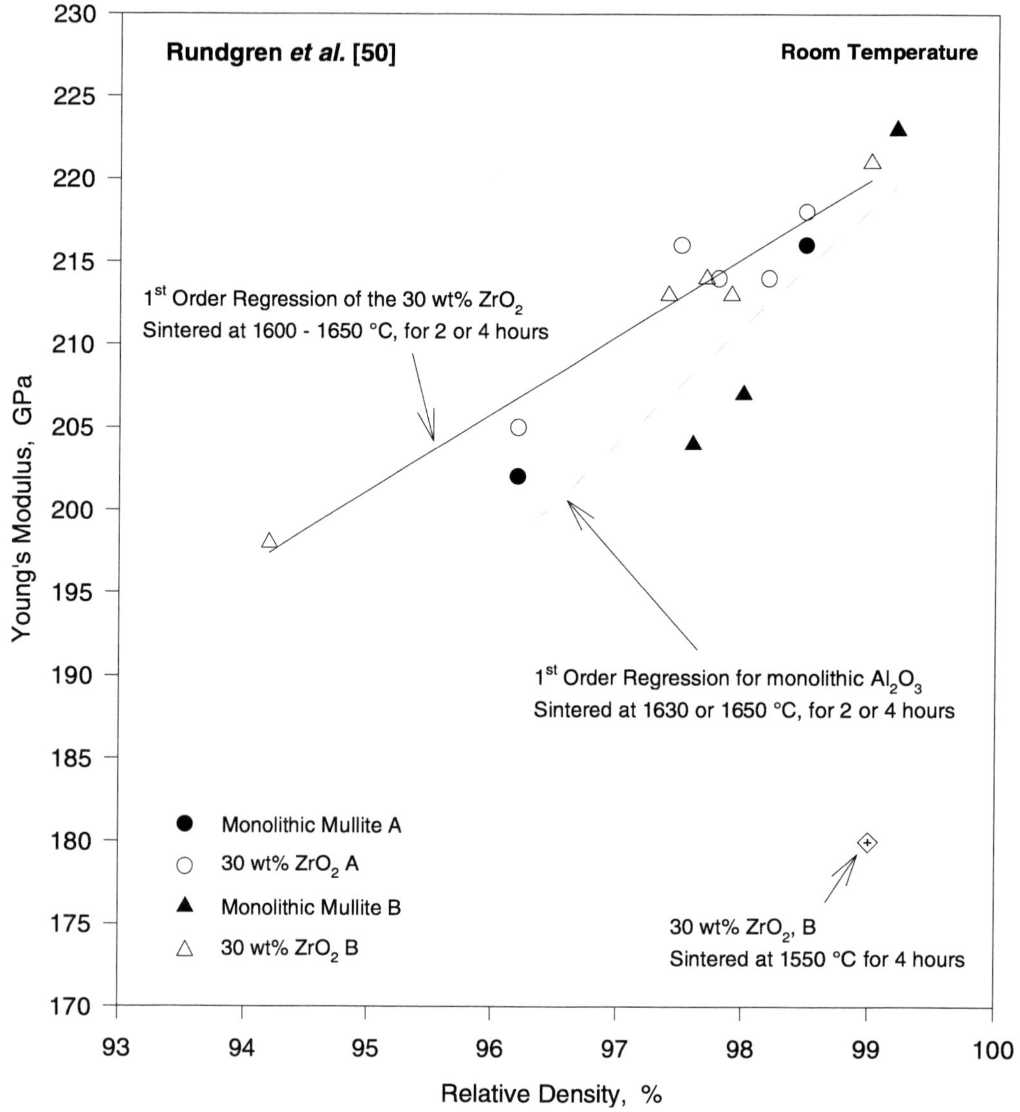

Discussion: The sintering conditions affected the composite density and microstructure, which in turn controls the modulus of the composite. Figure 92 shows the general trend of modulus vs. relative density; as the relative density increased, the Young's modulus increased. The one exception to this trend was the sample B composite sintered at 1550 °C, that reached near full density (99%), but had a very low modulus. This indicated that the microstructure of this composite was quite different from the microstructure of the other samples. The exact nature of the microstructure that caused the low modulus in this composite was not clear.

$SiC(p)/Al_2O_3$

Reference: 33

Composition
37 vol% SiC (with B_4C)

Composite Architecture
in-situ reinforcement combined particulate reinforcement

Processing Method
Conventionally Hot-Pressed or Reaction Hot-Pressed

Detailed Description
Powders were mixed, in the appropriate stoichiometric ratios, according to the expected solid-state reactions, by wet ball milling with Si_3N_4 media. Isopropanol, used as the fluid medium, was subsequently removed by drying under vacuum at 50 °C. The dried powders were loaded into graphite dies and hot-pressed under vacuum or Ar at temperatures from 1600-1800 °C and pressures of 7-28 MPa for times up to 1 hour.

Reinforcement Description
B_4C and SiC particulates

Supplier Designation
B_4C: Advanced Ref. Tech.
SiC: Buelher

Reinforcement Typical Properties
grain size (SiC):
reaction HP = 2 μm
conventional HP = 3 μm

Matrix Description
α-Al_2O_3

Matrix Properties
powder particle size ~ 0.3 μm
grain size:
reaction HP = 4 μm
conventional HP = 6 μm

Test Type
Pulse-Echo Resonance Technique

Specimen Preparation
Samples were machined either by conventional diamond machining or EDM/

Specimen Dimensions
2 x 3 mm cross-section
span = 12.7 mm

Test Description
crosshead speed = 0.5 mm/min

Test Temperature
RT

Reinforcement Content (volume percent)	Young's Modulus (GPa)	Relative Density (percent)	Comments
37	342	96	*reaction HP*
37	370	99	*conventionally HP*

Remarks/Comments
The sonic modulus values of 342 and 370 GPa for the reaction hot-pressed and conventionally hot-pressed materials, respectively, were less than the porosity corrected Voight and Reuss limits.

$SiC(p)/Al_2O_3$

Reference: 88

Composition
0 - 30 vol% SiC

Processing Method
Hot-Pressed

Detailed Description
First, ball-milling was done to the alumina for 24 hours to break up the agglomerates in a deflocculated, water-based slurry (Darvan 821-A). Next, the slurry was passed through a 325- mesh sieve to remove any large agglomerates. Afterwards, SiC was added, and the mixture was milled for one hour. The material was then dried, dry-pressed at 12 MPa, and then pre-fired to 600 °C for one hour to remove any residual volatile species. The pre-fired discs were then inductively hot-pressed in a nitrogen atmosphere for one hour at 1600 °C. Pressure was applied in a gradual rate beginning at 400 °C and held at 15 MPa.

Reinforcement Description
SiC particulates

Supplier Designation
Lonza LS-13

Reinforcement Typical Properties
density = 3.2 g cm^{-3}
CTE = 4.3 $x10^{-6}$ $°C^{1}$
E = 470 GPa
fracture toughness = 3-5 MPa $m^{1/2}$

Reinforcement Dimensions
powder particle size = 13 μm

Matrix Description
Al_2O_3

Matrix Designation
Ferro A-310

Matrix Properties
ave. grain size = 0.35 μm
density = 3.96 g cm^{-3}
CTE = 7.7-8.3 $x10^{-6}$ $°C^{-1}$
E = 380 GPa
fracture toughness = 2-3 MPa $m^{1/2}$

Test Type
Ultrasonic Technique

Specimen Preparation
Samples were surface polished to 1 μm. Tensile edges were chamfered (tensile side was perpendicular to the hot-pressing direction).

Specimen Dimensions
3 x 4 x 50 mm

Test Temperature
RT

Reinforcement Content (volume percent)	Young's Modulus (GPa)	Density (g cm^{-3})
0	353	3.96
10	383	3.91
20	288	3.81
30	289	3.73

Remarks/Comments
With 10 vol% of SiC particulate, the modulus was increased relative to that of the pure alumina. With 20 vol% SiC additions the elastic modulus dropped below the value for pure alumina. The drop in modulus at 20 and 30 vol% SiC reinforcement content indicated microcracking.

SiC(p)/Al_2O_3

Reference: 112

Composition
~55 vol% SiC

Processing Method
Slip-Cast/DIMOX

Detailed Description
Commercially available SiC(p)/Al_2O_3 (Du Pont Lanxide composite Inc., USA) was obtained. This material was prepaired using the DIMOX method. An SiC preform was slip-cast and then an Al alloy was infiltrated and oxidized in the furnace, resulting in the composite.

Reinforcement Description
SiC particulates

Reinforcement Properties
ave. particle diameter = 5-20 μm

Matrix Description
aluminum powder

Test Type
4-pt. Flexural

Specimen Dimensions
6 x 3 x 50 mm

Test Temperature
RT

Temperature (°C)	Young's Modulus (GPa)	Poisson's Ratio (dimensionless)
25	320	0.27

Figure 93: Young's Modulus of SiC(p)/Al_2O_3

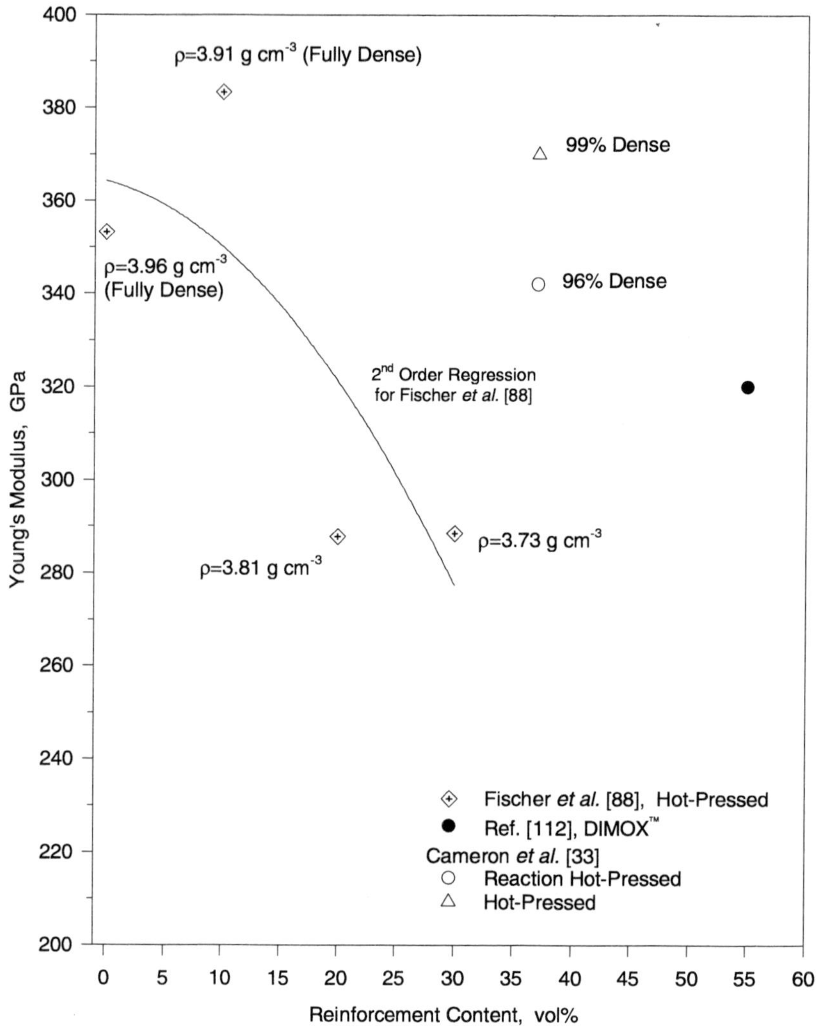

Discussion: Cameron *et al.* [33] found that the conventionally hot-pressed sample had a higher Young's modulus than the reaction hot-pressed composite. This difference in modulus was due to microstructural differences and the difference in density. Fischer *et al.* [88] found that the monophase alumina sample had a greater Young's modulus than most of the SiC(p)/Al_2O_3 composites. This was a result of the defects caused by internal stresses associated with the second phase. Because SiC particulates had a lower thermal expansion coefficient than the Al_2O_3 matrix, there was a radial compressive stress resulting in a circumferential tensile stress in the matrix. When the particulate exceeded a critical size, the stress caused microcracks that resulted in a decrease of the elastic modulus. The 10 vol% SiC(p)/Al_2O_3 sample had the highest modulus due to the SiC reinforcement combined with the lack of microcracks and high density.

In comparison, the Du Pont Lanxide DIMOX™ composite with 55 vol% SiC reinforcement had a higher modulus than the composites tested by Fischer *et al.* [88] with 20 or 30 vol% SiC that contained microcracks. Microcracking is generally not a problem with DIMOX™ composite because processing temperatures are much lower and because residual ductile aluminum would allow for some stress redistribution. The DIMOX™ composites may have a lower modulus compared to the ref. [33] composites due to some residual aluminum content.

$TiC(p)/ZrO_2.Al_2O_3$

Reference: 15

Composition
10 vol% ZrO_2
1 - 10 vol% TiC

Processing Method
Sintered

Detailed Description
The starting powders were mixed by ball-milling using plastic containers and balls with acetone. After drying, the composite powders were uniaxially pressed at 300 MPa to form pellets 5 mm in diameter and 5 mm thick. These pellets were sintered at 1500 °C for 10 hours in an argon atmosphere.

Reinforcement Description
0Y-ZrO_2 or 3Y-ZrO_2 and TiC particulates

Matrix Description
Al_2O_3 powder

Matrix Designation
AKP-20

Test Type
Knoop Indentation Technique

Test Description
Indenter load = 20 and 50 N

TiC Content (volume percent)	Young's Modulus (GPa)	Relative Density (percent)	Stabilizer Content (atomic percent)
1	324	96.8	0
3	310	95.6	0
6	348	94.7	0
10	309	92.5	0
1	309	97.2	3
3	291	96.3	3
5	296	96.9	3
10	315	95.9	3

Remarks/Comments
The elastic modulus increased with the addition of 1 vol% nonoxide, then slightly decreased with increasing TiC content. This was due to the decrease in density with increasing nonoxide content.

$SiC(p)/ZrO_2.Al_2O_3$

Reference: 15

Composition
10 vol% ZrO_2
0 - 10 vol% SiC

Processing Method
Sintered (some HIPed)

Detailed Description
The powders were ball-milled in plastic containers with acetone. After drying, the powders were uniaxially pressed at 300 MPa to form pellets 5 mm in diameter and 5 mm thick. These pellets were sintered at 1500 °C for 10 hours in an argon atmosphere. Some composites were HIPed at 1500 °C and 200 MPa for 30 minutes.

Reinforcement Description
0Y-ZrO_2 or 3Y-ZrO_2 and SiC particulates

Matrix Description
Al_2O_3 powder

Matrix Designation
AKP-20

Test Type
Knoop Indentation Technique

Test Description
Indenter load = 20 and 50 N

SiC Content (volume percent)	Young's Modulus (GPa)	Relative Density (percent)	Stabilizer Content (atomic percent)	Comments
0	342	98.8	0	*as-sintered*
1	361	98.7	0	
3	333	96.0	0	
6	306	95.4	0	
10	285	93.4	0	
0	326	98.3	3	*as-sintered*
1	297	96.8	3	
3	302	96.0	3	
5	317	96.6	3	
10	300	95.6	3	
1	320	97.6	3	*sintered and HIPed*
5	336	96.7	3	
10	330	96.7	3	

Remarks/Comments
The elastic modulus increased with the addition of 1 vol% nonoxide then slightly decreased with increasing SiC content. This was due to the decrease in density with increasing nonoxide content.

TiC(p)/Al_2O_3

Reference: 47

Composition
0 - 76 wt% TiC

Processing Method
Hot-Pressed

Detailed Description
The composites were wet-mixed for 20 hours in a ball-mill. Specimens were hot-pressed in a graphite die at 1500-1750 °C and 20 MPa for 1 hour. Applied pressure for the temperature range from room temperature to the hold temperature was one-half the final pressure. Hot-pressed specimens were removed from the die after cooling to room temperature.

Sintering Additives
TiO_2 and MgO powder

Reinforcement Description
TiC powder

Reinforcement Typical Properties
S < 400 ppm, Fe < 200 ppm, Ni < 20 ppm, Cr < 10 ppm

Reinforcement Dimensions
ave. particle size = 1.0 μm

Matrix Description
α-Al_2O_3 powder

Matrix Properties
Si < 50 ppm, Na < 10 ppm, Mg < 10 ppm, Cu < 10 ppm, Fe < 20 ppm

Matrix Dimensions
ave. particle size = 0.8 μm

Test Type
3-pt. Flexural

Specimen Preparation
Specimens were diamond cut from hot-pressed products. Surfaces were ground, and the edges of the tensile surface were beveled with grinding paper.

Specimen Dimensions
36 x 4 x 3 mm

Test Description
crosshead speed = 0.5 mm/min

Test Temperature
RT

Reinforcement Content (volume percent)	Young's Modulus (GPa)
0	393
8.3	399
17	405
26	410
35	415
55	425
76	434

Remarks/Comments
Proper units were calculated by multiplying (9.8066 x 0.001) and the given values.

TiC(p)/Al_2O_3

Reference: 84

Composition
0 and 30 vol% TiC

Processing Method
Hot-Pressed

Detailed Description
Al_2O_3 powders were homogenized with 30 vol% TiC in plastic jars with alumina balls and isobuthyl alcohol. The composites were hot-pressed in a vacuum at 1600-1650 °C and 30 MPa.

Reinforcement Description
TiC particulates

Matrix Description
Al_2O_3

Test Type
Resonance Frequency Method

Test Temperature
RT

Reinforcement Content (volume percent)	Young's Modulus (GPa)	Relative Density (percent)
0	396	100
30	410	99.4

Remarks/Comments
Young's modulus increased in the composites in comparison to the base-line Al_2O_3, according to the rule of mixtures.

TiN(p)/Al_2O_3

Reference: 84

Composition
0 and 30 vol% TiN

Processing Method
Hot-Pressed

Detailed Description
Al_2O_3 powders were homogenized with 30 vol% TiN in plastic jars with alumina balls and isobuthyl alcohol. The composites were hot-pressed in a vacuum at 1600-1650 °C and 30 MPa.

Reinforcement Description
TiN particulates

Matrix Description
Al_2O_3

Test Type
Resonance Frequency Method

Test Temperature
RT

Reinforcement Content (volume percent)	Young's Modulus (GPa)	Relative Density (percent)
0	396	100
30	415	99.3

Remarks/Comments
Young's modulus increased in the composites in comparison to the base-line Al_2O_3, according to the rule of mixtures.

TiB_2(p)/Al_2O_3

Reference: 84

Composition
0 and 30 vol% TiB_2

Processing Method
Hot-Pressed

Detailed Description
Al_2O_3 powders were homogenized with 30 vol% TiB_2 in plastic jars with alumina balls and isobuthyl alcohol. The composites were hot-pressed in a vacuum at 1600-1650 °C and 30 MPa.

Reinforcement Description
TiB_2 particulates

Matrix Description
Al_2O_3

Test Type
Resonance Frequency Method

Test Temperature
RT

Reinforcement Content (volume percent)	Young's Modulus (GPa)	Relative Density (percent)
0	396	100
30	424	98.6

Remarks/Comments
Young's modulus increased in the composites in comparison to the base-line Al_2O_3, according to the rule of mixtures.

$TiB_2(p)/Al_2O_3$

Reference: 100

Composition
30 and 100 vol% particulates

Processing Method
Hot-Pressed

Detailed Description
Monolithic TiB_2 was hot-pressed at 1850 °C and 30 MPa for 1 hour. The composite was hot-pressed at 1600 °C and 30 MPa for 30 minutes.

Reinforcement Description
TiB_2 powder

Reinforcement Supplier
Starck

Supplier Designation
Grade F

Matrix Description
Al_2O_3 particulates

Matrix Supplier
ALCOA

Matrix Designation
A16 S. G.

Test Type
Elastic Modulus Tests

Reinforcement Content (volume percent)	Young's Modulus (GPa)
Monolithic TiB_2	554
30	424

Remarks/Comments
Young's modulus decreased when reinforcement content decreased.

Figure 94a: Young's Modulus of Hot-Pressed TiB_2, TiC, and TiN Particulate Reinforced Alumina

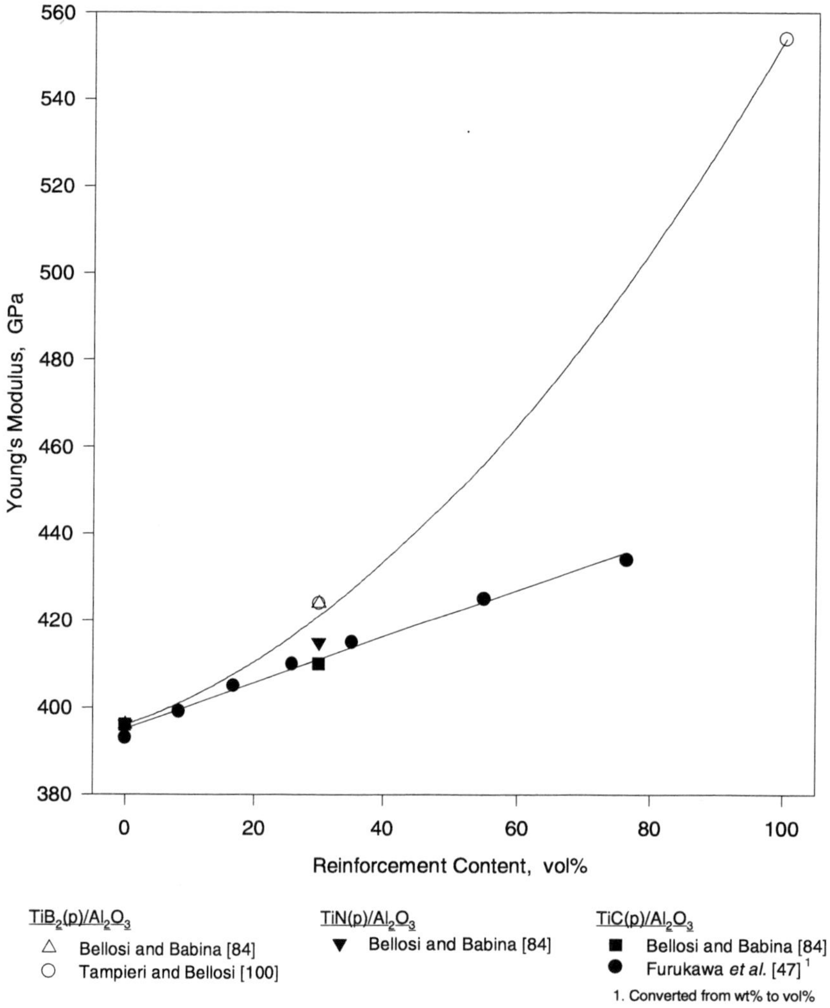

Discussion: As more reinforcement was added, the Young's modulus increased. The TiB_2 and TiN particulates had approximately the same Young's modulus, thus, the same effect on the modulus of Al_2O_3. As the volume fractions of the reinforcement increased, the elasticity of the composites increased according to the rule of mixtures.

Figure 94b: Young's Modulus of SiC or TiC Particulate Reinforced Zirconia-Alumina

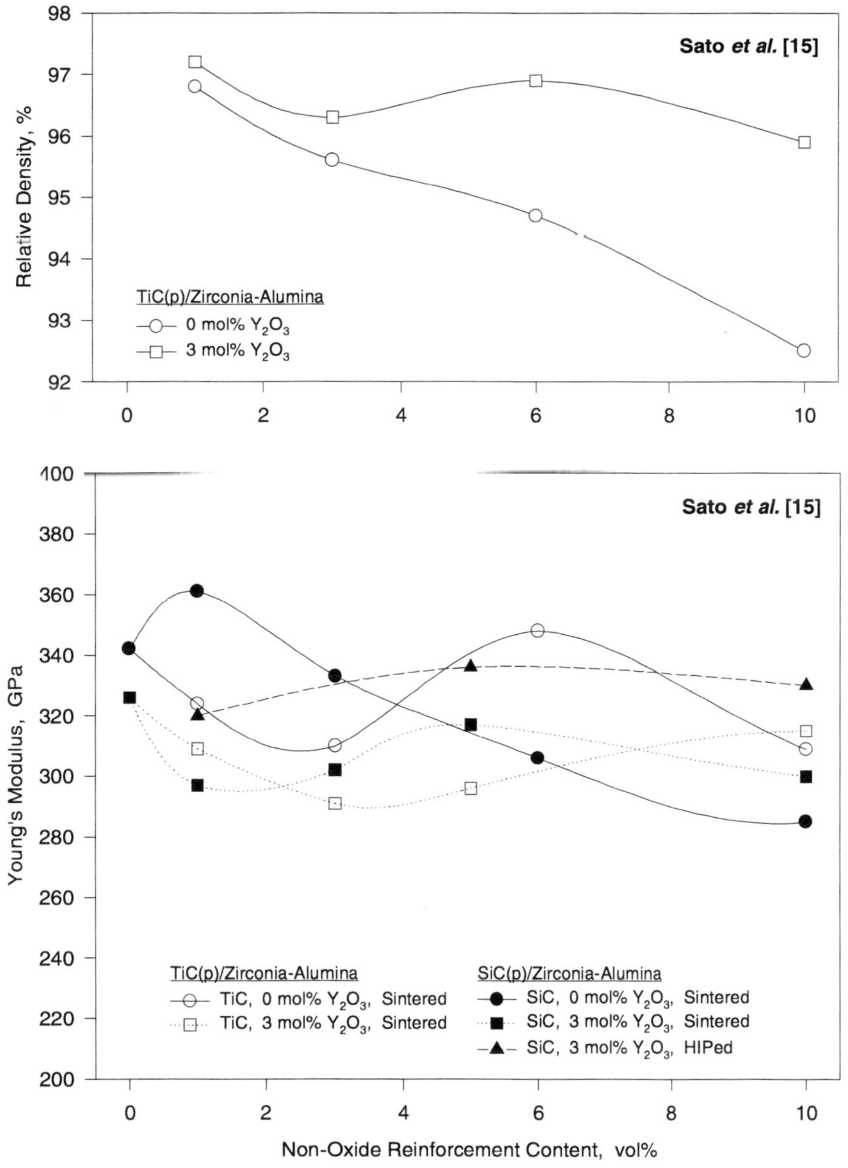

Discussion: Sato *et al.* [15] added 0-10 vol% TiC or SiC reinforcement to zirconia-alumina composites, with or without yttria stabilizers. There appears to be considerable scatter in the data presented in figure 94b. Microcracking accounted for some of this scatter. Trend analysis showed that the 3 mol% yttria stabilizers had a significant affect on the modulus of these composites. Without yttria, the modulus of the zirconia-alumina composites without non-oxide reinforcement was higher. However, the modulus dropped with the addition of TiC or SiC non-oxide reinforcement because of a drop in density. With yttria stabilizers, the relative density of the zirconia-alumina composites was above 96 % regardless of the non-oxide content. For this reason, the drop in modulus for the yttria containing composites with increasing non-oxide content was not as severe. HIPing of the SiC(p)/3Y-zirconia-alumina composite appeared to slightly increase relative density and heal microcracks, which resulted in an increase in modulus of 20-30 GPa.

$ZrO_2.HfO_2(p)/Al_2O_3.Cr_2O_3$

Reference: 13

Composition
10 and 30 mol% HfO_2
5 and 20 mol% Cr_2O_3

Processing Method
CIPed/Sintered

Detailed Description
The starting powders were prepared by coprecipitation. Aqueous solutions of aluminum nitrate, chromium nitrate, zirconium oxychloride, and hafnium oxychloride were mixed together. Precipitates were dried and calcined at 1050 °C. The powders were then isostatically pressed and sintered in an induction furnace using a graphite susceptor. Particulate and matrix powders were mixed in a ball-mill with water. Dried mixtures were granulated (~60 mesh) and CIPed at 28 MPa. The bars were then sintered for 2 hours at 1550 °C.

Reinforcement Description
$ZrO_2.HfO_2$ particulates

Matrix Description
$Al_2O_3.Cr_2O_3$ (Cr_2O_3-5,20 mol%)

Test Type
Sonic Method

Specimen Dimensions
5 mm diameter
60 mm length

Test Temperature
RT

Reinforcement Content (volume percent)	Young's Modulus (GPa)	Comments
0	390	*5 mol% Cr_2O_3, 10 mol% HfO_2*
2.5	385	
5.0	380	
7.5	374	
10.0	366	
12.5	349	
15.0	296	
0	390	*5 mol% Cr_2O_3, 30 mol% HfO_2*
2.5	385	
5.0	380	
7.5	374	
10.0	366	
12.5	352	
15.0	333	
0	367	*20 mol% Cr_2O_3, 10 mol% HfO_2*
2.5	365	
5.0	362	
7.5	357	
10.0	349	
12.5	332	
15.0	299	
0	367	*20 mol% Cr_2O_3, 30 mol% HfO_2*
2.5	365	
5.0	362	
7.5	357	
10.0	349	
12.5	337	
15.0	324	

Remarks/Comments
Hafnia was added to the ZrO_2 reinforcement to increase the t→m transformation temperature. Cr_2O_3 was added to the aluminum matrix to decrease the thermal conductivity. Young's modulus decreased as the amount of dispersed phase increased following the rule of mixtures.

Figure 95: Young's Modulus vs. ZrO_2 Content of Sintered $ZrO_2(p).HfO_2(p)/Al_2O_3.Cr_2O_3$

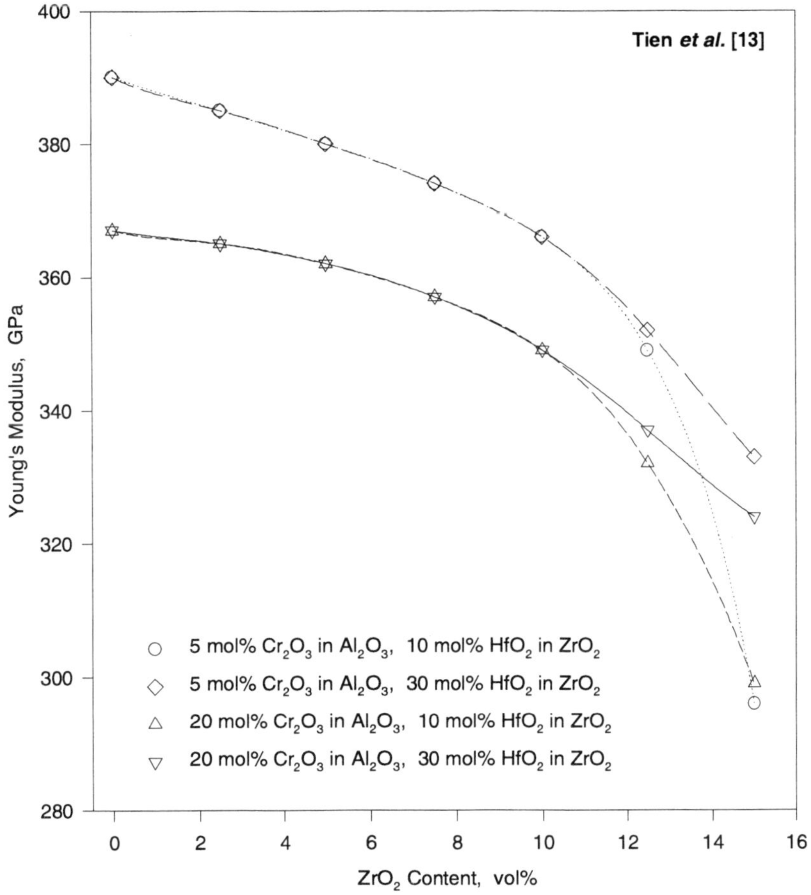

Discussion: The Young's modulus of the composite decreased as the ZrO_2 content increased, following the rule of mixtures up to 10 vol% ZrO_2 content. A considerable drop in the modulus was observed in the high ZrO_2 containing composites. This sharp drop indicated microcracks in the composite resulting from the volume change of zirconia during the tetragonal to monoclinic phase change. Hafnia inhibited the t-m phase change. As a result, the composites with 30 mol% HfO_2 in zirconia were not as severely micro-cracked and had a higher modulus at high zirconia (> 10 vol%) contents. From 0 to 10 vol% zirconia content, the effect of microcracking was not as evident. In this range, the modulus of the composites with 20 mol% Cr_2O_3 in Al_2O_3 was 15 to 20 GPa lower than the composites with only 5 mol% Cr_2O_3. This was due to the lower modulus of chromia compared to alumina.

SiC(p)/SiAlON

Reference: 44

Composition
10 vol% SiC

Processing Method
Hot-Pressed (some HIPed)

Detailed Description
The mixed powders were ball-milled for 24 hours with a sufficient amount of ethanol. The slurry was dried in a rotary evaporator and sieved to a particle size of <210 μm. The mixed powders were then hot-pressed at 1720 °C and 24.5 MPa for 30 minutes in a 0.1 MPa nitrogen atmosphere. Some were also HIPed at 1850-1950 °C under 10-100 MPa nitrogen pressure.

Reinforcement Description
β-SiC particulates

Reinforcement Supplier
Ibiden Co., Ltd

Supplier Designation
UF30

Reinforcement Dimensions
ave. particle size = 0.28 μm

Matrix Description
α-SiAlON and Si_3N_4 powders

Matrix Supplier
Ube Kosan Co., Ltd.

Matrix Designation
Si_3N_4: E10

Matrix Properties
60 wt% Si_3N_4 (particle size = 0.5 μm)
40 wt% SiAlON

Test Type
4-pt. Flexural

Specimen Dimensions
4 x 3 x 40 mm
inner span = 10 mm
outer span = 30 mm

Test Description
crosshead speed = 0.5 mm/min

Test Temperature
RT

HIP Temperature (°C)	Young's Modulus (GPa)	Density (g cm^{-3})	Range (GPa)	Comments
-----	323	3.20	± 2	*Tested as-hot-pressed*
1,850	332	3.27	± 2	*HIPed in 10 MPa nitrogen for 1 hour*
1,850	330	3.27	± 2	*HIPed in 100 MPa nitrogen for 1 hour*
1,950	330	3.27	± 2	*HIPed in 100 MPa nitrogen for 1 hour*

Remarks/Comments
Young's modulus increased slightly at a HIPing temperature of 1850 °C

TiC(p)/Si_3N_4

Reference: 27

Composition
0 and 5 vol% TiC

Processing Method
Cold-Pressed/Reaction Bonded

Detailed Description
TiC and Si_3N_4 powders were cold pressed then nitrided in a dry nitrogen atmosphere in a temperature range of 1350-1380 °C.

Reinforcement Description
TiC particulates

Reinforcement Supplier
H.C. Stark, Germany

Matrix Description
Si powder

Matrix Supplier
Indian Metals and Ferro Alloys, India

Test Type
4-pt. Flexural

Specimen Dimensions
4.5 x 3.5 x 45 mm
inner span = 10.04 mm
outer span = 40.1 mm

Test Description
load rate = 1.25 N/min
$E = [P*d*(3L^2-4d^2)] / (4y*B*W^3)$
where:
P = load at fracture
d = bending arm length (10.04 mm)
B = sample breadth
W = sample width
L = major span (40.10 mm)
y = deflection of central point of sample with respect to the outer loading points

Test Temperature
RT

Test Equipment
High Temperature Bending Strength Tester (Model 422S, Netzsch, F.R.G.)

Reinforcement Content (volume percent)	Young's Modulus (GPa)	Range (GPa)
0	67	±3
5	64	+6/-3

TiC(p)/Si_3N_4

Reference: 38

Composition
0 and 15 vol% TiC

Processing Method
Hot-Pressed

Detailed Description
The starting powders were homogenized and mixed together. The mixtures were filter pressed at 2.07 MPa for ~45 minutes over a 0.02 µm polymer filter. The mixtures were dried for 2 weeks. The resulting green bodies were hot-pressed at 1800 °C and 41 MPa for 2 hours.

Sintering Additives
6 wt% CeO_2
(Cerac Inc., Milwaukee, WI)

Reinforcement Description
TiC particulates

Reinforcement Supplier
Hermann C. Starck, Inc., New York, NY

Supplier Designation
c.a.s. Grade

Reinforcement Dimensions
ave. particle size = 1.25 µm

Matrix Description
α-Si_3N_4

Matrix Supplier
Hermann C. Starck, Inc., New York, NY

Matrix Designation
Grade LC12

Matrix Properties
ave. particle size = 0.55 µm

Test Type
4-pt. Flexural

Specimen Dimensions
50.8 x 6.35 x 3.18 mm
inner span = 22.2 mm
outer span = 44.4 mm

Test Description
crosshead speed = 0.51 mm/sec
Selected specimens were strain gauged to obtain the elasticity.

Test Temperature
RT

Reinforcement Content (volume percent)	Young's Modulus (GPa)
0	329
15	328

$TiC(p)/Si_3N_4$

Reference: 95

Composition
0 and 20 vol% TiC

Processing Method
Hot-Pressed or Pressureless-Sintered

Detailed Description
The powders were mixed wet or attrition-milled under isoproponal or hexane. Sintering was done in a graphite resistance furnace under a positive pressure of nitrogen (10 mm of Hg) and at a temperature of 1750 °C. Hot-pressing was done in a vacuum hot-press in a BN coated graphite die and punched under a pressure of around 30 MPa at 1650 °C. The pressureless-sintered materials were in the form of billets initially isostatically-pressed in the green state (67 x 17 x 17 mm size).

Sintering Additives
Y_2O_3, AlN, and quartz powder

Reinforcement Description
TiC particulates

Reinforcement Supplier
H.C. Starck, Goslar, Germany

Reinforcement Dimensions
surface area = 4.5 m^2/g
dia. = 50-70 μm
length = 10 μm

Matrix Description
Si_3N_4 powder

Matrix Properties
surface area = 8.78 m^2/g

Test Type
Elastic Modulus Test

Specimen Dimensions
dia. = 25 mm
thickness = 8-10 mm

Reinforcement Content (volume percent)	Young's Modulus (GPa)	Comments
0	133	*11 % glassy phase, pressureless-sintered*
0	130	*15 % glassy phase, pressureless-sintered*
0	390	*hot-pressed*
20	437	

Remarks/Comments
TiC converted to TiN during sintering in the presence of nitrogen. This conversion was less in the case of hot-pressed products. Higher amounts of sintering aid with Si_3N_4 had hardly any effect on Young's modulus. The TiC reinforced composite showed an improved Young's modulus compared to the monolith and SiC platelet reinforced composite.

Figure 96: Young's Modulus of TiC(p)/Si_3N_4

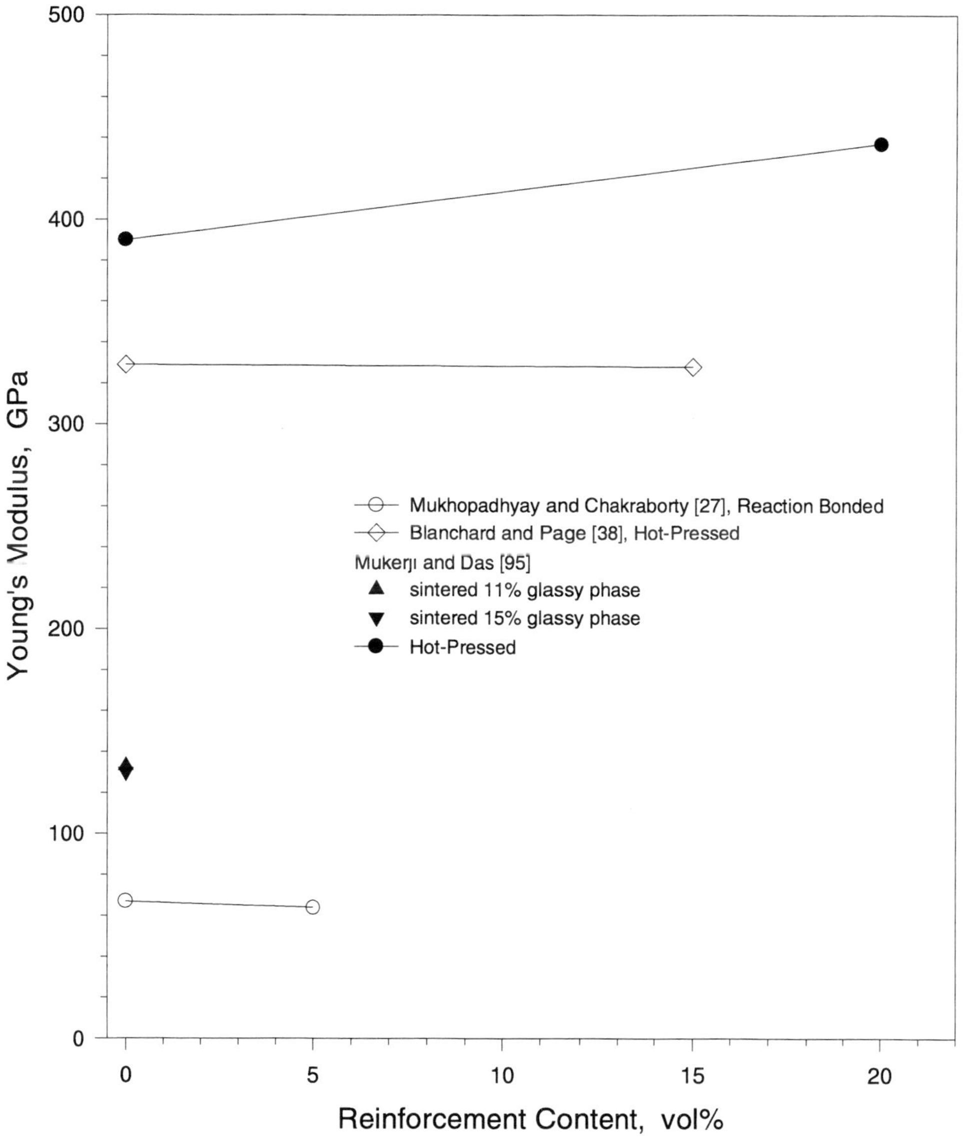

Discussion: The Young's modulus for the composites from references [38] and [27] that were hot-pressed and reaction bonded did not change with the addition of TiC particulates to the monolithic composition. However, the hot-pressed Si_3N_4 composite tested by Mukerji and Das [95] showed an increase of about 50 GPa with the addition of 20 vol% TiC particulates. Figure 96 shows that hot-pressing produced Si_3N_4 materials with higher elastic moduli than the sintered materials and much higher than the reaction bonded silicon nitride. The two sintered composites that had approximately 11 and 15 percent glassy phase had nearly the same elastic modulus. The moduli of these materials was relatively low because of the high glassy phase content [95].

Figure 97: Young's Modulus vs. HIP Temperature of SiC(p)/SiAlON

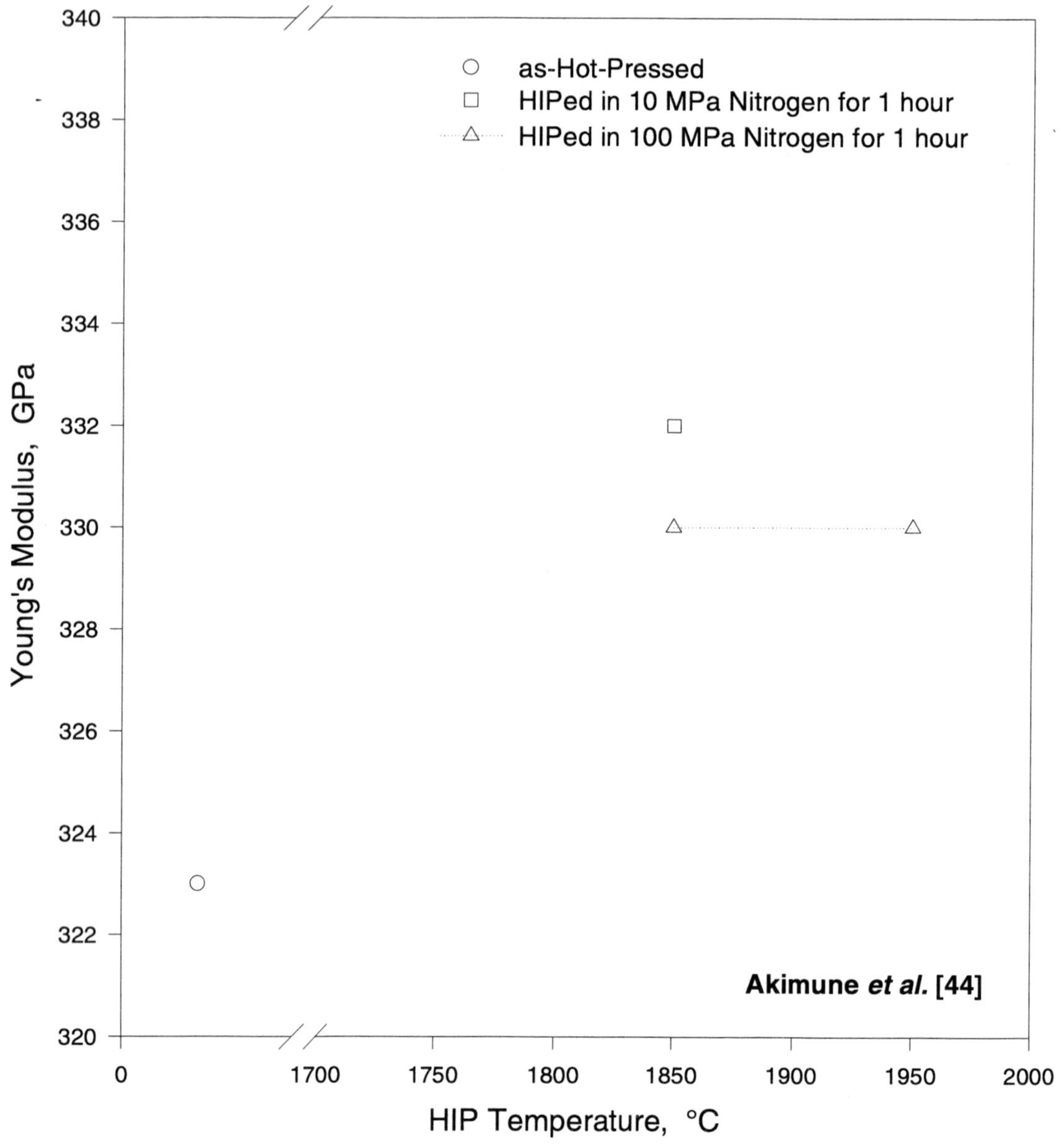

Discussion: HIPing of the as-hot-pressed SiC(p)/SiAlON composites slightly increased its modulus by 7-10 GPa. At a HIP temperature of 1850 °C, the Young's modulus for the SiC(p)/SiAlON composite decreased approximately 2 GPa when the HIP pressure increased from 10 to 100 MPa. For the SiC(p)/SiAlON composite that was sintered with 100 MPa pressure, increasing the HIPing temperature from 1850 °C to 1950 °C did not change the Young's modulus.

SiC(p)/Si_3N_4

Reference: 27

Composition
0 - 20 vol% SiC

Processing Method
Reaction Bonded

Detailed Description
SiC and Si_3N_4 powders were cold pressed then nitrided in a dry nitrogen atmosphere in a temperature range of 1350-1380 °C.

Reinforcement Description
SiC particulates

Reinforcement Supplier
Grindwell Norton

Matrix Description
Si_3N_4

Test Type
3-pt Flexural

Test Description
E = fracture load*bending arm length*(3*major span^2 - 4* bending arm length) / (4*deflection*sample length*sample width)

Test Temperature
RT-1400 °C

Test Environment
Air

Test Equipment
High Temperature Bending Strength Tester (Model 4225, Netzsch, F.R.G.)

Temperature (°C)	Young's Modulus (GPa)	Reinforcement Content (volume percent)	Relative Density (percent)
30	72	0	72
1,000	58	0	72
1,200	50	0	72
1,400	40	0	72
30	119	5	77
1,000	97	5	77
1,200	89	5	77
1,400	90	5	77
30	113	10	81
1,000	112	10	81
1,200	97	10	81
1,400	87	10	81
30	97	15	73
1,000	87	15	73
1,200	86	15	73
1,400	70	15	73
30	104	20	76
1,000	92	20	76
1,200	89	20	76
1,400	82	20	76

Remarks/Comments
Young's modulus decreased with increasing temperature. The low relative density resulted in composites with low modulus.

$SiC(p)/Si_3N_4$

Reference: 35

Composition

12 wt% Si-rich SiC
10 wt% C-rich SiC
20 wt% C-rich SiC

Processing Method

Colloidal Pressed/Reaction Bonded

Detailed Description

Four sample groups were fabricated. The first group, the monolithic RBSN, was fabricated either in methanol with CIPing or in octanol without CIPing. The remaining RBSN composite samples were fabricated in octanol without CIPing. Octanol was substituted for methanol to reduce the probability of differential sedimentation between the dissimilar Si and SiC powders before samples were pressed. A high-intensity ultrasonic probe was used to break up agglomerates in the dispersed powder mixture rather than the low-intensity ultrasonic bath used for the methanol processed baseline group. Si and SiC powders were ultrasonically dispersed separately, mixed together by hand to visual uniformity, and again sonicated. Total sonication time was 5-8 minutes at up to 40 W. Oxygen exposure was avoided except during pressing. Samples were colloidally pressed at 17 MPa to yield 1.27 diameter x 0.2 cm thick samples. A 0.4 °C/min. ramp followed by a 24 hour, 200 °C soak in nitrogen was used to dry the samples. Nitridation was accomplished in ultra-high purity nitrogen at atmospheric pressure. The firing schedule for the samples was:

(A) RBSN (methanol):
1 °C/min. to 1200 °C
0.5 °C/min. to 1400 °C; 1 hour hold

(B) RBSN (octanol):
1 °C/min. to 1250 °C; 4 hour hold
0.5 °C/min. to 1400 °C; 7 hour hold

(C) RBSN + 12 % Si-rich SiC:
2 hours at 1250-1400 °C; 4 hour hold
10 hours at > 1400 °C

(D) RBSN + 10 % C-rich SiC:
3 °C/min. to 1250 °C; 4 hour hold
0.5 °C/min. to 1400 °C; 10 hour hold

(E) RBSN + 20 % C-rich SiC:
3 °C/min. to 1250 °C; 4 hour hold
0.5 °C/min. to 1400 °C; 10 hour hold

Reinforcement Description

SiC particulates

Reinforcement Dimensions

ave. Si dia. = 0.23 μm
ave. C-rich SiC dia. = 0.04 μm
ave. Si-rich SiC dia. = 0.05 μm

Matrix Description

Si powder

Matrix Designation

F93 S

Test Type

Knoop Indentation Method

Specimen Preparation

Samples were ground and polished to 1 μm on the tensile surface.

Specimen Dimensions

dia. = 12.43-12.67 mm
thickness = 1 mm

Type of Specimen

Disk

Reinforcement Content (weight percent)	Young's Modulus (GPa)	Relative Density (percent)	Range (GPa)	Comments
0	156	77	±23	*A*
0	160	77	±23	
0	182	76	±13	*B*
12	150	73	±10	*C, Si-rich SiC*
10	166	76	±14	*D, C-rich SiC*
20	115	73	±10	*E, C-rich SiC*

Remarks/Comments

Modulus was estimated from the modulus/hardness ratios measured by Knoop indentation. With the same density, the octanol-processed RBSN modulus was higher than the methanol-processed RBSN.

SiC(p)/Si_3N_4

Reference: 74

Composition
0 - 38 vol% SiC

Processing Method
Colloidal Pressed/Reaction Bonded

Detailed Description
Si and SiC powders were colloidally pressed from an octanol dispersion. The bodies were then reaction bonded to form the composite.

Reinforcement Description
SiC particulates

Reinforcement Dimensions
dia. = 15 or 50 μm

Matrix Description
Si powder

Matrix Properties
dia. = 0.2-0.3 μm

Test Type
Knoop Indentation

Specimen Preparation
Samples were polished to 9 μm surface finish.

Specimen Dimensions
dia. = 12 mm
thickness = 1.5 mm

Type of Specimen
Disk

Test Temperature
RT

Reinforcement Content (volume percent)	Young's Modulus (GPa)	Comments
0	147	*type A*
22	163	
38	176	
0	142	*type B*
21	182	

Remarks/Comments

Type B was processed with a higher solids loading in the colloidal slurry to prevent segregation of SiC particulates. A large degree of segregation was found in the type A composite, and some agglomeration was seen in the type B composite. The modulus was calculated from the Knoop indentation technique.

$SiC(p)/Si_3N_4$

Reference: 37

Composition
0 and 10 vol% SiC

Processing Method
CIPed/Sintered or Sinter-HIPed

Detailed Description
The starting powders were ball-milled, dried, and sieved. The mixture was CIPed at 400 MPa and pressureless-sintered at 1750 °C under 0.1 MPa of nitrogen for 3 hours. Half of the samples were HIPed at 1850 °C and 100 MPa in nitrogen for 1 hour.

Sintering Additives
10 wt% Y_2O_3
5 wt% Al_2O_3

Reinforcement Description
SiC particulates

Reinforcement Supplier
Ibiden Co., Ohgaki, Japan

Reinforcement Typical Properties
specific density = 3.21 g cm^{-3}

Reinforcement Dimensions
ave. particle dia. = 0.26 μm

Matrix Description
Si_3N_4

Matrix Supplier
Ube Kosan Co., Japan

Matrix Designation
E-10

Matrix Properties
specific density = 3.18 g cm^{-3}
ave. particle dia. = 0.5 μm

Test Type
Resonance Frequency Method

Specimen Preparation
Samples were surface ground.

Specimen Dimensions
4 x 3 x 40 mm

Type of Specimen
Bar

Test Temperature
RT

Reinforcement Content (volume percent)	Young's Modulus (GPa)	Relative Density (percent)	Std. Dev. (GPa)	Comments
0	289	98	5	*Sintered*
10	292	97	4	
0	298	99	4	*HIPed*
10	303	99	4	

Remarks/Comments
Five specimens were tested for each composition.

$SiC(p)/Si_3N_4$

Reference: 40

Composition
0 and 10 vol% SiC

Processing Method
Hot-Pressed

Detailed Description
The starting powders were mixed, dried, and sieved. The powders were hot-pressed at 1700-1780 °C and 30 MPa under 0.1 MPa nitrogen.

Sintering Additives
5 wt% Y_2O_3

Reinforcement Description
β-SiC particulates

Reinforcement Supplier
Ibiden Co., Ltd., Oogaki-city, Gifu-prefecture, Japan

Supplier Designation
β-SiC

Reinforcement Typical Properties
specific density = 3.21 g cm^{-3}

Reinforcement Dimensions
ave. particle dia. = 0.26 μm
surface area = 15.1 m^2 g^{-1}

Matrix Description
Si_3N_4

Matrix Supplier
Ube Industries., Ltd., Japan

Matrix Designation
COA

Matrix Properties
specific density = 3.18 g cm^{-3}
ave. particle dia. = 0.5 μm
surface area = 10.3 m^2 g^{-1}
contains (wt%): 5 Y_2O_3 and 5 Al_2O_3
composition (wt%): N > 38, O = 1.27, C < 0.3, Al < 0.05, Ca < 0.05, Fe < 0.01

Test Type
Resonance Frequency Method

Specimen Preparation
Samples were polished with 6 and 3 μm diamond paste.

Specimen Dimensions
6 x 3 x 50 mm

Type of Specimen
Bar

Test Temperature
RT

Reinforcement Content (volume percent)	Young's Modulus (GPa)
0	284
10	295

Remarks/Comments
The Young's modulus increased by 11 GPa with the addition of SiC reinforcement.

SiC(p)/Si_3N_4

Reference: 67

Composition
24 wt% SiC

Processing Method
Polymer Pyrolysis/CIPed/Sintered

Detailed Description
The amorphous Si, N, and C powders were prepared by pyrolysis of a polymethylsilazane at 1000 °C in argon. 80 grams of the polymethylsilazane was mixed with 10.5 wt% Al_2O_3 and 4.5 wt% Y_2O_3 were dissolved in 300 ml of THF under an argon atmosphere. The THF was evaporated in a vacuum at 0.1 Pa and room temperature. The dried residue was heated at 550 °C for 1 hour then heated at 1000 °C for 1 hour. It was then attrition milled for 3 hours at 1000 rpm and cold-pressed at 630 MPa into green compacts of 35 x 17 x 15 mm. Reference samples were made from α– Si_3N_4, β– SiC, 10.5 wt% Al_2O_3, and 4.5 wt% Y_2O_3. They were attrition milled, dried, sintered, and hot-pressed in the same manner as pyrolyzed sample. Both the as-pyrolyzed sample and the reference sample were sintered in a graphite resistivity heated furnace at 1750 °C and 1850 °C under 0.1 MPa nitrogen.

Sintering Additives
10.5 wt% Al_2O_3 (A16, Alcoa, Pittsburg, PA)
4.5 wt% Y_2O_3 (Ventron Alta Produckte, Karlstuhe, FRG)

Reinforcement Description
SiC particulates

Reinforcement Supplier
H.C. Stark, Berlin, Germany

Supplier Designation
B10

Matrix Description
Si_3N_4 powder

Matrix Supplier
H.C. Stark, Berlin, Germany

Matrix Designation
LC10

Matrix Properties
Polymethylsilazane powder A:
7.1 wt% C, 22.5 wt% N, 10.4 wt% O
Reference powder B:
9 wt% C, 20.6 wt% N, 9.2 wt% O

Test Type
Resonance Frequency Method

Type of Specimen
Bar

Test Temperature
RT

Reinforcement Content (weight percent)	Young's Modulus (GPa)	Relative Density (percent)	Comments
24	286	95	*sample A*
24	251	93	*sample B (Reference)*

Remarks/Comments
Pyrolyzed sample had a smaller grain size, smaller grain size distribution, and a higher Young's modulus.

SiC(p)/Si_3N_4

Reference: 93

Composition
0 - 20 vol% SiC

Processing Method
Hot-pressed

Detailed Description
These powders were mixed by ball-milling with Al_2O_3 ball using ethanol solvent for 24 hours in a polyethylene vessel. Hot-pressing was carried out at 1800 °C and 30 MPa for 1 hour in a nitrogen atmosphere.

Sintering Additives
8 wt% Y_2O_3

Reinforcement Description
SiC powders

Reinforcement Supplier
Ibiden Co., Ltd.

Supplier Designation
UF grade

Reinforcement Dimensions
ave. particle size = 0.3 μm

Matrix Description
Si_3N_4 powder

Matrix Supplier
Ube Industries, Ltd., Japan

Matrix Designation
E-10 grade

Matrix Properties
ave. particle size = 0.2 μm

Test Type
Resonance Frequency Method

Specimen Dimensions
3 x 4 x 40 mm

Type of Specimen
Bar

Test Temperature
RT

Reinforcement Content (volume percent)	Young's Modulus (GPa)	Relative Density (percent)
0	292	99.9
5	291	99.9
10	300	98.4
20	314	98.7

Remarks/Comments
Young's modulus was constant up to 5 vol% SiC. There were two types of dispersion configurations: nano-meter scale particles uniformly distributed and submicron particles forming aggregates at grain boundaries of the matrix. Grains changed to a rod-like structure as SiC content increased. Young's modulus reached a maximum value at 20 vol% SiC.

Figure 98: Young's Modulus of Reaction Bonded SiC(p)/Si_3N_4

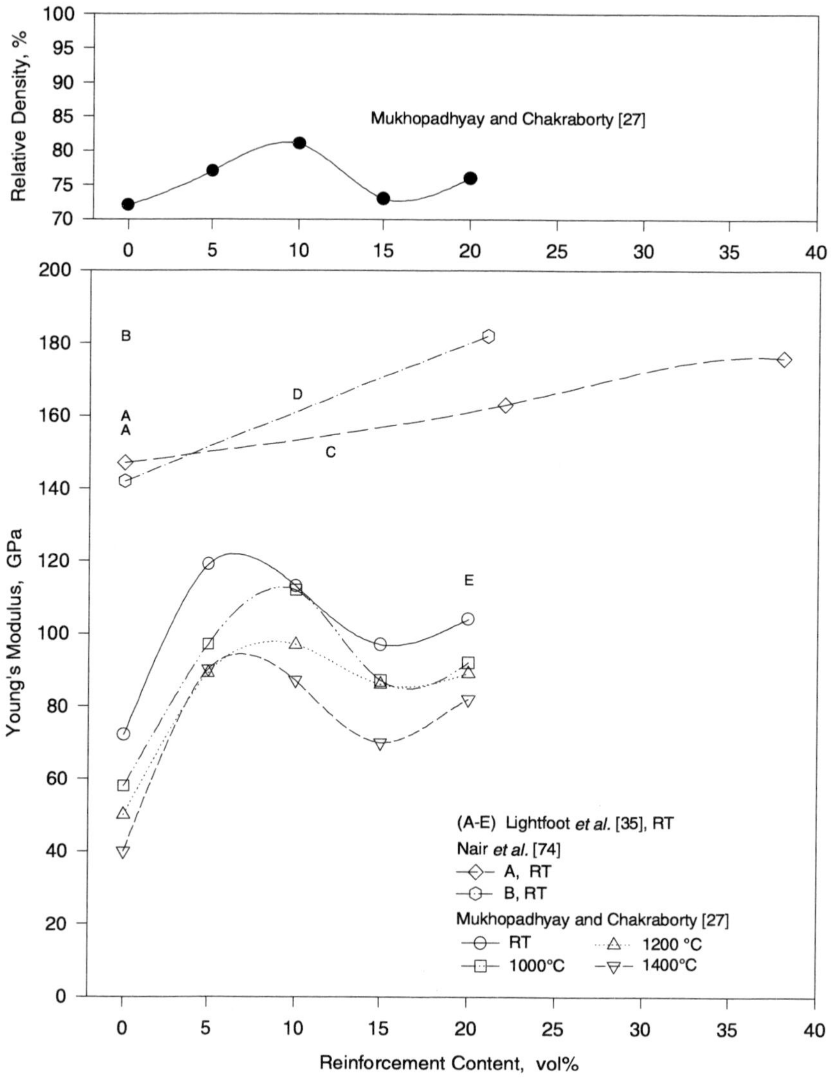

Discussion: The overall trend found by ref. [27] was due to density (maximum at ~ 5-10 vol%). With reaction bonded composites there was a maximum density that compensated for the volume change. When this density was reached, the further addition of SiC particulates caused stresses that resulted in microcracking. This could have reduced the Young's modulus. After the microcracks had occurred, the matrix would be free to add more particulates without further cracking. Then the composite, increasing in the phase that had the higher Young's modulus, had its Young's modulus dominated by the rule of mixtures.

The modulus of the RBSN materials tested by Lightfoot *et al.* [35] and Nair *et al.* [74] were much higher than the composites tested by ref. [27]. Nair *et al.* [74] found that the modulus increased with increasing SiC content. The modulus of the materials tested by Lightfoot *et al.* [35] was likely more dependent on processing conditions than reinforcement content. It was therefore difficult to discern the effect of SiC content on the modulus.

Figure 99: Young's Modulus vs. Temperature of Reaction Bonded SiC(p)/Si_3N_4

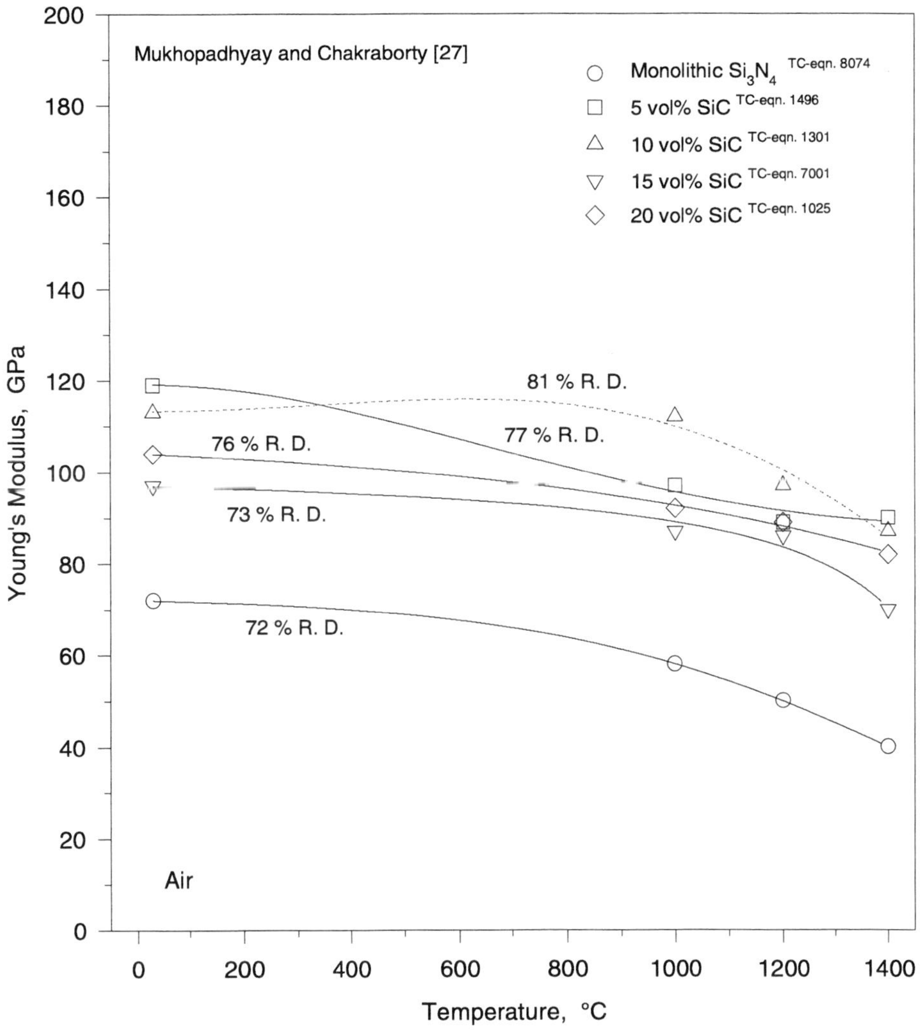

Discussion: When the data from reference [27] is plotted vs. temperature, the results showed that modulus decreased with increasing temperature. However, the modulus of these composites was primarily dependent on relative density.

Figure 100: Young's Modulus of SiC(p)/Si_3N_4

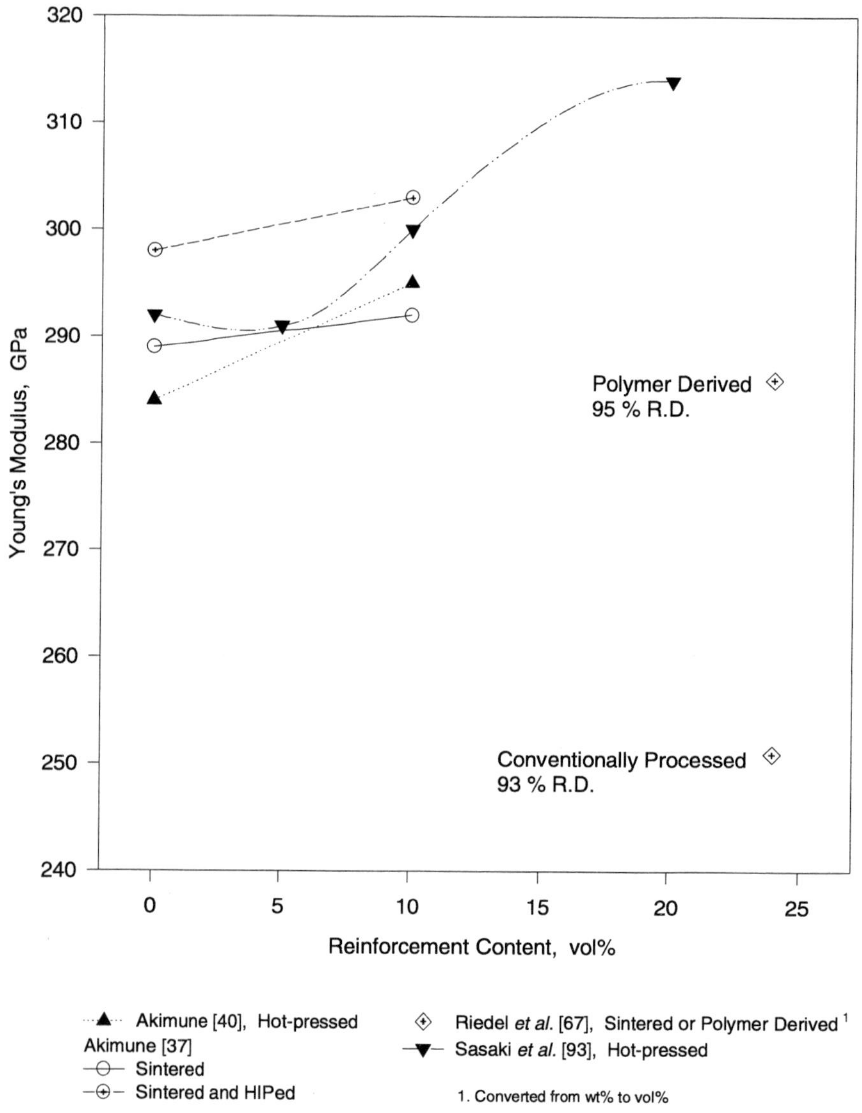

Discussion: The composites tested by Sasaki *et al.* [93] approached the rule of mixtures. Deviations from this rule can possibly be due to the presence of disorder at the SiC-Si_3N_4 grain boundaries. A large amount of disorder in the composite occurred at the interface between the matrix and the reinforcement. Because there was greater surface area between phases in a nano-composite than a micro-composite, there was more disorder. This was probably one of the reasons why Young's modulus did not increase up to 5 vol% SiC [93]. The lower modulus of the conventionally sintered composite than the polymer derived composite was at least partially due to the lower density of the conventionally sintered composite [67]. The composite tested by Akimune [37 and 40] showed a slight increase in modulus with the addition of 10 vol% SiC reinforcement.

$BN(p)/Si_3N_4$

Reference: 27

Composition
0 and 10 vol% BN

Processing Method
Cold-Pressed/Reaction Bonded

Detailed Description
BN and silicon powders were cold pressed then nitrided in a dry nitrogen atmosphere in a temperature range of 1350-1380 °C.

Reinforcement Description
BN particulates

Reinforcement Supplier
Grindwell Norton

Matrix Description
Si powder

Test Type
3-pt. Flexural

Test Description
E = fracture load*bending arm length*(3*major span^2 - 4* bending arm length) / (4*deflection*sample length*sample width)

Test Temperature
RT-1400 °C

Test Environment
Air

Test Equipment
High Temperature Bending Strength Tester (Model 422S, Netzsch, Germany)

Temperature (°C)	Young's Modulus (GPa)	Reinforcement Content (volume percent)	Relative Density (percent)
25	71	0	72
1,000	59	0	72
1,200	49	0	72
1,400	40	0	72
25	87	10	80
1,000	71	10	80
1,200	68	10	80
1,400	65	10	80

Remarks/Comments
Young's modulus decreased with increasing temperature. Density was 2.21 g/cm^3.

TiN(p)/Si_3N_4

Reference: 54

Composition
0 - 40 vol% TiN

Processing Method
Hot-Pressed

Detailed Description
Si_3N_4 powder and sintering aids were homogenized in water for 48 hours and freeze dried. The second phase (TiN) was added to the matrix mix and mixed in isobutyl alcohol for 48 hours. The materials were then uniaxially hot-pressed at 1800 °C and 30 MPa for 10 minutes to near full density.

Sintering Additives
3 wt% Al_2O_3
8 wt% Y_2O_3

Reinforcement Description
TiN particulates

Reinforcement Supplier
Starck, Germany

Supplier Designation
Grade C

Reinforcement Dimensions
ave. size = 0.3-3 μm

Matrix Description
Si_3N_4 powder

Matrix Supplier
Starck, Germany

Matrix Designation
LC12

Test Description:

Test Type
Resonance Frequency Method

Test Description
A crossed strain gauge measured strains.

Test Temperature
RT

Reinforcement Content (volume percent)	Young's Modulus (GPa)	Relative Density (percent)
0	340	100
20	331	99.4
30	351	99.6
40	359	99.4

Remarks/Comments
Young's modulus increased with 30 and 40 vol% additions of TiN reinforcement and was maximized at 40 vol% with a value of 359 GPa; however, these values were not much higher than the monolithic value.

TiN(p)/Si_3N_4

Reference: 115

Composition
10 vol% TiN

Processing Method
CIPed/Pressureless-Sintered

Detailed Description
The powders were mixed in a vibratory mill using methanol and followed by drying. Then, the powders were dispersed in water at pH ~10 by means of ultrasonification. The mixed slurry was washed with acetone and filtered off on the funnel. Then, the obtained powder cake was dispersed again in toluene, and residual water was removed by utilizing azeotropy between water and toluene. After drying in nitrogen- 5 % hydrogen atmosphere, the powders were CIPed at 490 MPa and then sintered at 1850 °C for 4 hours in a 0.9 MPa nitrogen atmosphere.

Sintering Additives
5 wt% Y_2O_3, 2 wt% Al_2O_3 (Hokko Chem. Co., Ltd.)
2 wt% MgO (Iwatani Chem. Ind. Co.)

Reinforcement Description
TiN particulates

Reinforcement Supplier
Nippon Shin-Kinzoku Co.

Matrix Description
α-Si_3N_4 powder

Matrix Supplier
Denki Kaguku Co.

Matrix Designation
SNP21 C3 grade

Matrix Properties
specific surface area = 11.2 $mm^2 g^{-1}$
92.2 % α– phase
oxygen content = 1.57 wt%

Test Type
Pulse-Echo Method

Test Description
Density was measured by Archimedes' method.

Test Temperature
RT

Grain Size (μm)	Young's Modulus (GPa)	Relative Density (percent)
0.3	316	98.0
1.0	320	98.5
1.8	320	98.3
3.7	312	98.2
7.3	303	98.0
13.6	287	98.1

Remarks/Comments

The Young's modulus decreased gradually as the TiN particulate size increased.

Figure 101: Young's Modulus vs. Temperature of Reaction Bonded BN(p)/Si_3N_4

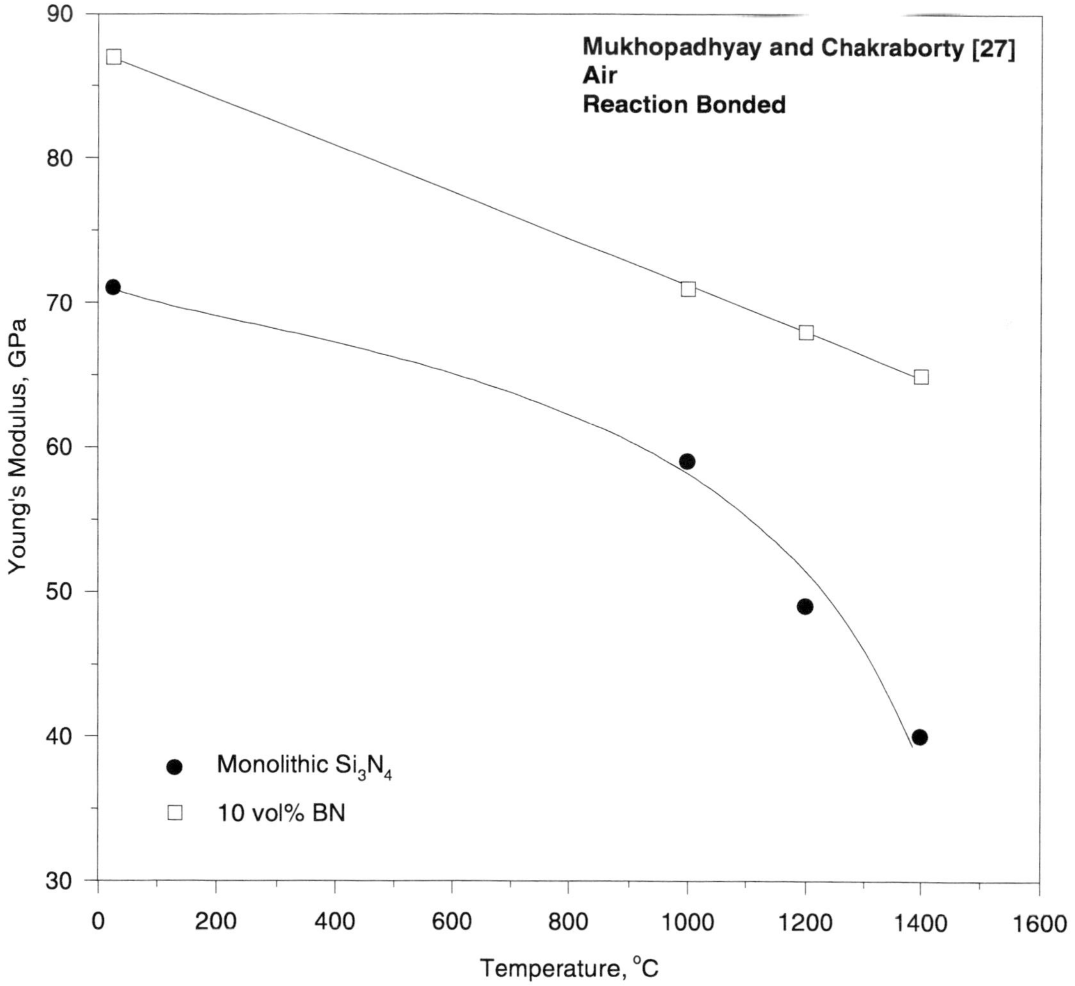

Discussion: The modulus of BN-RBSN was higher than the RBSN. The modulus of RBSN dropped rapidly above 1000 °C, whereas the addition of BN stabilized the decrease in modulus.

Figure 102: Young's Modulus of Hot-Pressed $TiN(p)/Si_3N_4$

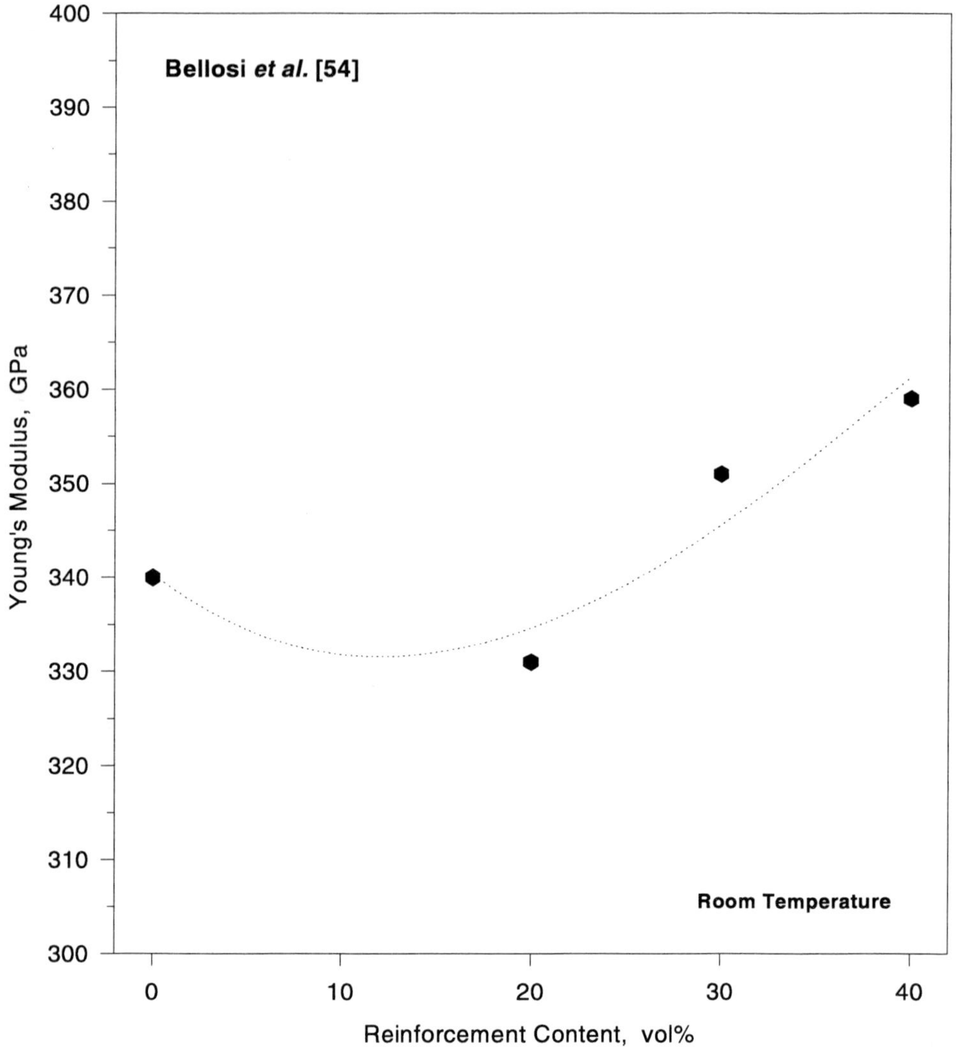

Discussion: Figure 102 presents the Young's modulus of hot-pressed $TiN(p)/Si_3N_4$. The Young's modulus initially decreased slightly with the addition of 20 vol% TiN, perhaps due to grain boundary disorder. The modulus increased with continued addition of TiN reinforcement up to 40 vol% because of the much higher modulus of TiN compared to Si_3N_4 [54].

Figure 103: Young's Modulus vs. Grain Size of Sintered TiN(p)/Si_3N_4

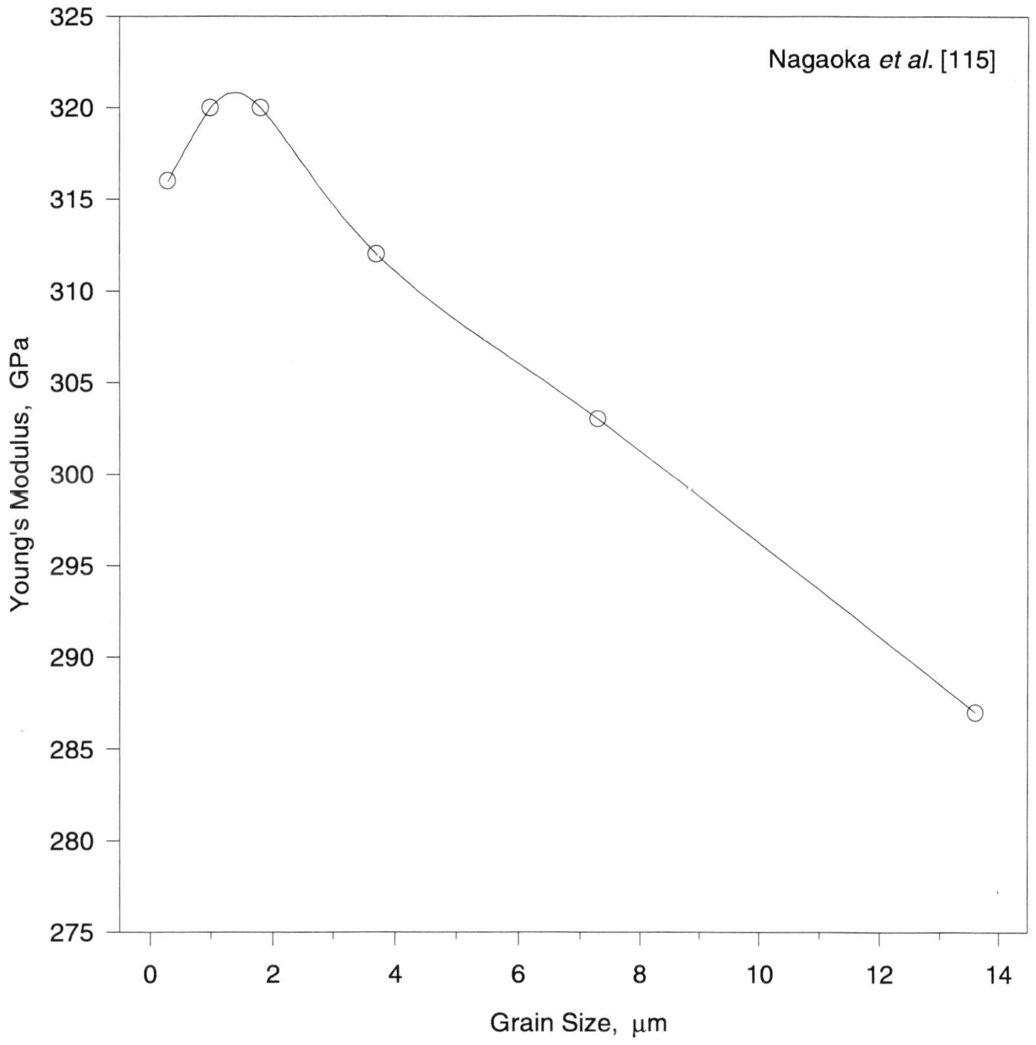

Discussion: As the grain size increased in sintered TiN(p)/Si_3N_4, residual stresses due to thermal mismatch of the phases increased. This produced microcracks that caused the composite to decrease in elastic modulus [115].

BN(p)/SiC

Reference: 17

Composition
0 - 85 wt% BN

Processing Method
Hot-Pressed

Detailed Description
The powders were mixed in a plastic ball-mill with Si_3N_4 balls for 4 hours in isopropyl alcohol. After drying, the powders were pelletized, and specimens were uniaxially hot-pressed at 2100 °C and 35 MPa under a vacuum for 1 hour in graphite dies lined with graphite foil.

Reinforcement Description
BN particulates

Reinforcement Supplier
Research Organic/Inorganic Chemical Corp., Bellville, NJ

Supplier Designation
Grade B-27

Reinforcement Dimensions
particle size = 15 μm

Matrix Description
β-SiC powder

Matrix Supplier
H.C. Starck, Goslar, Germany

Matrix Designation
Lot S 1626

Matrix Properties
particle size < 1 μm

Test Type
Resonant-Sphere Technique

Specimen Preparation
Specimens were ground into a sphere.

Type of Specimen
Sphere

Test Temperature
RT

Reinforcement Content (weight percent)	Young's Modulus (GPa)	Relative Density (percent)
0	437	100
6.4	310	97.8
13	317	96.9
21	255	96.0
29	241	96.2
38	243	93.3
48	239	91.5
59	198	89.7
71	167	87.3
85	185	85.5

Remarks/Comments
With increasing BN content, the Young's modulus dropped sharply to 30 wt% BN, decreased more gradually to 70 wt%, and then dropped with a slightly greater slope to 90 wt% BN.

TiB_2(p)/SiC

Reference: 69

Composition
0 - 30 vol% TiB_2

Processing Method
Reaction Sintered (some HIPed)

Detailed Description
Fine TiB_2 particles were prepared by milling the TiB_2 agglomerates produced by the following carbothermal reaction at 2000 °C:

$$TiO_2 + 0.5\ B_4C + 1.5\ C \rightarrow TiB_2 + 2\ CO$$

The removal of the CO gas from the system was important to enhance the reaction at lower temperatures. The TiO_2 particle size was 0.4 µm, the B_4C powder size was 1.5 µm, and the carbon black powder size was 0.018 µm. The amounts of B_4C and carbon that were added were more than the stoichiometrically required amount. The mixing ratio of SiC and TiO_2 was designed so that the volume ratio of SiC matrix to TiB_2 particulates formed by the carbothermal reaction should be 8:2. The starting materials were mixed with water by ball milling for 20 hours, using a polyethylene pot and nylon-coated iron balls. The slurries were filtrated by suction, and dried at 100 °C. Screened powders under 32-mesh were molded in dies at a pressure of 20 MPa and isostatically pressed at 300 MPa. The green compacts in a graphite container were heated in a vacuum at 1500 °C for 4 hours. After the vacuum soaking treatment, argon gas was introduced into the furnace and the compacts were sintered in an argon atmosphere at 2150 °C for 1 hour. Some composites were also HIPed in a graphite container at 1750-1900 °C for 1 hour in argon at 200 MPa.

Sintering Additives
B_4C: Denki Kagaku Kogyo
C: Mitsubishi Chemical

Reinforcement Description
TiB_2 particulates

Reinforcement Supplier
Fuji Titan Kogyo

Supplier Designation
TM-1

Matrix Description
α-SiC powder

Matrix Supplier
Showa Denko Co

Matrix Properties
ave. particle size = 0.48 µm

Test Type
Composite Bar Method

Specimen Dimensions
4 x 3 x 35 mm

Test Temperature
RT

Reinforcement Content (volume percent)	Young's Modulus (GPa)	Density ($g\ cm^{-3}$)	Comments
0	420	3.15	*reaction-sintered*
10	436	3.28	
20	452	3.40	
30	462	3.53	
0	432	3.19	*reaction sintered/HIPed*
10	456	3.32	
20	468	3.43	
30	480	3.56	

Remarks/Comments
Young's modulus was averaged for at least 2 sets of values. The reaction sintered composite with each composition attained approximately 99 % of the density of the hot-pressed composite. The Young's modulus increased linearly with increasing TiB_2 content.

TiC(p)/SiC

Reference: 70

Composition
0 - 100 wt% TiC

Processing Method
Hot-Pressed

Detailed Description
Powders of β-SiC and TiC were mixed with B_4C powder and carbon black as sintering aids for SiC. Well-mixed powders were attained by ball-milling in n-hexane for 24 hours using a polyamide media. After drying, the powders were loaded into a graphite die with id = 90 mm and hot-pressed at 2150 °C and 40 MPa for 2 hours in an argon gas atmosphere. Hot-pressed discs of SiC-TiC composites were machined into pieces.

Sintering Additives
B_4C: Denki Kagaku Kogyo KK (1500A)
ave. particle size = 1.2 μm
C (carbon black): Mitsubishi Chem. Ind. (600MA)
ave. particle size = 0.02 μm
purities of > 99 %

Reinforcement Description
TiC powders

Reinforcement Supplier
Kojundo Chem. Lab. Co., Ltd

Reinforcement Typical Properties
ave. particle size = 0.4 μm
purity > 99 %

Matrix Description
β-SiC powders

Matrix Supplier
Ibiden Co., Ltd

Matrix Designation
Ultrafine

Matrix Properties
ave. particle size = 0.3 μm
purity > 98 %

Test Type
Pulse Echo Overlap Method

Specimen Dimensions
3 x 4 x 38 mm

Test Temperature
RT

Reinforcement Content (weight percent)	Young's Modulus (GPa)	Poisson's Ratio (dimensionless)
0	384	0.15
10	415	0.16
30	425	0.15
50	433	0.16
70	418	0.16
90	462	0.17
100	450	0.19

Remarks/Comments
The Poisson's ratio increased linearly with addition of TiC from 0.15 for SiC to 0.19 for TiC. The elastic modulus increased linearly with addition of TiC from 390 GPa for SiC to 450 GPa for TiC.

TiB_2(p)/SiC

Reference: 83

Composition
0 and 16 vol% TiB_2

Processing Method
Colloidal Processed / Pressureless-Sintered

Detailed Description
The commercially available SiC-TiB_2 (Hexoloy ST) CMC was fabricated by mixing presized TiB_2 and submicrometer-size SiC powders using a nonaqueous blending and drying procedure. The homogeneous mixture, combined with a binder, was formed into green compacts that were then pressureless-sintered in an inert atmosphere at temperatures of over 2000 °C.

Reinforcement Description
TiB_2 particulates

Reinforcement Typical Properties
melting point = 2790 °C
E = 531 GPa
Poisson ratio = 0.28
CTE = 4.6x10^{-6} °C^{-1}

Reinforcement Dimensions
ave. particle size = 2 μm

Matrix Description
α-SiC

Matrix Properties
ave. grain size = 4 μm
melting point = 2540 °C
E = 410 GPa
Poisson ratio = 0.19
CTE = 4.02x10^{-6} °C^{-1}

Test Type
Elastic Modulus Test/ Poisson Ratio Test

Test Temperature
RT

Reinforcement Content (volume percent)	Young's Modulus (GPa)	Poisson's Ratio (dimensionless)	Density (g cm^{-3})
0	410	0.14	3.1
16	427	0.15	3.3

Remarks/Comments
The Young's modulus increased with the addition of TiB_2 particles. The Poisson's ratio increased slightly with the addition of TiB_2 particles.

Figure 104: Young's Modulus of BN(p), TiB_2(p), and TiC(p) in SiC Matrix

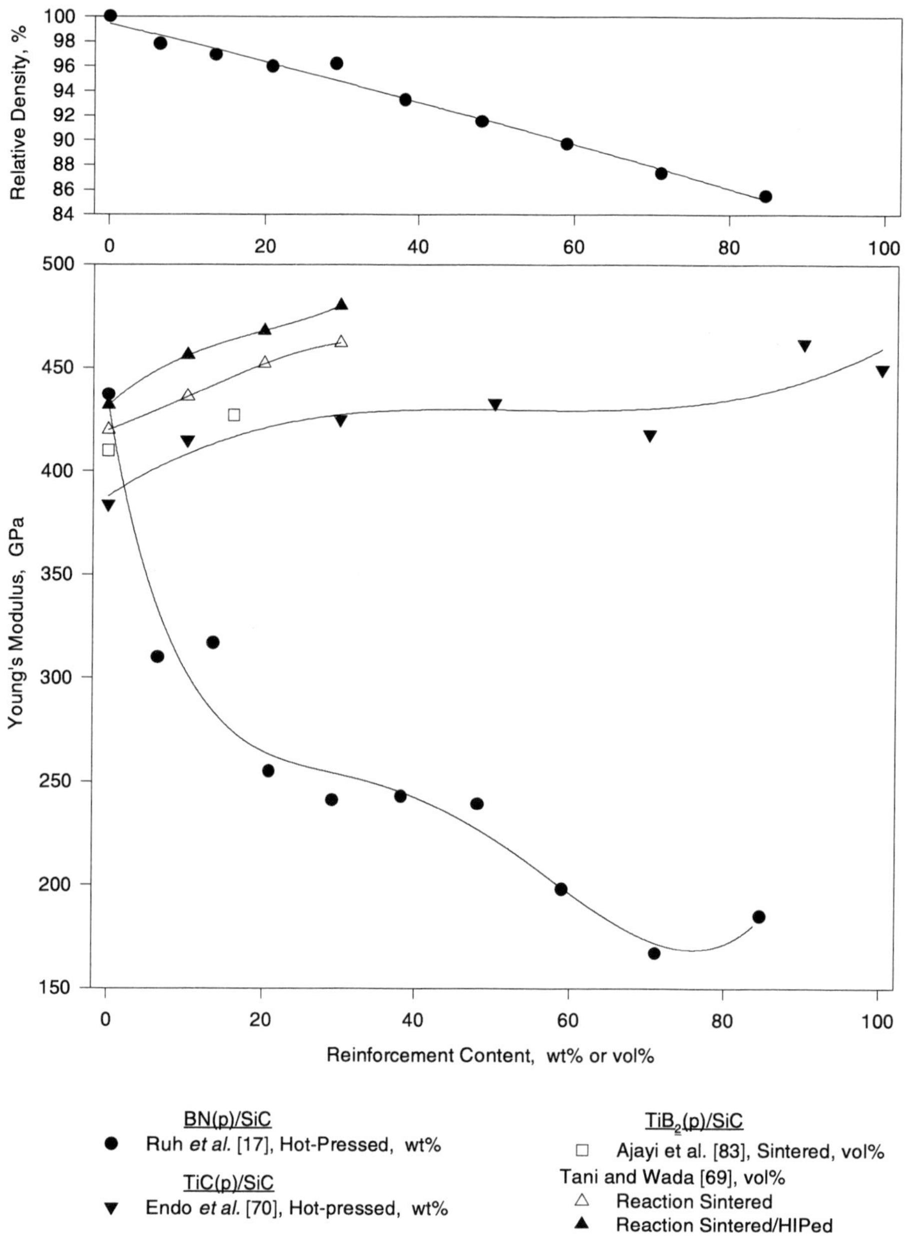

Discussion: BN has a lower elastic modulus than SiC, so the elastic modulus of the SiC composite decreased with increased BN content, according to the rule of mixtures. The decrease in modulus for the BN(p)/SiC composites was also partly the result of decreased relative density. The composites TiB_2(p)/SiC and TiC(p)/SiC showed similar increases in modulus with an increase in reinforcement content. This was also due to the rule of mixtures; TiB_2 and TiC have higher Young's modulus than the SiC matrix.

Figure 105: Poisson's Ratio of TiB_2(p) and TiC(p) in SiC Matrix

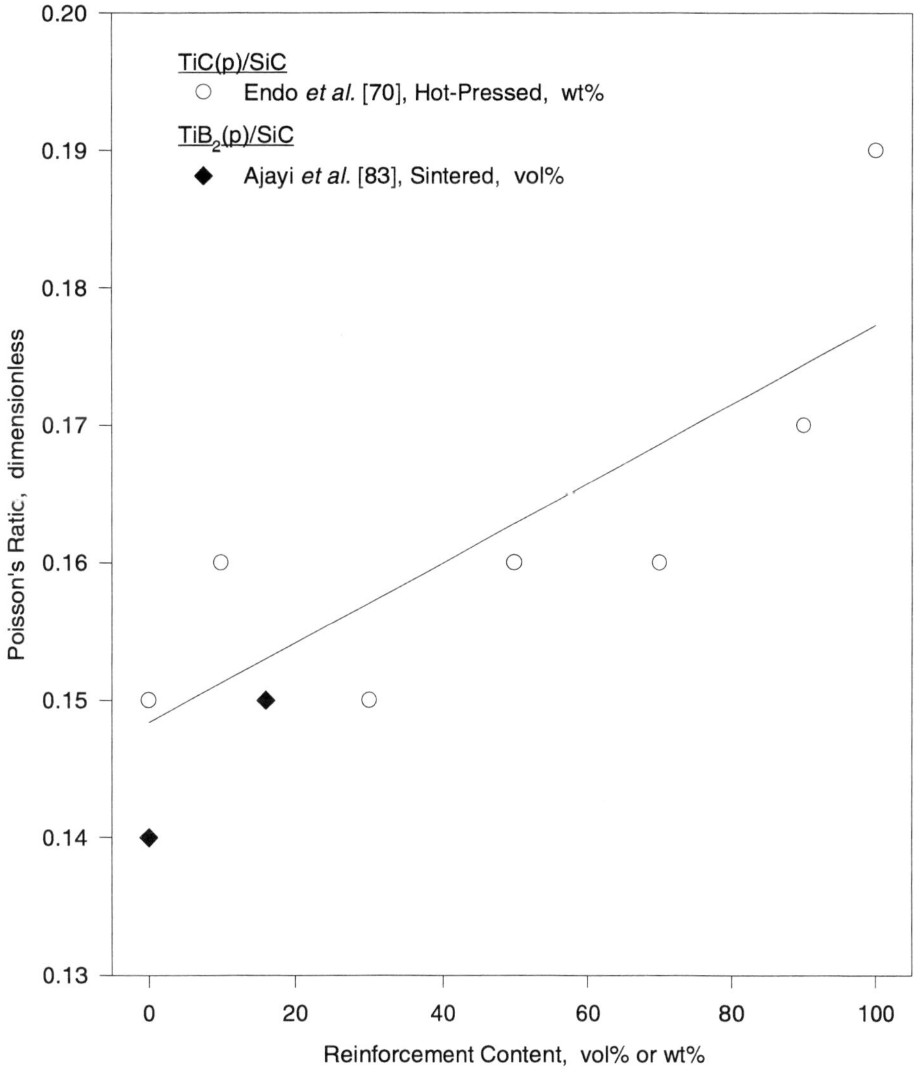

Discussion: As reinforcement content increased, Poisson's ratio increased for the TiB_2(p)/SiC and TiC(p)/SiC composites. The fit of a first order regression to the data suggested that the law of mixtures was observed.

Zr(p).ZrB_2(pl)/ZrC

Reference: 28

Composition
0 - 30 vol% Zr particles
ZrB_2/ZrC ratio = 58:42 by volume

Composite Architecture
multi-reinforcement

Processing Method
Lanxide Reaction Bonded

Detailed Description
The material was formed by reaction of molten zirconium with a preform of boron carbide particulates, which were substantially consumed in the process. These materials were being commercialized as Lanxide NX-3400 series platelet-reinforced ceramics.

Reinforcement Description
Zr particulates
ZrB_2 platelets

Matrix Description
ZrC particulates

Test Type
Resonance Frequency Method

Specimen Dimensions
5 x 6 x 50 mm

Test Temperature
RT

Residual Metal Content (volume percent)	Young's Modulus (GPa)	Shear Modulus (GPa)
2.0	422	175
2.5	396	167
3.1	436	169
6.0	382	162
8.0	356	144
10	362	158
12	334	141
20	308	122
21	308	129
22	289	113
24	290	120
26	287	101
30	290	114

Remarks/Comments
As the Zr content rose, the shear modulus fell slightly. As Zr content rose to 20 vol%, the Young's modulus decreased. Thereafter, the Young's modulus remained relatively constant with increasing Zr content.

$SiC(p)/ZrB_2$

Reference: 92

Composition
0 and 20 vol% SiC

Composite Architecture
particulate reinforcement

Processing Method
Hot-Pressed

Detailed Description
All materials were fabricated by hot-pressing and contain some degree of anisotropy.

Reinforcement Description
SiC particulates

Reinforcement Supplier
Superior Graphite

Reinforcement Dimensions
ave. size = 3 μm

Matrix Description
ZrB_2 powder

Test Type
Pulse-Echo Technique

Type of Specimen
Bar

Test Description
Mechanical properties were tested in the transverse direction.

Reinforcement Content (volume percent)	Young's Modulus (GPa)	Poisson's Ratio (dimensionless)	Density ($g\ cm^{-3}$)
0	500	0.11	6.10
20	515	0.14	5.52

Remarks/Comments
Poisson's ratio was higher for the composite than for the monolith and was greater for the SiC platelet reinforced composite than the SiC particulate reinforced composite. Young's modulus was slightly higher for the SiC particulate and platelet reinforced composites.

$SiC(p)/ZrB_2$

Reference: 109

Composition
10 wt% SiC

Composite Architecture
particulate reinforcement

Processing Method
Hot-Pressed

Reinforcement Description
SiC particulates

Matrix Description
ZrB_2 powder

Test Type
Vickers Indentation

Specimen Preparation
Test specimens were produced by wire EDM.

Reinforcement Content (weight percent)	Young's Modulus (GPa)	Density ($g\ cm^{-3}$)
10	460	5.4

Figure 106: Young's Modulus of $Zr(p).ZrB_2(pl)/ZrC$ and $SiC(p)/ZrB_2$

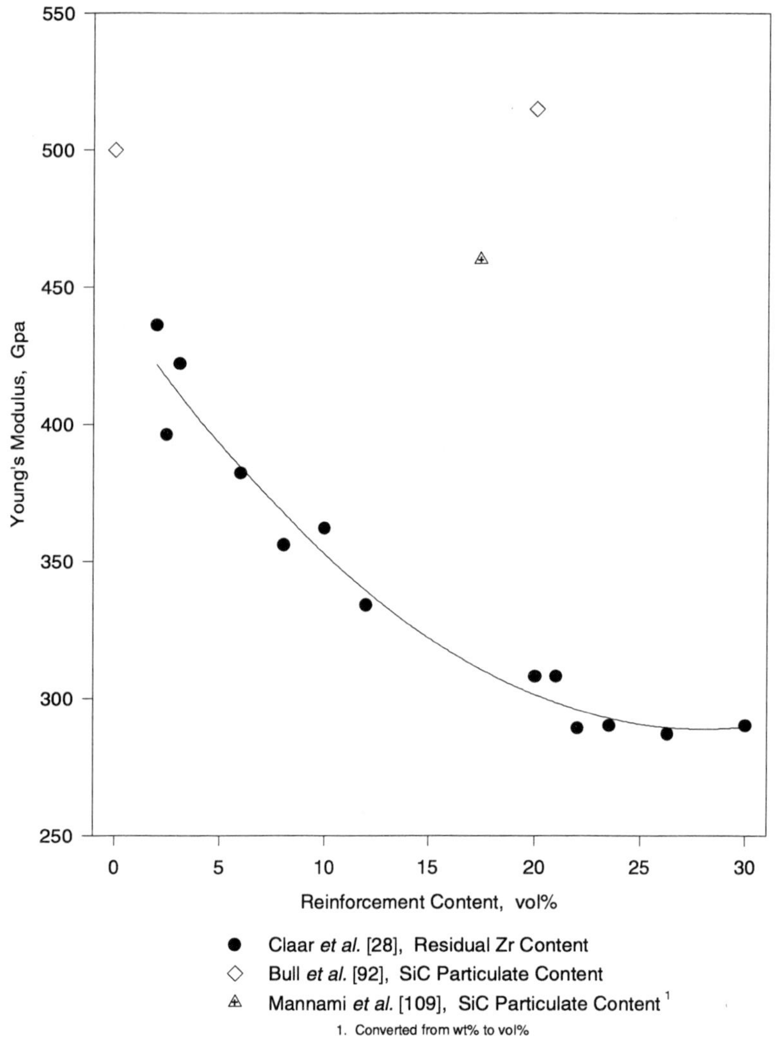

Discussion: The Young's modulus for the $Zr(p).ZrB_2(pl)/ZrC$ composite decreased with increasing residual metal content. The modulus of this composite was lower because of the presence of the lower modulus residual metal phase. The $SiC(p)/ZrB_2$ composite had a much higher elastic modulus, by about 200 GPa, than did the $Zr(p).ZrB_2(pl)/ZrC$ composite at 20 vol% reinforcement content. The Young's modulus of the $SiC(p)/ZrB_2$ composite increased only a small amount, by 15 GPa, with the addition of 20 vol% reinforcement content compared with the monolithic ZrB_2. The composite tested in Ref. [92] was tested in the transverse direction and contained some degree of anisotropy.

Figure 107: Shear Modulus of $Zr(p).ZrB_2(pl)/ZrC$

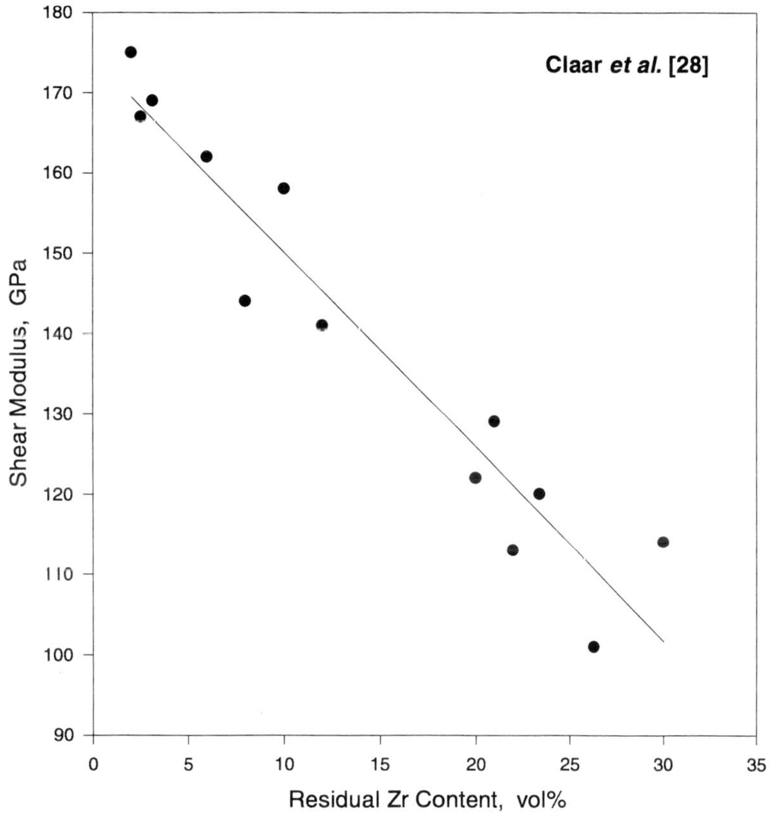

Discussion: The shear modulus of the $Zr(p).ZrB_2(pl)/ZrC$ composite decreased as the reinforcement content increased. The shear modulus was the greatest for the monolithic material and decreased with each addition of reinforcement because of the residual metal phase present in the reinforcement.

Figure 108: Poisson's Ratio for $SiC(p)/ZrB_2$

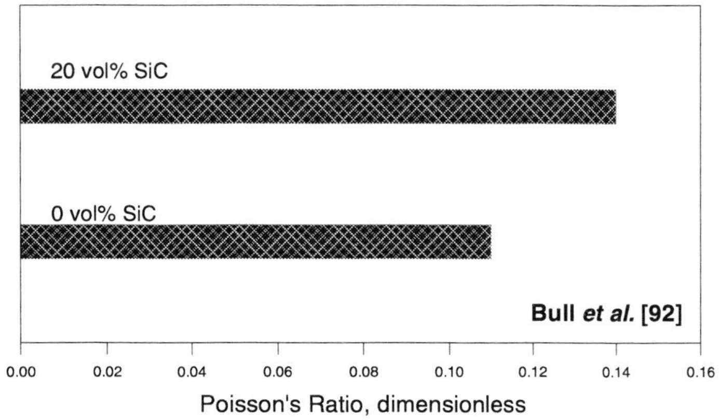

Discussion: The Poisson's ratio of the 20 vol% $SiC(p)/ZrB_2$ composite was much greater, by 0.03, than that of the monolithic ZrB_2 composite. Zirconium diboride composites with moderate amounts of carbon or SiC particulates offer the best combination of mechanical properties and ablation resistance [92].

13. Hardness
(Section IV)

SiC(p)/Al_2O_3

Reference: 33

Composition
37 vol% SiC (with trace B_4C)

Processing Method
Conventionally Hot-Pressed or Reaction Hot-Pressed

Detailed Description
Composite powders were mixed together, then reaction hot-pressed or conventionally hot-pressed under vacuum or Ar at temperatures from 1600-1800 °C and pressures of 7-28 MPa for up to 1 hour.

Reinforcement Description
B_4C and SiC particulates

Supplier Designation
B_4C-Advanced Ref. Tech.
SiC-Buclhcr

Reinforcement Typical Properties
grain size (SiC): reaction HP = 2 μm
conventional HP = 3 μm

Matrix Description
Al_2O_3

Matrix Properties
grain size: reaction HP = 4 μm
conventional HP = 6 μm

Test Type
Vickers Indentation

Test Description
Indenter load = 1 kg

Test Temperature
RT

Relative Density (percent)	Hardness, Vickers (GPa)	Comments
96	20.09	*reaction HP*
99	21.56	*conventionally HP*

Remarks/Comments
The reaction used in the reaction hot-pressing technique was unspecified. The hardness averaged 20.09 and 21.56 GPa for the reaction HP and conventionally HP materials, respectively.

SiC(p)/Al_2O_3

Reference: 77

Composition
0 - 17 vol% SiC

Processing Method
Hot-Pressed

Detailed Description
The composites containing various volume fraction of SiC particles were mixed by ball milling in highly pure toluene with Al_2O_3 balls for 24 hours. The dried mixtures were hot-pressed in N_2 atmosphere at 1500-1900 °C at 28 MPa.

Reinforcement Description
β-SiC particulates

Reinforcement Supplier
Ibiden Co., Ltd.

Reinforcement Dimensions
ave. particle size < 0.3 μm

Matrix Description
α-Al_2O_3 particulates

Matrix Supplier
Asahi Chemical Co.

Matrix Properties
ave. particle size = 0.3 μm

Test Type
Vickers Indentation

Test Description
load = 4.9 N, loading duration = 20 sec
Density measured by an Archimedes method using toluene.

Test Temperature
RT- 1300 °C

Test Environment
Vacuum

Test Equipment
Nikon, QM-high temperature Vickers hardness tester

Temperature (°C)	Hardness, Vickers (GPa)	Reinforcement Content (volume percent)
22	18.3	0
400	14.4	0
600	10.1	0
800	7.6	0
1,000	5.5	0
1,190	4.2	0
1,310	2.0	0
22	19.5	5
200	16.9	5
400	14.9	5
600	12.3	5
800	10.6	5
1,000	8.8	5
1,190	6.9	5
1,310	3.8	5
22	22.3	17
200	19.3	17
400	16.8	17
600	14.6	17
800	12.2	17
1,000	9.9	17
1,190	7.6	17
1,310	6.7	17

Remarks/Comments
Hardness decreased as temperature increased. The composite with 17 vol% SiC reinforcement had a higher hardness than that with 5 vol% SiC reinforcement. Both were higher than the monolithic material.

Figure 109: Vickers Hardness vs. Temperature of $SiC(p)/Al_2O_3$

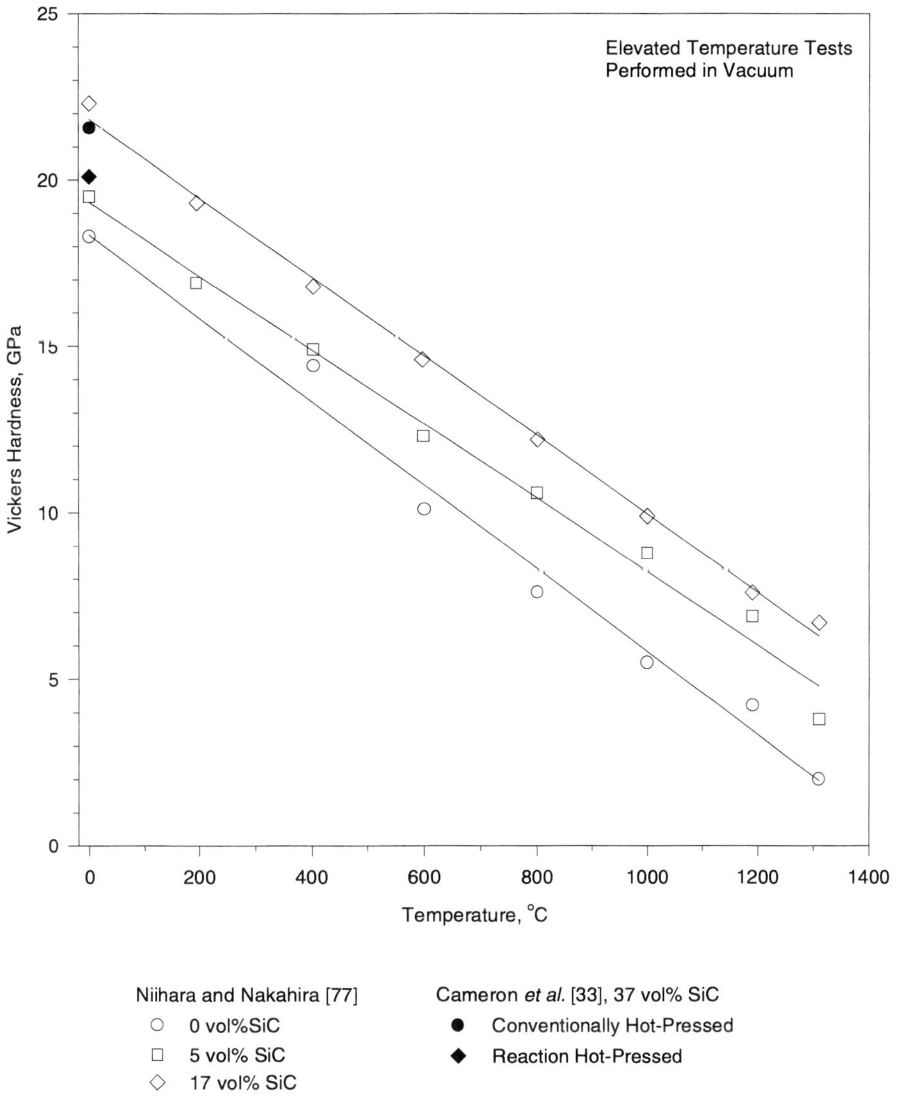

Discussion: Niihara and Nakahira [77] examined the temperature dependence of hardness of a $SiC(p)/Al_2O_3$ nanocomposite. For all samples, the Vickers hardness decreased monotonically with increasing temperature. The Vickers hardness was greater for the 17 vol% $SiC(p)/Al_2O_3$ composite versus the 5 vol% $SiC(p)/Al_2O_3$ and monolithic Al_2O_3 at all temperatures. The slope of the strength degradation was similar for all three materials, which suggests that the drop in hardness of the SiC reinforcement was proportional to the drop in hardness for the Al_2O_3 matrix in the temperature range of RT to 1300 °C. It may equally imply that there are two effects that cancel each other out with increasing temperature: (1) residual thermal stresses are lost as the temperature approached processing temperatures, which would result in a decrease in hardness for the composites, and (2) SiC is harder than alumina at elevated high temperatures, which would result in an increase in hardness for the composites.

Cameron *et al.* [33] found that the room temperature hardness and relative density were greater for the conventionally hot-pressed $SiC(p)/Al_2O_3$ composite than for the reaction hot-pressed composite. Both composites were within the range of room temperature hardness values obtained for the nanocomposites tested by Niihara and Nakahira [77].

$TiB_2(p)/Al_2O_3$

Reference: 29

Composition
0 - 50 wt% TiB_2

Processing Method
CIPed/Sintered

Detailed Description
Al_2O_3 and 0-50 wt% TiB_2 were mixed for 5 hours in an alumina ball mill. After drying the mixtures were pressed into rods 8 x 20 mm under 30 GPa. These rods were sintered at 1600-1750 °C for 15-20 minutes under 0.001-0.01 Pa.

Reinforcement Description
TiB_2 particulates

Reinforcement Supplier
H. C. Starck Berlin, Gosler, Germany

Reinforcement Typical Properties
Specific gravity = 4.38 g cm^{-3}
Microhardness = 33.7 GPa
Thermal conductivity = 122 W/(m K)

Reinforcement Dimensions
ave. dia. = 20 μm

Matrix Description
Al_2O_3 powder

Matrix Supplier
Fujimi Kenmazai Kagyo Ltd. Inazawa, Japan

Matrix Properties
ave. particle size = 0.2 μm

Test Type
Vickers Indentation

Specimen Preparation
Specimens had their surfaces polished prior to testing.

Test Description
Load = 300 g

Test Temperature
RT

Reinforcement Content (weight percent)	Hardness, Vickers (GPa)	Relative Density (percent)	Mean Grain Size (μm)
0	18.4	97.2	38
10	25.9	98.8	16
20	30.7	99.5	14
30	26.2	99.0	9
40	25.7	98.7	7
50	23.9	97.8	5

Remarks/Comments
The addition of TiB_2 to Al_2O_3 increased the hardness from 18.4 GPa to a maximum of 30.7 GPa. This was due to the increase in relative density as well as the higher hardness of TiB_2.

$TiC(p)/Al_2O_3$

Reference: 23

Composition
25 vol% TiC

Processing Method
CIPed/Pressureless Sintered

Detailed Description
The powders were mixed by ball milling with ethanol for 24 hours. After drying, the mixtures were uniaxially pressed and finally isostatically pressed at 200 MPa. Sintering of the compact was done in a radio-frequency induction heating furnace. Specimens were set in an SiC crucible to prevent reaction with the graphite susceptor, and were fired at 1600-1800 °C for 1 hour under an argon gas flow at atmospheric pressure.

Sintering Additives
MgO (0.5 wt% of matrix)

Reinforcement Description
TiC particulates

Reinforcement Supplier
H.C. Stark

Supplier Designation
c.a.s. grade

Reinforcement Dimensions
ave. grain size = 1.5 μm

Matrix Description
Al_2O_3

Matrix Supplier
Shinkosha

Matrix Properties
ave. grain size = 0.2 μm

Test Type
Vickers Indentation

Test Temperature
RT

Reinforcement Content (volume percent)	Hardness, Vickers (GPa)
25	22.5

TiC(p)/Al_2O_3

Reference: 58

Composition
30 wt% TiC

Processing Method
Cold Pressed/ Sintered (some HIPed)

Detailed Description
Powders were dispersed and milled in a 15 cm diameter stainless steel mill with hexane for 24 hours with a ball to charge ratio of 10 to 1. Larger batches were attrition milled in deionized water and a binder was added prior to spray drying. Powders were dry screened to -80 mesh and wet screened to -325 mesh. Powders were uniaxially pressed and sintered in N_2 at 1840 °C for 10 minutes. HIPing was performed at 1300-1700 °C and 100 or 200 MPa by stacking the parts on a graphite setter plate and heating for 1 hour while increasing the argon pressure and holding at that temperature for 15-30 minutes.

Sintering Additives
3.7 wt% TiH_2

Reinforcement Description
TiC powder

Reinforcement Supplier
Hermann C. Starck Inc., New York, NY

Supplier Designation
Grade CA

Matrix Description
Al_2O_3 powder

Matrix Supplier
ALCOA, Pittsburgh PA

Matrix Designation
A-16 Superground

Test Type
Vickers Indentation

Specimen Preparation
Specimens were ground to a 30 μm surface finish.

Specimen Dimensions
4 x 5 x 50 mm

Test Description
Indenter Load = 50-75 N

Test Temperature
RT

Density (g cm^{-3})	Hardness, Vickers (GPa)	Range (GPa)	HIPing Temperature (°C)	HIPing Pressure (°C)	Comments
4.22	17.3	±0.2	----	---	*as-sintered*
4.24	17.8	±0.4	1300	100	*sintered and HIPed*
4.33	19.4	±0.5	1600	100	
4.31	19.7	±0.2	1700	200	

Remarks/Comments
Vickers hardness was greatest for the composite that had the highest HIP temperature and pressure (1700 °C and 200 MPa). Hardness was considerably greater for the composites that were sintered and HIPed compared to the composite that was tested as-sintered.

$TiC(p)/Al_2O_3$

Reference: 58

Composition
30 wt% TiC

Processing Method
Cold Pressed/Sintered (some HIPed) or Hot-Pressed

Detailed Description
Powders were dispersed and milled in a 15 cm diameter stainless steel mill with hexane for 24 hours with a ball to charge ratio of 10:1. Larger batches were attrition milled in deionized water and a binder was added prior to spray drying. Powders were dry screened to -80 mesh and wet screened to -325 mesh. Powders were uniaxially pressed, sintered in Ar at 1875 °C for 10 minutes. HIPing was performed at 1600 °C and 100 MPa by stacking the parts on a graphite setter plate and heating for 1 hour while increasing the argon pressure and holding at that temperature for 15-30 minutes. Some specimens were only hot-pressed at 1650 °C and 35 MPa for 10 minutes.

Sintering Additives
3.7 wt% TiH_2

Reinforcement Description
TiC powder

Reinforcement Supplier
Hermann C. Starck Inc.,
New York, NY

Supplier Designation
Grade CA

Matrix Description
Al_2O_3 powder

Matrix Supplier
ALCOA, Pittsburgh PA

Matrix Designation
A-16 Superground

Test Type
Vickers Indentation

Specimen Preparation
Specimens were ground to a 30 μm surface finish.

Specimen Dimensions
4 x 5 x 50 mm

Test Description
Indenter Load = 50-75 N

Test Temperature
RT

Relative Density (percent)	Hardness, Vickers (GPa)	Range (GPa)	Comments
94.9	18.7	±0.6	*sintered*
99.5	22.0	±1.1	*sintered and HIPed*
98.6	23.1	±0.7	*hot-pressed*

Remarks/Comments
The hardness value was the highest at 23.1 GPa when the specimen was hot-pressed.

$TiB_2(p)/Al_2O_3$

Reference: 84

Composition
0 and 30 vol% TiB_2

Processing Method
Hot-Pressed

Detailed Description
Al_2O_3 powders were homogenized with 30 vol% TiB_2 in plastic jars with alumina balls and isobuthyl alcohol. The composites were hot-pressed in a vacuum at 1600-1650 °C and 30 MPa.

Reinforcement Description
TiB_2 particulates

Matrix Description
Al_2O_3 powder

Test Type
Vicker's Indentation Test

Specimen Preparation
The surface of the specimen was polished prior to testing.

Test Description
load = 0.5 kg

Test Temperature
RT

Reinforcement Content (volume percent)	Hardness, Vickers (GPa)	Relative Density (percent)	Mean Grain Size (microns)	Range (GPa)
0	18.7	100	-	±0.6
30	19.2	98.6	1.3	±0.9

Remarks/Comments
Hardness values were greater for the composites in comparison to the Al_2O_3 monolith, due to the second phase being harder than the matrix and alumina grain refinement.

TiN(p)/Al_2O_3

Reference: 84

Composition
0 and 30 vol% TiN

Processing Method
Hot-Pressed

Detailed Description
Al_2O_3 powders were homogenized with 30 vol% TiN in plastic jars with alumina balls and isobuthyl alcohol. The composites were hot-pressed in a vacuum at 1600-1650 °C and 30 MPa.

Reinforcement Description
TiN particulates

Matrix Description
Al_2O_3 powder

Test Type
Vickers Indentation

Specimen Preparation
The surface of the specimens were polished prior to testing.

Test Description
load = 0.5 kg

Test Temperature
RT

Reinforcement Content (volume percent)	Hardness, Vickers (GPa)	Relative Density (percent)	Mean Grain Size (microns)	Range (GPa)
0	18.7	100	-	±0.6
30	19.0	99.3	0.8	±0.6

Remarks/Comments
Hardness values were slightly greater for the composites in comparison to the monolithic Al_2O_3. This was likely due to alumina grain refinement.

TiC(p)/Al_2O_3

Reference: 84

Composition
0 and 30 vol% TiN

Processing Method
Hot-Pressed

Detailed Description
Al_2O_3 powders were homogenized with 30 vol% TiC in plastic jars with alumina balls and isobuthyl alcohol. The composites were hot-pressed in a vacuum at 1600-1650 °C and 30 MPa.

Reinforcement Description
TiC particulates

Matrix Description
Al_2O_3 powder

Test Type
Vicker's Indentation

Specimen Preparation
The surface of the specimens were polished prior to testing.

Test Description
load = 0.5 kg

Test Temperature
RT

Reinforcement Content (volume percent)	Hardness, Vickers (GPa)	Relative Density (percent)	Mean Grain Size (microns)	Range (GPa)
0	18.7	100	-	±0.6
30	20.7	99.4	0.9	±0.9

Remarks/Comments
Hardness values were greater for the composites in comparison to the monolithic Al_2O_3, due to the second phase being harder than the matrix and alumina grain refinement.

TiN(p)/Al_2O_3

Reference: 94

Composition
43 and 50 vol% TiN

Processing Method
CIPed/Reaction Sintered (some Hot-Pressed)

Detailed Description
When the two powders reacted, they produced a mixture of Al_2O_3 and TiN according to the reaction:
$xTiO_2 + 2AlN = Al_2O_3 + xTiN + (1\text{-}x/2)N_2 + (x\text{-}3/2)O_2$.
The powders were intimately mixed in a ball-mill with acetone and were reacted in a alumina crucible under an overpressure of nitrogen and 1600 °C. 50 x 20 x 20 mm fired billets were made by initial cold-isostatic-pressing, followed by firing at 1600 °C in N_2 under a cover of AlN and with an overpressure on N_2 (10 mm). Hot-pressing was done for some specimens on 70 mm discs at 1550 °C using a pressure of 30 MPa in a BN coated graphite die and punch and using an N_2 overpressure.

Reinforcement Description
TiO_2

Reinforcement Supplier
Glaxo Labs, India

Supplier Designation
Laboratory grade

Matrix Description
AlN

Matrix Supplier
C.S. Starck, Berlin, Gaslav, Germany

Matrix Designation
Grade A

Test Type
Vickers Indentation

Test Description
load = 0-20 kg

Test Temperature
RT

Test Equipment
Lietz miniload hardness tester (wezler, FRG) or Wolpert Dia-tester 2 RC (Wilhelm Herm. Holm Hamburg, FRG)

Reinforcement Content (volume percent)	Applied Load (kg)	Hardness, Vickers (GPa)	Comments
43	4.89	15.2	*pressureless sintered*
43	0.90	19.2	*sintered and hot-pressed*
43	4.77	18.5	
43	10.2	17.2	
43	20.2	15.3	
50	0.08	18.5	*pressureless sintered*
50	0.96	14.9	
50	4.78	14.0	
50	4.82	14.8	*sintered and hot-pressed*

Remarks/Comments
The composite with less TiN showed better sintered density (97%) than the other composite which contained a higher percentage of nonsinterable TiN (93%). The composite with the lower TiN content had higher hardness values than the composite with the greater TiN content. Hot-pressed composites had higher hardness values than pressureless sintered composites.

$TiN(p)/Al_2O_3$

Reference: 94

Composition
50 vol% TiN

Processing Method
CIPed/Reaction Sintered/Hot-Pressed

Detailed Description
When the two powders reacted, they produced a mixture of Al_2O_3 and TiN according to the reaction:
$xTiO_2 + 2AlN = Al_2O_3 + xTiN + (1-x/2)N_2 + (x-3/2)O_2$.
The powders were intimately mixed in a ball-mill with acetone and were reacted in a alumina crucible under an overpressure of nitrogen (10 mm) and 1600 °C. 50 x 20 x 20 mm fired billets were made by initial cold-isostatic-pressing, followed by firing at 1600 °C in N_2 under a cover of AlN and with an overpressure on N_2 (10 mm). Hot-pressing was done on 70 mm discs at 1550 °C using a pressure of 30 MPa in a BN coated graphite die and punch and using 10 mm N_2 overpressure.

Reinforcement Description
TiO_2

Reinforcement Supplier
Glaxo Labs, India

Supplier Designation
Laboratory grade

Matrix Description
AlN

Matrix Supplier
C.S. Starck, Berlin, Gaslav, Germany

Matrix Designation
Grade A

Test Type
Vickers Indentation

Test Description
load = 5 kg

Test Temperature
RT

Test Equipment
Lietz miniload hardness tester (wezler, FRG) or Wolpert Dia-tester 2 RC (Wilhelm Herm. Holm Hamburg, FRG)

Relative Density (percent)	Hardness, Vickers (GPa)
93.7	13.0
95.7	14.6
97.0	15.8

Remarks/Comments
The Vickers hardness increased almost linearly as relative density increased. Extrapolation of this data to full density yields a hardness of 18.32 GPa.

$TiB_2(p)/Al_2O_3$

Reference: 100

Composition
30 vol% TiB_2

Processing Method
Hot-Pressed

Detailed Description
Monolithic TiB_2 was hot-pressed at 1850 °C for 1 hour and at 30 MPa. TiB_2-Al_2O_3, 30 vol% TiB_2, was produced by hot-pressing at 1600 °C for 30 minutes and at 30 MPa.

Reinforcement Description
TiB_2 powder

Reinforcement Supplier
Hermann C. Starck Inc.

Supplier Designation
Grade F

Matrix Description
Al_2O_3 particulates

Matrix Supplier
ALCOA

Matrix Designation
A16 S.G.

Test Type
Vickers Indentation

Test Temperature
RT

Reinforcement Content (volume percent)	Hardness, Vickers (GPa)	Range (GPa)
Monolithic TiB_2	24.4	±1.0
30	19.2	±0.8

Remarks/Comments
Vickers hardness decreased with increasing reinforcement content.

Figure 110: Hardness vs. Relative Density of TiB_2(p)/Al_2O_3, TiN(p)/Al_2O_3, and TiC(p)/Al_2O_3

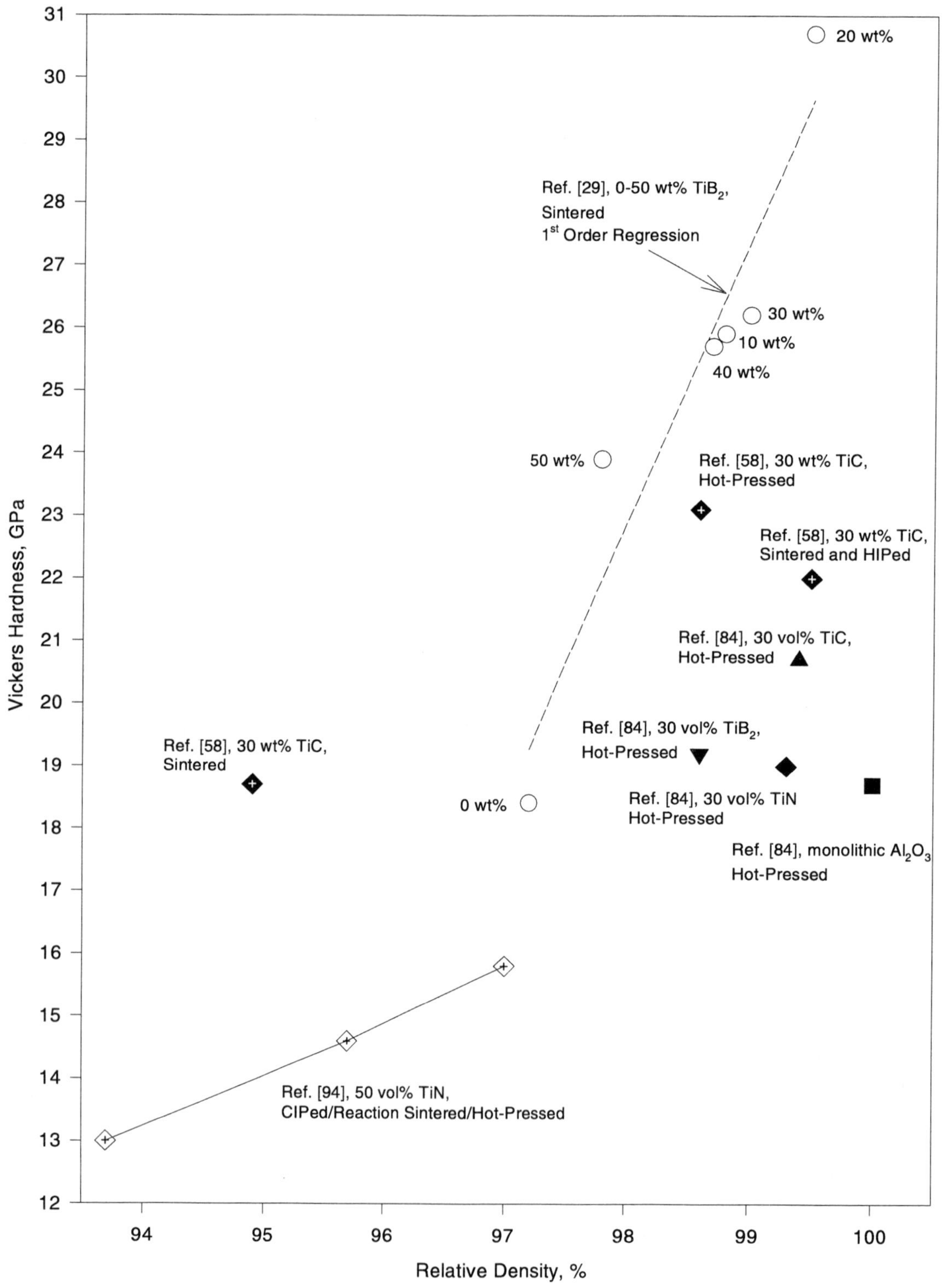

Discussion: The Vickers hardness of the composites was influenced by the material processing (sintering, HIPing, and hot-pressing), reinforcement type and content. Cutler *et al.* [58] found that high pressure processing improved the Vickers hardness of a TiC(p)/Al_2O_3 composite. Performing a post sintering HIPing step increased the hardness from 18.7 GPa to 22 GPa. This increase in hardness occurred primarily due to the removal of porosity; the relative density increased from 94.9 to 99.5 after the HIPing step. The hot-pressed composite was harder than the sinter-HIPed composite, despite having slightly lower relative density. The higher hardness of the hot-pressed composite was the result of a much finer grain size.

Bellosi and Babina *et al.* [84] tested hot-pressed TiN, TiC, or TiB_2 particulate reinforced alumina composites. The hardness of these composites was relatively low compared to the composites tested by the other references; the hardness was only around 19 GPa for the TiB_2 and TiN reinforced alumina, and around 21 GPa for the TiC(p)/Al_2O_3. The hardness of these composites was only slightly higher than the hardness of the monolithic alumina.

Kimura *et al.* [29] found that the hardness of TiB_2(p)/Al_2O_3 composites was highly dependent on relative density. Kimura *et al.* [29] also found that the role of TiB_2 in Al_2O_3 was to inhibit the rapid grain growth of Al_2O_3. Thus, TiB_2 acted as a sintering aid. The hardness was greater due to the smaller grain size of the alumina matrix, as well as due to the greater hardness of the TiB_2 reinforcement compared to the alumina matrix. The hardness of these composites was very high when the composite density was above 99%. At 20 wt% TiB_2, where the composite density reached 99.5 %, the hardness reached 30.7 GPa. This was much closer to the hardness of monolithic TiB_2 (33.7 GPa) than monolithic alumina (19.1 GPa). Further analysis of the Kimura *et al.* [29] composites is presented in figure 111.

The hardness of the reaction sintered TiN(p)/Al_2O_3 composite tested by Mukerji and Biswas [94] was considerably lower than the other composites shown in figure 110. This was partly due to low densification. The hardness increased monotonically with increasing relative density. If the hardness is extrapolated to full density, a hardness of 18.3 GPa would be reached. This is still relatively low for a 50 vol% TiN(p)/Al_2O_3 composite. This low hardness was also the result of the microstructure formed by the reaction sintering processing method.

Figure 111: Vickers Hardness of $TiC(p)/Al_2O_3$, $TiN(p)/Al_2O_3$, and $TiB_2(p)/Al_2O_3$

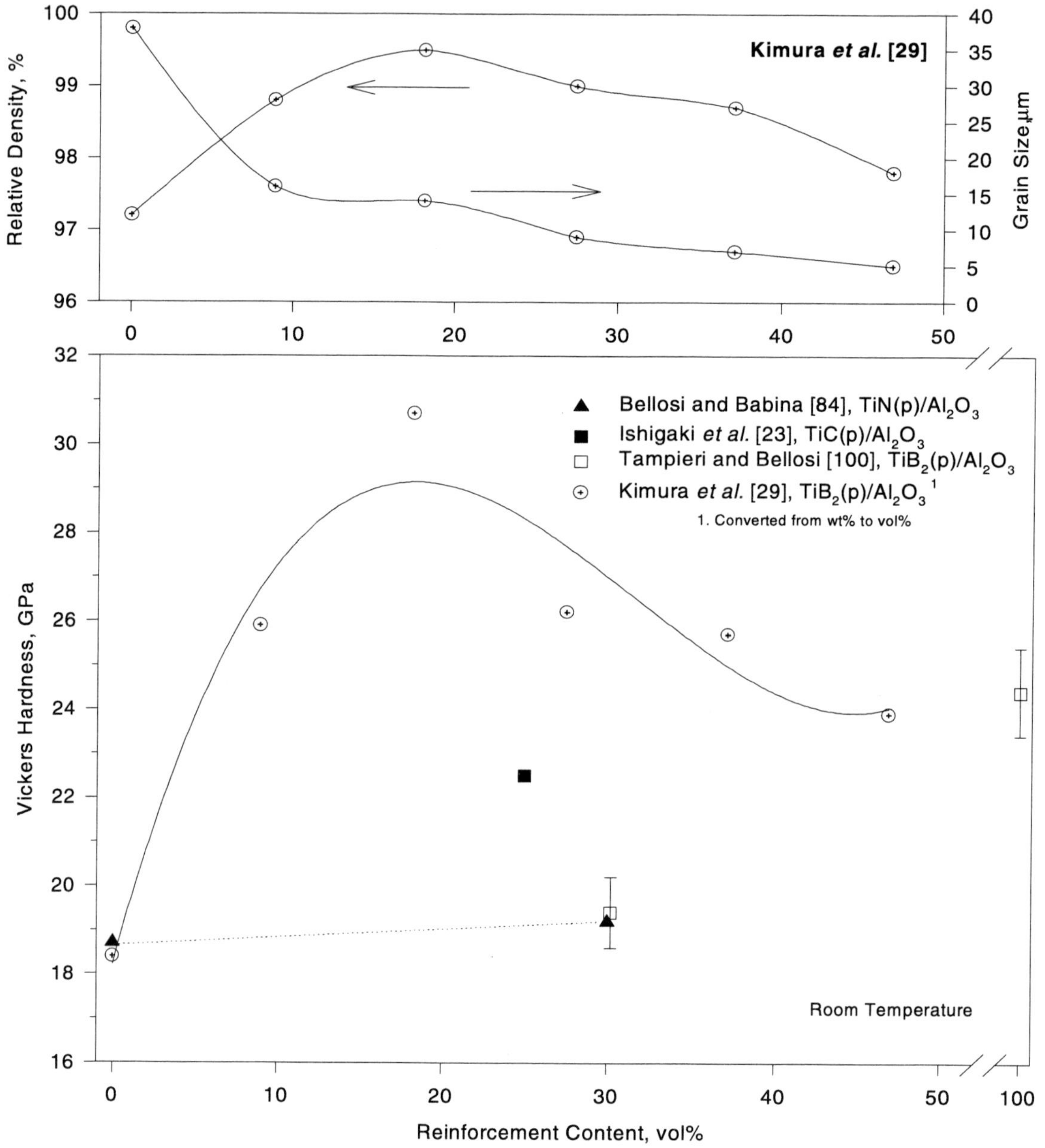

Discussion: The Vickers hardness for the $TiB_2(p)/Al_2O_3$ composites was substantially greater than the hardness for either the $TiC(p)/Al_2O_3$ or $TiN(p)/Al_2O_3$ composites. The addition of 30 vol% TiC to the Al_2O_3 did not increase nor decrease the hardness. However, the addition of 10 vol% TiB_2 particulates to the monolithic Al_2O_3 ceramic substantially improved the Vickers hardness. Hardness continued to increase up to 20 vol% TiB_2. However, subsequent additions of the TiB_2 particulates over 20 vol% resulted in a decrease in the Vickers hardness. The role of the TiB_2 particulates was to inhibit rapid grain growth of Al_2O_3. Thus, the TiB_2 behaved as a sintering aid. As the TiB_2 particulate content increased, the mean grain size of the Al_2O_3 in the composite decreased.

The relative density for the TiB_2/Al_2O_3 composite of ref. [29] increased to a maximum at 20 vol% reinforcement content and then decreased gradually with further increases of TiB_2 reinforcement. The grain size of the $TiB_2(p)/Al_2O_3$ composite was the greatest for the monolithic composition and decreased with each addition of TiB_2 reinforcement [29].

The hardness of the composite was primarily dependent on relative density. The maximum hardness of 30.7 GPa occurred at 20 vol% TiB_2 content. This corresponded to a maximum bending strength of 310 MPa.

Figure 112: Effect of Post Sinter HIPing on the Vickers Hardness of $TiC(p)/Al_2O_3$

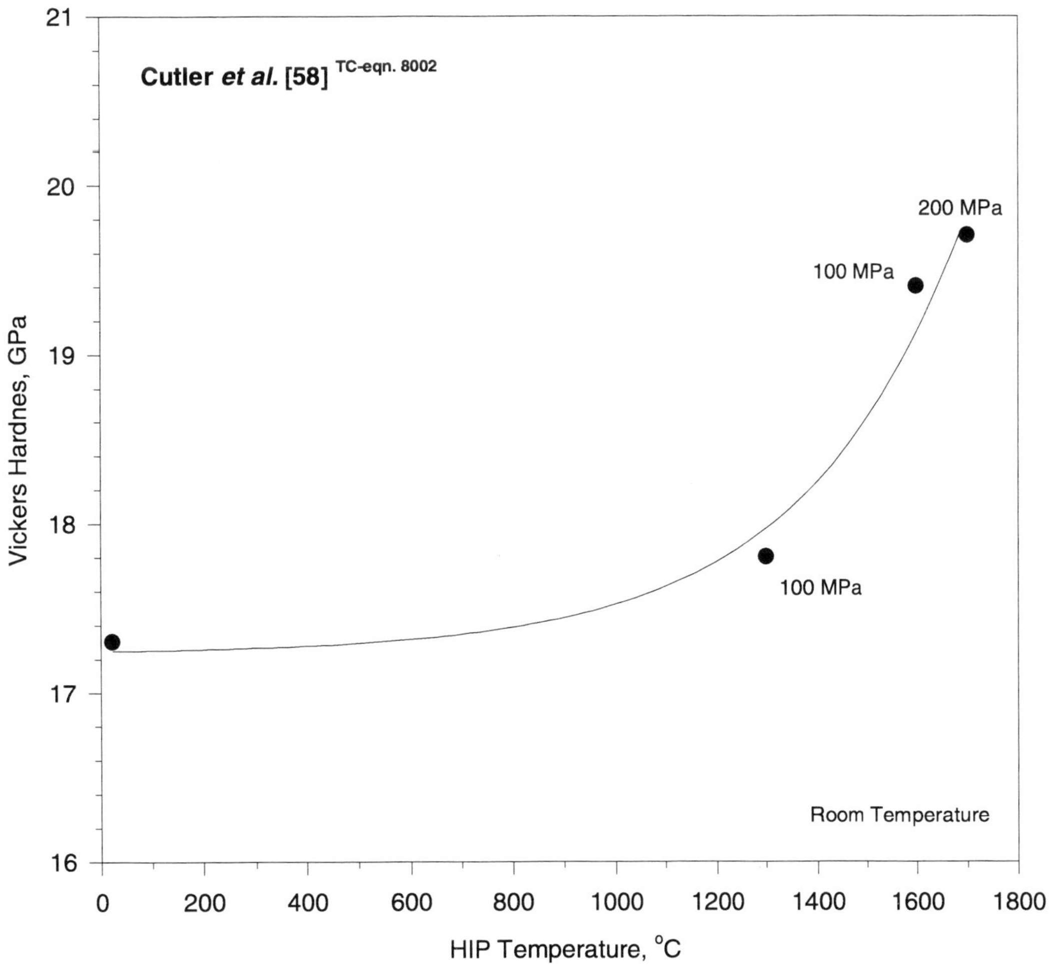

Discussion: Cutler *et al.* [58] examined the effect of a post-sinter HIPing step on a $TiC(p)/Al_2O_3$ composite. As the HIPing temperature of the $TiC(p)/Al_2O_3$ composite increased, the Vickers hardness also increased. The as-sintered composite contained some open porosity. This porosity was only partially removed by sintering at 1300 °C; however, HIPing at 1600 °C or above completely removed all open porosity and significantly improved composite hardness.

Figure 113: Vickers Hardness of TiN(p)/Al_2O_3

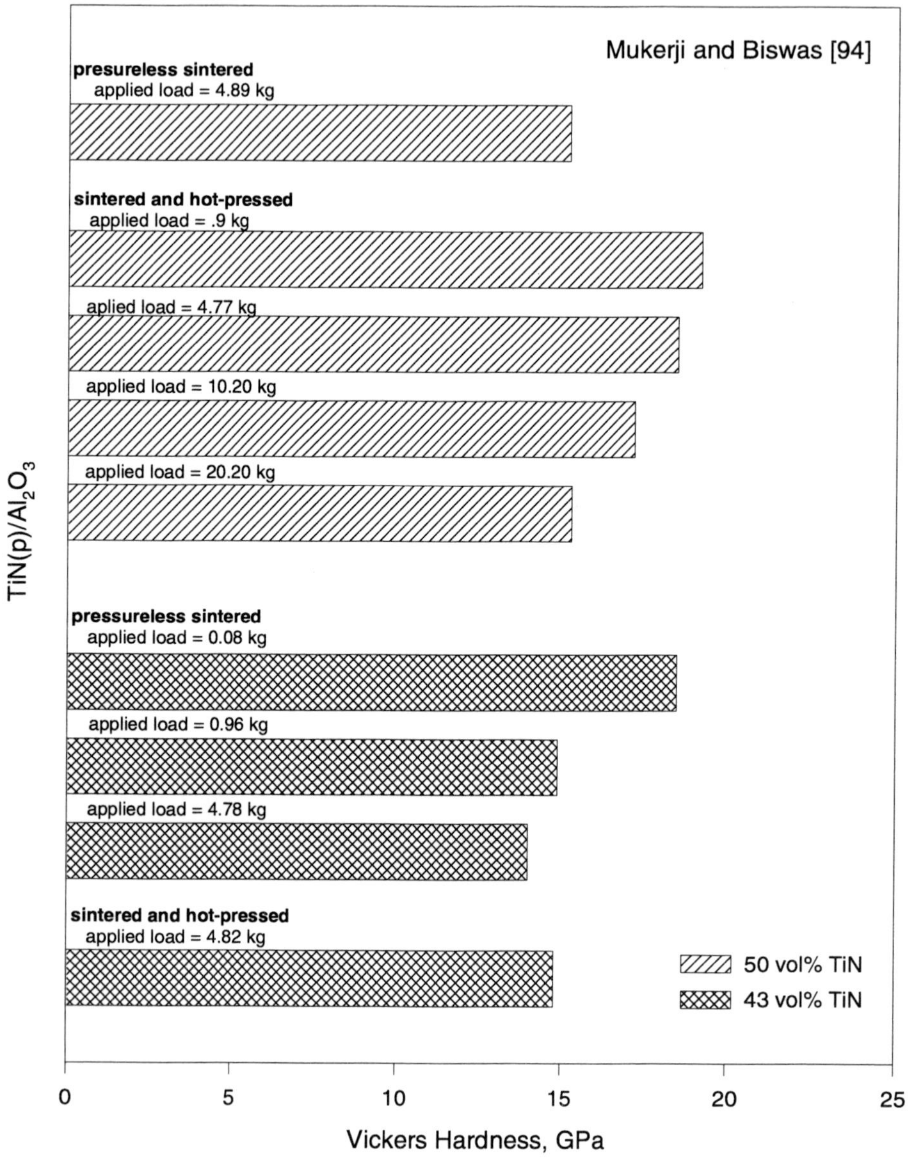

Discussion: As the applied load increased for the hot-pressed and pressureless sintered TiN(p)/Al_2O_3 composites, the Vickers hardness decreased. Overall, the hot-pressed composites were harder than the pressureless sintered composites. The composites with a reinforcement content of 43 vol% were slightly harder than those with a reinforcement content of 50 vol%.

$ZrO_2(p)/Al_2O_3$

Reference: 7

Composition
0 - 100 vol% ZrO_2

Processing Method
Hot-Pressed

Detailed Description
Sub-micron composite powders, including yttrium nitrate (a Y_2O_3 precursor), were mixed by ball-milling with methanol and Al_2O_3 balls in a plastic container. The powders were dried, and those with yttrium nitrate were calcined at 400 °C for 4 hours. Most Y_2O_3 containing composites were hot-pressed at 1600 °C for 2 hours. Composites containing 80 and 100 vol% ZrO_2 were hot-pressed at 1400 °C in order to achieve a smaller grain size which allowed the retention of tetragonal ZrO_2. Non-yttria composites were hot-pressed at 1500 °C, again to achieve smaller grain size. The pure Al_2O_3 end-member was hot-pressed at 1400 °C to achieve a grain size comparable to the two-phase materials.

Reinforcement Description
0, 2, 7.5 Y-ZrO_2 particulates

Reinforcement Supplier
Zircas Corp.

Matrix Description
Al_2O_3 powder

Matrix Supplier
Lindy B, Union Carbide Corp.

Test Type
Vickers Indentation

Test Description
Load = 20 kg

Reinforcement Content (volume percent)	Hardness, Vickers (GPa)	ZrO_2 Phase
0	17.6	-
0 mol% Y_2O_3		% Tetragonal
7.5	17.2	~90
10	15.8	~80
13	16.9	~70
15	17.3	~50
20	10.1	<20
2 mol% Y_2O_3		% Tetragonal
6	16.8	100
12	15.9	100
18	16.1	100
24	16.4	100
30	15.7	100
45	15.1	100
60	13.7	~95
80	12.6	~85
100	11.6	~80
7.5 mol% Y_2O_3		
20	15.8	100 % Cubic
40	15.9	
60	15.0	
80	14.3	
100	11.4	

Remarks/Comments
A linear rule of mixtures exist for the 2 mol% composite. The 20 vol% ZrO_2/Al_2O_3 composite was shown to be highly micro-cracked, which gave a much smaller hardness value. This small value was due to the indenter pushing the microcracked material aside as it extended into the surface.

ZrO_2(p)/Al_2O_3

Reference: 15

Composition
0 - 100 vol% ZrO_2

Processing Method
Sintered

Detailed Description
The starting powders were mixed by ball milling using plastic containers with acetone. After drying, the composite powders were uniaxially pressed at 300 MPa to form pellets 5 mm in diameter and 5 mm thick. These pellets were sintered at 1600 °C for 10 hours in an argon atmosphere.

Reinforcement Description
0, 3, 6Y-ZrO_2 particulates

Matrix Description
Al_2O_3 powder

Matrix Designation
AKP-20

Test Type
Vickers Indentation

Test Description
Indenter load = 10 and 20 N

Reinforcement Content (volume percent)	Hardness, Vickers (GPa)	ZrO_2 Phase
0 mol% Y_2O_3		
0	18.2	48.5 % t, 51.5 % m
5	17.9	37.7 % t, 65.3 % m
10	17.3	21.2 % t, 78.8 % m
15	16.2	8.7 % t, 91.3 % m
20	14.2	3.9 % t, 96.1 % m
30	12.1	fully monoclinic
40	9.2	fully monoclinic
60	8.0	fully monoclinic
80	6.4	fully monoclinic
3 mol% Y_2O_3		
20	18.0	fully tetragonal
40	15.7	
60	14.3	
80	12.9	
100	11.9	
6 mol% Y_2O_3		
5	17.9	fully cubic
20	17.3	
40	16.1	
60	14.9	
80	13.9	

Remarks/Comments
All samples were approximately 99 % dense. The addition of ZrO_2 to Al_2O_3 reduced the hardness of the material. As the 0Y-ZrO_2 became monoclinic, microcracks formed in the matrix, significantly reducing the hardness of the composite. The hardness decreased with the addition of fully tetragonal or fully cubic ZrO_2 according to the rule of mixtures.

$ZrO_2(p)/Al_2O_3$

Reference: 64

Composition
0 - 25 vol% ZrO_2

Processing Method
Hot-Pressed

Detailed Description
Al_2O_3 powder mixed with 10, 15, 20, and 25 vol% ZrO_2 by ball milling for 24 hours in anhydrous alcohol. Mixtures were green compacted at 300 MPa, then hot-pressed at 1650 °C and 25 MPa.

Reinforcement Description
2Y-ZrO_2 particulates

Reinforcement Supplier
Toso Co. Ltd.

Reinforcement Typical Properties
54 % tetragonal ZrO_2 & 46 % monoclinic ZrO_2

Reinforcement Dimensions
particle size = 0.06 μm

Matrix Description
alumina matrix

Matrix Properties
0.1 μm particle size

Test Type
Vickers Indentation

Specimen Dimensions
thickness = 5 mm
dia. = 20 mm

Test Description
load = 500 N (HV50)

Test Temperature
RT

Reinforcement Content (volume percent)	Hardness, Vickers (HV (GPa))	Alumina Grain Size (μm)	Zirconia Grain Size (μm)	t-ZrO_2 Content (percent)	Relative Density (percent)	Comments
0	1,640 [16.1]	35.9	-	-	95.3	+12/-11
10	1,750 [17.2]	31.5	3.09	79.9	96.1	+16/-17
15	1,630 [16.0]	29.8	2.97	86.2	99.6	+11/-10
20	1,600 [15.7]	27.8	3.74	92.1	99.8	±10
25	1,610 [15.8]	24.8	4.78	88.7	99.9	±11

Remarks/Comments
Highest hardness was observed at 10 vol% ZrO_2.

Figure 114a: Vickers Hardness of $ZrO_2(p)/Al_2O_3$

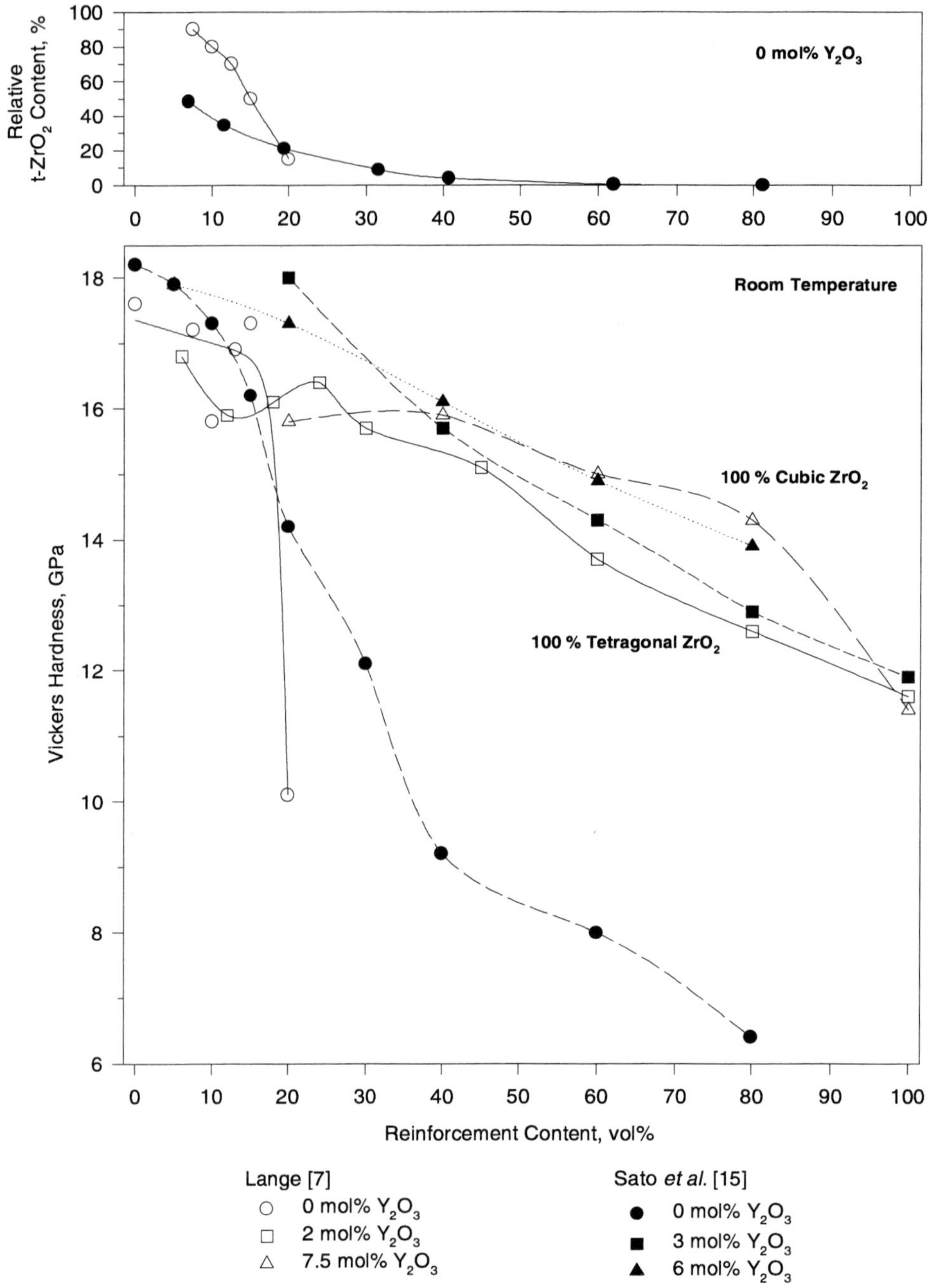

Discussion: The mechanical properties of zirconia reinforced ceramic composites are highly dependent on the phase of the zirconia reinforcement. It is typically desirable to stabilize the zirconia in the tetragonal phase to increase strength and toughness. A high modulus matrix, such as alumina, constrains the zirconia reinforcement in the tetragonal phase. Phase analysis of the pure zirconia reinforced alumina showed that the zirconia was partially stabilized at low reinforcement contents. However, the t-ZrO_2 content decreased rapidly with increasing reinforcement content to nearly 0 % at 40 vol% zirconia. This decrease occurred because there was less and less matrix constraining the reinforcement at higher reinforcement contents. The addition of 2-3 mol% yttria stabilized the zirconia reinforcement to 100 % tetragonal at all reinforcement contents. The addition of greater than 6 mol% Y_2O_3 stabilized the ZrO_2 to 100 % cubic phase [7, 15].

The hardness of 0Y-ZrO_2(p)/Al_2O_3 dropped rapidly as the zirconia content was increased above 15 vol% of the composite and as the relative t-ZrO_2 content dropped below 50 %. The presence of monoclinic zirconia suggests a t-m phase transformation occurred during processing. This phase transformation and corresponding volume expansion during processing caused a great deal of matrix microcracking. This resulted in the composites with more than 15 vol% zirconia content having significantly lower hardness [7].

Due to the lower hardness of zirconia compared to alumina, the addition of fully stabilized t-ZrO_2 and cubic zirconia both showed a decrease in hardness with increasing reinforcement content according to the rule of mixtures. The hardness of cubic zirconia composites was slightly higher than the hardness of t-ZrO_2 composites.

Figure 114b: Vickers Hardness of $ZrO_2(p)/Al_2O_3$

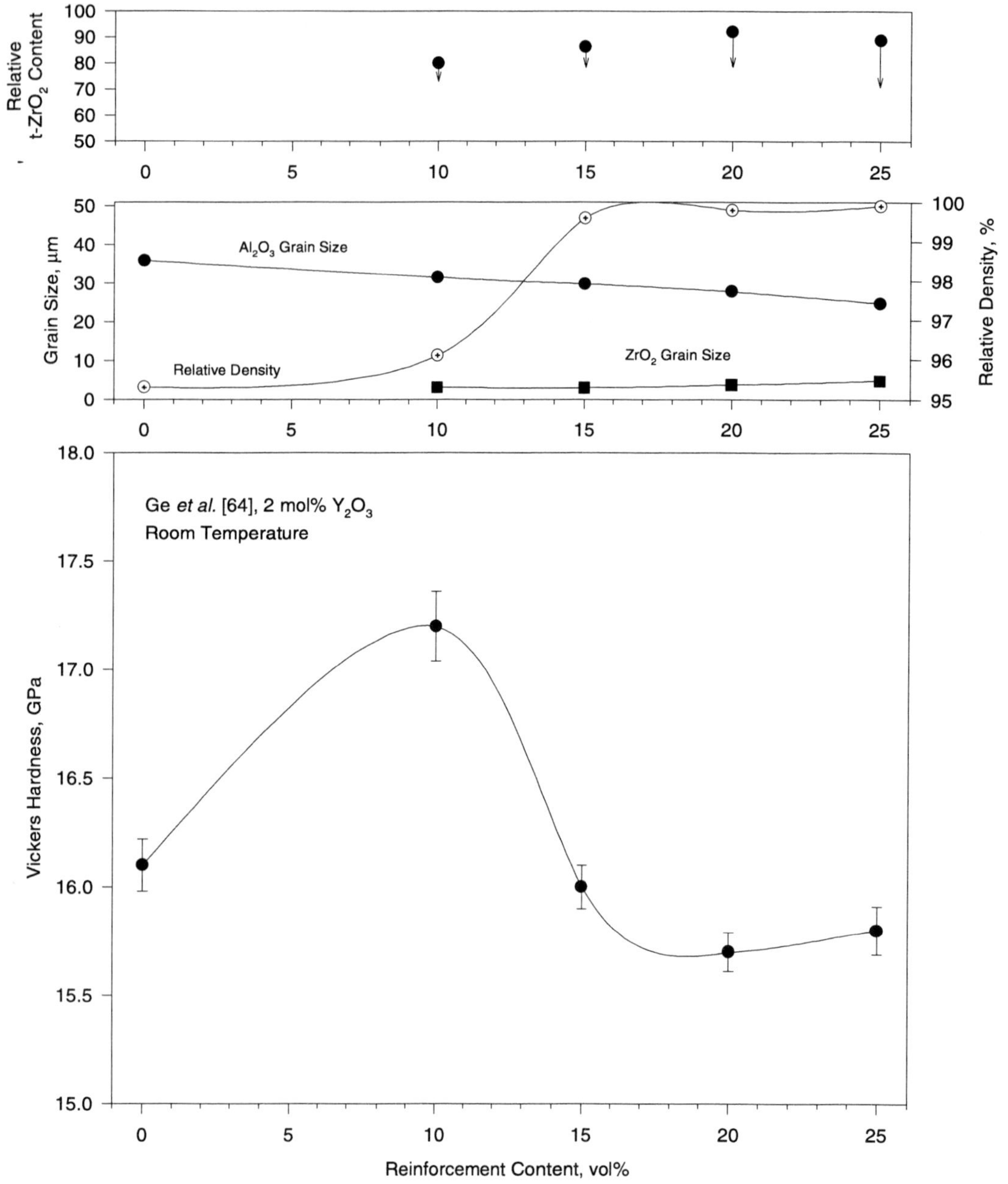

Discussion: Figure 114b shows that the hardness of the specimens tested by Ge *et al.* [64] peaked at 10 vol% reinforcement content. Due to the lower hardness of zirconia, it was expected that the hardness of the composite would decrease with increasing ZrO_2 content according to the rule of mixtures; however, the hardness at 10 vol% ZrO_2 composite benefited from the alumina grain refinement and the presence of fine grain zirconia particulates. Above 10 vol% ZrO_2, microcracks resulting from the t→m transformation during processing decreased the hardness of the composite [64]. The arrows represent the decrease in tetragonal ZrO_2 content during fracture.

$SiC(p)/ZrO_2 \cdot Al_2O_3$

Reference: 15

Composition
0 - 10 vol% SiC

Processing Method
Sintered (some HIPed)

Detailed Description
The starting powders were mixed by ball milling using plastic containers and balls with acetone. After drying, the composite powders were uniaxially pressed at 300 MPa to form pellets 5 mm in diameter and 5 mm thick. These pellets were sintered at 1500 °C for 10 hours in an argon atmosphere. Some samples were also HIPed at 1500 °C and 200 MPa for 30 minutes.

Reinforcement Description
SiC particulates

Matrix Description
Al_2O_3 and 10 vol% 0Y- or 3Y-ZrO_2 powder

Matrix Designation
AKP-20

Test Type
Vickers Indentation

Test Description
Indenter load = 10 and 20 N

Test Temperature
RT

Reinforcement Content (volume percent)	Hardness, Vickers (GPa)	Relative Density (percent)	Y_2O_3 Content (atomic percent)	Comments
0	17.2	98.8	0	*as-sintered*
1	18.7	98.7	0	
3	18.0	96.0	0	
5	16.5	95.4	0	
10	15.0	93.4	0	
0	15.9	98.3	3	*as-sintered*
1	15.8	96.8	3	
3	16.5	96.0	3	
5	16.3	96.6	3	
10	16.6	95.6	3	
1	17.3	95.8	3	*sintered and HIPed*
5	17.4	96.7	3	
10	17.7	96.7	3	

Remarks/Comments
The hardness increased with the addition of 1 vol% nonoxide then slightly decreased with increasing SiC content. This was due to the decrease in density with increasing nonoxide content.

TiC(p)/$ZrO_2 \cdot Al_2O_3$

Reference: 15

Composition
1 - 10 vol% TiC

Processing Method
Sintered

Detailed Description
The starting powders were mixed by ball milling using plastic containers with acetone. After drying, the composite powders were uniaxially pressed at 300 MPa to form pellets 5 mm in diameter and 5 mm thick. These pellets were sintered at 1500 °C for 10 hours in an argon atmosphere.

Reinforcement Description
TiC particulates

Matrix Description
Al_2O_3 and 10 vol% 0Y- or 3Y-ZrO_2 powder

Matrix Designation
AKP-20

Test Type
Vickers Indentation

Test Description
Indenter load = 10 and 20 N

Test Temperature
RT

Reinforcement Content (volume percent)	Hardness, Vickers (GPa)	Relative Density (percent)	Y_2O_3 Content (mole percent)
0	17.2	98.8	0
1	17.7	96.8	0
3	16.8	95.6	0
5	17.2	94.7	0
10	14.4	92.5	0
0	15.9	98.3	3
1	16.6	97.2	3
3	15.9	96.3	3
5	15.7	96.9	3
10	16.2	95.9	3

Remarks/Comments
The hardness increased with the addition of 1 vol% nonoxide then slightly decreased with increasing TiC content. This was due to the decrease in density with increasing nonoxide content.

TiC(p)/$ZrO_2 \cdot Al_2O_3$

Reference: 91

Composition
20 vol% TiC

Processing Method
Hot-Pressed

Reinforcement Description
TiC particulates

Matrix Description
Al_2O_3 and 4Y-ZrO_2 powder

Test Type
Vickers Indentation

Test Temperature
RT

Reinforcement Content (volume percent)	Hardness, Vickers (GPa)	Density ($g\ cm^{-3}$)
20	15.7	4.32

Figure 115: Vickers Hardness of $SiC(p)/ZrO_2 \cdot Al_2O_3$ and $TiC(p)/ZrO_2 \cdot Al_2O_3$

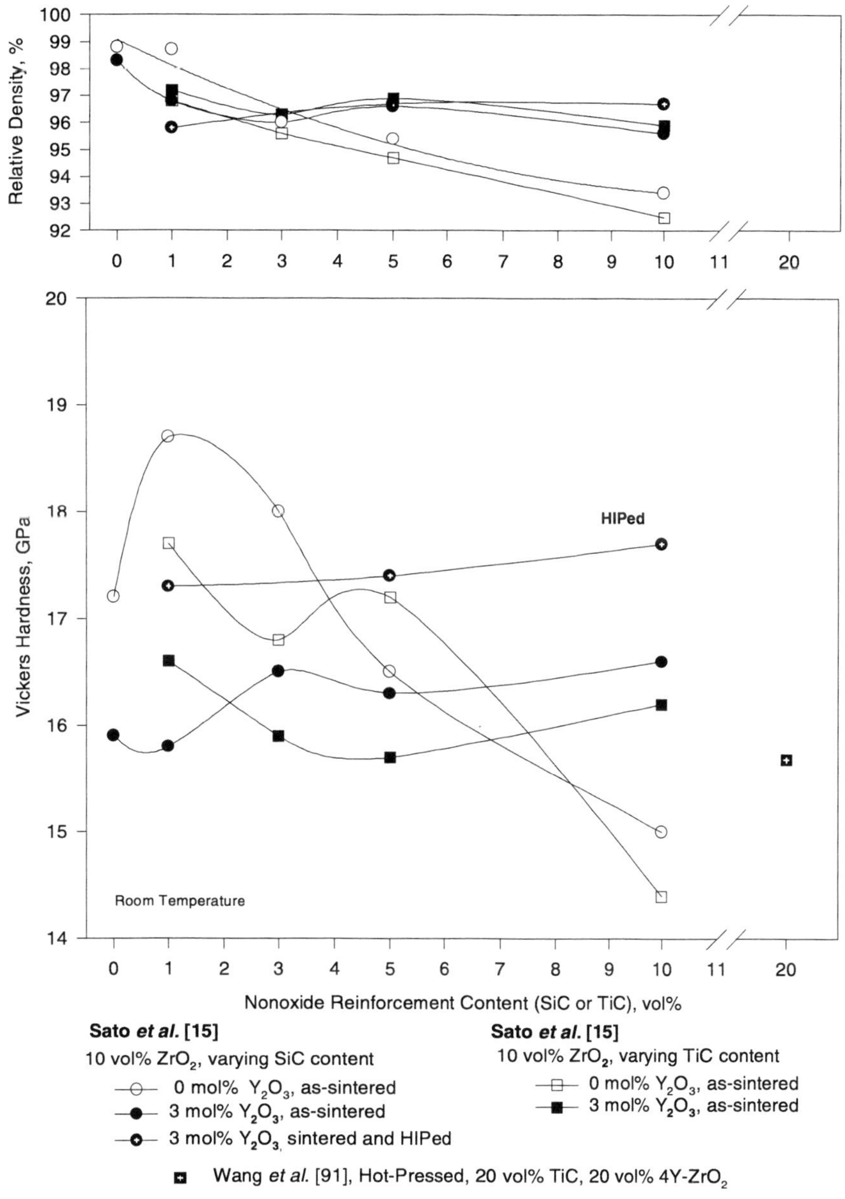

Discussion: Sato *et al.* [15] tested zirconia-alumina composites with either TiC or SiC reinforcement and with or without 3 mol% yttria stabilizer in the zirconia. Both the $SiC(p)/ZrO_2 \cdot Al_2O_3$ and $TiC(p)/ZrO_2 \cdot Al_2O_3$ composites without yttria stabilizer showed a higher hardness than the yttria containing composites in the range of 0 to 2 vol% SiC or TiC. On the other hand, these composites also showed a substantial decrease in hardness with carbide additions greater than 2 vol% due to inhibited densification. The hardness of the composites that contained the yttria stabilizer remained approximately constant with the addition of carbide reinforcement. The HIPed $SiC(p)/ZrO_2 \cdot Al_2O_3$ composite had greater hardness than the as-sintered composite with 3 mol% Y_2O_3 at all reinforcement contents. The HIPing process helped to remove internal porosity which resulted in a greater densification and consequently a higher hardness. Overall, the $SiC(p)/ZrO_2 \cdot Al_2O_3$ and $TiC(p)/ZrO_2 \cdot Al_2O_3$ composites showed similar behavior.

Figure 116: The Effects of Processing and Sintering Aids on $TiC(p)/ZrO_2{\cdot}Al_2O_3$ and $SiC(p)/ZrO_2{\cdot}Al_2O_3$

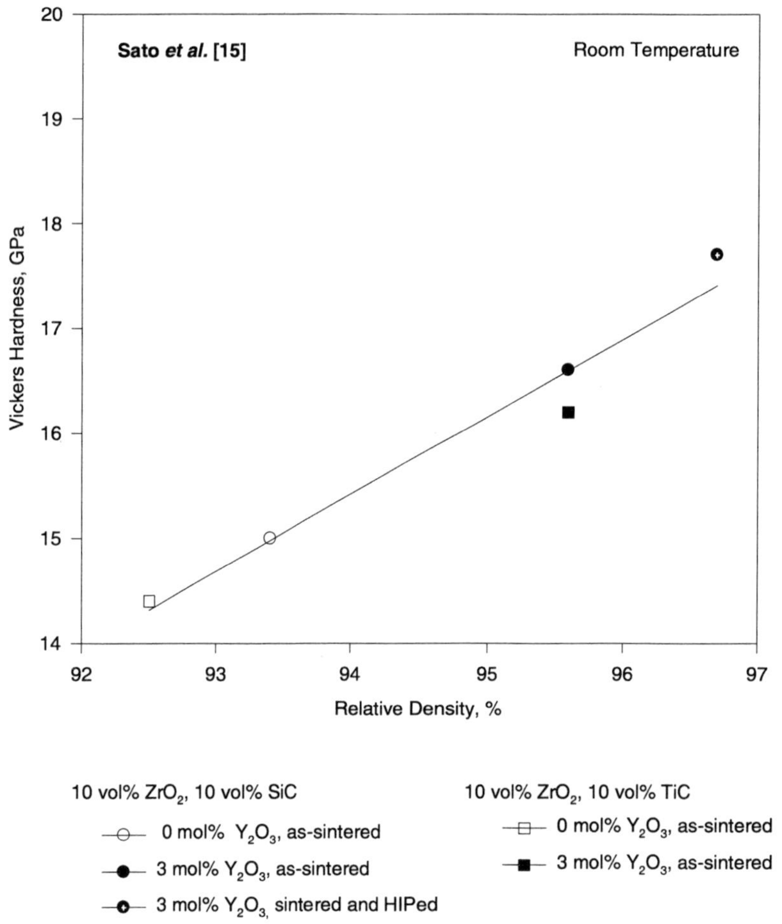

Discussion: The Vickers hardness was greatest for the composites that contained the Y_2O_3 stabilizer. As the relative density increased for the composites, the hardness also increased. The relative density was a consequence of the material processing and composition. The HIPed $SiC(p)/ZrO_2{\cdot}Al_2O_3$ composite resulted in the greatest densification and thus the highest hardness. Without yttria content, the SiC particulates produced a greater densification and higher hardness than the TiC particulate reinforced composites. Density of the two composites containing yttria was identical, but the hardness of the $SiC(p)/ZrO_2{\cdot}Al_2O_3$ composite was still slightly higher.

SiC(p)/Si_3N_4

Reference: 35

Composition
12 wt% Si-rich SiC
10 and 20 wt% C-rich SiC

Processing Method
Colloidal Pressed/Reaction Bonded

Detailed Description
Four sample groups were fabricated. Group one, monolithic RBSN, was fabricated either in methanol with CIP or in octanol without CIP. The remaining RBSN samples were fabricated in octanol without CIP. A high-intensity ultrasonic probe was used to break up agglomerates. Si and SiC powders were ultrasonically dispersed separately, mixed together by hand, and again sonicated. Total sonication time was 5-8 minutes at up to 40 W. Samples were colloidally pressed at 17 MPa to yield samples. The samples were then dried and nitrided. The firing schedule for the samples are:
(**A**) RBSN (methanol): 1 °C/min. to 1200 °C; 0.5 °C/min. to 1400 °C; 1 hour hold
(**B**) RBSN (octanol): 1 °C/min. to 1250 °C; 4 hour hold; 0.5 °C/min. to 1400 °C; 7 hour hold
(**C**) RBSN + 12 % Si-rich SiC: 2 hours at 1250-1400 °C; 4 hour hold; 10 hours at > 1400 °C
(**D**) RBSN + 10 % C-rich SiC: 3 °C/min. to 1250 °C; 4 hour hold; 0.5 °C/min. to 1400 °C; 10 hour hold
(**E**) RBSN + 20 % C-rich SiC: 3 °C/min. to 1250 °C; 4 hour hold; 0.5 °C/min. to 1400 °C; 10 hour hold

Reinforcement Description
SiC particulates

Reinforcement Dimensions
ave. C-rich SiC dia. = 0.04 μm
ave. Si-rich SiC dia. = 0.05 μm

Reinforcement Designation
C-rich SiC: L61SC
Si-rich SiC: L62SC

Matrix Description
Si powder

Matrix Properties
ave. dia. = 0.23 μm

Matrix Designation
F93 S

Test Type
Vickers Indentation

Specimen Preparation
Samples were ground and polished to 1 μm on the tensile surface.

Specimen Dimensions
dia. = 12.7 mm
thickness = 1 mm

Relative Density (percent)	Hardness, Vickers (GPa)	Range (GPa)	Comments
77	8.9	±0.5	*A*
77	9.1	±0.4	*A*
76	10.2	±0.4	*B*
73	7.8	±0.5	*C*
76	9.7	±0.7	*D*
73	7.4	±0.4	*E*

Remarks/Comments
With the same density, the octanol-processed RBSN is significantly harder than the methanol-processed RBSN.

SiC(p)/Si_3N_4

Reference: 37

Composition
0 and 10 vol% SiC

Processing Method
CIPed/Sintered (Some HIPed)

Detailed Description
The starting powders were ball-milled, dried and sieved. This mixture was cold isostatically pressed at 400 MPa and pressureless sintered at 1750 °C under 0.1 MPa of N_2 for 3 hours. Half of the samples were HIPed for 1 hour at 1850 °C under 100 MPa of N_2.

Sintering Additives
10 wt% Y_2O_3
5 wt% Al_2O_3

Reinforcement Description
SiC particulates

Reinforcement Supplier
Ibiden Co., Ohgaki, Japan

Reinforcement Typical Properties
specific density = 3.21 g cm^{-3}

Reinforcement Dimensions
ave. particle dia. = 0.26 μm

Matrix Description
Si_3N_4

Matrix Supplier
Ube Kosan Co., Japan

Matrix Designation
E-10

Matrix Properties
specific density = 3.18 g cm^{-3}
ave. particle dia. = 0.5 μm

Test Type
Vickers Indentation

Specimen Preparation
Samples were polished using 3 and 6 μm paste.

Specimen Dimensions
4 x 3 x 40 mm

Test Description
Indenter load = 300 g

Reinforcement Content (volume percent)	Hardness, Vickers (GPa)	Relative Density (percent)	Standard Deviation (GPa)	Comments
0	16.1	98	±0.9	*Pressureless sintered*
10	17.3	97	±0.9	
0	16.4	99	±0.8	*HIPed*
10	14.5	99	±0.4	

Remarks/Comments
Fifteen measurements were made for each composition.

$SiC(p)/Si_3N_4$

Reference: 39

Composition
0 - 50 vol% SiC

Processing Method
Hot-Pressed

Detailed Description
To make powders from SHS reactants were mixed in isopropyl alcohol for 3 hours, dried and put into graphite containers for combustion experiments performed in high pressure stainless steel vessel under a nitrogen pressure of 30 atm. Products from the SHS were attrition milled in isopropyl alcohol for 24 hours using Si_3N_4 grinding media. After adding 4 wt% of Y_2O_3 and drying, the samples were hot-pressed at 1700 °C and 3000 psi (about 21 MPa) under a 30 atm nitrogen atmosphere.

Sintering Additives
4 wt% Y_2O_3

Reinforcement Description
Si powder
carbon black

Reinforcement Supplier
Si: Elkem Metals Co., Niagara Falls, NY; C: Thermax, R. T. Vanderbilt Co., Norwalk, CT

Reinforcement Typical Properties
ave. size = 3.9 μm for Si powder
ave. size = 0.25 μm for C powder

Matrix Description
β-Si_3N_4 powder

Matrix Designation
SHS

Matrix Properties
ave. size = 4 μm

Test Type
Vickers Indentation

Test Description
Load = 100 N for 60 seconds

Test Temperature
RT

Reinforcement Content (volume percent)	Hardness, Vickers (GPa)	Relative Density (percent)
0	16.5	99.8
5	16.8	99.7
10	17.0	99.8
17	17.5	99.8
34	16.5	92.6
50	9.0	84.8

Remarks/Comments
Powders derived from self-propagating high-temperature synthesis (SHS) tended to be mixed better then conventional mixing.

SiC(p)/Si_3N_4

Reference: 39

Composition
16 vol% SiC

Processing Method
Hot-Pressed

Detailed Description
To make powders from SHS reactants were mixed in isopropyl alcohol for 3 hours, dried and put into graphite containers for combustion experiments performed in high pressure stainless steel vessel under a nitrogen pressure of 30 atm. Products from the SHS were attrition milled in isopropyl alcohol for 24 hours using Si_3N_4 grinding media. After adding 4 wt% of Y_2O_3 and drying, the samples were hot-pressed at 1700 °C and 3000 psi (about 21 MPa) under a 30 atm nitrogen atmosphere.

Sintering Additives
4 wt% Y_2O_3

Reinforcement Description
β-SiC powder

Reinforcement Supplier
Lonza Ltd., Basel, Switzerland

Supplier Designation
Lonza

Reinforcement Typical Properties
ave. size = 1.00 μm

Matrix Description
Si_3N_4 powder (α or β)

Matrix Designation
β-Si_3N_4 (SHS)
α-Si_3N_4 (Cookson)

Matrix Properties
ave. size = 4 μm (β-SiC)
ave. size = 1.5 μm (α-SiC)

Test Type
Vickers Indentation

Test Description
Load = 100 N for 60 seconds

Test Temperature
RT

Relative Density (percent)	Hardness, Vickers (GPa)	Hot-Press Time (hours)	Comments
99.8	18.1	1	*in situ SHS β-Si_3N_4-SiC*
78.8	6.0	1	*Mechanically mixed α-Si_3N_4-SiC*
91.9	9.2	1	*Mechanically mixed β-Si_3N_4-SiC*
91.6	9.4	2	*Mechanically mixed α-Si_3N_4-SiC*
100	15.4	2	*Mechanically mixed β-Si_3N_4-SiC*

Remarks/Comments
Commercial synthesized composites exhibited lower hardness values than the *in situ* composite as a result of high residual porosity after hot-pressing.

SiC(p)/Si_3N_4

Reference: 40

Composition
0, 10 vol% SiC

Processing Method
Hot-Pressed

Detailed Description
The starting powders were mixed, dried and sieved. The powders were hot-pressed at 30 MPa under 0.1 MPa nitrogen at 1700 °C and 1780 °C.

Sintering Additives
5 wt% Y_2O_3 (in addition to the 5 wt% Y_2O_3 and 5 wt% Al_2O_3 already in the Si_3N_4 powder)

Reinforcement Description
β-SiC particulates

Reinforcement Supplier
Ibiden Co. Ltd., Oogaki-city, Gifu-Prefecture, Japan

Reinforcement Typical Properties
specific density = 3.21 g cm^{-3}

Reinforcement Dimensions
ave. particle dia. = 0.26 µm
surface area = 15.1 m^2g^{-1}

Matrix Description
Si_3N_4

Matrix Supplier
Ube Industries., Ltd., Japan

Matrix Designation
COA

Matrix Properties
specific density = 3.18 g cm^{-3}
ave. particle dia. = 0.5 µm
surface area = 10.3 m^2g^{-1}
contains 5 wt% Y_2O_3 and 5 wt% Al_2O_3
composition (wt%): N > 38, O = 1.27, C < 0.3, Al < 0.05, Ca < 0.05, Fe < 0.01

Test Type
Vickers Indentation

Specimen Preparation
Specimens were polished with 6 and 3 µm diamond paste.

Specimen Dimensions
4 x 3 x 40 mm

Test Description
Indenter load = 300 g

Test Temperature
RT

Reinforcement Content (volume percent)	Hardness, Vickers (GPa)
0	18.9
10	20.2

Remarks/Comments
The hardness increased with the addition of particulates.

SiC(p)/Si_3N_4

Reference: 67

Composition
24 wt% SiC

Processing Method
CIPed/Pressureless Sintered (some Polymer Pyrolyzed)

Detailed Description
Polymethylsilazane was dissolved in tetrahydrofuran (THF) in an argon atmosphere. 10.5 wt% Al_2O_3 and 4.5 wt% Y_2O_3 were added, and the THF was evaporated in a vacuum at 0.1 Pa and room temperature. The dried THF residue was heated at 550 °C for 1 hour then heated at 1000 °C for 1 hour. The mixture was then attrition milled for 3 hours at 1000 rpm, sieved, and finally cold isostatically pressed at 630 MPa into green compacts of 35x17x15 mm. Reference samples made from α–Si_3N_4, β–SiC, 10.5 wt% Al_2O_3 and 4.5 wt% Y_2O_3. They were attrition milled, dried, sintered and hot pressed in same manner as pyrolyzed sample. Both the as-pyrolyzed sample and the reference sample were sintered in a graphite resistivity heated furnace at 1750 °C and 1850 °C under 0.1 MPa N_2.

Sintering Additives
10.5 wt% Al_2O_3 (A16, Alcoa, Pittsburg, PA)
4.5 wt% Y_2O_3 (Ventron Alta Produckte, Karlstuhe, Germany)

Reinforcement Description
β-SiC particulates

Reinforcement Supplier
H.C. Stark, Berlin, Germany

Supplier Designation
B10

Matrix Description
β-Si_3N_4 powder

Matrix Supplier
H.C. Stark, Berlin, Germany

Matrix Designation
LC10

Matrix Properties
Polymethylsilazane powder A:
(7.1 wt% C, 22.5 wt% N, 10.4 wt% O)
Reference powder B:
(9 wt% C, 20.6 wt% N, 9.2 wt% O)

Test Type
Vickers Indentation

Test Temperature
RT

Grain Size (μm)	Hardness, Vickers (GPa)	Relative Density (percent)	Comments
0.25	14.5	95	*sample A, polysilazane derived*
0.41	11.3	93	*sample B, conventionally processed*

Remarks/Comments
Pyrolyzed sample has smaller grain size, smaller grain size distribution, and a higher hardness.

SiC(p)/Si_3N_4

Reference: 93

Composition
0-20 vol% SiC

Processing Method
Hot-Pressed

Detailed Description
These powders were mixed by ball-milling with Al_2O_3 balls using ethanol solvent for 77.8 hours in a vessel made by polyethylene. Hot-pressing was carried out at 1800 °C for 1 hour in nitrogen atmosphere under a pressure of 30 MPa.

Sintering Additives
8 wt% Y_2O_3

Reinforcement Description
β-SiC powders

Reinforcement Supplier
Ibiden Co., Ltd.

Supplier Designation
UF grade

Reinforcement Dimensions
ave. particle size = 0.3 μm

Matrix Description
α-Si_3N_4 powder

Matrix Supplier
Ube Industries, Ltd., Japan

Matrix Designation
E-10 grade

Matrix Properties
ave. particle size = 0.2 μm

Test Type
Vickers Indentation Microfracture

Test Description
Density measured by Archimedes' method.

Test Temperature
RT

Reinforcement Content (volume percent)	Hardness, Vickers (GPa)	Relative Density (percent)	Range (GPa)
0.0	12.9	99.9	± .05
5.0	13.2	99.9	± .12
10	13.1	98.4	± .08
20	14.3	98.7	± .20

Remarks/Comments
Two types of dispersion configurations were evident: nano-meter scale particles were uniformly distributed, and submicrometer particles forming aggregates at grain boundaries of the matrix. Grains changed to rod-like structure as SiC content increased. Fracture toughness reached a maximum at 5 vol% SiC-Si_3N_4.

Figure 117a: Vickers Hardness of SiC(p)/Si_3N_4 (Reaction Bonded)

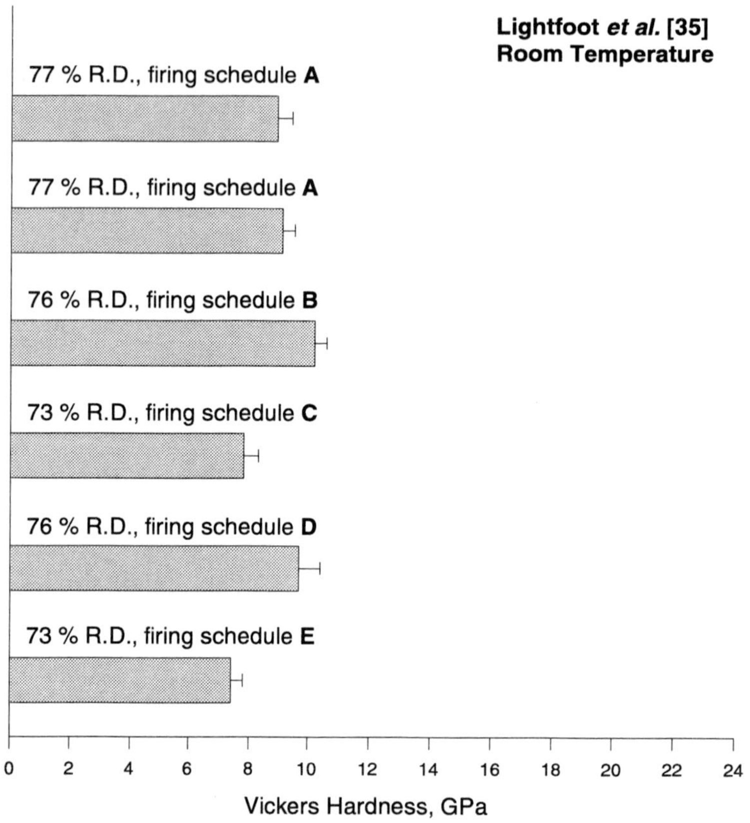

Discussion: Figures 117a-c examined the use of several different techniques and additives to produce a range of hardness for SiC(p)/Si_3N_4. Figure 117a shows the hardness of reaction bonded composites, Lightfoot *et al.* [35]. Figure 117b shows the hardness of composite fabricated by conventional powder techniques, Akimune [37 and 40] and Sasaki *et al.* [93]. Figure 117c presents data for SHS and polymer pyrolysis derived composites, Agrafiotis *et al.* [39] and Riedel *et al.* [67]. For the samples tested by Lightfoot *et al.* [35], synthesis in methanol or octanol was followed by different firing schedules. These samples were less dense than conventional materials. This was partially due to the usage of uniform-diameter Si powders. The usage of these powders resulted in voids between Si particles that were not effectively filled by the SiC. The low density and high porosity of these samples make them subject to crushing under the indenter load rather than plastic deformation leading to low hardness measurements [35]. Schedule B, synthesized in octanol, was the hardest at 10.2 GPa.

Akimune [40 and 37] has tested SiC(p)/Si_3N_4 composites for use in advanced gas turbine and internal combustion engines for the Nissan Motor Co. These composites were fabricated by conventional powder techniques, i.e., they were either hot-pressed, sintered, or HIPed. Figure 117b shows that a different hardness was achieved by different conventional sintering processes. Akimune [37] examined the effects of CIPed/sintered and sintered-HIPed techniques on monolithic and 10 vol% SiC. The added HIP processing resulted in 99% dense samples, while the samples that were not HIPed were only 98 and 97 % dense. Analysis of the 10 vol% SiC(p)/Si_3N_4 composites showed that the sinter-HIPed sample reached a higher density. Despite this fact, the sinter-HIPed materials was not as hard as the sintered sample that was not HIPed. The sintered 10 vol% SiC(p)/Si_3N_4 sample had the lowest relative density but possessed the highest

Figure 117b: Vickers Hardness of $SiC(p)/Si_3N_4$ (Conventional Methods)

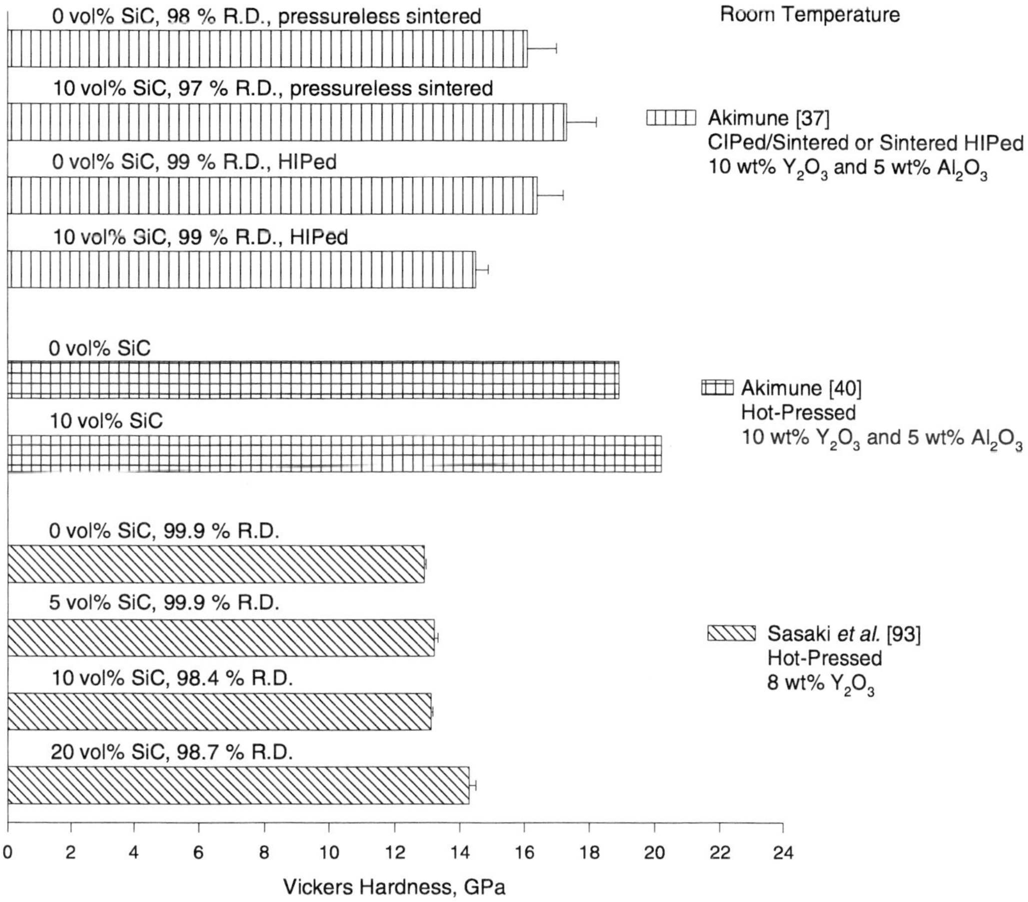

hardness of 17.3 GPa for any of the composites tested by ref. [37]. The composite was softer after HIPing. Possible reasons for this may have been that HIPing reduced the deformation resistance through relaxation of the particulate induced stresses and/or grain growth of silicon nitride grains.

Akimune [40] examined the effect of hot-pressing on hardness. The samples had the same sintering aids as the samples in ref. [37]. The 10 vol% sample produced here was the hardest in the comparison of the composites shown in figures 117a-c at 20.2 GPa. No other reference (except ref. [37]) used such high temperatures and pressures during processing. This could have caused the high hardness.

Sasaki *et al.* [93] examined the hardness of hot-pressed $SiC(p)/Si_3N_4$ samples with Y_2O_3 added as a sintering aid. The density of these samples compared to those of ref. [37], but the hardness was not as high as ref. [37].

Figure 117c: Vickers Hardness of SiC(p)/Si_3N_4 (Reaction Processed and Polymer Derived)

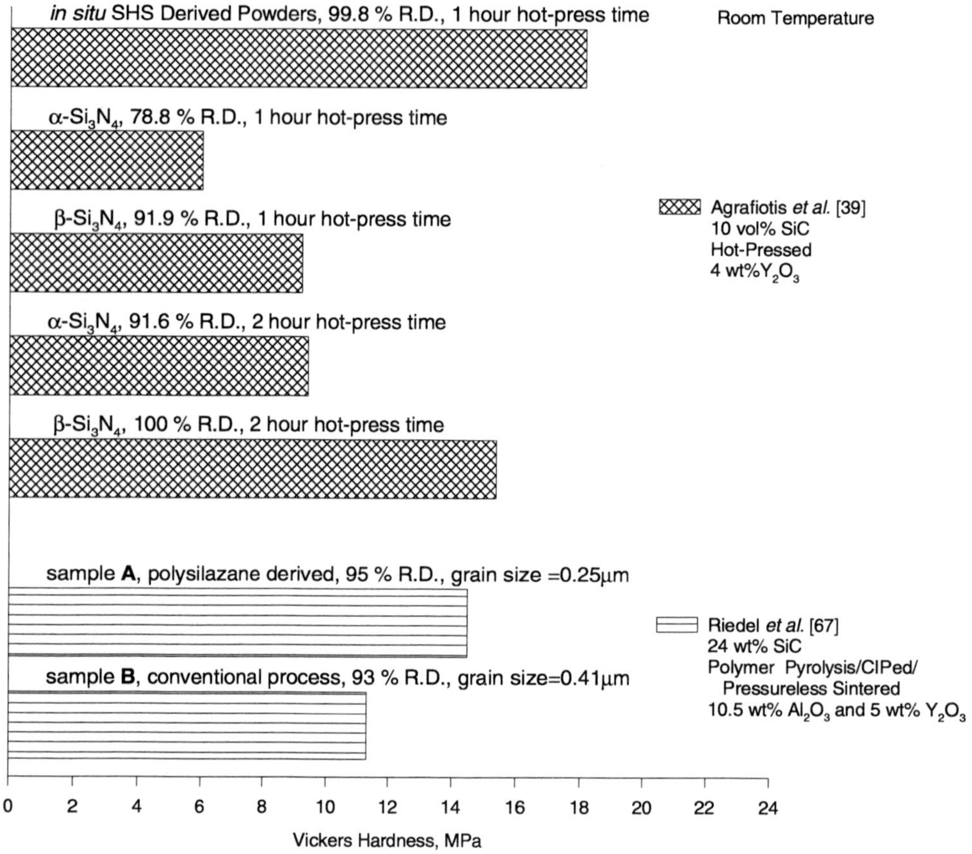

Agrafiotis *et al.* [39] examined combustion synthesis of SiC(p)/Si_3N_4. Although the increased sintering time increased density for all hot-pressed samples which led to a higher hardness, no samples were harder than the *in situ* β-Si_3N_4 with only 1 hour of hot-pressing time. This was due to the combustion synthesis of the material produced from the chemical equation of:

$$(3(1 - x)+y)Si + xSi_3N_4 + yC + 2(1 - x)\ N_2 \rightarrow Si_3N_4 + ySiC.$$

This led to a self propagating chemical reaction that resulted in the starting powders. This type of synthesis leads to well-dispersed particles that results in a hard composite.

Riedel *et al.* [67] tested composites sintered from polymer derived powders. The polysilazane derived sample, A, had a higher hardness than the conventionally processed composite. This was due to the as-pyrolyzed polymer derived powder made from amorphous silicon carbo-nitride, while the reference sample was fabricated from crystalline powders. The amorphous polymer derived powders sintered much faster than the crystalline powders and resulted in smaller grain size and a higher relative density. The small grain size of the polysilazane composite and higher density led to a higher hardness.

Figure 118: Vickers Hardness of SiC(p)/Si_3N_4 Composites from SHS Derived Powders

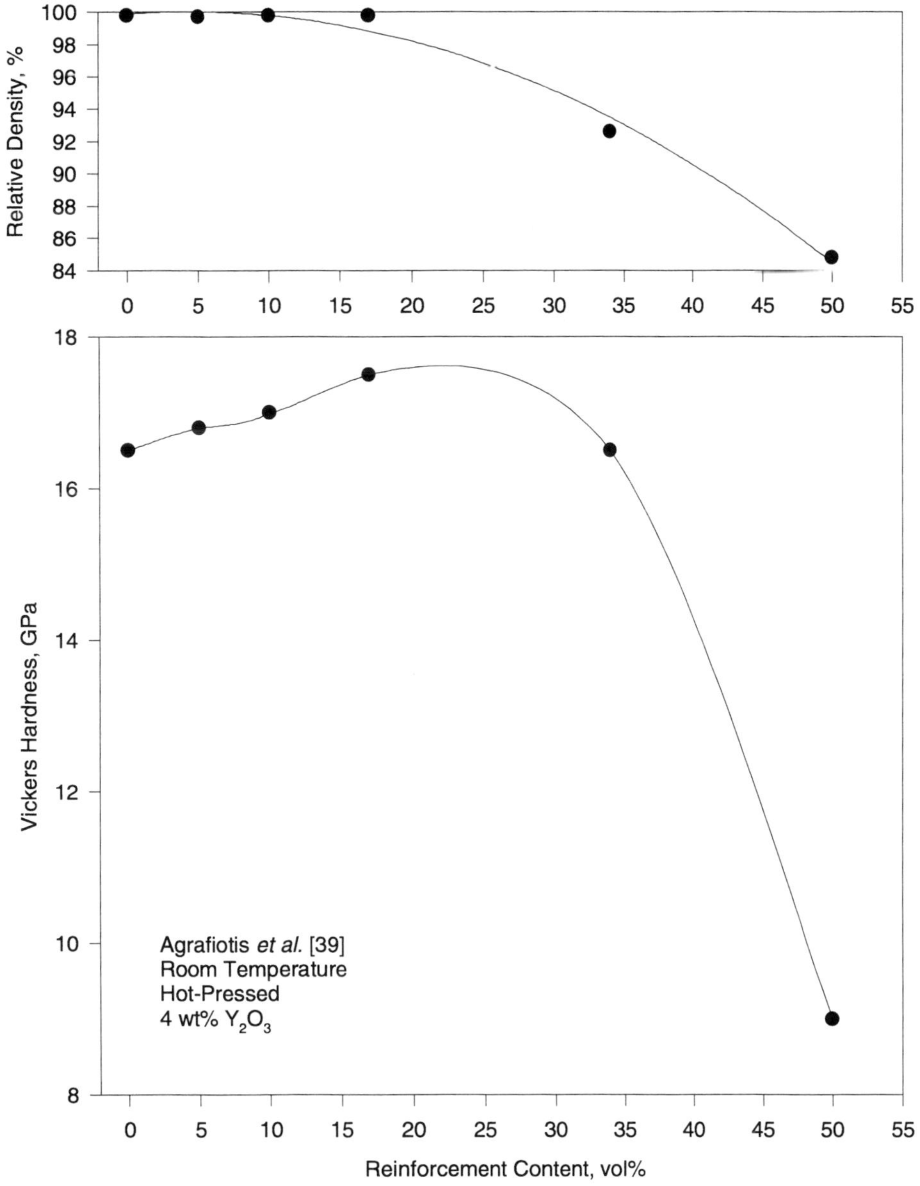

Discussion: Figure 118 shows the hardness of a sample from SHS reaction derived powders. For near fully dense samples with less than 17% reinforcement content, hardness increased with the addition of SiC particulates from 16.5 to 17.5 GPa. For samples that contained over 17 % added SiC, full density could not be obtained by hot-pressing. The decrease in density resulted in a decrease in hardness.

$TiC(p)/Si_3N_4$

Reference: 21

Composition
0 and 15 vol% TiC

Processing Method
Colloidal-Processed/Hot-Pressed

Detailed Description
The starting powders were homogenized and mixed together. The mixtures were filter pressed at 2.07 MPa for approximately 45 minutes over a 0.02 μm polymer filter. The mixtures were dried for 2 weeks. The resulting green bodies were hot-pressed under 41 MPa at 1800 °C for 2 hours.

Sintering Additives
6 wt% CeO_2 from Cerac Inc., Milwaukee, WI

Reinforcement Description
TiC particulates

Reinforcement Supplier
Hermann C. Starck, Inc., New York, NY

Supplier Designation
c.a.s. Grade

Reinforcement Dimensions
ave. particle size = 1.25 μm

Matrix Description
α-Si_3N_4

Matrix Supplier
Hermann C. Starck, Inc., New York, NY

Matrix Designation
Grade LC12

Matrix Properties
ave. particle size = 0.55 μm

Test Type
Vickers Indentation

Specimen Preparation
The specimen surfaces were polished prior to testing.

Test Description
Indenter load = 10 kg

Test Temperature
RT

Reinforcement Content (vol%)	Hardness, Vickers (GPa)
0	17.8
15	22.8

Remarks/Comments
Hardness increased with the addition of TiC particulates.

$TiC(p)/Si_3N_4$

Reference: 95

Composition
0 - 20 vol% TiC

Processing Method
Hot-Pressed or Gas Pressure Sintered

Detailed Description
The powders were mixed wet or attrition-milled under isopropanal or hexane. Sintering was done in a graphite resistance furnace under positive pressure of N_2 (10 mm) and at a temperature of 1750 °C. Hot-pressing was done in a vacuum hot press in BN coated graphite die and punched under a pressure of around 30 MPa at 1650 °C. The pressureless sintered materials were in the form of billets initially isostatically-pressed in the green state (67 x 17 x 17 mm size).

Sintering Additives
Y_2O_3
quartz powder

Reinforcement Description
TiC particulates

Reinforcement Supplier
H.C. Starck, Goslar, Germany

Reinforcement Dimensions
surface area = 4.5 $m^2\ g^{-1}$

Matrix Description
Si_3N_4 powder

Matrix Properties
surface area = 8.78 $m^2\ g^{-1}$

Test Type
Vickers Indentation Test

Specimen Dimensions
dia. = 25 mm
thickness = 8 - 10 mm

Test Description
load = 10 kg

Test Equipment
Wolpert hardness tester

Reinforcement Content (volume percent)	Hardness, Vickers (GPa)	Comments
0	14.2	*11 % glassy phase, pressureless sintered*
0	15.1	*15 % glassy phase, pressureless sintered*
0	18.1	*hot-pressed*
10	17.1	*hot-pressed*
15	17.8	
20	17.5	

Remarks/Comments

TiC converts to TiN during sintering in the presence of N_2. This conversion is less in the case of hot-pressed products. Higher amounts of sintering aid can increase the Vickers microhardness. Hot-pressed materials had considerably higher hardness values than the sintered samples.

TiN(p)/Si_3N_4

Reference: 54

Composition
0 - 40 vol% TiN

Processing Method
Hot-Pressed

Detailed Description
Si_3N_4 powder and sintering aids were homogenized in water for 48 hours and freeze dried. The second phase (TiN) was added to the matrix mix and mixed in isobutyl alcohol for 48 hours. The materials were then uniaxially hot-pressed at 1800 °C and 30 MPa for 10 minutes to nearly full density.

Sintering Additives
3 wt% Al_2O_3
8 wt% Y_2O_3

Reinforcement Description
TiN particulates

Reinforcement Supplier
Starck, Germany

Supplier Designation
Grade C

Reinforcement Dimensions
ave. size = 0.3-3 μm

Matrix Description
Si_3N_4 powder

Matrix Supplier
Starck, Germany

Matrix Designation
LC12

Test Type
Vickers Indentation

Test Description
500 g load
10 sec indentation time

Test Temperature
RT

Reinforcement Content (volume percent)	Hardness, Vickers (GPa)	Relative Density (percent)
0	20.7	100
20	17.8	99.4
30	19.2	99.6
40	15.2	99.4

Remarks/Comments

The hardness decreased with increasing TiN reinforcement, with the exception of 30 vol% TiN, where the hardness value for the composites increased to 19.2 GPa. This was at least partially due to the increase in relative density.

$TiN(p)/Si_3N_4$

Reference: 65

Composition
12 - 40 wt% TiN

Processing Method
Hot-Pressed

Detailed Description
Specimens were hot-pressed in graphite dies. The pressure was applied uniaxially. Heating was done by an inductor from a high-frequency generator.

Sintering Additives
Y_2O_3

Reinforcement Description
TiN particulates

Matrix Description
Si_3N_4

Matrix Properties
furnace synthesized:
porosity = 3-5%
Hv = 12.2 ± 0.5 GPa
plasma-chemical synthesized:
porosity = 0%
Hv = 18.9 ± 0.8 GPa

Test Type
Vickers Indentation

Specimen Dimensions
12.7 x 12.7 x 3.5 mm

Specimen Preparation
Specimens were heat treated for 2 hours at 1450 °C prior to testing.

Test Description
Vickers pyramid indentation. Loads of 50 N and 200 N were used.

Test Temperature
RT

Reinforcement Content (weight percent)	Hardness, Vickers (GPa)	Comments
12	20.3	*50 N load*
30	19.9	
40	20.9	
12	16.8	*200 N load*
30	16.2	
40	16.2	

Remarks/Comments
Hardness did not change before or after heat treatment in a reducing atmosphere. The hardness values were higher at the 50 N load, compared to the tests performed at 200 N.

Discussion: Hardness data for hot-pressed TiN and TiC particulate reinforced silicon nitrides are presented in figure 119. Two pressureless sintered silicon nitride ceramics are also presented as a comparison. It was not expected that the $TiN(p)/Si_3N_4$ composites would show a strong dependence on reinforcement content since titanium nitride and silicon nitride have close to the same hardness. The $TiN(p)/Si_3N_4$ composites tested by Bellosi *et al.* [54] did show that the hardness of the composite decreased, increased, then decreased again with the addition of 20, 30, and 40 vol% TiN reinforcement, respectively. This trend closely followed the relative density trend, which suggested that the hardness of these composites was dependent on relative density. The $TiN(p)/Si_3N_4$ composites tested by Gnesin *et al.* [65] did not show a strong dependence on reinforcement content; however, there was a 2 - 4 GPa difference in the hardness as measured with a 50 N load or a 200 N load.

Since TiC is harder than silicon nitride, it might be expected that the hardness of silicon nitride would be increased with the addition of TiC particulates. Blanchard and Page [21] found that the hardness of silicon nitride increased with the addition of 15 vol% TiC. On the other hand, Mukerji and Das [95] found that the $TiC(p)/Si_3N_4$ did not show a strong dependence on TiC content. This difference was likely due to the differences in the hot-pressing conditions. The composites tested by Blanchard and Page [21] were hot-pressed at 1800 °C and 41 MPa, compared to only 1650 °C and 30 MPa for the composites tested by Mukerji and Das [95]. Higher hot-pressing temperatures and pressures likely led to decreased porosity, resulting in higher hardness for the $TiC(p)/Si_3N_4$ composites.

Figure 119: Vickers Hardness of $TiN(p)/Si_3N_4$ and $TiC(p)/Si_3N_4$ Composites

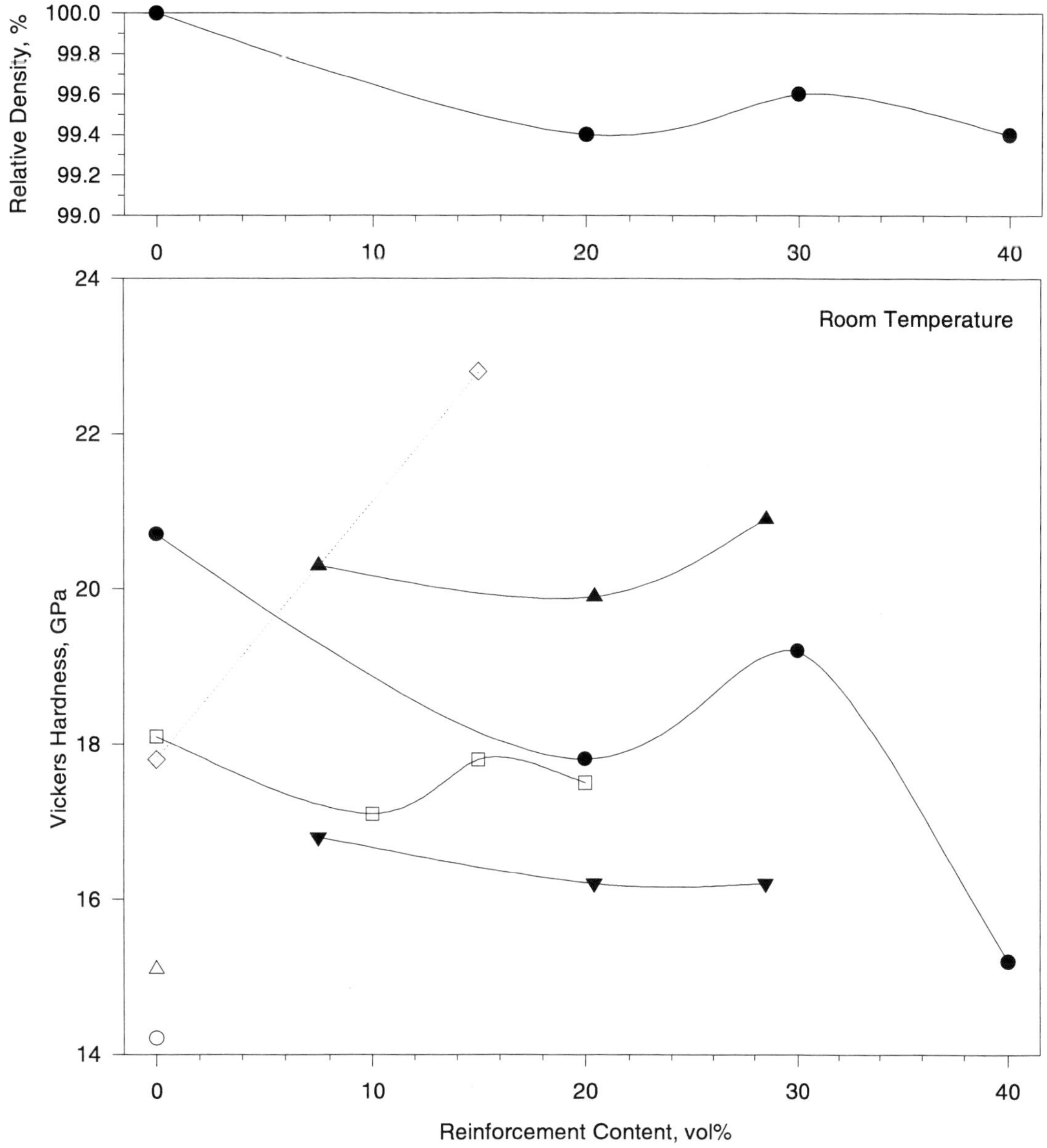

◇ Blanchard and Page [21], Hot-Pressed, $TiC(p)/Si_3N_4$, Sintering Aid: 6 wt% CeO_2

—●— Bellosi *et al.* [54], $TiN(p)/Si_3N_4$, Hot-Pressed, Sintering Aid: 3 wt% Al_2O_3 and 8 wt% Y_2O_3

Mukerji and Das [95], Sintering Aid: Y_2O_3 and Quartz

—□— $TiC(p)/Si_3N_4$, Hot-Pressed

○ 11% glassy phase, Pressureless Sintered Si_3N_4

△ 15% glassy phase, Pressureless Sintered Si_3N_4

Gnesin *et al.* [65], $TiN(p)/Si_3N_4$ [1]

Hot-Pressed, Sintering Aid: Y_2O_3

—▲— 50 N load

—▼— 200 N load

1. Converted from wt% to vol%

TiC(p)/AlN

Reference: 96

Composition
5 wt% TiC

Processing Method
Hot-Pressed

Detailed Description
The powders were milled in a planetary mill and then hot-pressed in a graphite mold.

Sintering Additives
2 wt% Al

Reinforcement Description
TiC powder

Matrix Description
AlN powder

Matrix Properties
ave. particle size = 4 μm
composition: 64.6 wt% Al; 30.9 wt% N; 0.7 wt% Fe; 0.7 wt% O
balance: Mg, Mn, Ti, Cr

Test Type
Vickers Indentation

Test Description
load = 1 N

Test Temperature
RT

Reinforcement Content (weight percent)	Hardness, Vickers (GPa)
5	18.9

Remarks/Comments
Self-reinforcement occurs by the intragranular crystal-oriented transition "2H to multilayer polytypes" to produce plate-like grains.

TiN(p)/AlN

Reference: 96

Composition
5 and 20 wt% TiN

Processing Method
Hot-Pressed

Detailed Description
The powders were milled in a planetary mill and then hot-pressed in a graphite mold.

Reinforcement Description
TiN powder

Matrix Description
AlN powder

Matrix Properties
ave. particle size = 4 μm
composition: 64.6 wt% Al; 30.9 wt% N; 0.7 wt% Fe; 0.7 wt% O
balance: Mg, Mn, Ti, Cr

Test Type
Vickers Indentation

Test Description
load = 1 N

Test Temperature
RT

Reinforcement Content (weight percent)	Hardness, Vickers (GPa)
5	15.5
20	17.5

Remarks/Comments
Self-reinforcement occurs by the intragranular crystal-oriented transition "2H to multilayer polytypes" to produce plate-like grains.

TiO_2(p)/AlN

Reference: 96

Composition
5 and 20 wt% TiO_2

Processing Method
Hot-Pressed

Detailed Description
The powders were milled in a planetary mill and then hot-pressed in a graphite mold.

Reinforcement Description
TiO_2 powder

Matrix Description
AlN powder

Matrix Properties
ave. particle size = 4 μm
composition: 64.6 wt% Al; 30.9 wt% N; 0.7 wt% Fe; 0.7 wt% O
balance: Mg, Mn, Ti, Cr

Test Type
Vickers Indentation

Test Description
load = 1 N

Test Temperature
RT

Reinforcement Content (weight percent)	Hardness, Vickers (GPa)
5.0	18.1
20	18.6

Remarks/Comments
Self-reinforcement occurs by the intragranular crystal-oriented transition "2H to multilayer polytypes" to produce plate-like grains.

Figure 120: Vickers Hardness of TiN(p)/AlN, TiO_2(p)/AlN, and TiC(p)/AlN

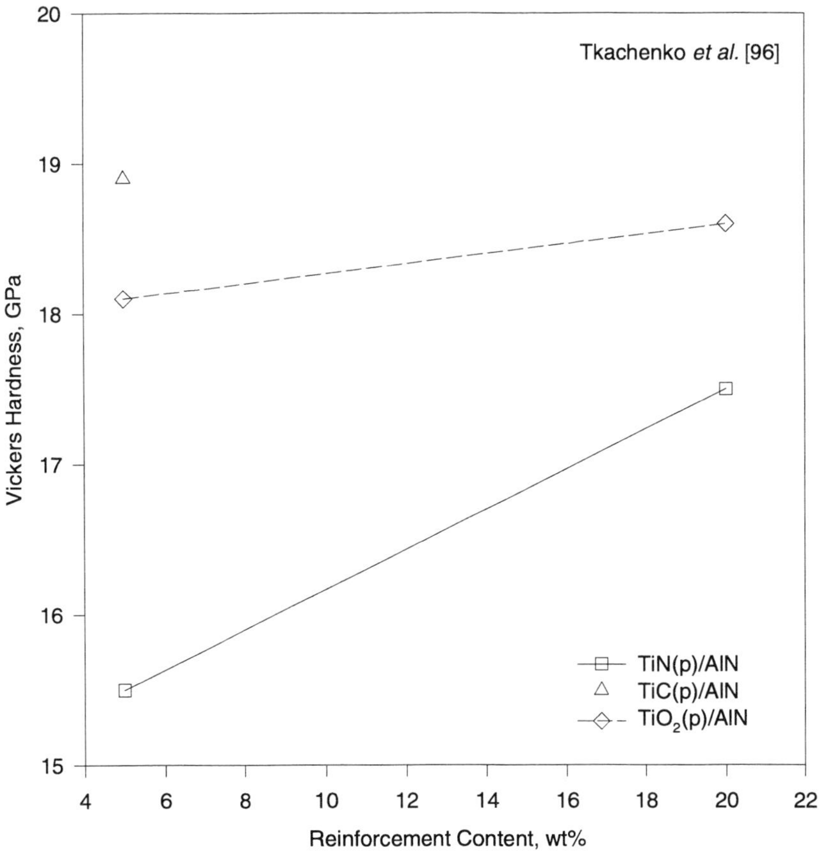

Discussion: The Vickers hardness for the TiO_2 (p)/AlN and the TiN(p)/AlN composites increased with an addition of reinforcement content. The addition of only 5 wt% TiC particulates resulted in the hardest composite tested, whereas the addition of TiN particulates resulted in a less hard material.

TiB_2(p)/SiC

Reference: 83

Composition
0 and 16 vol% TiB_2

Processing Method
Colloidal Processed/ Pressureless Sintered

Detailed Description
The commercially available SiC-TiB_2 (Hexoloy ST) CMC was fabricated by mixing presized TiB_2 and submicrometer-size SiC powders using a nonaqueous blending and drying procedure. The homogeneous mixture, combined with a binder, was formed into green compacts that were then pressureless-sintered in an inert atmosphere at temperatures of over 2000 °C.

Reinforcement Description
TiB_2 particulates

Reinforcement Typical Properties
melting point = 2790 °C
E = 531 GPa
Poisson ratio = 0.28
CTE = 4.6×10^{-6} °C^{-1}

Reinforcement Dimensions
ave. particle size = 2 μm

Matrix Description
α-SiC

Matrix Supplier
Sohio Engineered Materials CO., Niagara Falls, NY

Matrix Designation
Hexoloy ST

Matrix Properties
ave. grain size = 4 μm
melting point = 2540 °C
E = 410 GPa
Poisson ratio = 0.19
CTE = 4.02×10^{-6} °C^{-1}

Test Type
Vickers Indentation

Test Temperature
RT

Reinforcement Content (volume percent)	Hardness, Vickers (GPa)	Density (g cm^{-3})
0	27.5	3.1
16	27.5	3.3

Remarks/Comments
The addition of 16 vol% TiB_2 to SiC made no significant change in hardness compared to monolithic SiC.

TiC(p)/SiC

Reference: 70

Composition
0 - 100 wt% TiC

Processing Method
Hot-Pressed

Detailed Description
Powders of β-SiC and TiC were mixed with B_4C powder and carbon black as sintering aids for SiC. Well-mixed powders were attained by ball-milling in n-hexane for 24 hours using a polyamide media. After drying, the powders were loaded into a graphite die with inside dia. = 90 mm and hot-pressed at 2150 °C for 2 hours under 40 MPa in an Ar gas atmosphere. Hot-pressed discs of SiC-TiC composites were machined into pieces.

Sintering Additives
B_4C: ave. particle size = 1.2 μm
purity > 99%
Denki Kagaku Kogyo KK
carbon black:
ave. particle size = 0.02 μm
purity > 99%
Mitsubishi Chem. Ind.

Reinforcement Description
TiC powders

Reinforcement Supplier
Kojundo Chem. Lab. Co. Ltd

Reinforcement Typical Properties
ave. particle size = 0.4 μm
purity > 99 %

Matrix Description
β-SiC powders

Matrix Supplier
Ibiden Co. Ltd

Matrix Designation
Ultrafine

Matrix Properties
ave. particle size = 0.3 μm
purity > 98 %

Test Type
Vickers Indentation

Specimen Dimensions
3 x 4 x 38 mm

Test Description
load = 0.5 kg

Test Temperature
RT

Reinforcement Content (weight percent)	Hardness, Vickers (kg/mm^2 (GPa))	Range (kg/mm^2)
0	2,830 [27.7]	±40
10	2,730 [26.8]	±45
30	2,780 [27.2]	±85
50	2,700 [26.5]	±45
70	2,380 [23.3]	±75
90	2,340 [22.9]	±55
100	2,660 [26.1]	±200

Remarks/Comments
For all the compositions, the composites exhibited quite high hardness.

Figure 121: Vickers Hardness of TiC and TiB_2(p)/SiC

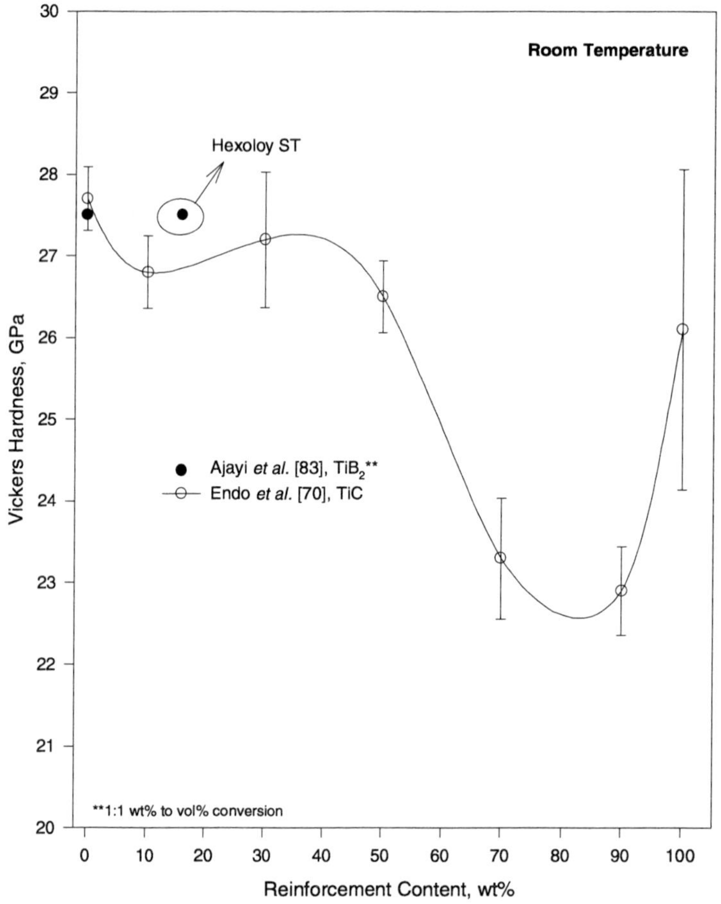

Discussion: Endo *et al.* [70] tested SiC-TiC ceramics as a continuous function of composition from monolithic SiC to monolithic TiC. The monolithic SiC had the highest hardness of any of the compositions. The hardness of all the compositions was above 26 GPa, except at 70 and 90 % TiC. It was suggested that the lower hardness at these two compositions was caused by the residual stress fields generated in the sintered composite due to the thermal expansion mismatch, causing higher microcrack density [70]. The SiC grains in the TiC(p)/SiC composite, in comparison to the SiC grains in the monolithic silicon carbide, were smaller and more uniform due to grain boundary pinning by the TiC during sintering. A comparison of the two monolithic ceramics showed that the silicon carbide was slightly harder than the titanium carbide.

Ajayi *et al.* [83] tested Hexaloy ST, which is a commercially available 20 vol% TiB_2(p)/SiC composite. The Hexaloy ST composite showed identical hardness to the monolithic silicon carbide also tested in ref. [83]; and it also showed approximately the same hardness as a similar TiC(p)/SiC composite tested in Endo *et al.* [70].

B_4C(p)/TiB_2

Reference: 101

Composition
0 - 30 vol% B_4C

Processing Method
Cold-Pressed/Pressureless Sintered

Detailed Description
Powder compacts formed by cold-pressing were fabricated by pressureless sintering at 1600 °C in an argon atmosphere.

Sintering Additives
8.6 mol% Ni, 1.4 mol% C

Reinforcement Description
B_4C powder

Matrix Description
TiB_2 powder

Test Type
Vickers Indentation

Specimen Dimensions
3 x 4 mm cross-section

Test Temperature
RT

Reinforcement Content (volume percent)	Hardness, Vickers (GPa)	Range (GPa)	Relative Density (percent)
0	15.0	+0.7/-1.6	94.3
10	13.5	+1.6/-0.9	94.7
20	13.1	+1.1/-0.6	93.7
30	9.95	±0.5	88.3

Remarks/Comments
Hardness decreased with the addition of B_4C.

ZrO_2(p)/TiB_2

Reference: 34

Composition
20 wt% YSZ

Processing Method
CIPed/HIPed

Detailed Description
Starting powders were ball-milled in ethanol. The mixed powder was dried, uniaxially pressed at 9.8 MPa and CIPed at 300 MPa. The green bodies were HIPed at 1600 °C and 200 MPa for 2 hours in Ar gas by the glass-encapsulation method.

Sintering Additives
0-5 wt% SiC

Reinforcement Description
2Y-ZrO_2 particulates

Reinforcement Dimensions
particle dia. = 1-3 μm

Matrix Description
TiB_2

Matrix Properties
particle dia. = 2-4 μm

Test Type
Vickers Indentation

Specimen Dimensions
3 x 4 x 36 mm

Test Temperature
RT

Stabilizer Content (weight percent)	Hardness, Vickers (GPa)
0	22.8
2	23.2
5	23.4

Remarks/Comments
The addition of SiC to the composite improved the Vickers hardness slightly for up to 5 wt% SiC due to grain size reduction.

ZrO_2(p)/TiB_2

Reference: 34

Composition
0 - 40 wt% ZrO_2

Processing Method
CIPed/HIPed

Detailed Description
Starting powders were ball-milled in ethanol. The mixed powder was dried, uniaxially pressed at 9.8 MPa and CIPed at 300 MPa. The green bodies were HIPed at 1600 °C and 200 MPa for 2 hours in Ar gas by the glass-encapsulation method.

Reinforcement Description
2Y-ZrO_2 particulates

Reinforcement Dimensions
particle dia. = 1-3 μm

Matrix Description
TiB_2

Matrix Properties
particle dia. = 2-4 μm

Test Type
Vickers Indentation

Specimen Dimensions
3 x 4 x 36 mm

Test Temperature
RT

Reinforcement Content (weight percent)	Hardness, Vickers (GPa)
0	24
5	25
10	24
20	21
40	19

Remarks/Comments
Vickers hardness decreased with increasing reinforcement content.

Figure 122: Vickers Hardness of 2Y-ZrO_2(p)/TiB_2 and B_4C(p)/TiB_2

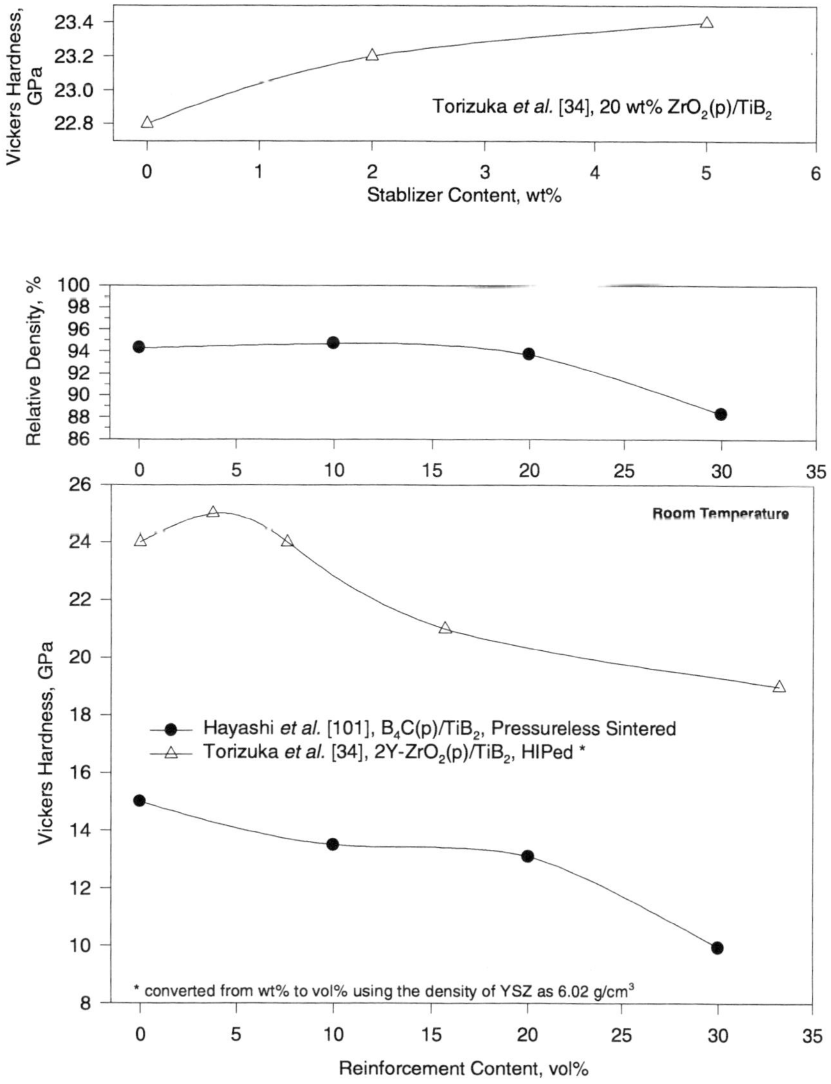

Discussion: The Vickers hardness for all composites decreased with the addition of either B_4C or YSZ particulates in TiB_2. The hardness for the HIPed TiB_2 composite tested by Torizuka *et al.* [34] was substantially greater than that for the pressureless-sintered TiB_2 composite from Hayashi *et al.* [101] at all reinforcement contents. The SiC sintering additive was effective in restraining the grain growth of the TiB_2 grains in the HIPed 2Y-ZrO_2(p)/TiB_2 composite, which resulted in higher hardness. The relative density for the B_4C(p)/TiB_2 composite remained approximately constant up to 20 vol% reinforcement content and then decreased at 30 vol% reinforcement. There was a significant drop off in hardness that resulted from the low relative density of the 30 vol% B_4C(p)/TiB_2 composite. X-ray diffraction revealed the presence of TiB_2, B_4C, and NiB phases present in the B_4C(p)/TiB_2 composite [101].

14. Thermal Properties (Section V)

Diamond(p)/Cordierite

Reference: 102

Composition
15 vol% diamond

Processing Method
Hot-Pressed

Detailed Description
Diamond particles of various sized were used for the different composites. The cordierite powder and diamond particles were mixed then hot-pressed in graphite dies at 1300 °C and 32 MPa.

Reinforcement Description
diamond particulates

Reinforcement Supplier
Megadiamond Division of Smith International, Provo, UT
GE Superabrasives, Worthington, OH

Reinforcement Dimensions
powder particle radius =
0 - 0.25 µm, 0 - 0.5 µm,
0.5 - 1 µm, 0 - 5 µm, 4 - 8 µm,
12 - 22 µm, and 50 µm

Matrix Description
cordierite powders

Matrix Supplier
Baikowski International, Charlotte, NC

Test Type
Laser-Flash Technique

Specimen Preparation
Specimens were polished with 1 µm diamond paste, and ultrasonically cleaned to remove the polishing medium.

Specimen Dimensions
dia. = ~9 mm, thickness = 3 mm

Type of Specimen
disks

Test Description
The thermal diffusivity was measured using a glass-Nd laser. The specimen temperature was measured using a nitrogen-cooled infrared detector. Measurements were made directly on the as-hot-pressed samples, with heat flow parallel to the hot-pressing direction. From room temperature to about 200 °C the measurements were made in air in a specially constructed specimen holder. Above 300 °C, the specimens were positioned in a carbon-resistance furnace with nitrogen protective atmosphere. Transient temperature measurements were made by a InSb liquid-nitrogen-cooled detector.

Test Temperature
RT - 700 °C

Test Environment
RT-300: Air
300-700 °C: Nitrogen

Test Equipment
Nd-glass laser

Diamond Particle Radius (µm)	Thermal Conductivity ($W\ m^{-1}\ K^{-1}$)
0.0625	3.52
0.375	5.46
1.25	6.11
3	6.20
8.5	6.46
25	7.22

Remarks/Comments
The hot pressed composites were ~ 99% of theoretical density, using 2.669 g/cm^3 as the theoretical density. Thermal conductivity increased with increasing diamond particle size. Except for the smallest diamond particle size, the thermal conductivity decreased with increasing temperature.

Temperature, (°C)	Thermal Conductivity, (W m^{-1} K^{-1})	Diamond Particle Radius (μm)
20	7.18	25
200	6.26	25
300	5.35	25
400	4.90	25
500	4.80	25
600	4.37	25
700	4.20	25
20	6.40	8.5
200	5.35	8.5
300	5.12	8.5
400	4.87	8.5
500	4.50	8.5
600	4.09	8.5
700	3.83	8.5
20	6.22	3.0
200	5.09	3.0
300	4.88	3.0
400	4.70	3.0
500	4.37	3.0
600	4.18	3.0
700	4.04	3.0
20	6.07	1.25
300	4.94	1.25
400	4.72	1.25
500	4.61	1.25
600	4.24	1.25
700	4.10	1.25
20	5.46	0.375
200	4.70	0.375
300	4.25	0.375
400	4.17	0.375
500	3.93	0.375
600	3.68	0.375
700	3.61	0.375
20	3.45	0.0625
200	3.55	0.0625
300	3.49	0.0625
400	3.57	0.0625
500	3.58	0.0625
600	3.42	0.0625
700	3.34	0.0625
20	4.01	***Monolithic***
200	3.40	***Cordierite***
300	3.13	-
400	2.84	-
500	2.67	-
600	2.59	-
700	2.55	-

Diamond(p)/Cordierite

Reference: 102

Composition
0, 15 vol% diamond

Processing Method
Hot-Pressed

Detailed Description
Diamond particles of various sized were used for the different composites. The cordierite powder and diamond particles were mixed then hot-pressed in graphite dies at 1300 °C and 32 MPa.

Reinforcement Description
diamond particulates

Reinforcement Supplier
Megadiamond Division of Smith International, Provo, UT
GE Superabrasives, Worthington, OH

Reinforcement Dimensions
powder particle radius =
0 - 0.25 μm, 0 - 0.5 μm,
0.5 - 1 μm, 0 - 5 μm, 4 - 8 μm,
12 - 22 μm, and 50 μm

Matrix Description
cordierite powders

Matrix Supplier
Baikowski International,
Charlotte, NC

Test Type
Differential Scanning Calorimetry

Specimen Preparation
Specimens were polished with 1 μm diamond paste, and ultrasonically cleaned to remove the polishing medium.

Specimen Dimensions
dia. = ~9 mm, thickness = 3 mm

Type of Specimen
disks

Test Temperature
RT - 700 °C

Test Environment
Air

Temperature (°C)	Specific Heat ($J\,g^{-1}\,K^{-1}$)	Comments
25	0.798	***Monolithic Cordierite***
200	0.990	
300	1.047	
400	1.082	
500	1.140	
600	1.159	
700	1.171	
25	0.726	***15 vol% Diamond Particulate Reinforced Cordierite***
200	0.952	
300	1.056	
400	1.176	
500	1.220	
600	1.240	
700	1.284	

Remarks/Comments
The specific heat of the monolithic cordierite was greater than the diamond(p)/cordierite composite at room temperature. However, the specific heat of the composite increased more rapidly with increasing temperature and by around 300 °C the composite had a greater specific heat than the matrix.

Figure 123: Diamond Particle Size Dependence of Elevated Temperature Thermal Conductivity of a Diamond(p)/Cordierite Composite

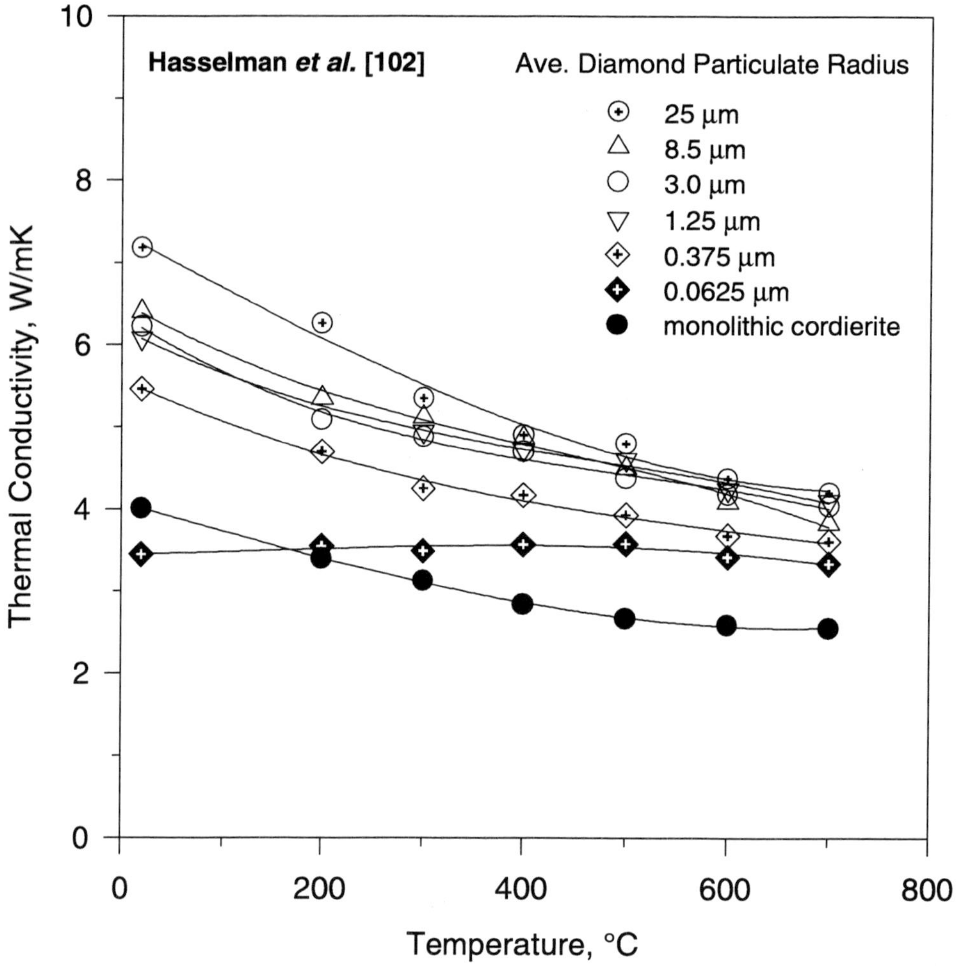

Discussion: Hasselman *et al.* [102] found that there was a significant diamond particle size effect on the thermal conductivity of diamond particulate reinforced cordierite composites at room temperature. To explain the size effect of the diamond particulates, the composites can be divided into three components in terms of thermal conductivity. The first is the cordierite matrix, and the second is the diamond reinforcement. Diamond particulates have a much higher thermal conductivity than the cordierite matrix, thus it was expected that the thermal conductivity would increase with the addition of 15 vol% diamond reinforcement, regardless of its particle size. The third component is the interfacial layer between the diamond particulates and cordierite matrix which acts as a thermal barrier to heat transfer. The effect of this third component increased dramatically as the diamond particle size decreased because of the surface/volume relationship. As a result, at room temperature the thermal conductivity of the 15 vol% diamond(p)/cordierite composite decreased with decreasing diamond particle size, as shown in figure 124. In fact, the thermal conductivity of the composite with very small 0.0625 μm diamond particles was actually lower than the conductivity of the cordierite matrix, despite the fact that it was expected that the thermal conductivity of the composite would be much higher.

The data showed that the thermal conductivity of the monolithic cordierite decreased with increasing temperature, and it is well known that diamond has a strong negative dependence on temperature, so it was not surprising that the thermal conductivity of the large diamond particle composites decreased with increasing temperature. What was surprising was that the thermal conductivity of the composite with the smallest diamond particles (0.0625 μm) showed no change in thermal conductivity with increasing temperature. This suggested that the interfacial layer showed the unique characteristic of increasing thermal conductivity with increasing temperature to offset the decrease in thermal conductivity of both the diamond and the cordierite. It was suggested that this temperature dependence could be the result of at least two effects. First of all, phonons within the diamond phase are scattered at the grain boundary, and it has been demonstrated that this effect lessens with increasing temperature. Secondly, thermal expansion mismatch stresses are lessened as temperatures approach processing temperatures. These thermal stresses can create significant lattice distortions, with elastic strains at or near the interface as high as 3%. Such lattice distortions can contribute significantly to phonon scattering [102].

Figure 124: Effect of Diamond Particle Radius on Thermal Conductivity of a 15 vol% Diamond(p)/Cordierite Composite

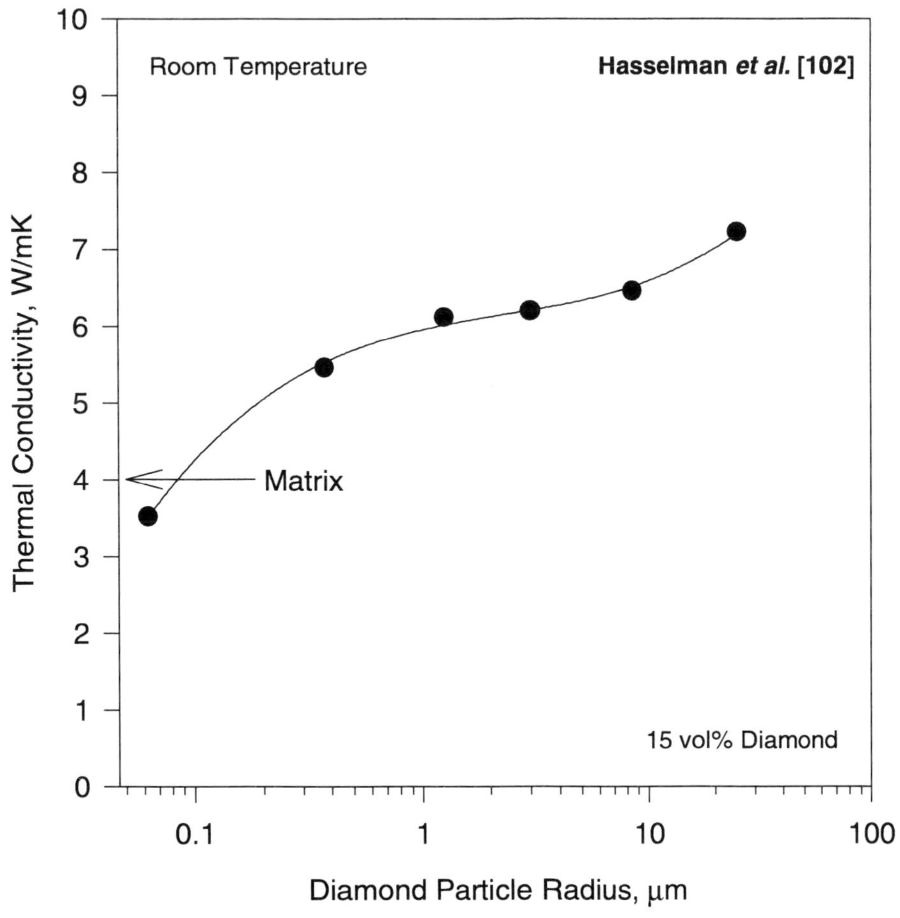

Figure 125: Specific Heat vs. Temperature of Cordierite and a 15 vol% Diamond(p)/Cordierite Composite

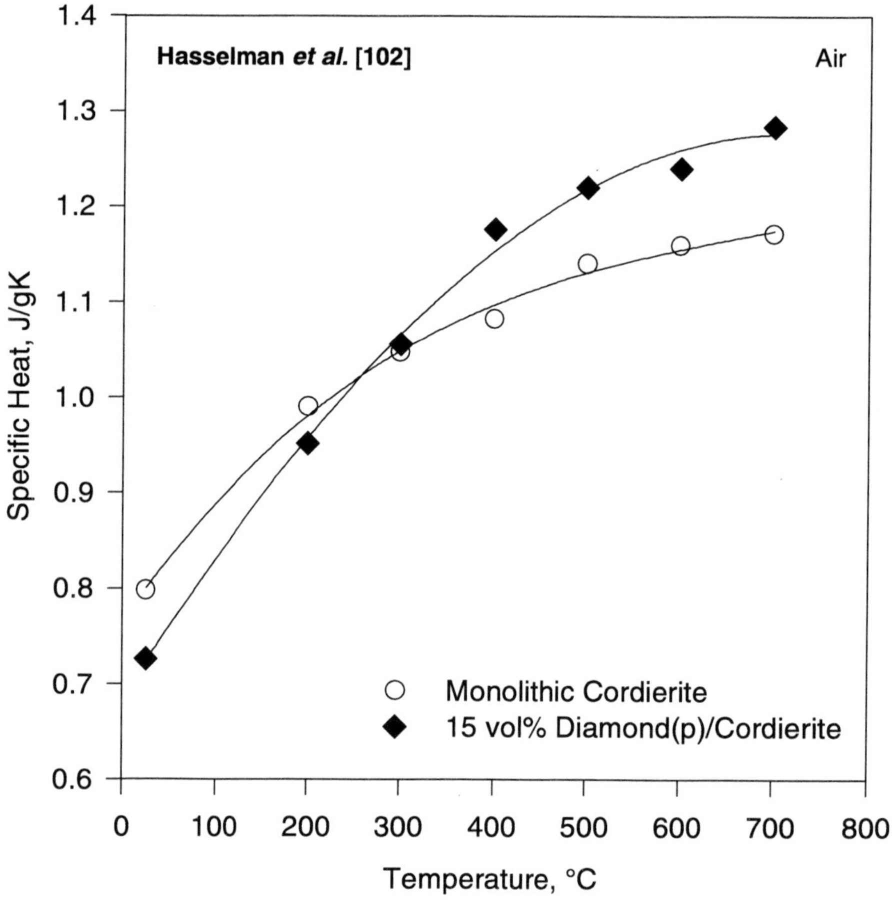

Discussion: The specific heat of the monolithic cordierite was greater than the diamond(p)/cordierite composite at room temperature. However, the specific heat of the composite increased more rapidly with increasing temperature and by around 300 °C the composite had a greater specific heat than the matrix.

$SiC(p)/Al_2O_3$

Reference: 55

Composition
0-87 wt% SiC

Processing Method
Hot-Pressed

Detailed Description
The composite powders were mixed and hot-pressed in graphite dies at an appropriate time, temperature and pressure. The hot-pressing schedule was precisely controlled such that samples with a wide range of porosities were obtained.

Reinforcement Description
SiC particulates

Reinforcement Typical Properties
The SiC particles were somewhat disc-shaped with a diameter of ~ 6 μm and a thickness of about 1-2 μm.
CTE = 5.1×10^{-6} °C^{-1} (20-1327 °C)
thermal conductivity = 168.0 W/m K
density = 3.21 g cm^{-3}
specific heat = 0.161 cal/g.°C
thermal diffusivity = 0.7770 cm^2/s

Matrix Description
Al_2O_3 powder

Matrix Properties
CTE = 8.9×10^{-6} °C^{-1} (20-1327 °C)
thermal conductivity = 36.4 W/m K
density = 3.986 g cm^{-3}
specific heat = 0.230 cal/g.°C
thermal diffusivity = 0.0950 cm^2/s
mean grain size = 1 μm

Test Type
Laser-Flash Technique

Specimen Dimensions
dia. = 12 mm
thickness = 2 mm

Type of Specimen
Disk

Test Description
The thermal diffusivity was measured using a glass-Nd laser with the specimen temperatures measured using a nitrogen-cooled infrared detector. The thermal diffusivity was multiplied by the density and the specific heat to obtain the thermal conductivity. The porosity was calculated from the measured density and the theoretical density.

Test Temperature
RT

Reinforcement Content (weight percent)	Thermal Conductivity (W m^{-1} K^{-1})
Experimental	
0.0	36.0
8.8	35.7
9.0	34.2
26	40.7
26	42.3
30	35.6
44	46.1
44	50.0
65	53.4
87	65.2
Theoretical	
0	36
100	167

Reinforcement Content (weight percent)	Thermal Diffusivity (cm^2 s^{-1})
Composite Theory	
0.0	0.093
20	0.146
50	0.276
70	0.411
100	0.662
Experimental	
0.0	0.0927
8.2	0.0967
26	0.124
46	0.170
66	0.220
88	0.305

Porosity (percent)	Relative Thermal Conductivity	SiC Content (weight percent)
0.289	0.825	8.2
0.527	0.801	8.2
0.663	0.684	25.7
0.986	0.662	25.7
3.12	0.536	30.0
0.45	0.588	44.6
2.27	0.535	44.6
6.79	0.457	65.3
8.77	0.423	87.9

Remarks/Comments

The addition of SiC increased the thermal conductivity through all content values. However, the experimental values were considerably lower than the calculated values. The difference can be attributed to the formation of microcracks due to the thermal expansion mismatch. Increasing porosity significantly increased the extent of microcracking in the composite and resulted in a substantial decrease in thermal diffusivity and conductivity.

SiC(p)/BeO

Reference: 9

Composition
0-100 wt% SiC

Processing Method
Hot-Pressed

Detailed Description
Specimens were received from the supplier, Ceradyne Inc., Santa Ana, CA. The composite designation was Code Ceralloy 2700 Series. The composite powders were mixed and hot-pressed in graphite dies at an appropriate time, temperature and pressure. The hot-pressing schedule was precisely controlled such that samples with a wide range of porosities were obtained.

Reinforcement Description
SiC particulates

Reinforcement Dimensions
grain size < 5 μm

Matrix Description
BeO powder

Matrix Properties
mean grain size = 1 μm

Test Type
Laser-Flash Technique

Test Temperature
300-1400 °C

Test Environment
Air

Temperature (°C)	Thermal Diffusivity ($cm^2 s^{-1}$)	Reinforcement Content (weight percent)
410	0.15	100
460	0.14	100
520	0.13	100
590	0.12	100
630	0.12	100
720	0.11	100
780	0.10	100
810	0.10	100
850	0.10	100
880	0.10	100
910	0.10	100
980	0.09	100
1,000	0.08	100
1,100	0.08	100
1,200	0.07	100
1,200	0.07	100
1,300	0.06	100
1,400	0.06	100
380	0.12	60
480	0.11	60
600	0.10	60
720	0.09	60
800	0.08	60
910	0.07	60
1,000	0.07	60
1,100	0.06	60
1,200	0.05	60
1,400	0.05	60

Temperature (°C)	Thermal Diffusivity (cm² s⁻¹)	Reinforcement Content (weight percent)
410	0.14	0.0
450	0.13	0.0
520	0.11	0.0
550	0.11	0.0
600	0.09	0.0
620	0.09	0.0
650	0.09	0.0
680	0.08	0.0
720	0.08	0.0
780	0.07	0.0
800	0.07	0.0
860	0.06	0.0
910	0.05	0.0
960	0.05	0.0
1,000	0.05	0.0
1,040	0.05	0.0
1,110	0.04	0.0
1,150	0.04	0.0
1,190	0.04	0.0
1,230	0.04	0.0
1,270	0.04	0.0
1,340	0.03	0.0
1,410	0.03	0.0
300	0.10	30
360	0.09	30
420	0.08	30
500	0.07	30
560	0.07	30
590	0.06	30
630	0.06	30
670	0.06	30
720	0.06	30
740	0.05	30
810	0.05	30
870	0.04	30
900	0.04	30
990	0.04	30
1,020	0.04	30
1,090	0.04	30
1,130	0.04	30
1,190	0.03	30
1,280	0.03	30
1,370	0.03	30

Reinforcement Content (weight percent)	Thermal Diffusivity (cm² s⁻¹)	Temperature (°C)
Composite Theory		
0	0.884	20
100	0.659	20
Experimental		
5	0.759	20
10	0.675	20
20	0.45	20
40	0.436	20
50	0.49	20
60	0.411	20
80	0.405	20
0	0.148	400
5	0.124	400
10	0.107	400
20	0.089	400
40	0.088	400
50	0.119	400
60	0.126	400
80	0.103	400
100	0.152	400
0.0	0.097	600
5	0.084	600
10	0.074	600
20	0.06	600
40	0.065	600
50	0.099	600
60	0.1	600
80	0.076	600
100	0.121	600
0	0.068	800
5	0.06	800
10	0.055	800
20	0.045	800
40	0.048	800
50	0.078	800
60	0.081	800
80	0.064	800
100	0.1	800
0	0.05	1000
5	0.046	1000
10	0.043	1000
20	0.04	1000
40	0.041	1000
50	0.059	1000
60	0.063	1000
80	0.051	1000
100	0.085	1000
0	0.043	1200
5	0.041	1200
10	0.039	1200
20	0.034	1200
40	0.034	1200
50	0.049	1200
60	0.053	1200
80	0.043	1200
100	0.073	1200

Porosity (percent)	Relative Thermal Conductivity	BeO Content (weight percent)
0.286	0.860	5
0.886	0.759	10
2.83	0.508	20
0.622	0.647	25
0.771	0.643	25
0.529	0.620	40
0.593	0.599	40
1.392	0.512	40
1.073	0.578	50
0.903	0.492	60

Remarks/Comments

The SiC(p)/BeO composite increased thermal diffusivity with increasing reinforcement content. However, the increase was less than expected due to microcracking caused by residual thermal stresses. The 30 wt% SiC composite had the lowest overall thermal diffusivity, with 60 wt% next and 100 wt% the highest. The monolith started very high then rapidly decreased with temperature to a low value comparable to the 30 wt% SiC composite. Increasing porosity significantly increased the extent of microcracking in the composite and resulted in a substantial decrease in thermal diffusivity and conductivity.

SiC(p)/MgO

Reference: 55

Composition

0-40 wt% SiC

Processing Method

Hot-Pressed

Detailed Description

The composite powders were mixed and hot-pressed in graphite dies at an appropriate time, temperature and pressure. The hot-pressing schedule was precisely controlled such that samples with a wide range of porosities were obtained.

Reinforcement Description

SiC particulates

Reinforcement Typical Properties

The SiC particles were somewhat disc-shaped with a diameter of ~ 6 μm and a thickness of about 1-2 μm.
CTE = 5.1×10^{-6} °C^{-1} (20-1327 °C)
thermal conductivity = 168.0 W/m K
density = 3.21 g cm^{-3}
specific heat = 0.161 cal/g.°C
thermal diffusivity = 0.7770 cm^2/s

Matrix Description

MgO powder

Matrix Properties

CTE = 14.4×10^{-6} °C^{-1} (20-1327 °C)
thermal conductivity = 48.4 W/m K
density = 3.576 g cm^{-3}
specific heat = 0.222 cal/g.°C
thermal diffusivity = 0.1457 cm^2/s
mean grain size = 1 μm

Test Type

Laser-Flash Technique

Specimen Dimensions

dia. = 12 mm
thickness = 2 mm

Type of Specimen

Disk

Test Description

The thermal diffusivity was measured using a glass-Nd laser with the specimen temperatures measured using a nitrogen-cooled infrared detector. The thermal diffusivity was multiplied by the density and the specific heat to obtain the thermal conductivity. The porosity was calculated from the measured density and the theoretical density.

Test Temperature

RT

Reinforcement Content (weight percent)	Thermal Diffusivity ($cm^2 s^{-1}$)
Composite Theory	
0.0	0.145
11	0.173
26	0.224
37	0.272
100	0.660
Experimental	
0.0	0.145
10	0.0917
20	0.0927
30	0.0930
40	0.105

Porosity (percent)	Relative Thermal Conductivity	MgO Content (weight percent)
0.111	0.851	1.5-2.25
0.240	0.873	1.5-2.25
0.468	0.845	1.5-2.25
0.804	0.806	1.5-2.25
0.798	0.787	1.5-2.25
1.06	0.772	1.5-2.25
1.46	0.858	1.5-2.25
1.68	0.768	1.5-2.25
1.19	0.515	10
1.42	0.549	10
1.42	0.572	10
1.82	0.538	10
1.86	0.430	20
2.38	0.437	20
2.39	0.418	20
3.36	0.421	20
3.88	0.412	20
4.89	0.413	20
8.91	0.348	20
2.18	0.397	30
2.84	0.355	30
3.20	0.518	30
4.53	0.338	30
1.41	0.453	40
2.06	0.326	40
2.39	0.338	40
2.60	0.343	40
4.18	0.329	40
4.79	0.326	40

Remarks/Comments:

The SiC(p)/MgO composite increased thermal diffusivity with increasing reinforcement content; however, the increase was less than expected due to microcracking caused by residual thermal stresses. Increasing porosity significantly increased the extent of microcracking in the composite and resulted in a substantial decrease in diffusivity.

Figure 126: Thermal Diffusivity of SiC(p)/MgO and Al_2O_3

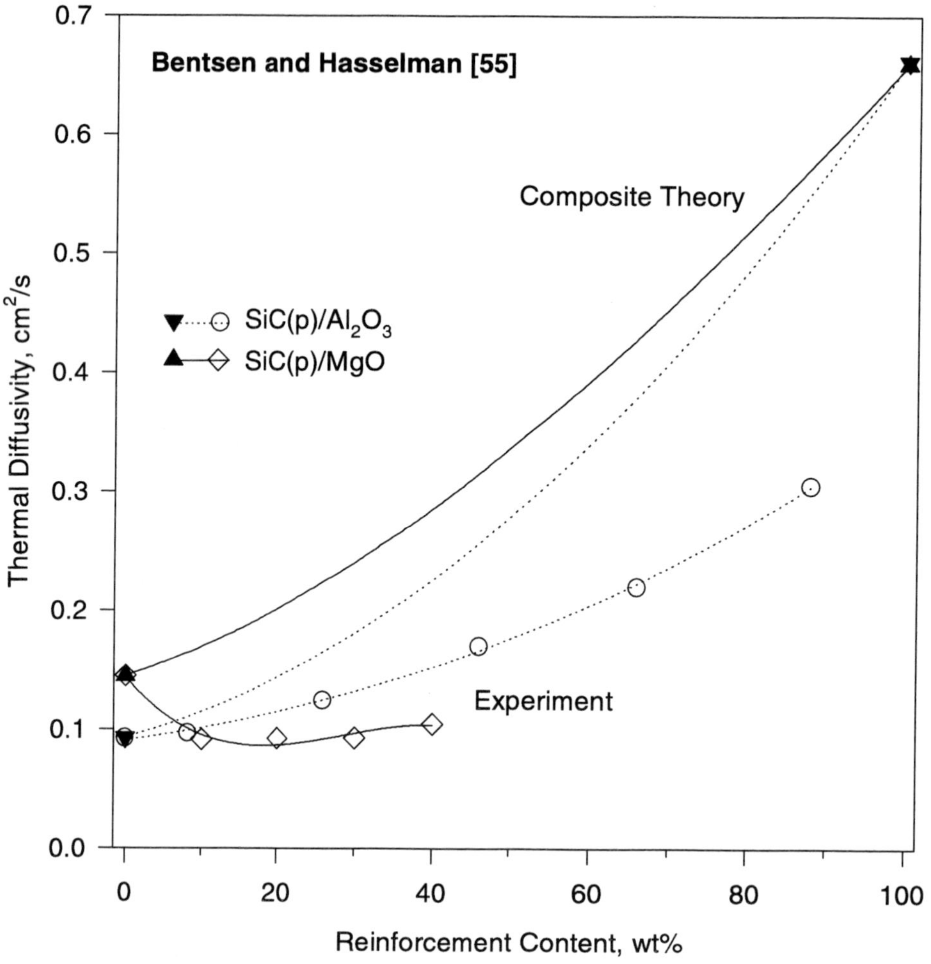

Discussion: Figures 126 and 127 present thermal diffusivity data for silicon carbide reinforced aluminum, magnesium and beryllium oxides with increasing SiC content. In each of these composite systems, the thermal expansion of the oxide phase is greater than the thermal expansion of the SiC reinforcement. This difference leads to significant thermal stresses and possible matrix microcracking when the composite is cooled from processing temperatures. Singh *et al.* [9] and Bentsen and Hasselman [55] examined the effect of microcracking on the heat transfer properties of these three composite systems.

Figures 126 shows the thermal diffusivity of SiC-particulate reinforced Al_2O_3 and MgO. According to the composite theory, the diffusivity of both the Al_2O_3 and MgO matrix composite should increase substantially with increasing SiC content. Yet, experiment showed that the increase was much less than expected from the calculated values. The difference was attributable to the formation of microcracks [55].

Figure 127: Thermal Diffusivity of SiC(p)/BeO

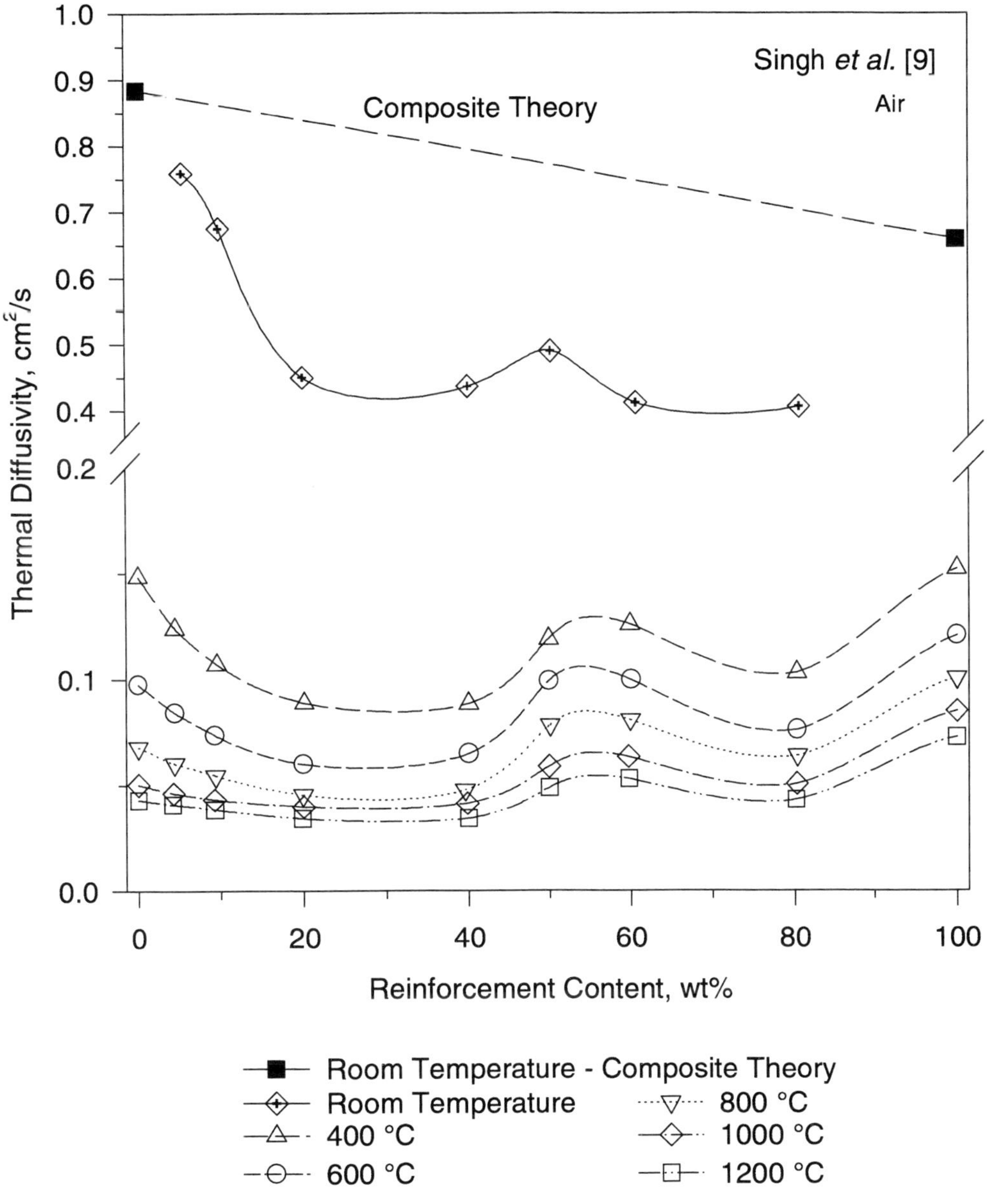

Discussion: Figure 127 shows the thermal diffusivity of SiC reinforced BeO from room temperature to 1200 °C. Like the Al_2O_3 and MgO matrix composites, the thermal diffusivity of the SiC(p)/BeO fell well below that of the calculated theoretical values, which was also due to matrix microcracking. At around 50 wt% SiC content, the composite showed a slight increase in diffusivity which corresponded with a decrease in porosity. This led researchers to further explore the relationship between porosity and thermal diffusivity. The results of this research are presented in figure 130. At higher temperatures, the thermal diffusivity decreased substantially, while the general shape of the curve with increasing SiC content remained the same.

Figure 128: Thermal Diffusivity vs. Temperature of SiC(p)/BeO

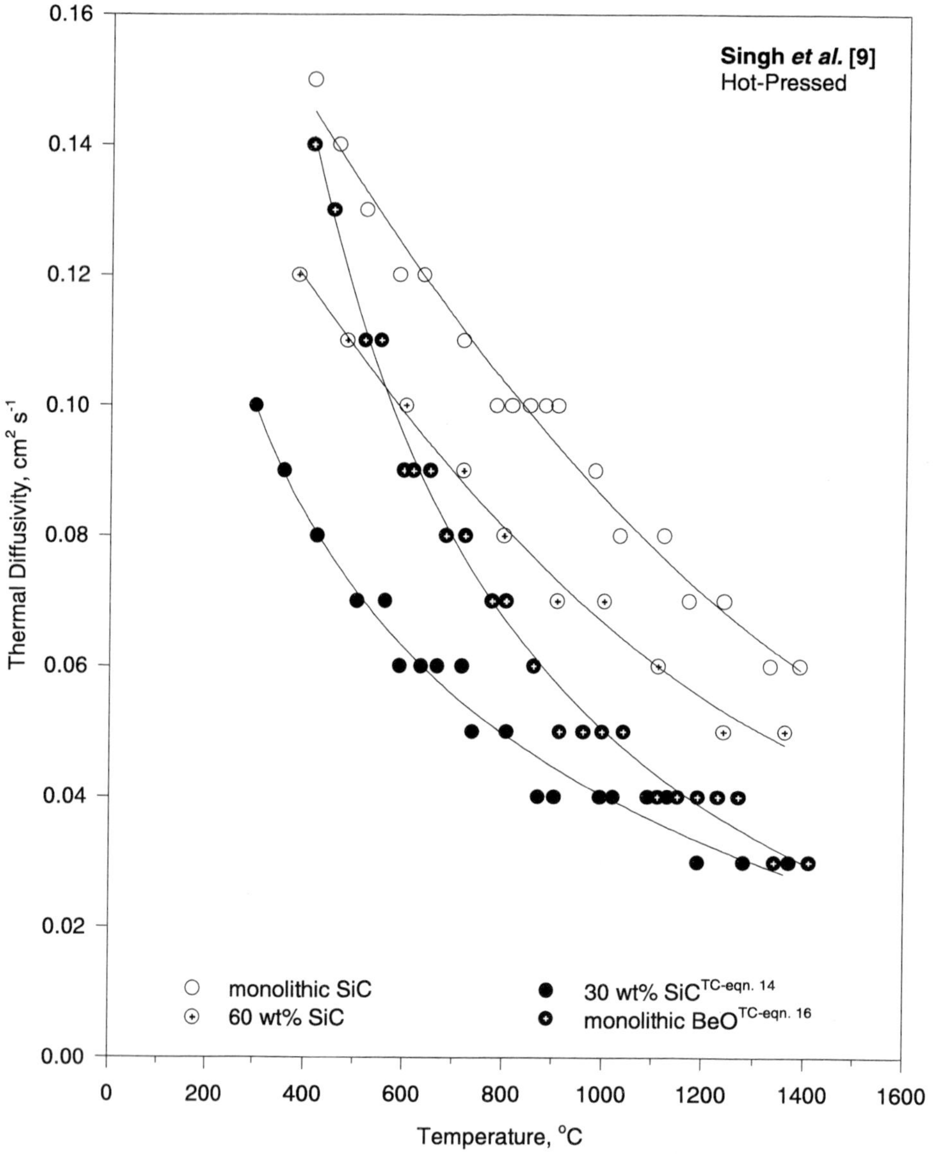

Discussion: The thermal diffusivity of the SiC(p)/BeO composite, monolithic SiC, and monolithic BeO decreased as the temperature increased. The data was obtained by heating the composites slowly from room temperature to 1400 °C. Upon cooling, a hysterisis effect occurred [9]. The monolithic SiC had the highest thermal diffusivity at all test temperatures. The two composites had lower thermal diffusivities at room temperature because of microcracking caused by thermal expansion mismatch stresses. At lower temperatures, the monolithic BeO showed a relatively high thermal diffusivity, similar to that of monolithic SiC. However, the thermal diffusivity of BeO was much more dependent on temperature and showed the greatest drop in diffusivity of any of samples tested [9]. The diffusivity of the monolithic beryllia fell below that of the microcracked 60 wt% SiC/BeO composite because of the low diffusivity of BeO at high temperatures.

Figure 129: Thermal Conductivity of SiC(p)/Al_2O_3

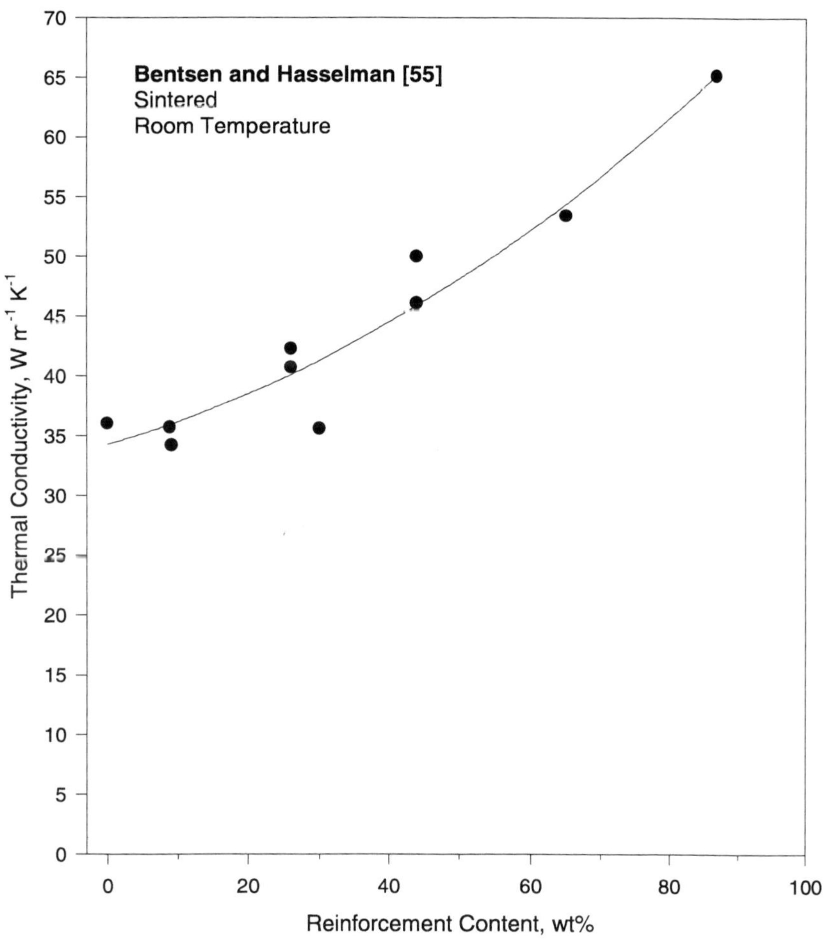

Discussion: The addition of the SiC particulates increased the thermal conductivity due the higher thermal conductivity of SiC compared to the alumina matrix. The thermal conductivity of silicon carbide, 168 W/mK, was much greater than 36.4 W/mK for alumina. The increase in conductivity corresponds to the increase in thermal diffusivity shown in figure 126. However, like the diffusivity, the experimental values for thermal conductivity were considerably lower than the theoretical values due to the formation of microcracks resulting from thermal expansion mismatch stresses.

Figure 130: Porosity Dependence of the Thermal Conductivity at Room Temperature of SiC Particulate Reinforced Oxides

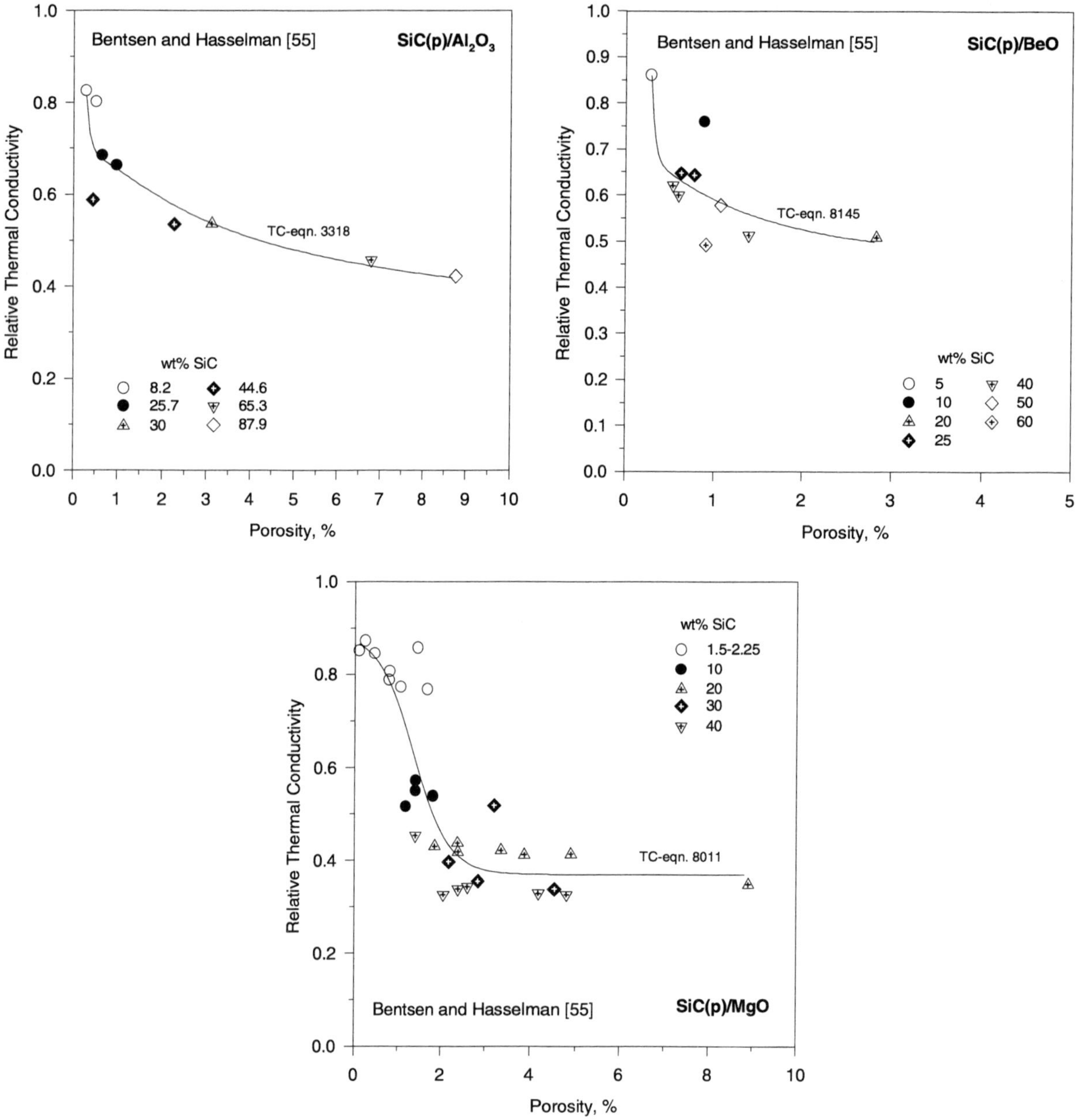

Discussion: Dividing the experimental value by the calculated theoretical value for any given silicon carbide content yields the relative thermal conductivity. Analysis of the relative thermal conductivity eliminates the reinforcement content as a variable in order to reveal the true dependence of thermal conductivity on microcracking. When relative thermal conductivity was plotted against porosity for all three oxides composites, it was found that the conductivity dropped substantially with only a small amount of porosity [55]. A pore content of only 2 % lowered the thermal conductivity by about 50%, which was far greater than about a 3 % decrease theoretically predicted by the Maxwell[1] equations for spherical pores. The strong dependence of thermal conductivity on porosity was thought to be the result of microcrack nucleation by the pores within the composite. High residual thermal stresses do not necessarily result in microcrack formation, since a crack initiation site of sufficient size to produce stress intensity equal to or greater than the critical stress intensity factor of the composite is also required. Pores at triple points or at grain boundaries can act as initiation sites. Therefore, a small amount of porosity can lead to extensive microcracking of the composite. The data presented in figure 130 indirectly shows the relationship between thermal conductivity and microcracking. Thermal conductivity dropped substantially with pore content, especially in the low porosity regime, because the pores initiated the extensive microcracking observed in the composites.[2]

---

1. J. C. Maxwell, A treatise on Electricity and Magnetism, 1, 3rd ed., Oxford University Press (1904)
2. D. P. H. Hasselman and K. Y. Donaldson, "Thermal Conductivity of Whisker- and Particulate-Reinforced Ceramic Matrix Composites," *Handbook on Discontinuously Reinforced Ceramic Matrix Composites*, (in press), 357 - 406

$SiC(p).ZrO_2(p)$/Mullite

Reference: 86

Composition
20 vol% SiC(p) and
20 vol% 2.8Y-ZrO_2

Composite Architecture
particulate reinforcement

Processing Method
Hot-Pressed

Detailed Description
The mullite, SiC, and Y-TZP powders were mixed with the desired volume ratio by wet milling with distilled water using Al_2O_3 balls in a polyethylene container for 6 hours, and dried. The obtained mixed powders were sieved to under 120 mesh, and were molded in a BN-coated graphite die for sintering. Hot-pressing was carried out at 1600 °C for 40 minutes at nitrogen atmosphere under 25 MPa for all composites.

Reinforcement Description
α-SiC and 2.8Y-ZrO_2

Reinforcement Typical Properties
SiC composition:
total C = 27.0 wt%; free C = 0.41 wt%
total Si = 68.5 wt%; free Si = 1.08 wt%
O = 2.15 wt%; Al_2O_3 = 0.13 wt%
Fe_2O_3 = 0.18 wt%; CaO = 0.02 wt%
Y-TZP composition: 2.8 mol% Y_2O_3

Reinforcement Dimensions
SiC ave. dia. = 0.93 μm
SiC specific surface = 7.53 $m^2.g^{-1}$

Matrix Description
mullite powder

Matrix Properties
ave. grain size = 2 μm
composition:
Al_2O_3 = 75.92 wt%
SiO_2 = 23.34 wt%
other oxides = 0.74 wt%

Test Type
Thermal Expansion Test

Specimen Preparation
Samples were cut and ground by a diamond wheel.

Specimen Dimensions
2.5 x 5 x 28 mm
span = 20 mm

Type of Specimen
Bar

Test Description
crosshead speed = 0.5 mm min^{-1}

Test Temperature
20-1100 °C

Test Environment
Air

Temperature (°C)	Mean Thermal Expansion Coefficient (10^{-6} K^{-1})
20	-
100	4.64
212	4.94
317	5.16
402	5.17
497	4.97
602	4.74
717	4.64
807	4.76
902	5.00
997	5.36
1,100	5.78

Remarks/Comments
From room temperature to 1273 K (1000 °C), the mean CTE of the composite was 5.35 $x10^{-6}$/°C, similar to that of monolithic mullite. At 673-973 K (400-700 °C), the thermal expansion coefficient decreased slightly with increasing temperature.

Figure 131: Temperature Dependence of the Mean Thermal Expansion Coefficient of SiC(p)/ZrO_2·Mullite

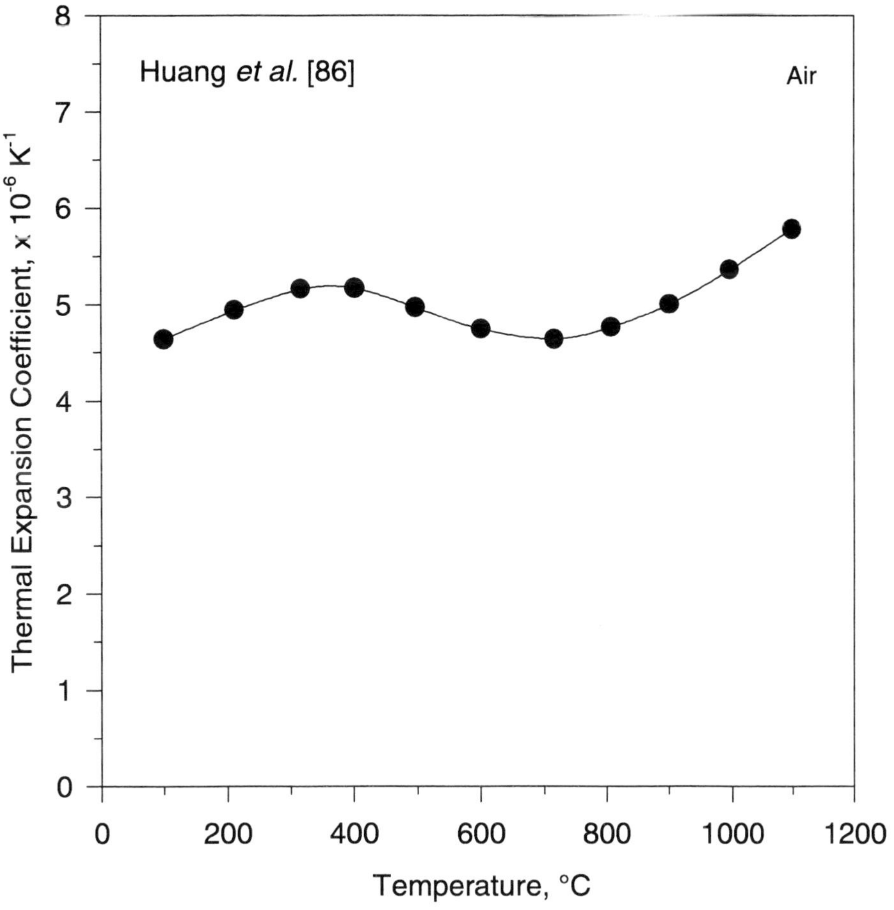

Discussion: The average CTE of SiC(p)/ZrO_2·mullite from room temperature to 1000 °C was 5.35x10^{-6} K^{-1}, similar to monolithic alumina. The coefficient of thermal expansion showed a slight drop from 400 °C to 700 °C. The drop in CTE in this temperature range was caused by the transformation of monoclinic zirconia to tetragonal zirconia, which involves a decrease in volume [86].

Al_2O_3(p)/Aluminum Titanate

Reference: 97

Composition
7.6 wt% Al_2O_3, 3.4 wt% TiO_2

Processing Method
DIMOX

Detailed Description
The aluminum alloy was blended with a stoichiometric ratio of Al_2O_3 and TiO_2 in a proprietary mixture incorporating stabilizers. The mixture was then reacted in air. The final composition was 72 wt% aluminum titanate, 7.6 wt% alumina, 3.4 wt% TiO_2, and the remainder consisting of the various stabilizers. The composite contain no residual metal phase.

Reinforcement Description
TiO_2 and Al_2O_3 powders

Matrix Description
aluminum alloy

Matrix Supplier
Du Pont Lanxide Composite Inc.

Matrix Designation
NX-1010

Test Type
Laser-Flash Technique

Specimen Dimensions
8 x 8 mm
thickness = 2 mm

Type of Specimen
square plate

Test Description
The thermal diffusivity was measured using a glass-Nd laser with the specimen temperature measured using a nitrogen-cooled infrared detector. From room temperature to about 200 °C the measurements were made in air in a specially constructed specimen holder. Above 300 °C, the specimen was positioned in a carbon-resistance furnace with nitrogen protective atmosphere. Transient temperature measurements were made by a InSb liquid-nitrogen-cooled detector. Data were obtained at approximately 100 °C intervals on both heating and cooling. During thermal cycling, the composites were held at the maximum temperature for 10 minutes.

Test Temperature
RT - 1200 °C

Test Environment
RT-300: Air
300-1200 °C: Nitrogen

Test Equipment
Nd-glass laser, Laser Applications Inc., Orlando, FL

Number of Thermal Cycles	Thermal Diffusivity ($cm^2 s^{-1}$)
0	0.01005
1	0.00714
2	0.00741
3	0.00818
4	0.00881
5	0.0087
6	0.0087
7	0.00867

Remarks/Comments
Density of the composites was around 2.6 g/cm^3. The composites were approximately 30 % porous.

Temperature-*Heating* (°C)	Thermal Diffusivity (cm²/s)
20	0.00933
133	0.00696
223	0.00624
330	0.00575
416	0.00541
516	0.00521
609	0.00518
20	0.00928
110	0.00704
214	0.00628
327	0.00595
408	0.00574
509	0.00562
604	0.00571
713	0.00616
802	0.00653
20	0.00947
123	0.00662
196	0.00625
308	0.00578
415	0.00533
501	0.00516
602	0.00583
687	0.00651
788	0.00669
895	0.00674
988	0.00689
20	0.00916
127	0.00652
211	0.00587
336	0.00551
416	0.00526
509	0.00547
606	0.00619
705	0.00643
804	0.0065
906	0.00625
1010	0.00614
1122	0.00605
20	0.00938
121	0.00694
188	0.00629
301	0.00596
416	0.00557
504	0.00542
596	0.0054
695	0.00579
793	0.006
902	0.00607
989	0.00613
1083	0.00625
1202	0.00621

Temperature-*Cooling* (°C)	Thermal Diffusivity (cm²/s)
533	0.00521
429	0.00549
354	0.00568
272	0.00618
20	0.00918
682	0.00688
564	0.00619
482	0.00573
363	0.0059
256	0.00596
20	0.00913
838	0.00702
743	0.00715
581	0.00722
506	0.00722
399	0.00732
298	0.00782
20	0.0096
1016	0.00634
916	0.00692
843	0.00774
726	0.00806
601	0.00834
486	0.00842
417	0.00863
379	0.00853
329	0.00862
20	0.00843
1021	0.00667
775	0.00719
638	0.00779
506	0.00815
399	0.00878
307	0.00878
274	0.00886
20	0.00814

Al_2O_3(p)/Aluminum Titanate

Reference: 97

Composition
7.6 wt% Al_2O_3, 3.4 wt% TiO_2

Processing Method
DIMOX

Detailed Description
The aluminum alloy was blended with a stoichiometric ratio of Al_2O_3 and TiO_2 in a proprietary mixture incorporating stabilizers. The mixture was then reacted in air. The final composition was 72 wt% aluminum titanate, 7.6 wt% alumina, 3.4 wt% TiO_2, and the remainder consisting of the various stabilizers. The composite contain no residual metal phase.

Reinforcement Description
TiO_2 and Al_2O_3 powders

Matrix Description
aluminum alloy

Matrix Supplier
Du Pont Lanxide Composite Inc.

Matrix Designation
NX-1010

Test Type
Differential Scanning Calorimetry

Specimen Dimensions
8 x 8 mm

thickness = 2 mm

Type of Specimen
square plate

Test Temperature
RT - 500 °C

Test Environment
Air

Temperature (°C)	Specific Heat ($J g^{-1} K^{-1}$)
25	0.698
50	0.785
100	0.835
150	0.880
200	0.918
250	0.945
300	0.969
350	0.966
400	0.974
450	1.001
500	1.032

Remarks/Comments
Density of the composites were around 2.6 g/cm^3. The composites were approximately 30 % porous. The specific heat increased with increasing temperature.

Figure 132: Thermal Expansion Behavior of Alumina-Aluminum Titanate Heated To and Cooled From 650 °C and 1250 °C

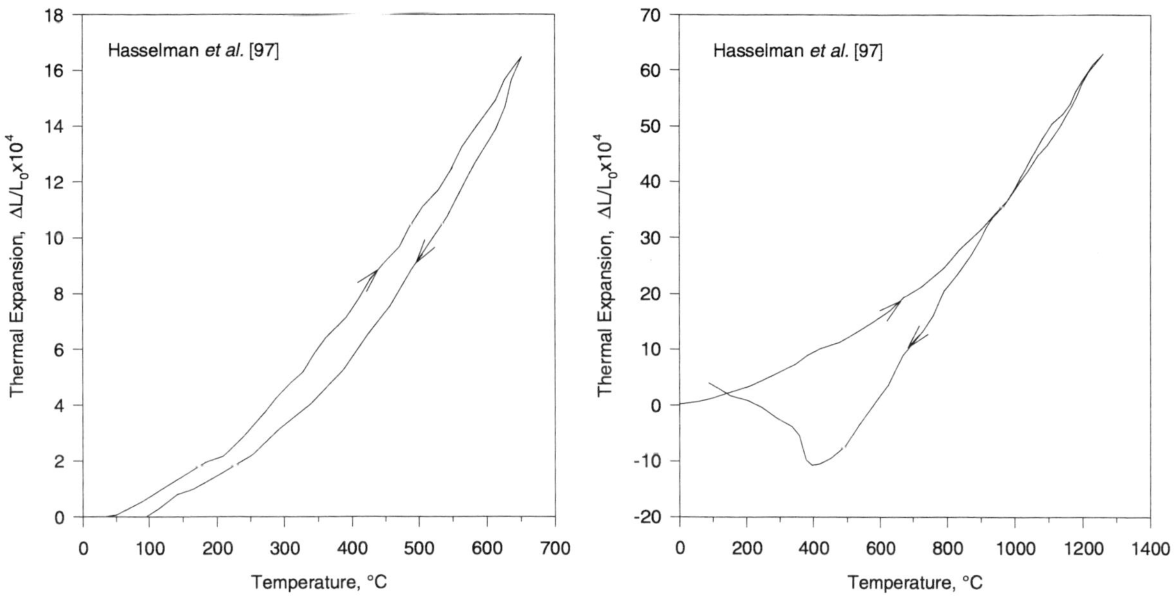

Discussion: Figure 132 shows the thermal expansion of an alumina-aluminum titanate composite heated to and cooled from 650 °C and another heated to and cooled from 1250 °C. The thermal expansion results were consistent with the microcracking theory presented to explain the thermal diffusivity behavior, figure 133. Upon thermal cycling to 650 °C, the composite showed little hysteresis in the thermal expansion, suggesting that there was no change in the microcrack density. However, there was considerable hysteresis upon cycling to 1250 °C, which showed that there was a change in the microcrack density. Above 1000 °C, the heating and cooling curves converge, suggesting that the healing process was complete at 1000 °C. From this data, it can be concluded that microcrack healing occurred during the heating cycle between 600 and 1150 °C and additional microcrack formation occurred during the cooling cycle from 1000 °C down to room temperature. At the end of the 1250 °C cycle, the specimen showed an increase in length compared to the as-processed sample, this was consistent with the conclusion that additional microcracking occurred above that which was already present in the as-processed sample [97].

Figure 133: Thermal Diffusivity of DIMOX™ Alumina-Aluminum Titanate Composites Heated to and Cooled from A) 600 °C, B) 800 °C, C) 1000 °C, D) 1100 °C, E) 1200 °C Hasselman *et al.* (97)

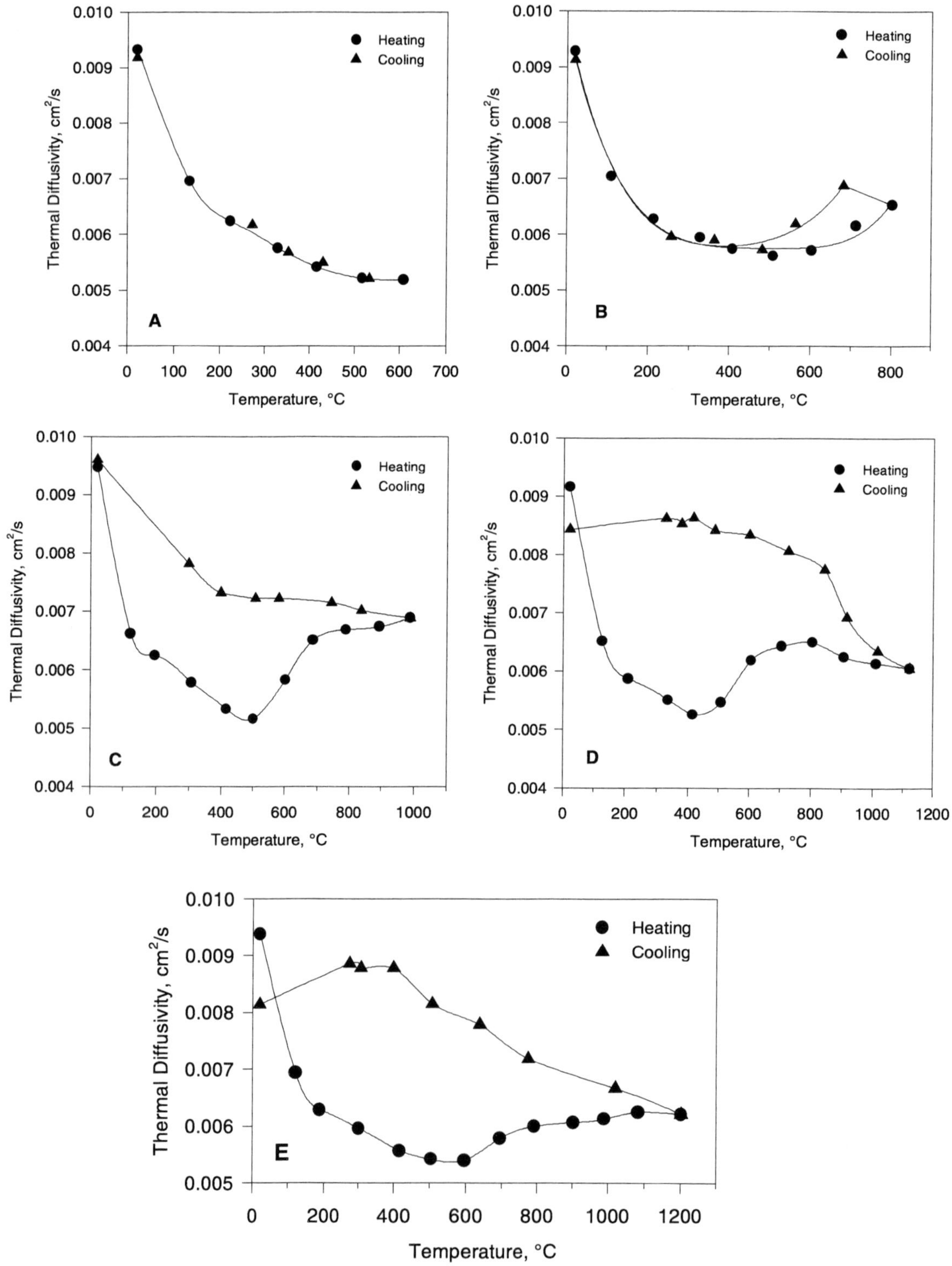

Discussion: Alumina-aluminum titanate shows the characteristic of microcracking when cooled from processing temperatures or exposed to a thermal gradient. This microcracking ability should give the composite decent thermal shock resistance despite low thermal conductivity. For these reasons, alumina-aluminum titanate shows promise as candidate materials for service conditions involving severe thermal shock coupled with the requirement of high thermal insulating ability, particularly in low stress applications. Hasselman *et al.* [97] examined the effect of thermal history on DIMOX™ processed, highly porous, alumina-aluminum titanate composites.

Figures 133a-e show the effect of heating and cooling an alumina-aluminum titanate composite in the temperature range of RT to 600, 800, 1000, 1100 and 1200 °C. It was expected, if microcracking were present, that the composites would show some hysteresis effect with heating and cooling as microcracks were healed and as new microcracks were formed. Figures 133a shows that there was no hysteresis effect upon heating to and cooling from 600 °C. This indicated that there was not a significant change in microcrack morphology. Heating to 800 °C showed some hysteresis. At 1000 °C, significant hysteresis was observed. This suggested significant crack healing at the higher temperatures of the thermal cycle; requiring cooling over a significant temperature range before the microcracks were reinitiated or repropagated [97]. Increasing the temperature range increased the hysteresis effect, as shown in figures 133d-e. It was noteworthy that the room temperature thermal diffusivity was lower after the thermal cycle than before, and heating to 1200 °C caused a greater drop than heating to 1100 °C. This suggested that further microcrack formation occurred during the cooling stage of the thermal cycle in addition to the microcracks that developed during processing, thus the thermal diffusivity decreased after thermal cycling. Figure 134 shows the effect of successive thermal cycling to 1200 °C on the thermal diffusivity of an alumina-aluminum titanate composite. The composite showed a significant drop in diffusivity after the first cycle, but increased with the second to the fourth cycle. This demonstrates the material's capability for self healing during multiple thermal cycling. After the fourth cycle, the process of healing and subsequent reformation of microcracks reached a steady-state, and the thermal diffusivity did not change with further thermal cycling.

Figure 134: Thermal Diffusivity of a DIMOX™ Alumina-Aluminum Titanate Composite After Thermal Cycling from RT to 1200 °C

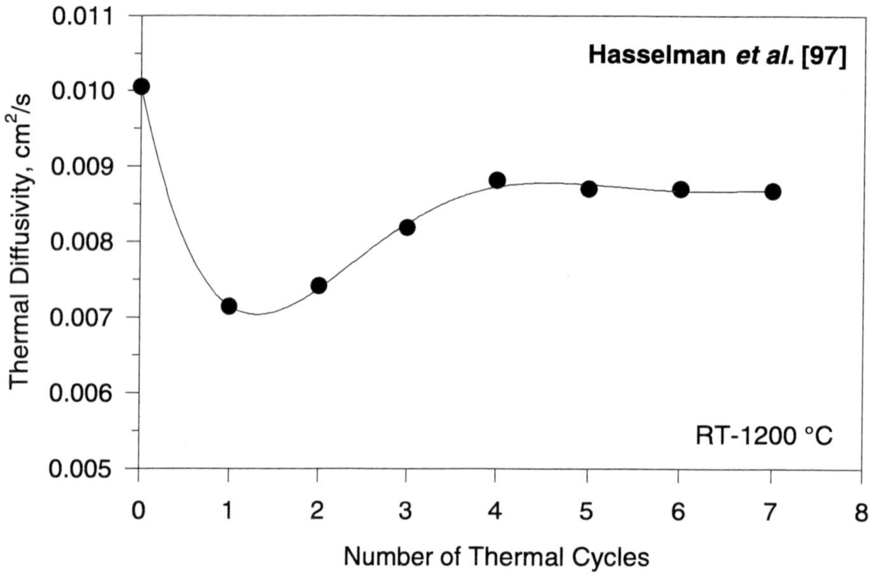

Figure 135: Specific Heat of DIMOX™ Alumina-Aluminum Titanate Composites

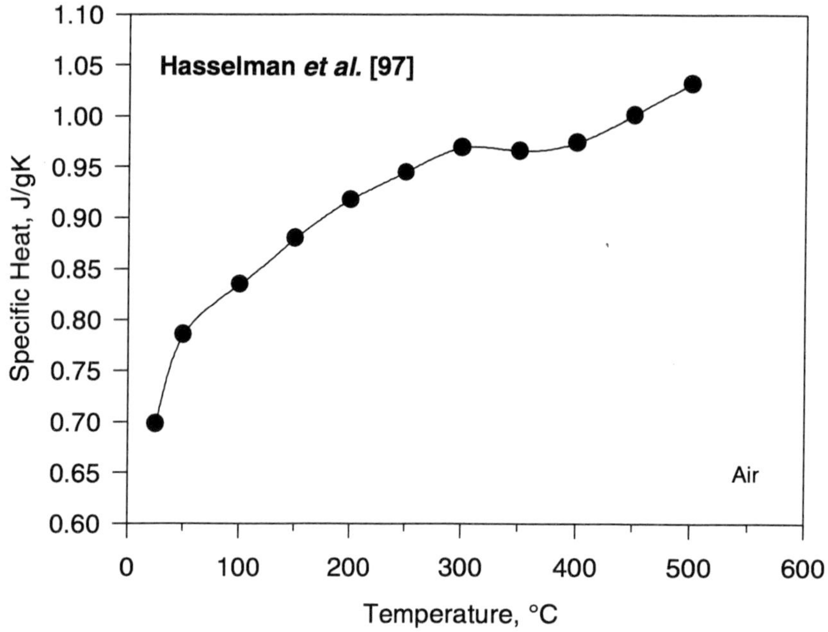

Discussion: The specific heat of the DIMOX™ alumina-aluminum titanate composite increased with increasing temperature [97]. Thermal conductivity of alumina-aluminum titanate composites showed the same trend as thermal diffusivity. However, the increasing specific heat caused the conductivity to show a stronger negative trend with increasing temperature than the thermal diffusivity.

Diamond(p)/Al_2O_3

Reference: 59

Composition
0-15 vol% diamond

Processing Method
Hot-Pressed

Detailed Description
The diamond powder was mixed with α-alumina and ball milled in methanol for 2 hours using alumina balls in a plastic jar. The resulting slurries were oven dried. The granulated powders were hot-pressed in a boron-nitride coated graphite die at 32 MPa and 1250 °C under an argon atmosphere.

Reinforcement Description
diamond powder

Reinforcement Supplier
Smith Mega Diamond, Inc., Provo, UT

Supplier Designation
S-diamond

Reinforcement Dimensions
0.5-1.0 μm
ave. particle size = 0.48 μm

Matrix Description
α-Al_2O_3 powder

Matrix Supplier
ALCOA Industrial Chemical, Bauxite, AR

Matrix Designation
A16SG

Matrix Properties
ave. particle size = 0.83 μm

Test Type
Dilatometry

Test Temperature
100-800 °C

Test Equipment
Orton 15BC-1 dilatometer

Reinforcement Content (volume percent)	Thermal Expansion Coefficient (10^{-6} K^{-1})	Relative Density (percent)
0.0	7.6	99.3
5.0	6.0	96.1
10	5.6	94.3
15	5.4	93.9

Remarks/Comments
The coefficient of thermal expansion decreased with the addition of diamond powder.

TiN(p)/Al_2O_3

Reference: 84

Composition
0, 30 vol% TiN

Processing Method
Hot-Pressed

Detailed Description
Al_2O_3 powders were homogenized with 30 vol% TiN in plastic jars with alumina balls and isobuthyl alcohol. The composites were hot-pressed in a vacuum at 1600-1650 °C and 30 MPa.

Reinforcement Description
TiN particulates

Matrix Description
Al_2O_3

Test Type
Dilatometry

Test Description
heating rate = 5 °C/min

Test Temperature
20-1000 °C

Reinforcement Content (volume percent)	Thermal Expansion Coefficient (10^{-6} K^{-1})	Relative Density (percent)
0.0	8.01	100
30	8.11	99.3

Remarks/Comments
The thermal expansion coefficient of the composite was related to the volume fraction of each phase; the dilatometric curves had no hysteresis which suggested the absence of microcracks in the sintered body.

TiB_2(p)/Al_2O_3

Reference: 84

Composition
0, 30 vol% TiB_2

Processing Method
Hot-Pressed

Detailed Description
Al_2O_3 powders were homogenized with 30 vol% TiB_2 in plastic jars with alumina balls and isobuthyl alcohol. The composites were hot-pressed in a vacuum at 1600-1650 °C and 30 MPa.

Reinforcement Description
TiB_2 particulates

Matrix Description
Al_2O_3

Test Type
Dilatometry

Test Description
heating rate = 5 °C/min

Test Temperature
20-1000 °C

Reinforcement Content (volume percent)	Thermal Expansion Coefficient (10^{-6} K^{-1})	Grain Size (μm)	Relative Density (percent)
0.0	8.01	-	100
30	7.82	1.3	98.6

Remarks/Comments
The thermal expansion coefficient of the composite was related to the volume fraction of each phase; the dilatometric curves had no hysteresis which suggested the absence of microcracks in the sintered body.

TiC(p)/Al_2O_3

Reference: 84

Composition
0, 30 vol% TiC

Processing Method
Hot-Pressed

Detailed Description
Al_2O_3 powders were homogenized with 30 vol% TiC in plastic jars with alumina balls and isobuthyl alcohol. The composites were hot-pressed in a vacuum at 1600-1650 °C and 30 MPa.

Reinforcement Description
TiC particulates

Matrix Description
Al_2O_3

Test Type
Dilatometry

Test Description
heating rate = 5 °C/min

Test Temperature
20-1000 °C

Reinforcement Content (volume percent)	Thermal Expansion Coefficient (10^{-6} K^{-1})	Relative Density (percent)
0.0	8.01	100
30	7.94	99.4

Remarks/Comments
The thermal expansion coefficient of the composite was related to the volume fraction of each phase; the dilatometric curves had no hysteresis which suggested the absence of microcracks in the sintered body.

Figure 136: Mean Thermal Expansion Coefficients of Diamond, TiN, TiB_2, and TiC Particulate Reinforced Alumina

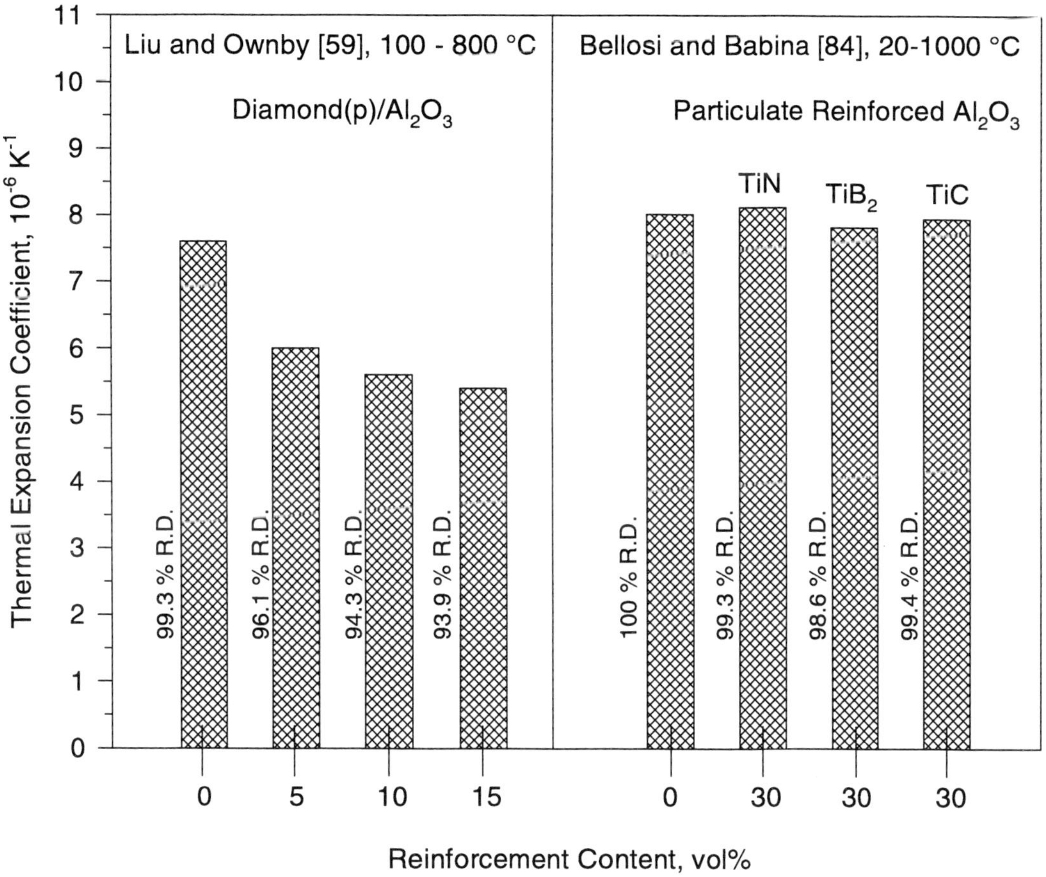

Discussion: Diamond has extremely high thermal conductivity and very low thermal expansion. Thus, it is expected that the addition of diamond particulates to a ceramic matrix could significantly improve thermal shock resistance. Liu and Ownby [59] measured the CTE of alumina composites with diamond particulate reinforcement. As expected, the CTE decreased with increasing diamond content. Some of the decrease in CTE could also be attributed to increasing porosity. Bellosi and Babina [84] measured the CTE of alumina composites with TiB_2, TiN, and TiC reinforcements. The addition of these electrically conductive ceramic particulates did not significantly affect thermal expansion.

$ZrO_2.HfO_2(p)/Al_2O_3.Cr_2O_3$

Reference: 13

Composition
0 and 15 vol% ZrO_2

Processing Method
Hot-Pressed

Detailed Description
The salts $Al(NO_3)_3.9H_2O$, $Cr(NO_3)_3.9H_2O$, $ZrOCl_2.3H_2O$, and $HfOCl_2.8H_2O$ were dissolved in water and were added slowly to an NH_4OH solution. The mixtures were kept at 70 °C for 30 minutes while stirring. The precipitates were then filtered, washed with dilute ammonia solution, and dried at 110 °C for 24 hours. The dried precipitates were calcined at 500 °C for 2 hours and ground for 10 hours in a ball mill. The co-precipitated powders were then hot-pressed to full density at 1500 °C for 30 minutes in BN-coated graphite dies under a pressure of 30 MPa.

Reinforcement Description
$ZrO_2.HfO_2$ particulates

Matrix Description
$Al_2O_3.Cr_2O_3$
(Cr_2O_3-0, 20, 50 mol%)

Test Type
Laser-Flash Technique

Test Description
Thermal diffusivity of specimens measured by the laser flash method and literature data on specific heat was used to calculate thermal conductivities.

Test Temperature
RT-1400 °C

Test Environment
RT: Air
>300 °C: Nitrogen

Temperature (°C)	Thermal Conductivity ($W m^{-1} K^{-1}$)	Comments
22	12.5	*monolithic Al_2O_3*
230	10.6	
470	9.10	
700	8.20	
920	7.50	
1,200	6.90	
22	10.6	*15 vol% ZrO_2*
180	9.10	
340	7.90	
560	6.60	
790	5.90	
1,100	5.40	
1,400	5.10	
22	9.60	*20 mol% Cr_2O_3, 15 vol% ZrO_2 fine grains*
160	8.10	
320	6.70	
610	5.40	
870	4.70	
1,200	4.30	
1,400	4.20	
22	6.70	*20 mol% Cr_2O_3, 15 vol% ZrO_2 coarse grains*
160	5.60	
350	4.50	
610	3.60	
850	3.10	
1,200	2.80	
1,400	2.70	
22	3.20	*50 mol% Cr_2O_3, 15 vol% ZrO_2 coarse grains*
180	2.30	
330	1.60	
470	1.10	
610	0.70	

Remarks/Comments

Thermal conductivity decreased with increasing temperature. The chromia was added to the composite to decrease thermal conductivity of the alumina matrix.

SiC(p)/ZrO_2

Reference: 85

Composition

15 wt% SiC

Processing Method

Sol-Gel/Hot-Pressed

Detailed Description

ZrO_2Cl_2-$8H_2O$, $Y(NO_3)_3$-$6H_2O$ and α–SiC were used as starting materials. Y-TZP and SiC(p)/Y-TZP were prepared by the coprecipitation method from the starting solution. The mixture was hot-pressed in a BN-coated graphite mold at 1600 °C and 25 MPa for 1 hour.

Reinforcement Description

α-SiC particulates

Matrix Description

2.8Y-ZrO_2

Test Type

Thermal Conductivity Test

Specimen Dimensions

2.5 x 5 x 26 mm

Type of Specimen

Bar

Test Temperature

330-1200 °C

Test Environment

Air

Temperature (°C)	Thermal Conductivity ($W m^{-1} K^{-1}$)
330	4.62
440	4.40
680	4.03
990	3.68
1,110	3.61
1,200	3.51

Remarks/Comments

Thermal conductivity decreased with increasing temperature.

$SiC(p).ZrO_2(p)$/Mullite

Reference: 86

Composition
20 vol% SiC(p) and 20 vol% 2.8Y-ZrO_2

Processing Method
Hot-Pressed

Detailed Description
The mullite, SiC, and Y-TZP powders were mixed with the desired volume ratio by wet milling with distilled water using Al_2O_3 balls in a polyethylene container for 6 hours, and dried. The obtained mixed powders were sieved to under 120 mesh, and were molded in a BN-coated graphite die for sintering. Hot-pressing was carried out at 1600 °C for 40 minutes at nitrogen atmosphere under 25 MPa for all composites.

Reinforcement Description
α-SiC and 2.8Y-ZrO_2 particulates

Reinforcement Typical Properties
SiC composition: total C = 27.0 wt%
free C = 0.41 wt%
total Si = 68.5 wt%
free Si = 1.08 wt%
O = 2.15 wt%
Al_2O_3 = 0.13 wt%
Fe_2O_3 = 0.18 wt%
CaO = 0.02 wt%

Reinforcement Dimensions
SiC ave. dia. = 0.93 μm
SiC specific surface = 7.53 $m^2.g^{-1}$

Matrix Description
Mullite powder

Matrix Properties
ave. grain size = 2 μm
composition: Al_2O_3 = 75.92 wt%
SiO_2 = 23.34 wt%
other oxides = 0.74 wt%

Test Type
Thermal Conductivity Test

Specimen Preparation
Specimens were cut and ground by a diamond wheel.

Specimen Dimensions
2.5 x 5 x 28 mm

Type of Specimen
Bar

Test Temperature
350-1200 °C

Test Environment
Air

Temperature (°C)	Thermal Conductivity ($W\,m^{-1}\,K^{-1}$)
350	6.69
450	6.61
640	5.98
750	5.86
950	5.31
1,100	5.10
1,200	4.81

Remarks/Comments
The thermal conductivity of the composite was relatively low.

$SiC(p)/Al_2O_3$

Reference: 112

Composition
~55 vol% SiC

Processing Method
Slip-Cast/DIMOX™

Detailed Description
Commercially available $SiC(p)/Al_2O_3$ (Du Pont Lanxide Composite Inc., USA) was obtained. This material was prepared using a DIMOX™ method. An SiC preform was slip cast and then an Al alloy was infiltrated and oxidized in the furnace, resulting in the composite.

Reinforcement Description
SiC particulates

Reinforcement Properties
ave. particle diameter = 5-20 μm

Matrix Description
aluminum powder

Test Type
Thermal Conductivity Test

Test Temperature
25-1550 °C

Test Environment
Air

Temperature (°C)	Thermal Conductivity ($W\ m^{-1}\ K^{-1}$)
25	65
1000	20
1550	5.5

ZrO_2(p)/Al_2O_3.Cr_2O_3

Reference: 114

Composition
0 and 15 vol% ZrO_2

Processing Method
Hot-Pressed

Detailed Description
The salts $Al(NO_3)_3.9H_2O$, $Cr(NO_3)_3.9H_2O$, $ZrOCl_2.3H_2O$, and $HfOCl_2.8H_2O$ were dissolved in water and were added slowly to an NH_4OH solution. The mixtures were kept at 70 °C for 30 minutes while stirring. The precipitates were then filtered, washed with dilute ammonia solution, and dried at 110 °C for 24 hours. The dried precipitates were calcined at 500 °C for 2 hours and ground for 10 hours in a ball mill. The co-precipitated powders were then hot-pressed to full density at 1500 °C for 30 minutes in BN-coated graphite dies under a pressure of 30 MPa.

Reinforcement Description
$ZrO_2.HfO_2$ particulates
(HfO_2 content was not specified)

Matrix Description
$Al_2O_3.Cr_2O_3$
(Cr_2O_3-0, 20, 50 mol%)

Test Type
Laser-Flash Technique

Specimen Preparation
Specimens were cut and ground by a diamond wheel.

Specimen Dimensions
8 x 8 x 2 mm

Type of Specimen
Bar

Test Description
Elevated temperature specimens were placed in a holder located in the center of a carbon resistance furnace with nitrogen atmosphere. Thermal diffusivity was determined during heating and cooling of the specimens. Thermal conductivity was calculated from the thermal diffusivity.

Test Temperature
RT-1400 °C

Test Environment
RT: Air
>300 °C: Nitrogen

CrO_2 Content (mol%)	Thermal Diffusivity (cm^2/s)	Comments
0.0	0.106	*0 % ZrO_2*
10	0.0483	
20	0.0400	
30	0.0337	
40	0.0280	
50	0.0246	
0.0	0.0252	*15 vol% ZrO_2*
10	0.0255	
20	0.0316	
30	0.0274	
40	0.0223	
50	0.0204	

Remarks/Comments
The thermal diffusivity decreased substantially with the addition of chromia to alumina, but the diffusivity of the composite did not change significantly with the addition of chromia. The addition of zirconia to alumina and alumina-chromia solid solutions decreased the thermal diffusivity as well. Thermal diffusivity decreased with increasing temperature.

Thermal Conductivity/Diffusivity vs. Temperature

Temperature - *Heating* (°C)	Thermal Diffusivity (cm^2/s)	Temperature - *Cooling* (°C)	Thermal Diffusivity (cm^2/s)	Comments
22	0.106	1200	0.0129	Al_2O_3
300	0.0371	1000	0.0143	
390	0.0277	820	0.0167	
500	0.0245	600	0.0210	
600	0.0211	22.0	0.105	
690	0.0188			
810	0.0165			
890	0.0153			
1000	0.0147			
1100	0.0139			
1200	0.0130			
1300	0.0127			
1400	0.0126			
22	0.0252	1200	0.00806	Al_2O_3 / 15 vol% ZrO_2
290	0.0112	1000	0.00614	
390	0.00750	800	0.00507	
500	0.00610	650	0.00564	
600	0.00513	22.0	0.0164	
700	0.00484			
800	0.00478			
900	0.00472			
1000	0.00545			
1100	0.00584			
1200	0.00539			
1300	0.00560			
1400	0.00748			
22	0.0478	22.0	0.0485	90 Al_2O_3 - 10 Cr_2O_3
300	0.0225	810	0.0131	
400	0.0197			
500	0.0176			
600	0.0152			
700	0.0138			
800	0.0134			
900	0.0124			
1000	0.0112			
1100	0.0106			
1200	0.0102			
1300	0.0102			
1400	0.00983			
22	0.0256	1050	0.0100	90 Al_2O_3 - 10 Cr_2O_3 / 15 vol% ZrO_2
300	0.0145	830	0.00784	
400	0.0117	600	0.00898	
500	0.0103	22.0	0.0214	
600	0.00950			
700	0.00936			
800	0.00887			
900	0.00860			
1000	0.00844			
1100	0.00846			
1200	0.00871			
1300	0.00884			
1400	0.00862			

Temperature - *Heating* (°C)	Thermal Diffusivity (cm^2/s)	Temperature - *Cooling* (°C)	Thermal Diffusivity (cm^2/s)	Comments
22	0.0396	105	0.0105	*80 Al_2O_3 - 20 Cr_2O_3*
360	0.0161	22.0	0.0381	
460	0.0150			
550	0.0143			
640	0.0128			
750	0.0126			
830	0.0118			
940	0.0115			
1030	0.0104			
1100	0.0106			
1200	0.0105			
1300	0.0103			
22	0.03.16	1030	0.0100	*80 Al_2O_3 - 20 Cr_2O_3 / 15 vol% ZrO_2*
360	0.0149	470	0.0129	
470	0.0133	22.0	0.0299	
550	0.0121			
640	0.0114			
740	0.0106			
820	0.0101			
940	0.00964			
1000	0.00938			
1100	0.00920			
1200	0.00871			
1400	0.00872			
22	0.0245	800	0.0114	*50 Al_2O_3 - 50 Cr_2O_3*
400	0.0143	22.0	0.0237	
500	0.0134			
700	0.0118			
800	0.0114			
900	0.0111			
1000	0.0108			
1100	0.0108			
1200	0.0109			
1300	0.0109			
1400	0.0106			
22	0.0200	1100	0.00842	*80 Al_2O_3 - 20 Cr_2O_3 / 15 vol% ZrO_2*
400	0.0109	800	0.00912	
500	0.0102	500	0.0106	
600	0.00952	22.0	0.0197	
700	0.00904			
800	0.00883			
900	0.00856			
1000	0.00858			
1100	0.00858			
1200	0.00852			
1300	0.00835			
1400	0.00835			

Figure 137: Thermal Conductivity vs. Temperature of DIMOX™ SiC(p)/Al_2O_3, SiC(p)/ZrO_2, SiC(p)/ZrO_2.Mullite, and ZrO_2.HfO_2(p)/Al_2O_3.Cr_2O_3

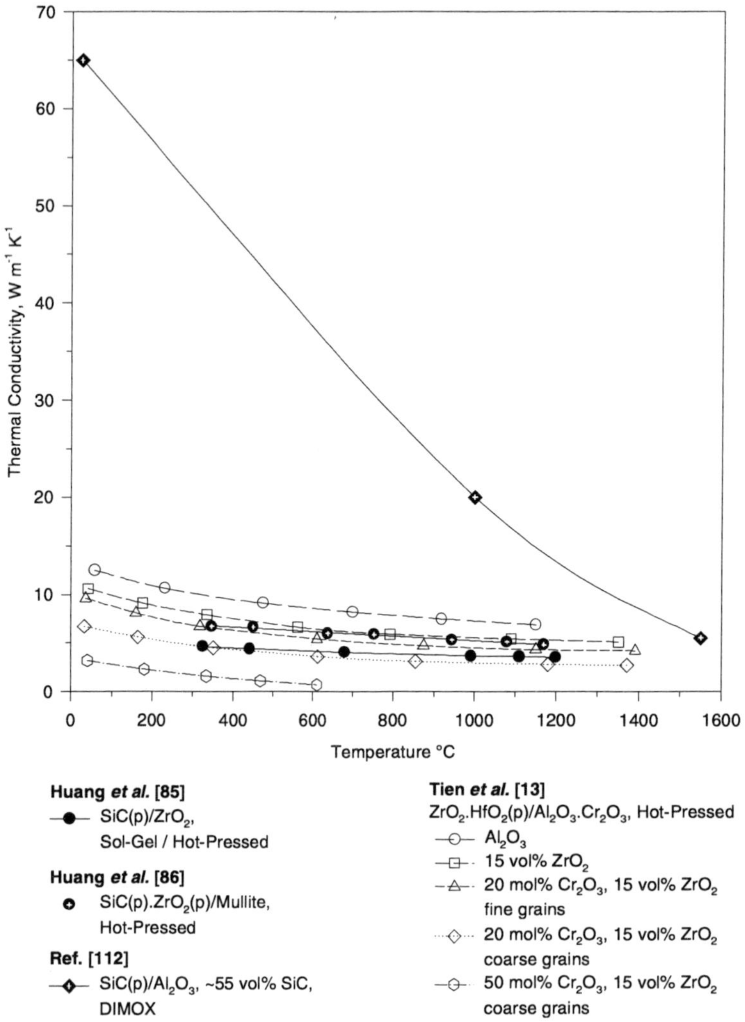

Discussion: The thermal conductivity of all the composites decreased with increasing temperature. The thermal conductivity of the sol-gel/hot-pressed SiC(p).ZrO_2/mullite composite tested by Huang *et al.* [86] showed the least dependence on temperature. The thermal conductivity of this composite was only slightly lower at 1200 °C than at room temperature.

The fine grain ZrO_2.HfO_2(p)/Al_2O_3.Cr_2O_3 composites tested by Tien *et al.* [13] exhibited better thermal conductivity at all temperatures than did the coarse grain composites. The added HfO_2 increased the transformation temperature of the zirconia. Chromia was added to the ZrO_2.HfO_2(p)/Al_2O_3.Cr_2O_3 composites to decrease thermal conductivity of the alumina matrix. The data shows that the added Cr_2O_3 decreased the thermal conductivity of the composite.

The DIMOX™ SiC(p)/Al_2O_3 composite showed the highest room temperature thermal conductivity of the composites presented in figure 137. The high thermal conductivity was due to the high silicon carbide content (~55 vol%) and residual aluminum content. At room temperature, the aluminum metal conducted very well; however, as the temperature increased, the residual aluminum softened and melted. The thermal conductivity dropped significantly as a result.

Figure 138: Thermal Diffusivity of Zirconia-Hafnia/Chromia-Alumina Composites vs. Chromia Content

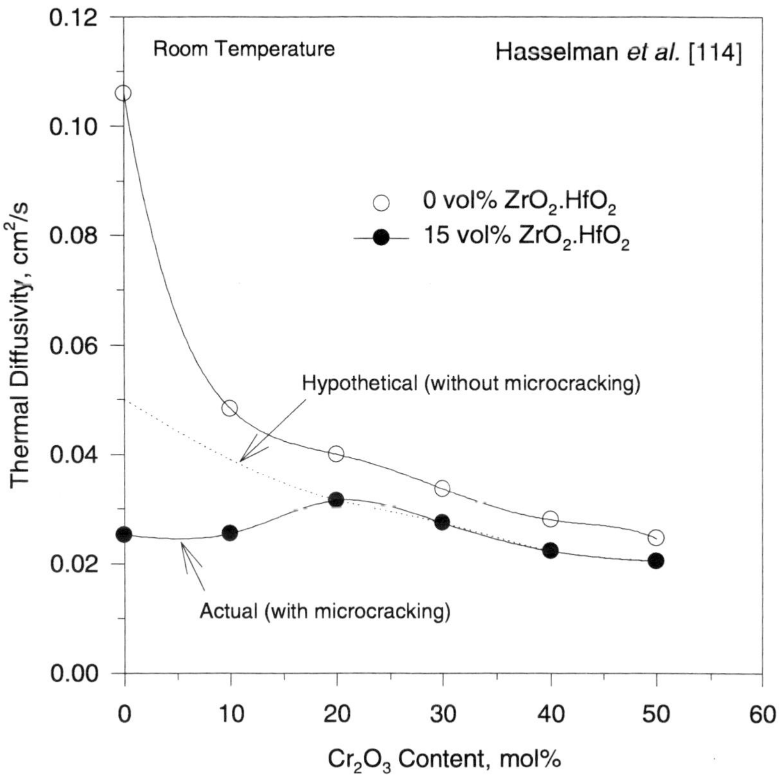

Discussion: The addition of chromia decreased the thermal diffusivity of monolithic alumina. The addition of chromia greatly increased phonon-phonon collisions and therefore decreased diffusivity [114]. The thermal diffusivity of zirconia was lower than that of the alumina or chromia-alumina solid solutions. Thus, the diffusivity of the composites with zirconia additions were lower than the samples without zirconia reinforcement. The anomalous low thermal diffusivity of the composites at low Cr_2O_3 contents was suspected to be due to microcracking at the grain boundaries. The microcracking developed below 20 vol% Cr_2O_3 content because of the extensive grain growth that occurred at low chromia contents [114]. If no microcracking had occurred, the thermal diffusivity could be extrapolated out as shown.

Figure 139: Thermal Diffusivity vs. Temperature of Alumina-Chromia Solid Solutions With or Without ZrO_2 Particulate Reinforcement

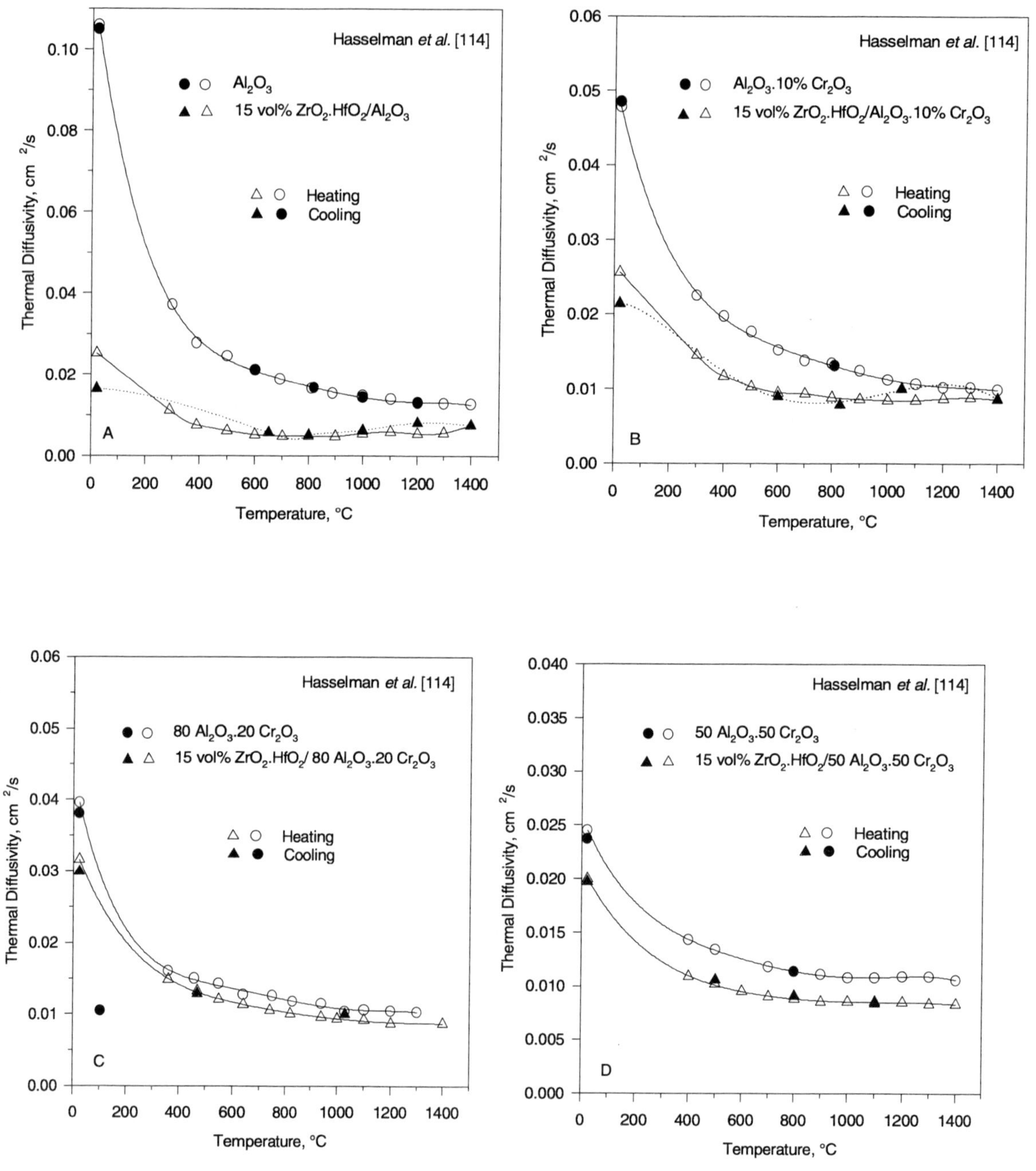

Discussion: Figures 139a-d shows the effect on thermal diffusivity of chromia solutions in alumina and the effect of zirconia reinforcement to the alumina-chromia solid solution. The addition of chromia greatly reduced the thermal diffusivity of the alumina and zirconia-alumina composites at room temperature. The thermal diffusivity of all samples decreased with increasing temperatures, until the diffusivity of all samples reached approximately the same level, around 0.01 cm^2/s, at 1400 °C. This showed that the chromia additions were effective in lowering the thermal diffusivity at low to intermediate temperatures, but were ineffective at high temperatures. In contrast to the chromia additions, the addition of zirconia resulted in only a slight decrease in thermal diffusivity at RT. However, this difference remained at all temperatures, which can be clearly seen by the parallel lines in figures 139c-d. The thermal diffusivity of both the alumina and 10% Cr_2O_3 alumina reinforced with zirconia showed a slight hysteresis in the thermal cycle. This hysteresis showed that some microcracking occurred. In both cases, the thermal diffusivity was lower after the thermal cycle than before. The 50 % chromia samples presented in figure 139d was observed to be relatively free of microcracks. This suggested that the diffusivity of the 50 % chromia samples did not display a hysteresis effect upon cooling [114].

TiO_2(p)/HfO_2

Reference: 116

Composition
25-50 mol% TiO_2

Processing Method
Sol-Gel/Sintering

Detailed Description
Samples were synthesized by hydrolysis and polycondensation of titanium alkoxide and hafnium dichloride oxide solutions. Powders obtained were cold-pressed (3 t/cm^2) and heat treated in air to promote sintering and crystallization. Material sintered and slowly cooled at 2 °C/min.

Reinforcement Description
TiO_2 particulates

Matrix Description
HfO_2 matrix

Test Type
Dilatometry

Type of Specimen
Bar

Test Description
Tests carried out in air on bar samples with heating and cooling speed of 3 °C/min. Expansion or contraction measured by a quartz push rod dilatometer.

Test Temperature
20-1000 °C

Test Environment
Air

Test Equipment
Quartz Push Rod Dilatometer

Reinforcement Content (mol%)	Thermal Expansion Coefficient (10^{-6} K^{-1})	Density (g cm^{-3})	Relative Density (%)
25.1	1.93		
30.1	0.39	8.14	94.7
36.5	0.90	7.39	92.5
40.2	1.74	7.36	94.3
50.2	3.70	10.0	97.0

Remarks/Comments
Ceramics with near zero CTE were produced in the TiO_2-HfO_2 system.

SiC(p)/ZrO_2

Reference: 85

Composition
15 wt% SiC

Processing Method
Sol-Gel/Hot-Pressed

Detailed Description
ZrO_2Cl_2-$8H_2O$, $Y(NO_3)_3$-$6H_2O$ and α–SiC were used as starting materials. Y-TZP and SiC(p)/Y-TZP were prepared by the coprecipitation method from the starting solution. The mixture was hot-pressed in a boron-coated graphite mold at 1600 °C and 25 MPa for 1 hour.

Reinforcement Description
SiC particulates

Matrix Description
2.8Y-ZrO_2

Test Type
Thermal Expansion Test

Specimen Dimensions
2.5 x 5 x 26 mm

Type of Specimen
Bar

Test Temperature
RT-1200 °C

Temperature (°C)	Mean Thermal Expansion Coefficient (10^{-6} K^{-1})
20	-
120	8.07
205	9.18
405	9.49
606	9.33
806	9.55
1,000	9.98
1,200	10.20

Remarks/Comments

The average coefficient of thermal expansion increased with increasing temperature.

Figure 140: Thermal Expansion Coefficient of TiO_2(p)/HfO_2

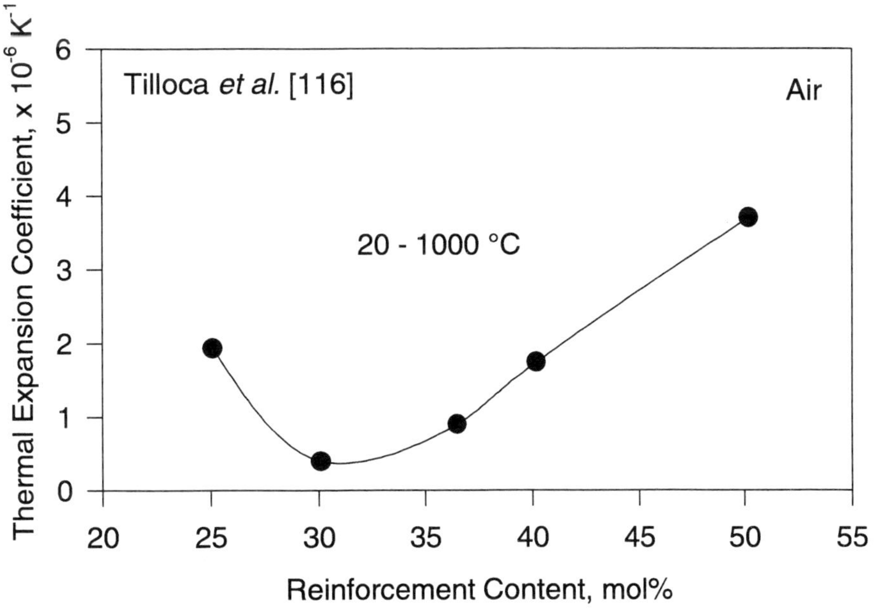

Discussion: Ceramics in TiO_2-HfO_2 system are of great interest because they have been found to possess ultra-low thermal expansion. The system is characterized by a wide solubility range based on the compound $HfTiO_4$, which exists from 35 to 53 mol% TiO_2. The $HfTiO_4$ phase has a high melting point (~1980 °C) and is quite stable, even under extremely reducing conditions. This shows that ceramics in the TiO_2-HfO_2 system have a great potential as a refractory material that also has high thermal shock resistance [116]. The thermal expansion coefficient as a function of TiO_2 content of a sol-gel derived TiO_2-HfO_2 composite is plotted in figure 140. The tests showed that the CTE nearly reached zero, only 0.39×10^{-6} K^{-1}, at 30 mol% TiO_2.

Repeated dilatometric heating and cooling trials showed a slight hysteresis effect. With increasing heating/cooling runs, the thermal expansion increased while the hysteresis diminished. By the fourth run, the CTE stabilized and the hysteresis had diminished. The low CTE of TiO_2-HfO_2 composites was partially due to the presence of a high density of microcracks which accommodated thermal stresses and resulted in the hysteresis effect observed in the TiO_2-HfO_2 system [116].

Figure 141: Temperature Dependence of the Mean Thermal Expansion Coefficient of $SiC(p)/Y\text{-}ZrO_2$

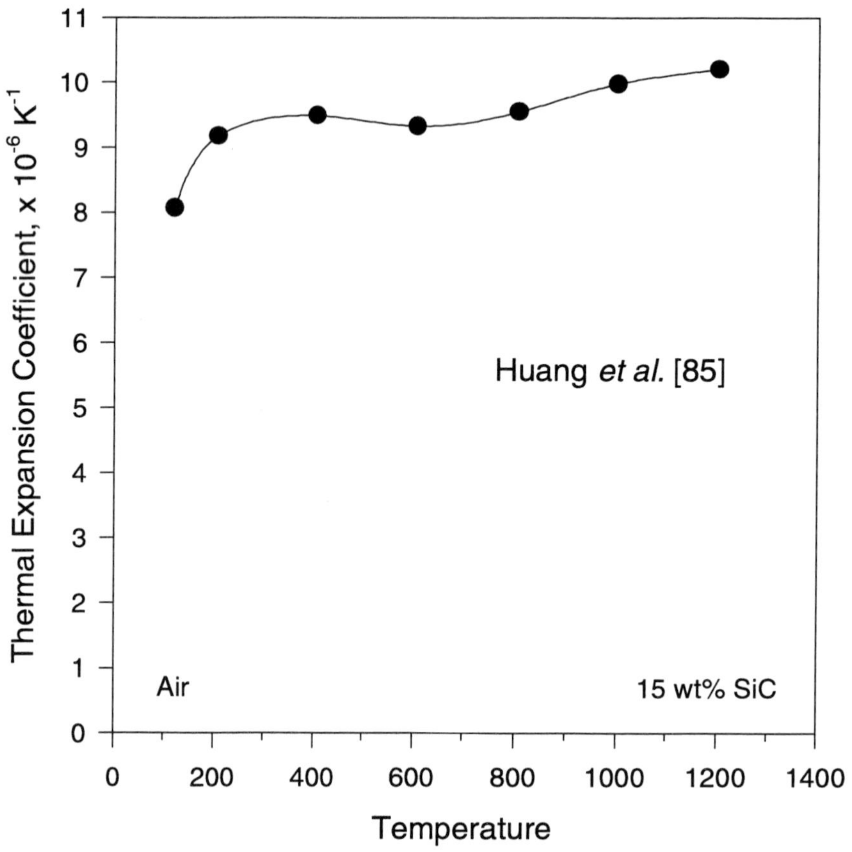

Discussion: Figure 141 shows the thermal expansion coefficient of a 15 vol% $SiC(p)/Y\text{-}ZrO_2$ composite. From 100 to 400 °C the CTE increased sharply, and from 800 to 1200 °C the CTE of the composite increased gradually. In the range 400 to 800 °C, the CTE decreased due to the transformation from monoclinic zirconia to tetragonal zirconia, which is associated with a decrease in volume [85].

$TiN(p)/Si_3N_4$

Reference: 36

Composition
0-100 vol% TiN

Processing Method
Reaction Bonded

Detailed Description
Suspended in alcohol, these particles were mixed for 24 hours in a pot mill together with Si_3N_4 balls. The mixtures were dried and kneaded with a thermoplastic resin binder, and then cooled and crushed. The crushed materials were heated to 160 °C in a die and pressed into a disk shape under 98 MPa. The samples were dewaxed and heated stepwise to 1350 °C in a nitrogen atmosphere for nitridation.

Reinforcement Description
TiN particulates (reaction-bonded)

Reinforcement Typical Properties
TiN particle size = 2.0 μm

Matrix Description
Si powder

Matrix Properties
particle size = 0.9 μm

Test Type
Thermal Expansion Test

Specimen Dimensions
dia. = 40 mm
thickness = 10 mm

Type of Specimen
Disk

Test Description
heating rate = 5 °C min^{-1}

Test Temperature
20-500 °C

Test Environment
Air

Reinforcement Content (volume percent)	Thermal Expansion Coefficient (10^{-6} K^{-1})	Comments
0.0	2.99	*pure Si_3N_4*
22.2	3.64	
40.2	4.52	
49.1	4.68	
61.3	5.42	
68.7	5.70	
100	7.46	*pure TiN*

Remarks/Comments
The measured results were located nearly on a straight line that connected the coefficients for Si_3N_4 and TiN. TiN and Al_2O_3 were both considered suitable because their thermal expansion coefficient were similar.

Al_2O_3(p)/Si_3N_4

Reference: 36

Composition
0-100 vol% Al_2O_3

Processing Method
Reaction Bonded

Detailed Description
Suspended in alcohol, these particles were mixed for 24 hours in a pot mill together with Si_3N_4 balls. The mixtures were dried and kneaded with a thermoplastic resin binder, and then cooled and crushed. The crushed materials were heated to 160 °C in a die and pressed into a disk shape under 98 MPa. The samples were dewaxed and heated stepwise to 1350 °C in a nitrogen atmosphere for nitridation.

Reinforcement Description
Al_2O_3 particulates

Reinforcement Typical Properties
Al_2O_3 particle size = 0.5 μm

Matrix Description
Si powder

Matrix Properties
particle size = 0.9 μm

Test Type
Thermal Expansion Test

Specimen Dimensions
dia. = 40 mm
thickness = 10 mm

Type of Specimen
Disk

Test Description
heating rate = 5 °C min^{-1}

Test Temperature
20-500 °C

Test Environment
Air

Reinforcement Content (volume percent)	Thermal Expansion Coefficient (10^{-6} K^{-1})	Comments
0.0	2.99	*pure Si_3N_4*
19.3	3.64	
30.0	4.03	
38.5	4.24	
50.5	5.05	
59.4	5.12	
70.3	6.04	
100	8.08	*pure Al_2O_3*

Remarks/Comments
The measured results were located nearly on a straight line that connected the coefficients for Si_3N_4 and Al_2O_3. . TiN and Al_2O_3 were both considered suitable because their thermal expansion coefficient were similar.

TiN(p)/Si_3N_4

Reference: 54

Composition
0-40 vol% TiN

Processing Method
Hot-Pressed

Detailed Description
Si_3N_4 powder and sintering aids were homogenized in water for 48 hours and freeze dried. The second phase (TiN) was added to the matrix mix and mixed in isobutyl alcohol for 48 hours. The materials were then uniaxially hot-pressed at 1800 °C and 30 MPa for 10 minutes to nearly full density.

Sintering Additives
3 wt% Al_2O_3
8 wt% Y_2O_3

Reinforcement Description
TiN particulates

Reinforcement Supplier
Starck, Germany

Supplier Designation
Grade C

Reinforcement Dimensions
ave. size = 0.3-3 μm

Matrix Description
Si_3N_4 powder

Matrix Supplier
Starck, Germany

Matrix Designation
LC12

Test Type
Thermal Expansion Test

Test Temperature
20-900 °C

Test Environment
Air

Reinforcement Content (volume percent)	Thermal Expansion Coefficient (10^{-6} K^{-1})	Relative Density (percent)
0.0	3.25	100
20	4.42	99.4
30	4.62	99.6
40	5.80	99.4

Remarks/Comments

As the TiN content increased, so did the thermal coefficient of expansion. Overall values of the thermal coefficient of expansion were greater for the composite materials than for the monolithic material.

Si_3N_4(p)/SiC

Reference: 56

Composition

10-50 % Si_3N_4

Processing Method

Reaction Bonded

Detailed Description

SiC and Si were mixed and pressed under 96.56 MPa. 1% polyvinyl alcohol was added as a binder. The green bodies were dried in a furnace at 200 °C for 2 hours. The green shapes were nitrided at 1350-1450 °C.

Reinforcement Description

in-situ Si_3N_4

Matrix Description

SiC

Test Type

Thermal Expansion Test

Type of Specimen

Bar

Test Temperature

20-1000 °C

Reinforcement Content (volume percent)	Thermal Expansion Coefficient (10^{-6} K^{-1})
10	4.37
20	3.98
30	3.68
40	3.47
50	3.16

Remarks/Comments

The CTE fell as the percentage of Si_3N_4 in the composite increased. The CTE of reaction-bonded silicon carbide was $5.9e^{-6}$/°C. Thermal expansion coefficients of all the composites was between the CTE of Si_3N_4 and SiC. The decrease of thermal expansion with increase in the percentage of Si_3N_4 in the composite was due to the lower thermal expansion of Si_3N_4 compared to that of SiC.

AlN(p)/BN

Reference: 79

Composition
25 wt% AlN

Processing Method
Reaction Bonded

Detailed Description
MBN-050(80 wt%), MBN-250(20 wt%), atomised Al powder, and the sintering and nitriding aids were mixed at a wt ratio of BN-AlN = 3:1 and then ball-milled. The mixed powder was dried, uniaxially pressed at 10 MPa and isostatically pressed at 700 MPa, until a 50 x 50 x 4 mm green body was obtained. The nitridation of aluminum was carried out in a mixed gas atmosphere N_2:H_2 = 3:1 in vol%. The temperature was initially kept at 150 °C for approximately 30 minutes, then increased at the rate of 6 °C/min until temperature reached 500 °C. Then increased at the rate of 3 °C/min till 600 °C. Temperature was maintained at 600 °C for 4 hours then increased at the rate of 6 °C/min till it reached 1900 °C, with an atmosphere change to pure N_2 at 1100 °C. The final densification was done at 1900 °C for 4 hours in nitrogen atmosphere. Then cooled at the rate of 10 °C/min.

Sintering Additives
Nitridation aid: CaF_2 = 1 wt%
Sintering Aids: Y_2O_3 = 4 and 8 wt%
$2CaO.3B_2O_3$ = 4 wt%
CaC_2 = 4 and 5 wt%
CaF_2 = 2 and 3 wt%

Reinforcement Description
Al powder (nitrided into AlN)

Reinforcement Supplier
Toyo Aluminum K. K.

Reinforcement Typical Properties
purity >= 99.3 %
grain size = 7-8 μm

Matrix Description
Hexagonal BN

Matrix Supplier
Mitsui Toatsu Chemicals Inc.

Matrix Designation
MBN-050; MBN-250

Matrix Properties
MBN-050: purity 97.0 percent
grain size = 0.1-0.2 μm
MBN-250: purity 92.8 %
ave. grain size = 0.01 μm

Test Type
Differential Method

Specimen Dimensions
50 x 50 x 4 mm

Type of Specimen
Bar

Test Temperature
20-400 °C

Test Equipment
Al_2O_3 Rod

Thermal Expansion Coefficient (10^{-6} K^{-1})	Density (g cm^{-3})	Comments
2.6	2.14	8 wt% Y_2O_3
2.6	2.08	4 wt% 2 $CaO{\cdot}3B_2O_3$
1.8	2.12	4 wt% CaC_2
2.2	2.11	2 wt% CaF_2

Remarks/Comments
The thermal expansion coefficient of AlN/BN was fairly low because the thermal expansion of h-BN is low.

Figure 142: Thermal Expansion Coefficient of Particulate Reinforced Silicon Nitride

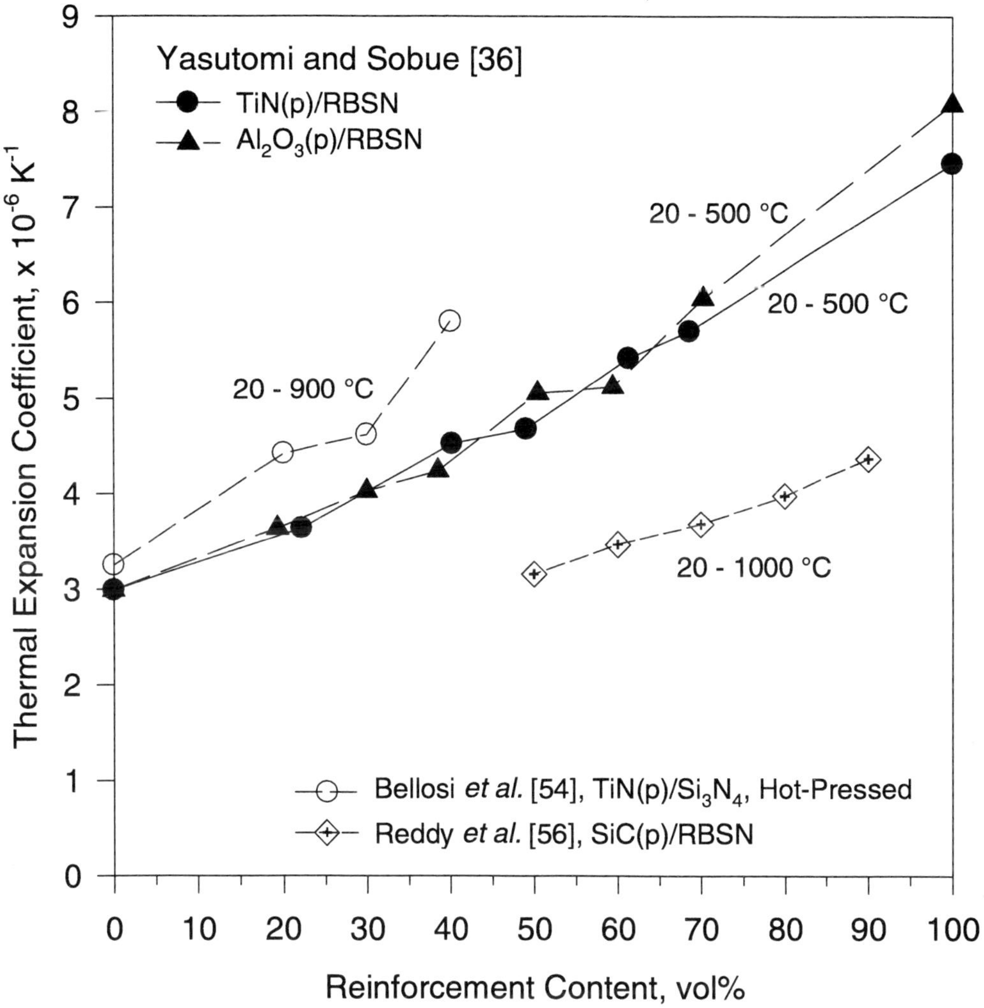

Discussion: Figure 142 shows thermal expansion coefficients of particulate reinforced silicon nitride composites. The thermal expansion of the hot-pressed TiN(p)/Si_3N_4 composite was greater than the TiN or Al_2O_3 particulate reinforced RBSN composites at the same reinforcement content. The hot-pressed composite had the higher CTE primarily because of the higher density reached by the hot-pressing method. The addition of sintering additives may also have significant effect on the CTE of a silicon nitride material. Yasutomi and Sobue [36] tested an alumina reinforced RBSN and a titanium nitride reinforced RBSN as an electrically resistive, and electrically conductive composite, respectively. Both reinforcements had nearly identical CTEs from RT to 500 °C. The reaction bonded matrix of the SiC(p)/RBSN composite was only semi-crystalline, somewhat porous and had a very low coefficient of thermal expansion, which was the cause of the low CTE of the composite [56].

Figure 143: Effect of Sintering Additives on the CTE of a Reaction Bonded AlN/BN Composite

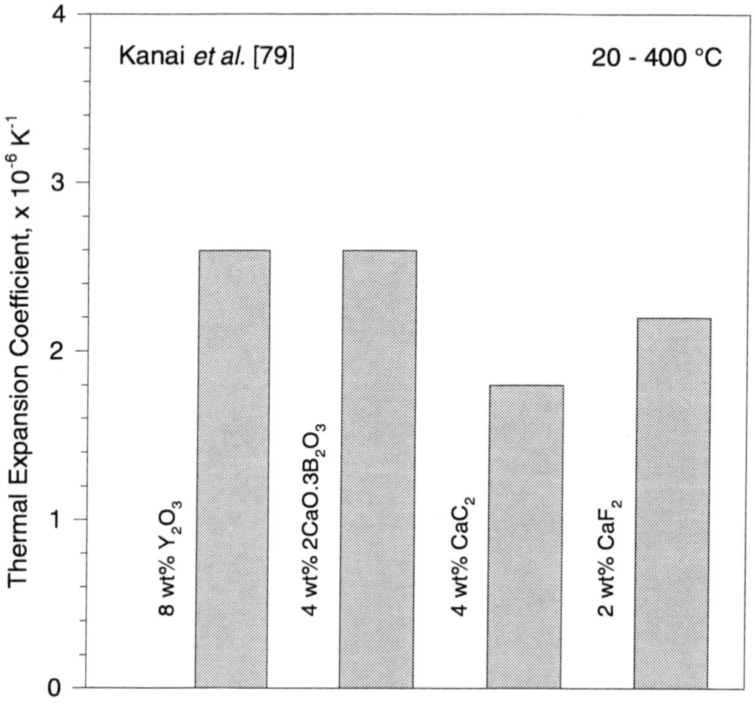

Discussion: The overall thermal expansion was low, 1.8 - 2.6 x 10^{-6} $°C^{-1}$. This was attributed to the low thermal expansion of hexagonal boron nitride. The effect of sintering additives was demonstrated. The yttria and $2CaO.3B_2O_3$ sintering additives produced a composites with higher thermal expansion, and the CaC_2 and CaF_2 produced composites with lower thermal expansion [79].

SiC(p)/AlN

Reference: 113

Composition
0 - 100 wt% SiC

Processing Method
Hot-Pressed

Detailed Description
Specimens were uniaxially hot-pressed in graphite dies lined with graphite foil. Hot-pressing was done in vacuum at 1700 to 2300 °C and 35 MPa.
Compaction was monitored by a micrometer dial gage attached to the ram.

Reinforcement Description
β-SiC

Reinforcement Supplier
Hermann C. Starck, Germany

Reinforcement Properties
Impurities (ppm):
5 Mg, 5 Mn, 2000 B, 1000 Al, 150 Fe, 10 Ni, 150 Ca, 20 Ti, <20 Cr, <200 W, 20 Mo, < 20 Co
Oxygen content: 0.30 wt%

Matrix Description
AlN

Matrix Supplier
Atomergic Chemetals Co., Div. of Gallard-Schlessinger Chemical Co., Carle Place, NY

Matrix Designation
D4744

Matrix Properties
Impurities (ppm):
200 Si, 10 Mg, 40 Mn, <20 B, 1000 Fe, 600 Ni, 40 Ca, 20 Ti, 600 Cr, 1000 W, 200 Mo, 1000 Co
Oxygen content: 0.98 wt%

Test Type
Laser-Flash Technique

Specimen Dimensions
2.5 cm dia., ~ 1 cm thick

Type of Specimen
disk

Test Description
The thermal diffusivity was measured using a glass-Nd laser with the specimen temperature measured using a nitrogen-cooled infrared detector. The thermal diffusivity was multiplied by the density and the specific heat to obtain the thermal diffusivity. The porosity was calculated from the measured density and the theoretical density.

Test Temperature
RT - 1423 °C

Reinforcement Content (weight percent)	Thermal Diffusivity ($cm^2 s^{-1}$)	Hot-Press Temperature (°C)	Comments
100	0.275	1700	*Two-Phase Mixture*
75	0.140	1850	
49	0.172	1900	
0	0.339	1950	
24	0.227	1950	
46	0.156	2000	----
0	0.391	2100	*Solid Solution*
0	0.335	2100	
10	0.142	2100	
34	0.083	2100	
49	0.092	2100	
64	0.092	2100	
78	0.119	2100	
100	0.202	2100	
59	0.079	2200	

Reinforcement Content (weight percent)	Thermal Conductivity ($W\ m^{-1}\ K^{-1}$)	Hot-Press Temperature (°C)	Comments
0	83.3	-	*Composite Mixture Theory*
100	46.6	-	

100	62.2	1700	*Two-Phase Mixture*
76	31.76	1850	
50	37.65	1900	
0.0	70.8	1950	
24	49.5	1950	
47	34.8	2000	

0	70.1	2100	*Solid Solution*
9	30.1	2100	
34	18.5	2100	
50	20.3	2100	
65	20.6	2100	
80	27.5	2100	
60	18.4	2200	

Temperature (°C)	Thermal Diffusivity ($cm^2\ s^{-1}$)	Comments
20	0.161	*two-phase mixture - HP-Temperature at 1900 °C*
700	0.0566	
800	0.0506	
900	0.0496	
990	0.0435	
1100	0.0420	
1200	0.0403	
1300	0.0393	
1400	0.0385	
20	0.0895	*solid solution HP-Temperature at 2100 °C*
300	0.0490	
422	0.0431	
500	0.0404	
600	0.0381	
693	0.0350	
800	0.0348	
895	0.0338	
992	0.0311	
1100	0.0283	
1200	0.0277	
1300	0.0297	
1423	0.0270	

Remarks/Comments

Full density was reached for all SiC and SiC-AlN samples, approximately 5 % porosity remained in the monolithic AlN sample. Below hot-pressing temperature of 2000 °C, the SiC-AlN existed as two-phase mixture, above 2000 °C HP temperature, the SiC and AlN formed a solid solution.

Figure 144a: Thermal Diffusivity of SiC/AlN Composites

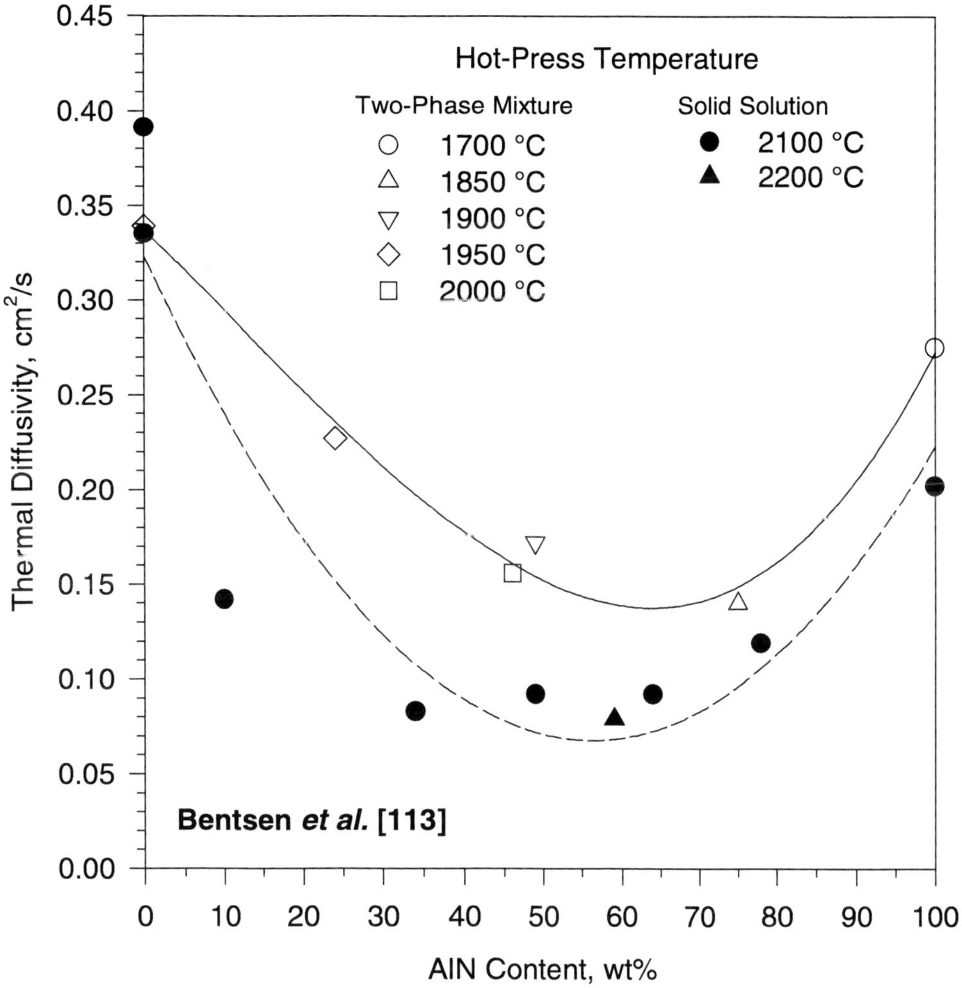

Discussion: Figures 144a-c present the data from Bentsen *et al.* [113] on the thermal conductivity and diffusivity of SiC/AlN composites. The composites were processed at varying temperatures such that the samples could be grouped according to the phases developed during processing. The first group, which were hot-pressed in the temperature range of 1700 -2000 °C, consisted primarily of two distinct phases of SiC and AlN. The second group was processed at a temperature at which interdiffusion constants were expected to be much higher, 2100 to 2200 °C. After processing at these higher temperatures the composite consisted primarily of a solid solution of SiC and AlN, which can be considered a SiC/AlN alloy. It was expected that thermal diffusivity and conductivity would be highly dependent on the phases present in the SiC/AlN composite, since previous experiments have found that foreign element inclusions typically negatively affect heat conduction.

Figure 144b: Thermal Conductivity of SiC/AlN Composites

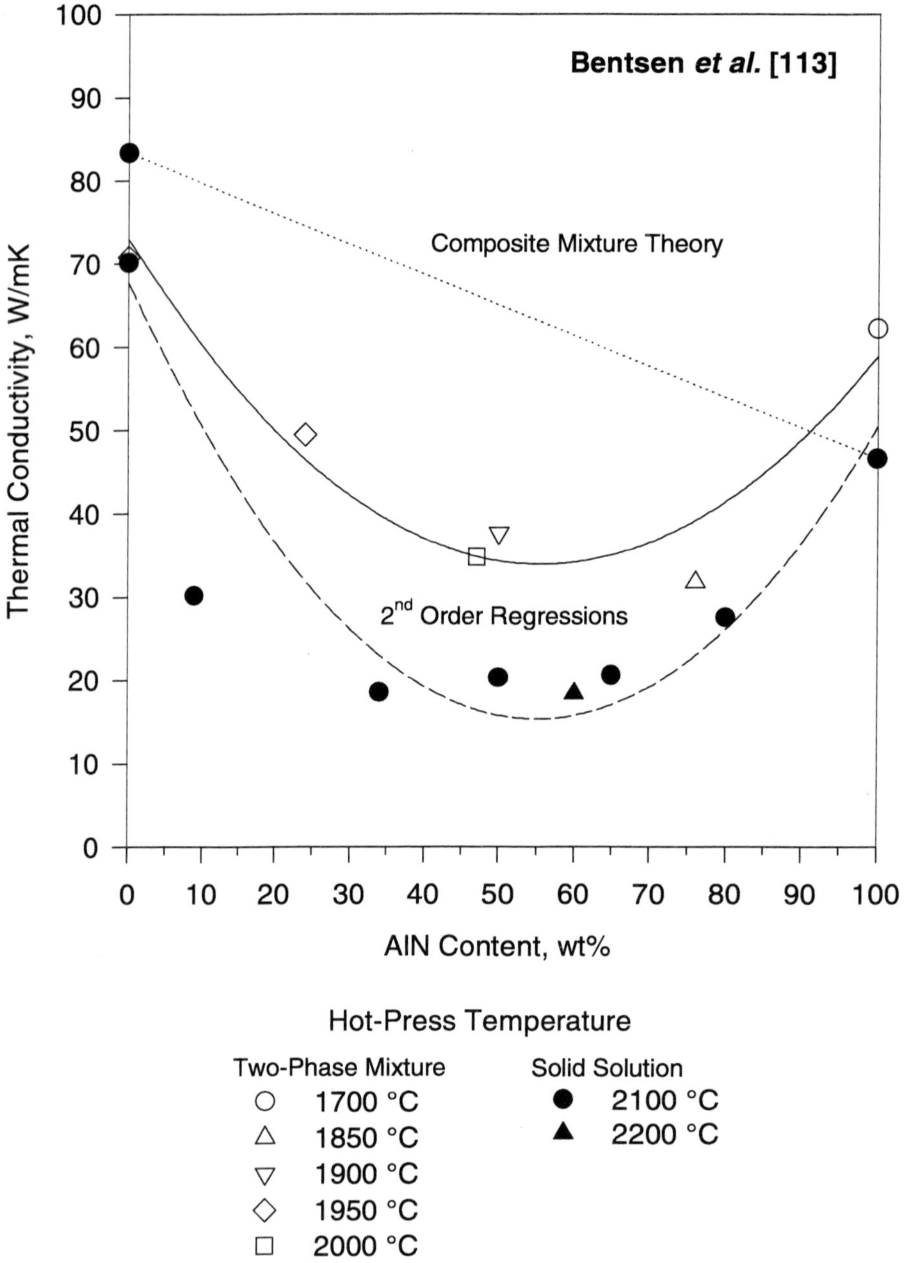

Discussion (cont.): Figure 144a shows that the thermal diffusivity of the composites decreased substantially compared to the monolithic SiC or AlN. Thermal conductivity was analogous to the thermal diffusivity. Figure 144b shows the results from the room temperature thermal conductivity tests on SiC/AlN composites compared to the thermal conductivity calculated by the composite mixture theory. Both the SiC/AlN composites with distinct phases and the SiC/AlN solid solution alloys showed a substantial drop in thermal conductivity compared to the theoretical conductivity, and the drop in conductivity for the solid solution alloy was approximately twice that of the SiC/AlN composite. In summary, increasing the hot-pressing temperature increased the formation of solid-solutions, which increased phonon scattering and decreased the thermal diffusivity and conductivity of the composite [113].

Figure 144c: Thermal Diffusivity vs. Temperature of SiC/AlN Composites

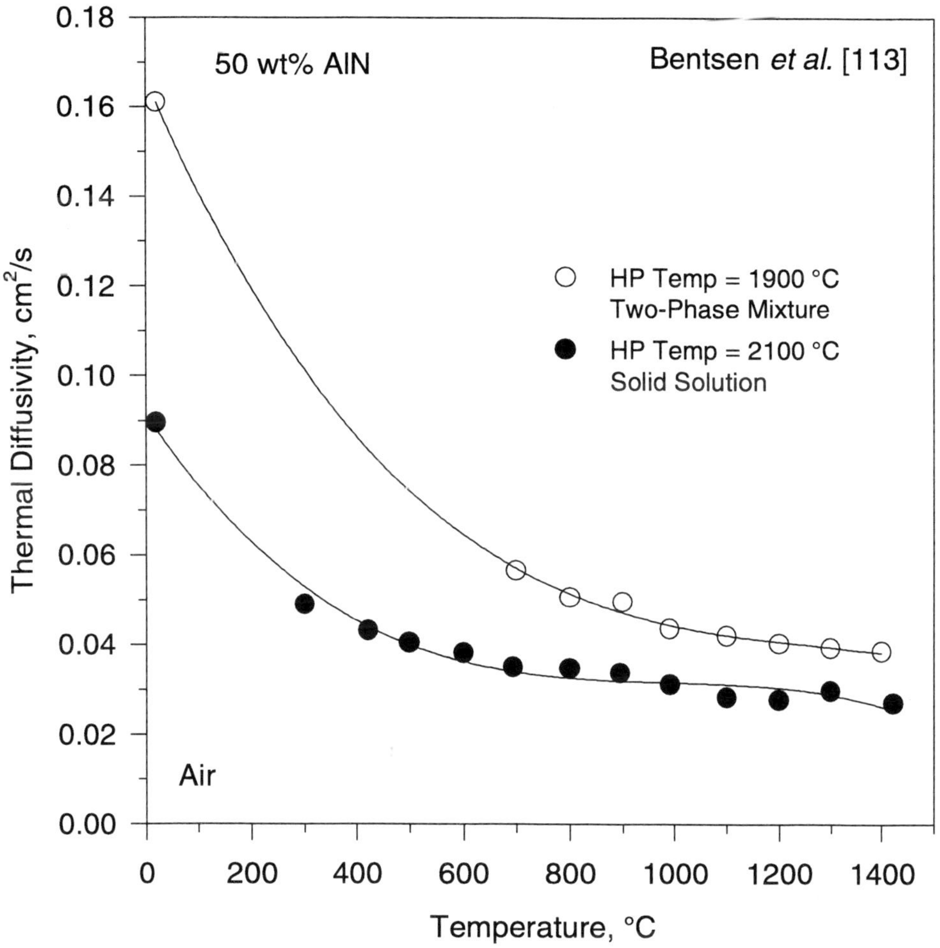

Discussion (cont.): Figure 144c shows the thermal diffusivity of the 50/50 wt% SiC/AlN composite and alloy with increasing temperature. As shown in the previous figures, the room temperature thermal diffusivity of the two-phase composite was nearly double that of the solid solution alloy. As temperature increased, the diffusivity of both samples decreased. On the other hand, the magnitude of this difference also decreased with increasing temperature. The reason for this was that phonon scattering increased at higher temperatures, and the effect of the distribution of impurity elements on thermal conductivity became less significant [113]. Nevertheless, above 1000 °C the difference in thermal diffusivity between the two samples remained approximately 0.01-0.02 cm^2/s, which was still a significant difference.

AlN(p)/BN

Reference: 52

Composition
25 wt% AlN

Processing Method
Hot-Pressed

Detailed Description
The mixed powders were ball milled, placed in a graphite mold, and hot pressed at 1800 °C for 2 hours under the uniaxial pressure of 40 MPa. Reagent grade CaC_2 was used as the sintering aid.

Sintering Additives
0-2 wt% CaC_2

Reinforcement Description
AlN powder

Reinforcement Supplier
Tokuyama Soda Co., Ltd.

Supplier Designation
F grade

Reinforcement Typical Properties
O_2 content <= 0.8 %
ave. grain size = 1.8 μm

Matrix Description
h-BN

Matrix Supplier
Denki Kagaku Kogyo K.K.

Matrix Designation
GP grade

Matrix Properties
purity >= 99.5 %O_2
content as B_2O_3 <= 0.1 %
grain size = 1-5 μm

Test Type
Laser-Flash Technique

Specimen Dimensions
8 x 2 mm cross-section

Type of Specimen
Pellet

Test Description
Measurements were made both parallel and perpendicular to the hot-pressing direction.

Test Temperature
RT

Sintering Aid Content (weight percent)	Thermal Conductivity ($W\ m^{-1}\ K^{-1}$)	Comments
0.09	10.3	***parallel***
0.34	12.3	
0.58	13.5	
0.86	20.5	
1.05	18.6	
1.53	17.8	
2.00	15.8	
0.10	33.4	***perpendicular***
0.34	53.2	
0.58	82.2	
0.86	246	
1.05	136	
1.53	36.6	
2.00	22.2	

Remarks/Comments
The composite was synthesized by uniaxial hot pressing with the expectation of producing high thermal conductivity due to the structural anisotropy of h-BN. The thermal conductivity for the direction perpendicular to the hot pressing axis had larger values than those parallel to the hot pressing axis. By optimizing the amount of added CaC_2 (0.8 wt%) as the sintering aid, a composite which had a high thermal conductivity of 246 W/(m K) in the perpendicular direction of the hot pressing axis was obtained. However, parallel to the hot pressing axis there was no distinguishable dependence on CaC_2 content and was equal to or less than 20 W/(m K) at all sintering aid contents.

AlN(p)/BN

Reference: 79

Composition
25 wt% AlN

Processing Method
Isostatically Pressed / Reaction Bonded

Detailed Description
MBN-050(80 wt%), MBN-250(20 wt%), atomised Al powder, and the sintering and nitriding aids were mixed at a wt ratio of BN-AlN = 3:1 and then ball-milled. The mixed powder was dried, uniaxially pressed at 10 MPa and isostatically pressed at 700 MPa, until a 50 x 50 x 4 mm green body was obtained. The nitridation of aluminum was carried out in a mixed gas atmosphere N_2:H_2 = 3:1 in vol%. The temperature was initially kept at 150 °C for approximately 30 minutes, then increased at the rate of 6 °C/min until temperature reached 500 °C. Then increased at the rate of 3 °C/min to 600 °C. Temperature was maintained at 600 °C for 4 hours then increased at the rate of 6 °C/min until it reached 1900 °C, with an atmosphere change to pure N_2 at 1100 °C. The final densification was done at 1900 °C for 4 hours in nitrogen atmosphere. Then cooled at the rate of 10 °C/min.

Sintering Additives
Nitridation aid: CaF_2 = 1 wt%
Sintering Aids: Y_2O_3 = 4 and 8 wt%
2 $CaO.3B_2O_3$ = 4 wt%
CaC_2 = 4 wt%
CaF_2 = 2 wt%

Reinforcement Description
Al powder (nitrided into AlN)

Reinforcement Supplier
Toyo Aluminum K.K.

Reinforcement Typical Properties
purity >= 99.3 %

Reinforcement Dimensions
grain size = 7-8 μm

Matrix Description
h-BN

Matrix Supplier
Mitsui Toastu Chemicals Inc.

Matrix Designation
MBN-050; MBN-250

Matrix Properties
MBN-050: purity 97.0 %
grain size = 0.1-0.2 μm
MBN-250: purity 92.8 %
ave. grain size = 0.01 μm

Test Type
Laser-Flash Technique

Specimen Dimensions
dia. = 8 mm
thickness = 2 mm

Type of Specimen
Pellet

Test Temperature
RT

Relative Density (percent)	Thermal Conductivity ($W m^{-1} K^{-1}$)	Density ($g cm^{-3}$)	Comments
87.1	66.0	2.14	8 wt% Y_2O_3
87.1	51.0	2.14	4 wt% Y_2O_3
84.7	40.0	2.08	4 wt% 2 $CaO*3B_2O_3$
86.3	53.0	2.12	4 wt% CaC_2
85.9	50.0	2.11	2 wt% CaF_2

Remarks/Comments
The highest thermal conductivity, 66 $W m^{-1}K^{-1}$ was attained for the 8 wt% Y_2O_3 composite. The lowest thermal conductivity was obtained from the 2 $CaO.3B_2O_3$ sintering additives. The overall thermal conductivity was found to be relatively high, in spite of low density.

Figure 145a: Thermal Conductivity vs. Sintering Aid Content of Hot-Pressed 25 vol% AlN(p)/BN

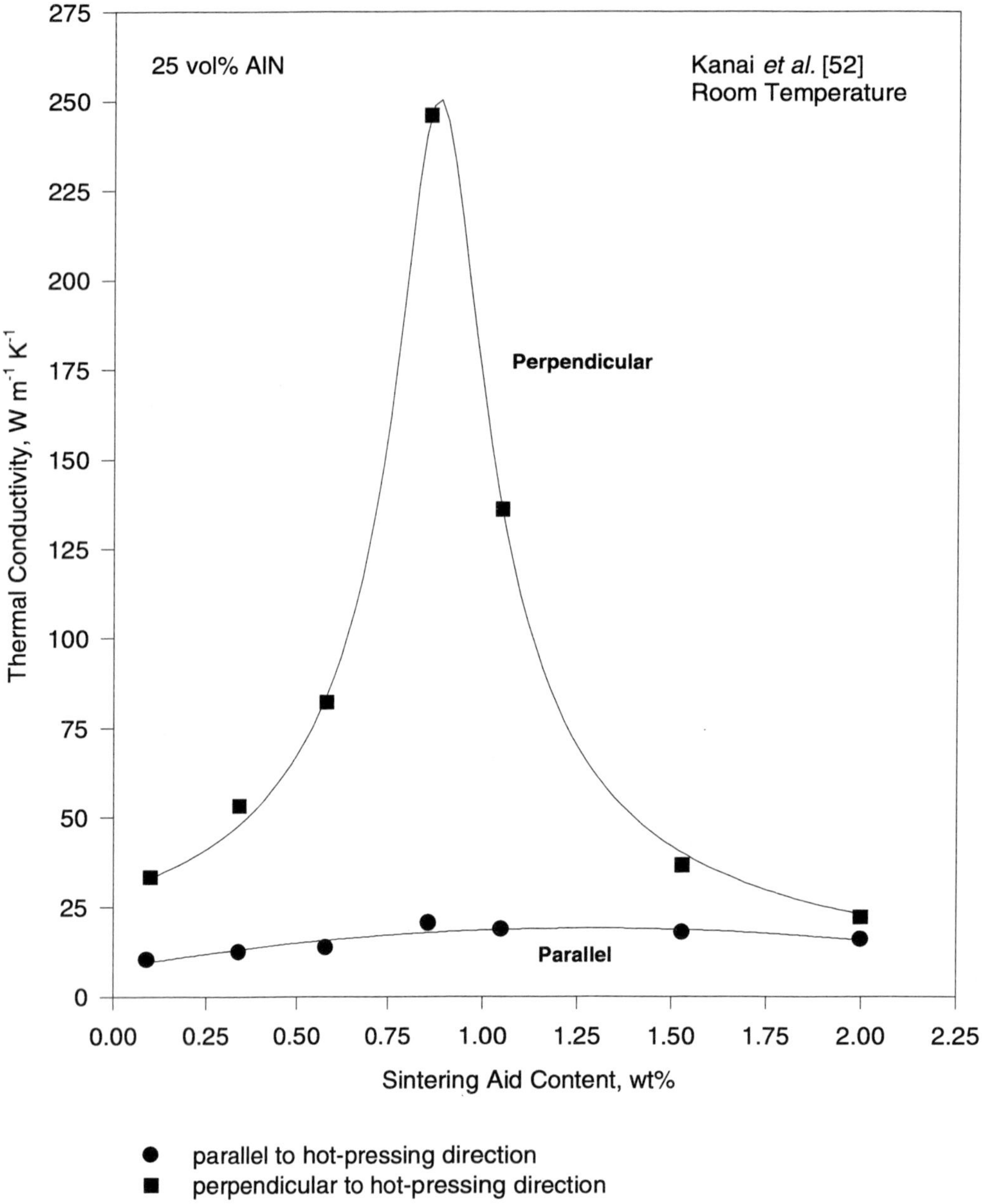

Discussion: Kanai *et al.* [52] examined the effect on thermal conductivity of varying types and quantities of sintering additives used in the processing of AlN/BN composites. Figure 145a shows that thermal conductivity of the AlN(p)/BN composite parallel to the hot-pressing axis remained relatively constant with increasing sintering aid content. However, a high thermal conductivity of 246 W/mK was achieved for the direction perpendicular to the hot-pressing axis through optimizing the amount of the calcium carbide (0.8 wt% CaC_2) sintering aid. A liquid phase, which formed by the reaction between CaC_2 and oxide on the surface of AlN particles, seemed to promote the orientation of hexagonal BN grains. This resulted in highly anisotropic thermal conductivity [52]. The AlN(p)/BN composite is most applicable for heat sink applications where it is desired to have directional heat flow through the material.

Figure 145b: Thermal Conductivity vs. Relative Density of Reaction Bonded 25 vol% AlN(p)/BN

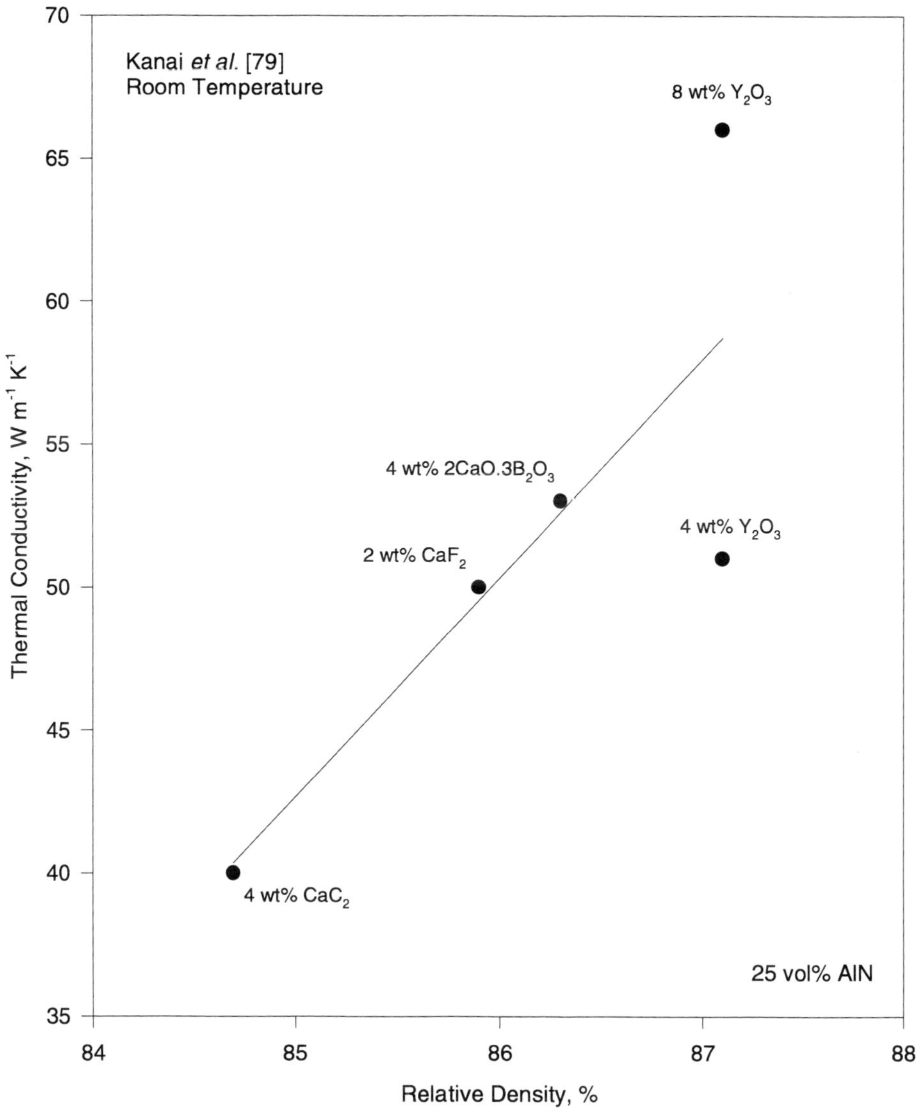

Discussion (cont.): Figure 145b shows AlN/BN composites processed by reaction bonding. Reaction bonding processing techniques are known to produce highly porous composites. Kanai *et al.* [79] studied the effect of various sintering additives on the reaction bonding of AlN/BN composites. The thermal conductivity of the AlN(p)/BN composites increased as the density increased for the different sintering aids except for the 4 wt% Y_2O_3 sintering aid, which saw a slight decrease. The low thermal conductivity of the 4 wt% Y_2O_3 composite was explained by the formation of a $Al_{27}O_{39}N$ phase, which was not observed in the other samples tested. Al-O-N compounds have low thermal conductivity. It was assumed that there is a good possibility for improving the thermal conductivity by preparing denser materials [79]. By using 8 wt% yttria as a sintering aid, a relatively high density of 87.1 % and a high thermal conductivity of 66 W/mK was obtained for the reaction-bonded material [79]. Under most circumstances, as porosity increases, the thermal conductivity decreases due to the fact that heat transfer across pores is slow.

BN(p)/B_4C

Reference: 89

Composition
0-100 vol% BN

Processing Method
Hot-Pressed

Detailed Description
A range of B_4C-BN compositions was prepared using a 99.9 % pure, 600-mesh B_4C powder and a 99.8 % pure hexagonal (graphitic) BN powder. The powders with appropriate compositions were mixed by dry-milling in a plastic jar with Si_3N_4 balls for 6-7 hours. Samples with a diameter of 2.5 cm and a thickness of 1 cm were uniaxially hot-pressed under vacuum in graphite dies lined with graphite foil at 2100 °C at a pressure of 35 MPa.

Reinforcement Description
h-BN particulates

Reinforcement Supplier
Cerac, Inc., Milwaukee, WI

Supplier Designation
No. 5717

Reinforcement Typical Properties
98.8 % pure

Reinforcement Dimensions
ave. dia. < 1 µm

Matrix Description
B_4C

Matrix Supplier
A. D. Mackay, New York

Matrix Properties
600 mesh, 99.9 % pure

Test Type
Laser-Flash Technique

Specimen Dimensions
8 x 8 x 2 mm

Type of Specimen
Plate

Test Description
A Nd-glass laser was used as a flash source for flash diffusivity. The test was performed in air from room temperature up to 200 °C and in a carbon resistance furnace with a protective atmosphere of nitrogen from 200 °C to 700 °C. A liquid-nitrogen-cooled IR detector was used to monitor transient specimen temperature. The laser pulse was passed through an aqueous solution of $CuSO_4$ to eliminate the effect of detector nonlinearity. The effect of pore content was normalized so that the data represents that of fully dense composites.

Test Temperature
RT- 700 °C

Test Environment
Air (<200 °C), and Nitrogen (>200 °C)

Test Equipment
Nd-glass laser

Reinforcement Content (volume percent)	Thermal Conductivity ($W\,m^{-1}\,K^{-1}$)	Comments
0.0	40.4	*perpendicular to hot-pressing direction*
38	58.7	
62	72.0	
80	82.8	
91	117	
100	143	
0.0	36.7	*parallel to hot-pressing direction*
21	25.7	
38	21.8	
62	20.0	
80	14.7	
90	14.1	
100	22.7	

Temperature (°C)	Thermal Conductivity ($W\ m^{-1}\ K^{-1}$)	Comments
22	132	*BN perpendicular to hot-pressing direction*
126	113	
208	101	
330	93.3	
417	87.8	
510	82.4	
606	78.1	
707	70.2	
22	90.4	*B_4C and 79.2 vol% BN perpendicular to hot-pressing direction*
127	72.0	
206	73.9	
326	67.9	
406	62.2	
525	59.7	
617	56.8	
700	51.9	
22	33.0	*B_4C parallel to hot-pressing direction*
127	26.4	
206	23.8	
311	21.8	
417	18.0	
518	17.7	
624	17.1	
706	16.5	
22	22.9	*BN parallel to hot-pressing direction*
128	22.6	
208	18.1	
332	15.1	
414	13.6	
511	13.1	
608	13.1	
707	12.4	
22	15.1	*B_4C + 79.2 vol% BN parallel*
129	16.1	
208	13.9	
329	14.4	
413	12.4	
524	11.4	
620	11.0	
708	10.5	

Remarks/Comments

For a B_4C matrix dispersed with h-BN particles, the thermal conductivity was significantly increased perpendicular to the hot-pressing direction, while leaving the thermal conductivity parallel to the hot-pressing direction relatively unaffected. This was expected since hot-pressing tends to align the hexagonal platelets with the plates oriented perpendicular to the hot-pressing direction. It was believed that the experimental values were less than the theoretical values since the hexagonal platelets are only partially reoriented, and because of the presence of interfacial thermal barriers.

BN(p)/SiC

Reference: 104

Composition
0-100 wt% BN

Processing Method
Hot-Pressed

Detailed Description
The powders were ball-milled in a plastic container, using Si_3N_4 balls for 4 hours in isopropyl alcohol. After drying, the powders were pelletized and uniaxially hot-pressed in vacuum for 1 hour at 2100 °C under 35 MPa pressure with graphite dies and graphite foil.

Reinforcement Description
h-BN

Reinforcement Supplier
Research Organic/Inorganic Chemical Corp., Belleville, NJ

Supplier Designation
B-27

Reinforcement Dimensions
ave. particle size = 15 μm

Matrix Description
β-SiC

Matrix Supplier
H.C. Stark, Goslar, Germany

Matrix Designation
Lot S 1626

Test Type
Laser-Flash Technique

Specimen Preparation
surfaces were ground

Specimen Dimensions
thickness = 2 mm

Type of Specimen
Plates

Test Description
A glass-Nd laser was used as the source and a remote optical sensor for monitoring the transient temperature response of the specimen. Room temperature measurements were done in air, but the high temperature measurements were done in a carbon resistance furnace in nitrogen. Measurements were made perpendicular and parallel to the hot-pressing direction. Density was determined by polishing the surfaces, measuring dimensions, and weighing.

Test Temperature
RT-1400 °C

Test Environment
Air (RT), Nitrogen (>RT)

Test Equipment
glass-Nd laser source, remote optical sensor detector

Parallel to Hot-Pressing Direction	
Reinforcement Content (weight percent)	Thermal Diffusivity ($cm^2\ s^{-1}$)
0.0	0.38
5.4	0.31
5.4	0.30
5.7	0.29
13	0.23
13	0.23
21	0.23
21	0.21
21	0.20
29	0.19
29	0.17
38	0.18
38	0.17
38	0.16
50	0.14
59	0.12
59	0.12
59	0.11
71	0.10
71	0.09
85	0.10
84	0.09

Perpendicular to Hot-Pressing Direction	
Reinforcement Content (weight percent)	Thermal Diffusivity ($cm^2\ s^{-1}$)
0.0	0.42
6.1	0.36
13	0.35
21	0.38
29	0.38
38	0.39
38	0.39
50	0.37
50	0.36
59	0.39
59	0.38
71	0.36
71	0.35
84	0.35
100	0.40

Remarks/Comments

The SiC matrices with BN particles appeared to be predominantly random in orientation though there was a significantly measurable variation in the thermal diffusivity.

Figure 146a: Thermal Conductivity of BN(p)/B_4C

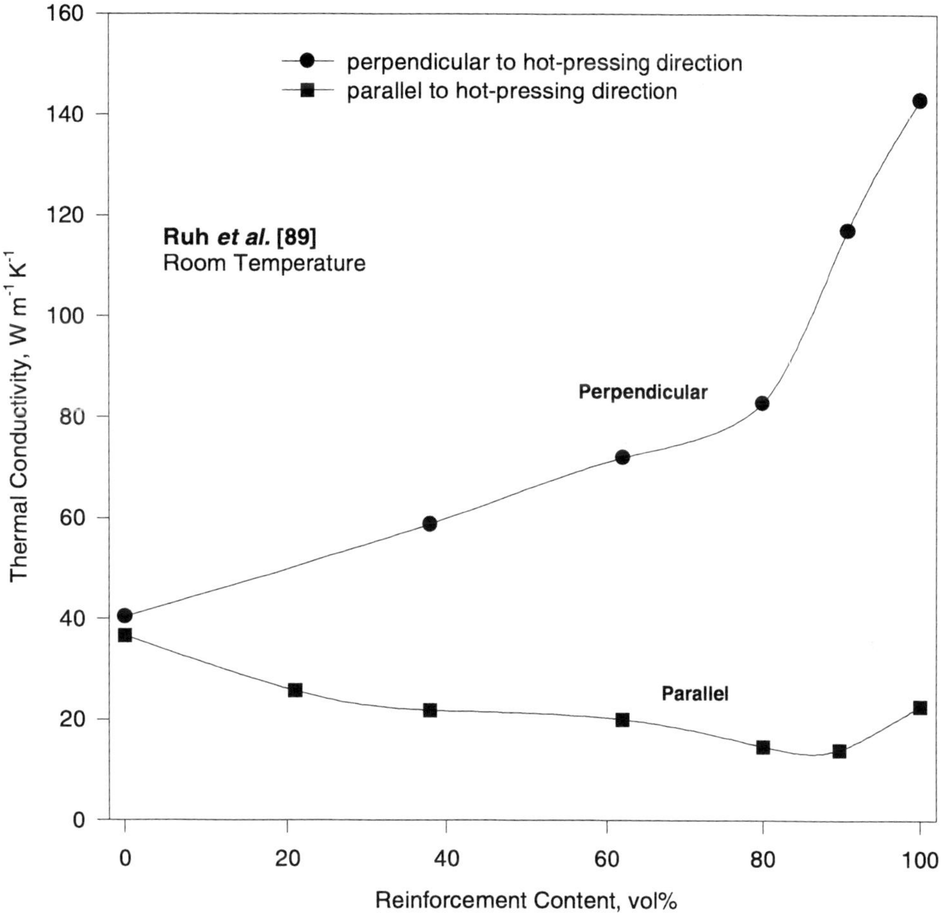

Discussion: Figure 146a illustrates the dependence of thermal conductivity on BN content in BN(p)/B_4C composites. The thermal conductivity increased with BN content perpendicular to the hot-pressing direction, and decreased with an addition of BN content parallel to the hot-pressing direction. Thus, the single phase BN exhibited a high degree of anisotropy. Thermal conductivity anisotropy is defined as the ratio of thermal conductivity measured perpendicular to the HP direction divided by the thermal conductivity measured parallel to the HP direction. The single phase BN exhibited a thermal conductivity ratio of 6 to 7 [89]. This was far less than the ratio of approximately 100 found by Simpson and Stuckes[1] (1971) for the thermal conductivity ratio of a highly oriented pyrolytic BN ceramic [89]. This implied that the BN ceramic measured by Ruh *et al.* [89] exhibited a preferred texture, but was far from perfectly oriented. The B_4C exhibited a slight anisotropy in thermal conductivity. This likely can be attributed to some preferred orientation of the B_4C grains or experimental scatter [89].

1. Simpson, A., and Stuckes, A.D., "The Thermal Conductivity of Highly Oriented Pyrolytic Boron Nitride", *Journal Phys. C: Solid State Phys.*, Vol. 4, pp. 1710-1717, (1971).

Figure 146b: Thermal Conductivity vs. Temperature of $BN(p)/B_4C$

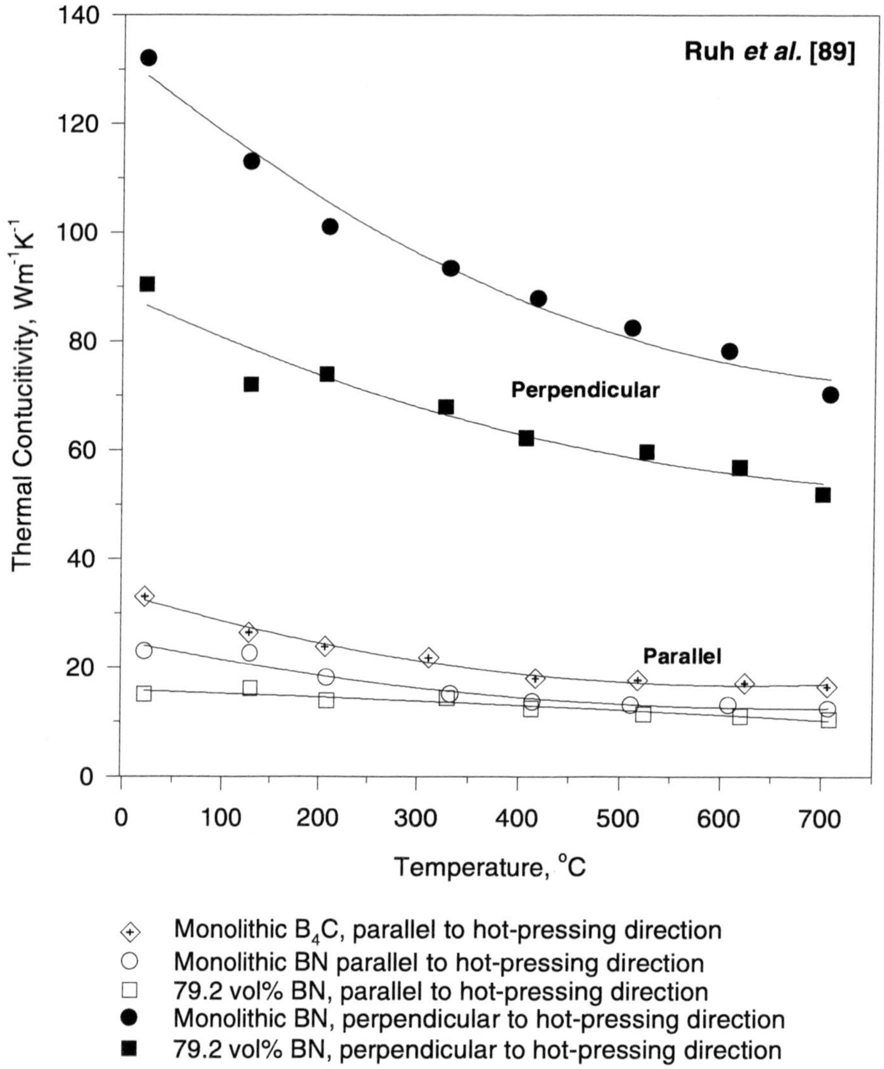

Discussion: Figure 146b illustrates the temperature dependence of thermal conductivity parallel and perpendicular to the hot-pressing direction for B_4C, BN, and a composite of B_4C with 79.2 vol% BN. The thermal conductivity decreased with increasing temperature for all the composites, as was expected for phonon conductors. Comparison of the data for BN and BN containing composites indicated that, at least up to the maximum temperature of 700 °C, the anisotropy ratio of the thermal conductivity maintained about the same value found at room temperature [89]. In summary, the results indicated that BN dispersions in a hot-pressed B_4C matrix served to significantly increase the thermal conductivity perpendicular to the hot-pressing direction, leaving the thermal conductivity parallel to the hot-pressing direction relatively unaffected [89].

The experimentally measured thermal conductivity was lower than the calculated theoretical thermal conductivity. This represents indirect evidence of thermal barriers at grain boundaries [89].

Figure 147: Thermal Diffusivity of BN(p)/SiC

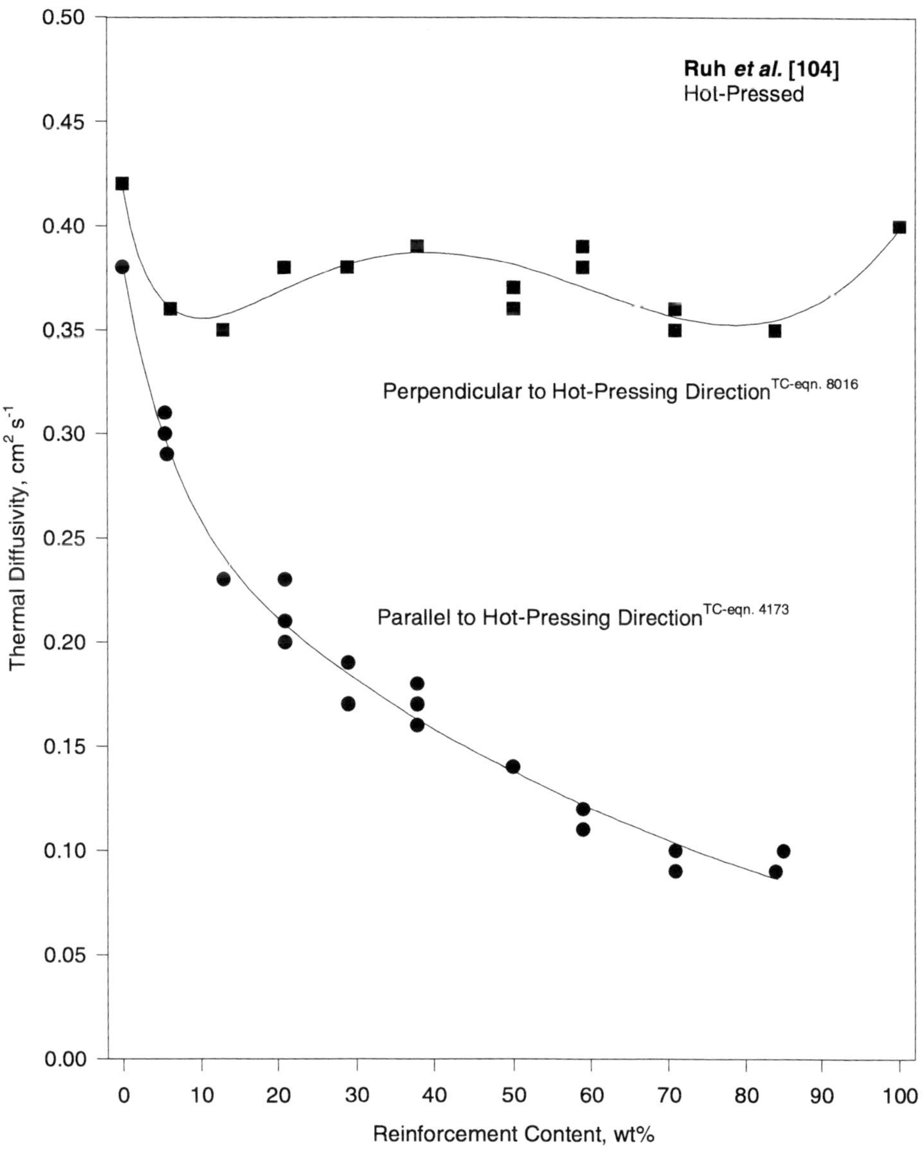

Discussion: The thermal diffusivity of the BN(p)/SiC composites was strongly influenced by the hot-pressing direction. The thermal diffusivity remained nearly constant perpendicular to the hot-pressing direction with increasing BN content. However, the thermal diffusivity rapidly decreased with increasing BN content parallel to the hot-pressing direction. Since the thermal diffusivity along the basal plane of BN was greater than that perpendicular to the basal plane, the anisotropy observed in the composites suggested that BN is preferentially oriented with the basal plane perpendicular to the direction of hot-pressing [104]. These results indicate that h-BN has approximately the same thermal diffusivity along the basal plane as silicon carbide.

BN(p)/SiC

Reference: 104

Composition
30 and 80 wt% BN

Processing Method
Hot-Pressed

Detailed Description
The powders were ball-milled in a plastic container, using Si_3N_4 balls for 4 hours in isopropyl alcohol. After drying, the powders were pelletized and uniaxially hot-pressed in vacuum for 1 hour at 2100 °C under 35 MPa pressure with graphite dies and graphite foil.

Reinforcement Description
h-BN

Reinforcement Supplier
Research Organic/Inorganic Chemical Corp., Belleville, NJ

Supplier Designation
B-27

Reinforcement Dimensions
ave. particle size = 15 μm

Matrix Description
β-SiC

Matrix Supplier
H.C. Stark, Goslar, Germany

Matrix Designation
Lot S 1626

Test Type
Laser-Flash Technique

Specimen Preparation
Sample surfaces were ground.

Specimen Dimensions
thickness = 2 mm

Type of Specimen
Plate

Test Description
A glass-Nd laser was used as the source and a remote optical sensor for monitoring the transient temperature response of the specimen. Room temp measurements were done in air, but the high temp measurements were done in a carbon resistance furnace in nitrogen. Measurements were made perpendicular and parallel to the hot-pressing direction. Density was determined by polishing the surfaces, measuring dimensions, and weighing.

Test Temperature
RT-1400 °C

Test Environment
Air (RT), Nitrogen (>RT)

Test Equipment
glass-Nd laser source, remote optical sensor detector

Temperature (°C)	Thermal Diffusivity ($cm^2 s^{-1}$)	Reinforcement Content (weight percent)	Comments
22	0.39	30	*perpendicular to hot-pressing direction*
300	0.20	30	
400	0.16	30	
500	0.14	30	
600	0.13	30	
700	0.11	30	
800	0.10	30	
900	0.10	30	
1,000	0.09	30	
1,200	0.07	30	
1,300	0.07	30	
1,400	0.07	30	
22	0.21	30	*parallel to hot-pressing direction*
300	0.10	30	
400	0.09	30	
500	0.08	30	
600	0.07	30	
700	0.06	30	
800	0.06	30	
1,000	0.05	30	
1,400	0.03	30	
22	0.36	80	*perpendicular to hot-pressing*
300	0.18	80	
400	0.15	80	
500	0.13	80	
600	0.12	80	
700	0.10	80	
800	0.09	80	
900	0.09	80	
1,000	0.08	80	
1,100	0.08	80	
1,200	0.07	80	
1,300	0.07	80	
1,400	0.06	80	
22	0.09	80	*parallel to hot-pressing*
300	0.05	80	
400	0.04	80	
500	0.04	80	
600	0.03	80	
700	0.03	80	
800	0.03	80	
900	0.03	80	
1,000	0.02	80	
1,100	0.02	80	
1,200	0.02	80	
1,300	0.02	80	
1,400	0.02	80	

Remarks/Comments

These SiC matrices with BN particles appear to be predominantly random in orientation though there was significant measurable variation in the thermal conductivity and diffusivity Thermal diffusivity dropped with increasing temperature due to phonon scattering..

B_4C(p)/Graphite

Reference: 99

Composition
0-30 mol% B

Processing Method
CIPed/HIPed

Detailed Description
The powders are mixed in an alcohol medium and the mixture is dried and the particles separated. The mixture undergoes cold-isostatic compacting prior to insertion in a metal capsule which is sealed under vacuum and placed in an HIP press. Densification is performed in an argon atmosphere at 190 MPa and 1600 °C for two hours. After densification, the metal clad is removed by machining and the test piece is characterized.

Reinforcement Description
B particles

Reinforcement Dimensions
ave. B_4C grain size = 1 μm

Matrix Description
graphite powder

Matrix Properties
ave. grain size = 4.1 μm

Test Type
Laser-Flash Technique

Specimen Preparation
Two large opposite faces of the test pieces were polished, and the edges beveled.

Test Temperature
RT-800 °C

Test Environment
Air

Temperature (°C)	Thermal Conductivity ($W m^{-1} K^{-1}$)	Comments
22	86.0	*0 mol% B*
300	48.1	
600	43.2	
800	39.6	
22	47.2	*10 mol% B*
300	31.6	
600	23.6	
800	27.8	
22	36.1	*20 mol% B*
300	27.6	
600	27.0	
800	20.8	
22	31.0	*30 mol% B*
300	24.2	
600	21.5	
800	20.5	

Remarks/Comments
Due to the low thermal conductivity of B_4C, the thermal conductivity of the 10, 20, and 30 mol% B composites will be lower than that of pure graphite. At 20 °C, thermal conductivity decreased by 45 % with the addition of 10 mol% B. Thermal conductivity decreased by 65 % with the addition of 30 mol% B. Above 600 °C, the thermal conductivity of the three composites was approx. the same, and about 50 % lower than that of pure graphite.

TiB_2(p)/SiC

Reference: 106

Composition
0, 16 vol% TiB_2

Processing Method
Pressureless Sintered

Detailed Description
Mixture combined with binders. The green compacts were pressureless sintered in an inert atmosphere at temperatures > 2000 °C.

Reinforcement Description
TiB_2 particulates

Reinforcement Dimensions
particle size = 1-5 μm

Matrix Description
SiC

Matrix Properties
grain size = 4-8 μm

Test Type
Thermal Diffusivity Tests

Test Temperature
RT-2000 °C

Test Environment
Air

Temperature (°C)	Thermal Diffusivity ($cm^2 s^{-1}$)	Reinforcement Content (volume percent)
30	0.39	0.0
300	0.17	0.0
400	0.16	0.0
500	0.14	0.0
600	0.12	0.0
700	0.11	0.0
800	0.10	0.0
900	0.09	0.0
1,000	0.09	0.0
1,100	0.08	0.0
1,200	0.07	0.0
1,300	0.07	0.0
1,400	0.06	0.0
1,500	0.06	0.0
1,600	0.05	0.0
1,700	0.05	0.0
1,800	0.05	0.0
1,900	0.04	0.0
2,000	0.04	0.0
30	0.41	16
300	0.19	16
400	0.15	16
500	0.13	16
600	0.12	16
700	0.12	16
800	0.11	16
900	0.10	16
1,000	0.10	16
1,100	0.09	16
1,200	0.08	16
1,300	0.08	16
1,400	0.07	16
1,500	0.07	16
1,600	0.06	16
1,700	0.06	16
1,800	0.05	16
1,900	0.06	16
2,000	0.05	16

Remarks/Comments
At 20 °C the diffusivity of composite was higher than SiC. Between 300 and 600 °C the thermal diffusivities were about equal. From 700-2000 °C the thermal diffusivity of the composite was less than the monolithic SiC. The differences in thermal diffusivities was attributed to the material's different heat capacities.

Figure 148: Thermal Conductivity vs. Temperature of B_4C(p)/Graphite

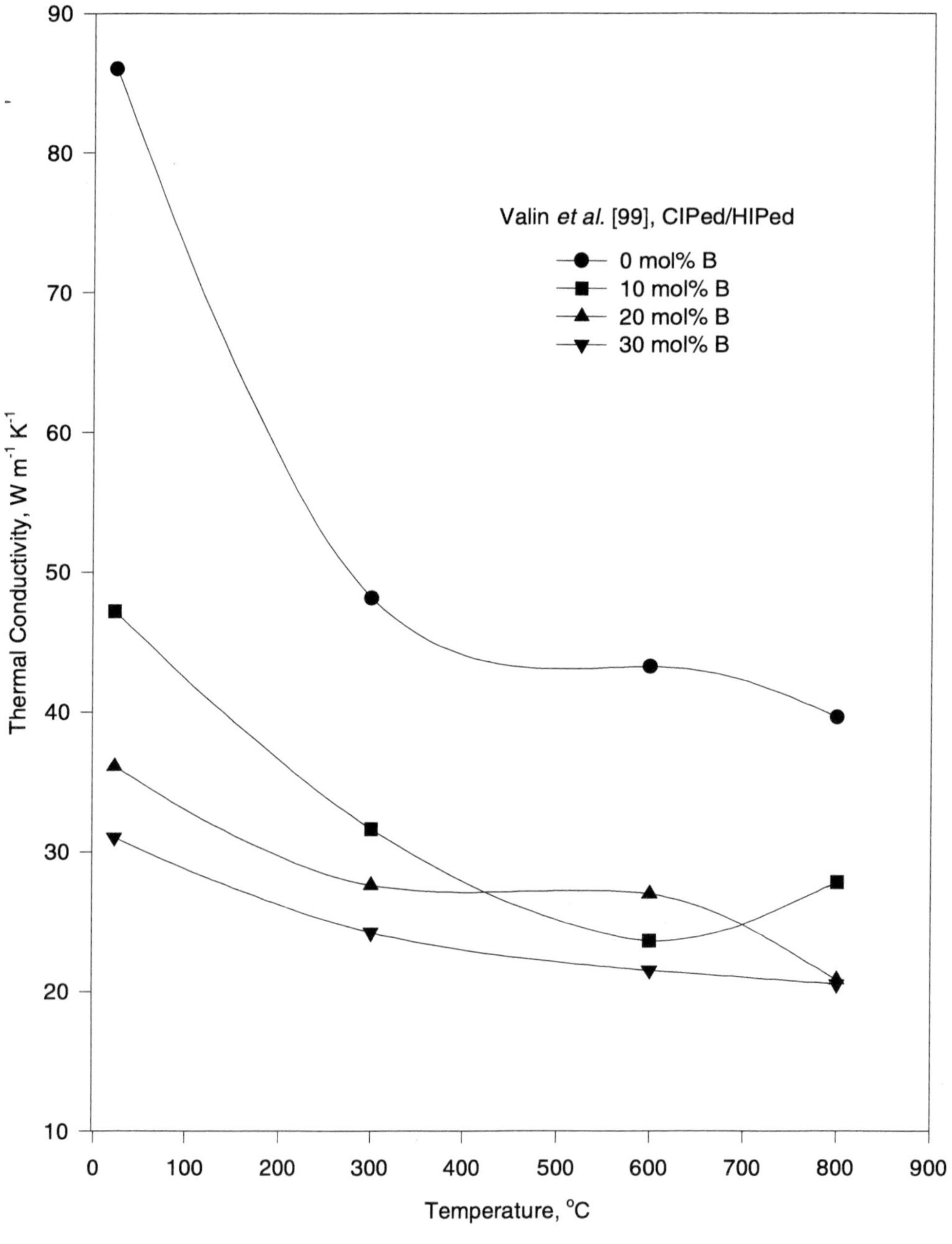

Discussion: The thermal conductivity for the B_4C(p)/Graphite composite increased with increasing mol% boron. Thermal conductivity of the B_4C(p)/Graphite composites decreased with increasing temperature [99].

Figure 149: Thermal Diffusivity vs. Temperature of BN(p)/SiC and TiB_2(p)/SiC

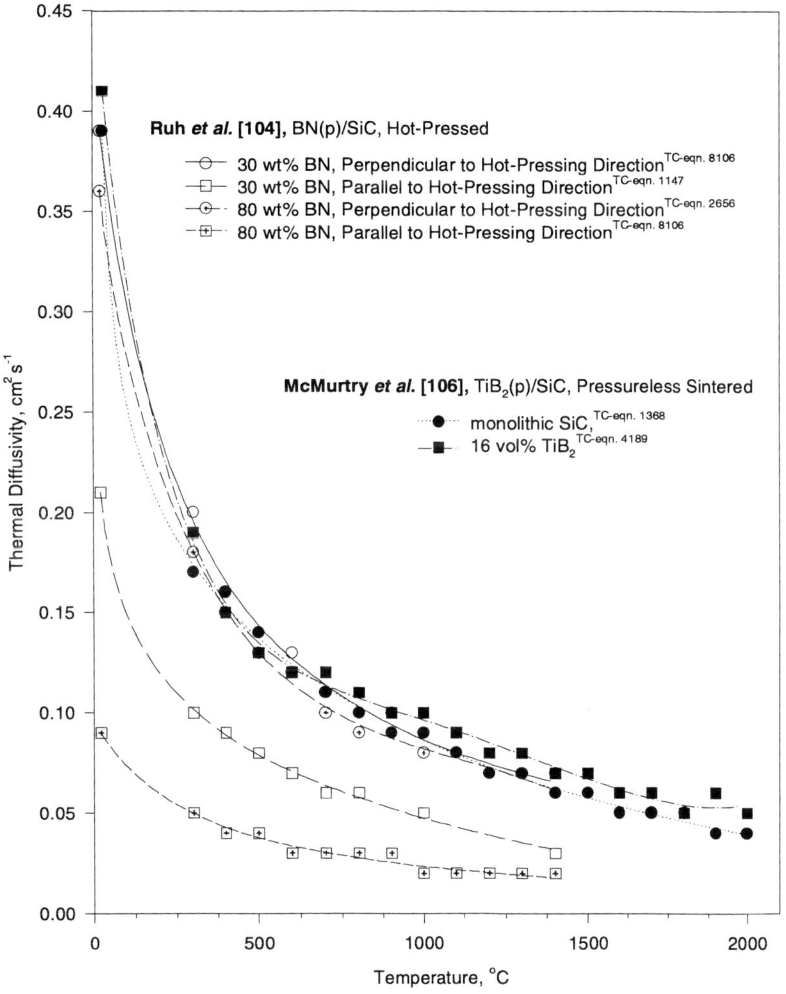

Discussion: The thermal diffusivity for both the BN(p)/SiC and TiB_2(p)/SiC composites decreased with increasing temperature. However, the rate of decrease in thermal conductivity dropped substantially after 600 to 700 °C. The thermal diffusivity for the pressureless sintered TiB_2(p)/SiC composites was greater than the BN(p)/SiC composites at nearly all elevated temperatures. The thermal diffusivity of the BN(p)/SiC composites tested by Ruh *et al.* [104] parallel to the hot-pressing direction decreased rapidly with increasing BN content. This resulted in a maximum anisotropy ratio of approximately four. Since the thermal diffusivity along the basal plane of BN is greater than that perpendicular to the basal plane, the anisotropy observed in the composites suggests that BN is preferentially oriented with the basal plane perpendicular to the direction of hot-pressing [104].

The measured thermal conductivity for the SiC-TiB_2 composite and the monolithic SiC at 20 °C were approximately equal to 112 and 125 W/mK, respectively [106]. Although there were slight differences in the absolute value and temperature dependence of the thermal diffusivity, the thermal conductivity of the composite material, SiC-TiB_2, was similar to monolithic SiC.

$Zr(p).ZrB_2(pl)/ZrC$

Reference: 28

Composition
12 vol% Zr particles
ZrB_2/ZrC ratio = 58:42 by volume

Processing Method
Lanxide Reaction Bonded

Detailed Description
The material is formed by the direct reaction of molten zirconium with a shaped preform of boron carbide particulates, which are substantially consumed in the process. These materials are currently being commercialized as the Lanxide NX-3400 series platelet-reinforced ceramics.

Reinforcement Description
Zr particulates and ZrB_2 platelets

Matrix Description
ZrC particulates

Test Type
Dilatometry

Specimen Dimensions
5 x 6 x 50 mm

Type of Specimen
Bar

Test Temperature
20-1200 °C

Test Environment
Air

Temperature (°C)	Mean Thermal Expansion Coefficient (10^{-6} K^{-1})
20	-
145	6.48
198	6.53
247	6.58
300	6.62
351	6.73
401	6.79
450	6.83
500	6.86
549	6.88
599	6.91
651	6.93
701	6.93
750	6.97
801	6.99
851	6.97
902	6.93
950	6.96
1,000	7.04
1,050	7.07
1,100	7.11
1,150	7.14
1,200	7.14

Remarks/Comments
As temperature increased, the coefficient of thermal expansion remained relatively constant for the composite.

SiC(p)/ZrB_2

Reference: 92

Composition
0 and 20 vol% SiC

Processing Method
Hot-Pressed

Detailed Description
All materials were fabricated by hot-pressing and contained some degree of anisotropy.

Reinforcement Description
SiC particulates

Reinforcement Supplier
Superior Graphite

Reinforcement Dimensions
ave. size = 3 μm

Matrix Description
ZrB_2 powder

Test Type
Thermal Expansion Test

Type of Specimen
Bar

Test Temperature
ZrB_2: 293-2273 K
SiC(p)/ZrB_2: 300 - 2400 K

Reinforcement Content (volume percent)	Thermal Expansion Coefficient (10^{-6} K^{-1})	Density (g cm^{-3})
0.0	8.30	6.10
20.0	7.60	5.47

Remarks/Comments
Thermal expansion coefficient was lower for the composite than for the monolith.

Al_2O_3(p)/TiB_2

Reference: 100

Composition
0 and 30 vol% particulates

Processing Method
Hot-Pressed

Detailed Description
Monolithic TiB_2 hot-pressed at 1850 °C for 1 hour and at 30 MPa. TiB_2-Al_2O_3, 30 vol% TiB_2, was produced by hot-pressing at 1600 °C for 30 minutes and at 30 MPa.

Reinforcement Description
TiB_2 powder

Reinforcement Supplier
Starck

Supplier Designation
Grade F

Matrix Description
Al_2O_3 particulates

Matrix Supplier
ALCOA

Matrix Designation
A16 S.G.

Test Type
Thermal Expansion Test

Type of Specimen
Billet

Test Temperature
unknown

Reinforcement Content (volume percent)	Thermal Expansion Coefficient (10^{-6} K^{-1})
0.0	7.45
30	7.82

Remarks/Comments
Thermal expansion coefficient increased with the addition of 30 vol% Al_2O_3.

$B_4C(p)/TiB_2$

Reference: 101

Composition
0-30 vol% B_4C

Processing Method
Pressureless Sintering

Detailed Description
Powder compacts formed by cold-pressing were fabricated by pressureless sintering at 1600 °C in an argon atmosphere.

Sintering Additives
8.6 mol% Ni, 1.4 mol% C

Reinforcement Description
B_4C powder

Matrix Description
TiB_2 powder

Test Type
Thermal Expansion Test

Specimen Dimensions
15 mm length

Type of Specimen
Beam

Test Temperature
RT-800 °C

Reinforcement Content (volume percent)	Thermal Expansion Coefficient (10^{-6} K^{-1})	Relative Density (percent)
0.0	7.83	94.3
10	7.80	94.7
20	7.62	93.7
30	7.45	88.3

Remarks/Comments
Thermal expansion coefficient decreased with the addition of B_4C.

$SiC(p)/ZrB_2$

Reference: 109

Composition
10 wt% SiC

Processing Method
Hot-Pressed

Reinforcement Description
SiC particulates

Matrix Description
ZrB_2

Test Type
Thermal Expansion Test

Specimen Preparation
Test specimens were produced by wire EDM.

Test Temperature
unknown

Reinforcement Content (weight percent)	Thermal Expansion Coefficient (10^{-6} K^{-1})
10	6.2

Figure 150: Mean Thermal Expansion Coefficient of Zr(p).ZrB_2(pl)/ZrC

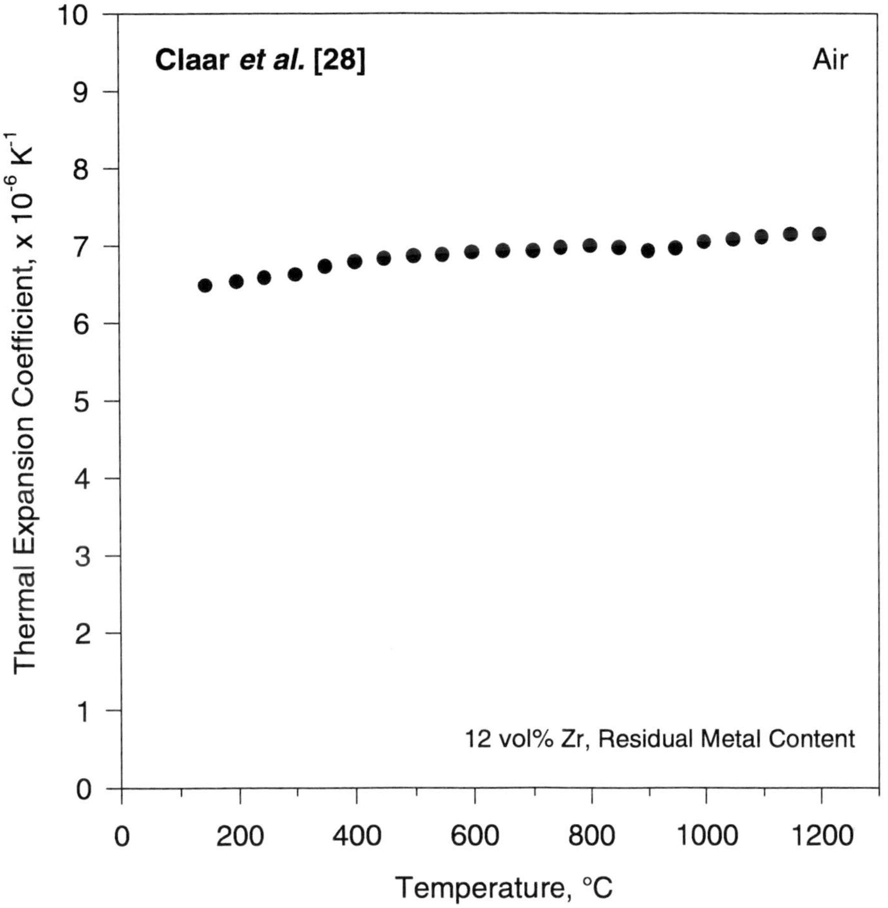

Discussion: The coefficient of thermal expansion for the material containing 12 vol% Zr increased from 6.2 x 10^{-6} K^{-1} at room temperature to 7.14 x 10^{-6} K^{-1} at 1200 °C [28].

Figure 151: Thermal Expansion Coefficients of Particulate Reinforced TiB_2 and ZrB_2

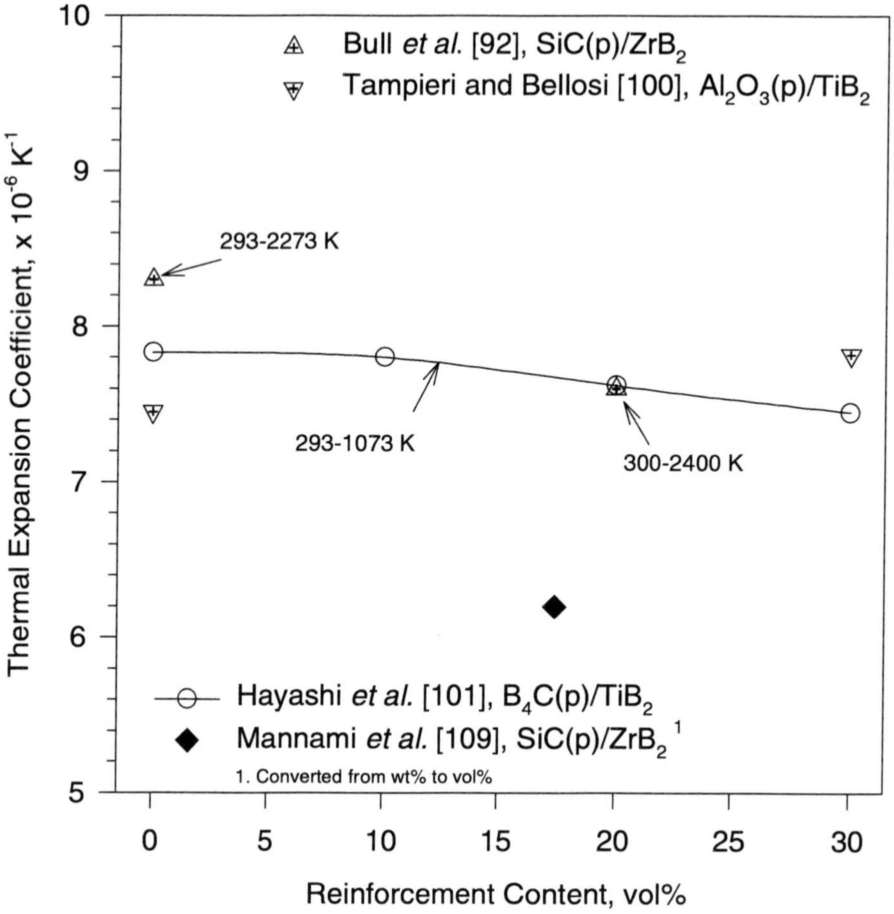

Discussion: Figure 151 shows the results of CTE measurements on selected particulate reinforced TiB_2 and ZrB_2 matrix composites. The CTE of a SiC(p)/ZrB_2 composite measured by Bull *et al.* [92] was considerably higher than the SiC(p)/ZrB_2 composite measured by Mannami *et al.* [109]. The CTE of the Al_2O_3(p)/TiB_2 and the B_4C(p)/TiB_2 samples were similar, both fell within 7-8 x10^{-6} K^{-1}, Tampieri and Bellosi [100] and Hayashi *et al.* [101].

SiC(p)/ZrB$_2$

Reference: 92

Composition
0 and 20 vol% SiC

Processing Method
Hot-Pressed

Detailed Description
All materials were fabricated by hot-pressing and contain some degree of anisotropy.

Reinforcement Description
SiC particulates

Reinforcement Supplier
Superior Graphite

Reinforcement Dimensions
ave. size = 3 μm

Matrix Description
ZrB$_2$ powder

Test Type
Laser-Flash Technique

Type of Specimen
Bar

Reinforcement Content (volume percent)	Thermal Conductivity (W m^{-1} K^{-1})	Density (g cm^{-3})
0.0	87	6.10
20	73	5.47

Remarks/Comments
Data taken from tests done at temperatures of 22-2000 °C for the monolith and 22-2100 °C for the composite. Thermal conductivity was lower for the composite than for the monolith.

Diamond(p)/SiC

Reference: 61

Composition
0-35 vol% diamond

Processing Method
Hot-Pressed

Detailed Description
The SiC powder and diamond particles were mixed in aqueous suspensions. The powder suspensions were evaporated during ultra-sonic vibration to provide thorough dispersion of the of the individual diamond particles in the dry powder. This composite was pressed at 5.5 GPa and 1500 °C in a press capable of operation in the diamond stability range.

Reinforcement Description
diamond particulates

Reinforcement Supplier
E.I. du Pont de Nemours & Co., Gibbstown, NJ

Supplier Designation
du Pont Mypolex

Reinforcement Dimensions
ave. particle size = 11 nm

Matrix Description
SiC powder

Matrix Supplier
Showa Denko K. K., Minato-Ku, Japan

Matrix Designation
DENSIC Type 2

Test Type
Laser-Flash Technique

Specimen Preparation
Specimens were polished with 1 μm diamond paste, and ultrasonically cleaned to remove the polishing medium.

Reinforcement Content (volume percent)	Thermal Conductivity (W m^{-1} K^{-1})
0.0	65.3
5.0	69.7
10	74.9
15	80.6
20	86.6
25	93.4
30	100
35	109

Remarks/Comments
Thermal conductivity of SiC was enhanced by the addition of diamond particles.

Figure 152: Thermal Conductivity of SiC(p)/ZrB_2, and Diamond(p)/SiC

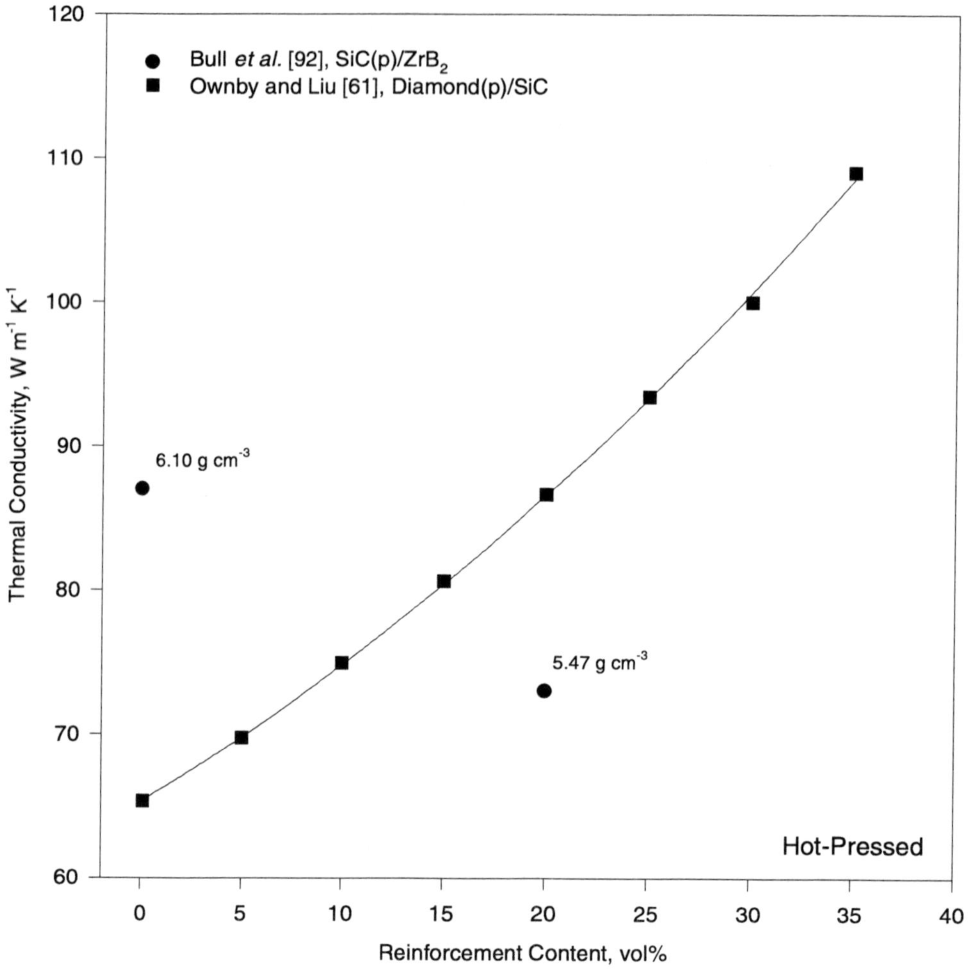

Discussion: The thermal conductivity for the diamond(p)/SiC composite increased according to the rule of mixtures with increasing diamond particulate reinforcement content. Diamond has a very high thermal conductivity (543 W/mK) whereas silicon carbide has a lower thermal conductivity (20-87 W/mK) [61]. The thermal conductivity and density for the SiC(p)/ZrB_2 composite decreased with the addition of SiC reinforcement to the monolithic ZrB_2 [92].

15. References
(Section VI)

References

1. R. W. Rice, "Ceramic Matrix Composite Toughening Mechanisms: An Update," *Ceram. Eng. Sci. Proc.* 6 (7-8): 589-607 (1985).

2. R. W. Rice, "Mechanisms of Toughening in Ceramic Matrix Composites," *Ceram. Eng. Sci. Proc.* 2 (7-8): 661-701 (1981).

3. G. D. Quinn and R. Morrell, "Design Data for Engineering Ceramics: A Review of the Flexure Test," *J. Am. Ceram. Soc.* 74 (9): 2037-66 (1991).

4. G. D. Quinn, "Fracture Toughness of Advanced Ceramics at Room Temperature: A VAMAS Round Robin," *Ceram. Eng. Sci. Proc.* 14 (7-8): 92-100 (1993).

5. K. E. Amin, "Toughness, Hardness, and Wear," *Ceramics and Glasses*, Vol. 4, Engineered Materials Handbook (Materials Park, OH: ASM International, 1991), 599-609.

6. K. D. Maglic, A. Cezairliyan, and V. E. Peletsky, *Survey of Measurement Techniques*, Vol. 1, Compendium of Thermophysical Property Measurement Methods (New York: Plenum Press, 1984).

7. F. F. Lange, "Transformation Toughening in the Al_2O_3/ZrO_2 Composite System," Technical Report 7, Document No. SC5117.7TR, Oct. 1979.

8. R. P. Wahi and B. Ilschner, "Fracture Behaviour of Composites Based on Al_2O_3-TiC," *J. Mater. Sci.* 15 (4): 875-85 (1980).

9. J. P. Singh, D. P. H. Hasselman, W. M. Su, J. A. Rubin, and R. Palicka, "Observations on the Nature of Micro-cracking in Brittle Composites," *J. Mater. Sci.* 16 (1): 141-50 (1981).

10. J. S. Moya and M. I. Osendi, "Microstructure and Mechanical Properties of Mullite/ZrO_2 Composites," *J. Mater. Sci.* 19 (9): 2909-14 (1984).

11. M. A. Janney, "Particulate Toughened Silicon Carbide," Oak Ridge National Laboratory, Technical Report, Conf-8410103-4, 8 pp., 1984.

12. R. C. Buchanan and W. W. Davison, "Influence of Al_2O_3 on Properties of Yttria Stabilized Zirconia-Al_2O_3 Composites," Report No. TR-13, AD-A167 328, 43 pp., April 1986.

13. T. Y. Tien, T. K. Brog, and A. K. Li, "Toughened Ceramics in the System Al_2O_3:Cr_2O_3/ZrO_2:HfO_2," *Int. J. High Tech. Ceram.* 2 (3): 207-19 (1986).

14. P. Vincenzini, A. Bellosi, and G. N. Babini, "Thermal Instability of Si_3N_4/ZrO_2 Composites," *Ceram. Int.* 12 (3): 133-45 (1986).

15. T. Sato, A. Shiratori, and M. Shimada, "Sintering and Fracture Behavior of Composites Based on Alumina-Zirconia (Yttria) - Nonoxides," *J. de Physique* 47 (2): 733-7 (1986).

16. W. Kladnig and G. Gritzner, "Bend Strength of Al_2O_3-ZrO_2 Composites," *J. Mater. Sci. Lett.* 6 (10): 1235-7 (1987).

17. R. Ruh, A. Zangvil, and R. R. Wills, "Phase and Property Studies of SiC-BN Composites," *Adv. Ceram. Mater.* 3 (4): 411-5 (1988).

18. E. Min-Haga and W. D. Scott, "Sintering and Mechanical Properties of ZrC-ZrO_2 Composites," *J. Mater. Sci.* 23 (8): 2865-70 (1988).

19. C. Martin, B. Cales, P. Vivier, and P. Mathieu, "Electrical Discharge Machinable Ceramic Composites," *Mater. Sci. Eng. A.* A109 (1-2): 351-6 (1989).

20. D. L. Jiang, J. H. Wang, Y. L. Li, and L. T. Ma, "Studies on the Strengthening of Silicon Carbide-based Multiphase Ceramics, I: The SiC-TiC System," *Mater. Sci. Eng. A.* A109 (1-2): 401-6 (1989).

21. C. R. Blanchard-Ardid and R. A. Page, "Improved Contact Damage Resistance of a Si_3N_4/TiC/SiC (w) Composite," *Ceram Eng. Sci. Proc.* 9 (9-10): 1443-52 (1988).

22. T. Ono, K. Nagata, M. Hashiba, E. Miura, Y. Nurishi, and T. Shimada, "Internal Friction, Crack Length of Fracture Origin and Fracture Surface Energy in Alumina-Zirconia Composites," *J. Mater. Sci.* 24 (6): 1974-8 (1989).

23. T. Ishigaki, K. Sato, and Y. Moriyoshi, "Pressureless Sintering of TiC- Al_2O_3 Composites," *J. Mater. Sci. Lett.* 8 (6): 678-80 (1989).

24. T. N. Tiegs, "Tailoring of Properties of SiC Whisker-Oxide Matrix Composites," Oak Ridge National Laboratory, Technical Report, Conf-881125-1, 18 pp., 1989.

25. W. Stadlbauer, W. Kladnig, and G. Gritzner, "Al_2O_3-TiB_2 Composite Ceramics," *J. Mater. Sci. Lett.* 8 (10): 1217-20 (1989).

26. W. C. Moffatt and H. K. Bowen, "Thermal Processing and Properties of Highly Homogeneous Alumina-Zirconia Composite Ceramics," *J. Mater. Sci.* 24 (11): 3984-90 (1989).

27. A. K. Mukhopadhyay and D. Chakraborty, "Influence of Some Parameters on the Strength and Fracture Toughness of Reaction-Bonded Silicon Nitride Composites," *Mater. Sci. Eng. A.* A122 (2): 173-82 (1989).

28. T. D. Claar, W. B. Johnson, C. A. Andersson, and G. H. Schiroky, "Microstructure and Properties of Platelet-Reinforced Ceramics Formed by the Directed Reaction of Zirconium with Boron Carbide," *Ceram. Eng. Sci. Proc.* 10 (7-8): 599-609 (1989).

29. I. Kimura, N. Hotta, Y. Hiraoka, N. Saito, and Y. Yokota, "Sintering and Characterization of Al_2O_3-TiB_2 Composites," *J. Eur. Ceram. Soc.* 5 (1): 23-7 (1989).

30. H. Suzuki and H. Saito, "Preparation and Sintering of Fine Composite Precursors of Mullite-Zirconia by Chemical Copolymerization of Metal Alkoxides," *J. Mater. Sci.* 25 (4): 2253-8 (1990).

31. H. Tanaka, P. Greil, and G. Petzow, "Sintering and Strength of Silicon Nitrided-Silicon Carbide Composites," *Int. J. High Tech. Ceram.* 1 (2): 107-18 (1985).

32. K.-H. Yang and A. S. Kobayashi, "Dynamic Fracture Responses of Alumina and Two Ceramic Composites," *J. Am. Ceram. Soc.* 73 (8): 2309-15 (1990).

33. C. P. Cameron, J. H. Enloe, L. E. Dolhert, and R. W. Rice, "A Comparison of Reaction vs Conventionally Hot-Pressed Ceramic Composites," *Ceram. Eng. Sci. Proc.* 11 (9-10): 1190-202 (1990).

34. S. Torizuka, J. Harada, and H. Nishio, "High Strength Titanium Diboride," *Ceram. Eng. Sci. Proc.* 11 (9-10): 1454-60 (1990).

35. A. Lightfoot, H. L. Ker, J. S. Haggerty, and J. E. Ritter, "Properties of RBSN and RBSN-SiC Composites," *Ceram. Eng. Sci. Proc.* 11 (7-8): 842-56 (1990).

36. Y. Yasutomi and M. Sobue, "Development of Reaction-Bonded Electro-Conductive Titanium Nitride-Silicon Nitride and Resistive Alumina-Silicon Nitride Composites," *Ceram. Eng. Sci. Proc.* 11 (7-8): 857-67 (1990).

37. Y. Akimune, "Impact Damage and Strength Degradation in a Silicon Carbide Reinforced Silicon Nitride Composite," *J. Am. Ceram. Soc.* 73 (10): 3019-25 (1990).

38. C. R. Blanchard and R. A. Page, "Effect of Silicon Carbide Whisker and Titanium Carbide Particulate Additions on the Friction and Wear Behavior of Silicon Nitride," *J. Am. Ceram. Soc.* 73 (11): 3442-52 (1990).

39. C. C. Agrafiotis, J. Lis, J. A. Puszynski, and V. Hlavacek, "Combustion Synthesis of Silicon Nitride-Silicon Carbide Composites," *J. Am. Ceram. Soc.* 73 (11): 3514-7 (1990).

40. Y. Akimune, "Spherical-Impact Damage and Strength Degradation in Silicon Nitride-Silicon Carbide Composites," *J. Mater. Sci.* 25 (8): 3439-48 (1990).

41. J. J. Petrovic and R. E. Honnell, "$MoSi_2$ Particle Reinforced-SiC and Si_3N_4 Matrix Composites," *J. Mater. Sci. Lett.* 9 (9): 1083-4 (1990).

42. S. T. Buljan, J. G. Baldoni, M. L. Huckabee, and G. Zilberstein, "Microstructure and Fracture Toughness of Silicon Nitride Composites," *Whisker Fiber-Toughened Ceram., Proc. Int. Conf. 1988*, eds. R. A. Bradley, D. E. Clark, D. C. Larsen and J. O. Stiegler (Metals Park, OH, ASM International, 1988), 113-23.

43. J. Liu and P. D. Ownby, "Enhanced Mechanical Properties of Alumina by Dispersed Titanium Diboride Particulate Inclusions," *J. Am. Ceram. Soc.* 74 (1): 241-3 (1991).

44. Y. Akimune, N. Hirosaki and T. Ogasawara, "Mechanical Properties of SiC-Particle/Sialon Composites," *J. Mater. Sci. Lett.* 10: 223-6 (1991).

45. J. Liu and P. D. Ownby, "Boron Carbide Reinforced Alumina Composites," *J. Am. Ceram. Soc.* 74 (3): 674-7 (1991).

46. Akihiro Sawaguchi, Kohji Toda, and Koichi Niihara, "Mechanical and Electrical Properties of Silicon Nitride-Silicon Carbide Nanocomposite Material," *J. Am. Ceram. Soc.* 74 (5): 1142-4 (1991).

47. Mitsuhiko Furukawa, Osamu Nakano, and Yoshio Takashima, "Fracture Toughness of Al_2O_3-TiC Ceramics," *Int. J. Refract. Hard Metals* 7 (1): 37-40 (1988).

48. Robert Ruh, K. S. Mazdiyasni, and M. G. Mendiratta, "Mechanical and Microstructural Characterization of Mullite and Mullite-SiC-Whisker and ZrO_2-Toughened-Mullite-SiC-Whisker Composites," *J. Am. Ceram. Soc.* 71 (6): 503-12 (1988).

49. Hideto Kamiaka, Chitake Yamagishi, and Junji Asaumi, "Mechanical Properties and Microstructure of Mullite-SiC-ZrO_2 Particulate Composite," *Ceramic Transactions 6, Mullite and Mullite Matrix Composites*, eds. Shigeyuki Somiya, Robert F. Davis, and Joseph A. Pask (Westerville, OH, American Ceramic Society, 1990), 509-18.

50. K. Rundgren, P. Elfving, H. Tabata, S. Kanzaki, and R. Pompe, "Microstructures and Mechanical Properties of Mullite-Zirconia Composites Made from Inorganic Sols and Salts," *Ceramic Transactions 6, Mullite and Mullite Matrix Composites*; eds. Shigeyuki Somiya, Robert F. Davis, and Joseph A. Pask (Westerville, OH, American Ceramic Society, 1990), 553-66.

51. Michael G. Jenkins, Jonathan A. Salem, and Srinivasa G. Seshadri, "Fracture Resistance of a TiB_2 Particle/SiC Matrix Composite at Elevated Temperature," *J. Comp. Mater.* 23 (1): 77-91 (1989).

52. Takao Kanai, Kei Tanemoto, and Hiroshi Kubo, "Hot-Pressed BN-AlN Ceramic Composites of High Thermal Conductivity," *Jpn. J. Appl. Phys., Part 1* 29 (4): 683-7 (1990).

53. H. Endo, M. Ueki, and H. Kubo, "Hot Pressing of SiC-TiC Composites," *J. Mater. Sci.* 25 (5): 2503-6 (1990).

54. A. Bellosi, A. Tampieri, and Yu-Zhen Liu, "Oxidation Behaviour of Electroconductive Si_3N_4-TiN Composites," *Mater. Sci. Eng. A.* A127 (1): 115-22 (1990).

55. L. D. Bentsen and D. P. H. Hasselman, "Role of Porosity in the Effect of Microcracking on the Thermal Conductivity of Brittle Ceramic Composites," *Thermal Conductivity 18*, eds. T. Ashworth and David R. (New York, Plenum Press, 1993), 485-98.

56. N. Kishan Reddy and J. Mukerji, "Preparation and Characterization of Silicon Nitride-Silicon Carbide Composites," *Bull. Mater. Sci.* 13 (3): 173-8 (1990).

57. A. Srinivasa Rao and W. Roger Cannon, "Alumina/Zirconia Composites Produced from Commercial Alumina and Synthesised Hydrous Zirconia," *Ceram. Int.* 15 (3): 179-88 (1989).

58. Raymond A. Cutler, Andrew C. Hurford, and Anil V. Virkar, " Pressureless-Sintered Al_2O_3-TiC Composites," *Mater. Sci. Eng. A.* A105-106 (1-2): 183-92 (1988).

59. Jenq Liu and P. Darrell Ownby, "Normal-Pressure Hot-Pressing of Alpha-Alumina-Diamond Composites," *J. Am. Ceram. Soc.* 74 (10): 2666-8 (1991).

60. Jenq Liu and P. Darrell Ownby, "Physical Properties of Alumina-Boron Carbide Whisker/Particle Composites," *Ceram. Eng. Sci. Proc.* 12 (7-8): 1245-53 (1991).

61. P. Darrell Ownby and Jenq Liu, "Nano-Diamond Enhanced Silicon Carbide Matrix Composites," *Ceram. Eng. Sci. Proc.* 12 (7-8): 1345-55 (1991).

62. Sergej T. Buljan, J. Gary Baldoni, and Marvin L. Huckabee, "Si_3N_4-SiC Composites," *Am. Ceram. Soc. Bull.* 66 (2): 347-52 (1987).

63. Youren Xu, Avigdor Zangvil, Martine Landon, and Francois Thevenot, " Microstructure and Mechanical Properties of Hot-Pressed Silicon Carbide-Aluminum Nitride Compositions," *J. Am. Ceram. Soc.* 75 (2): 325-33 (1992).

64. Q. L. Ge, T. C. Lei, and Y. Zhou, "Microstructure and Mechanical Properties of Hot Pressed Al_2O_3-ZrO_2 Ceramics Prepared from Ultrafine Powders," *Mater. Sci. Technol.* 7 (6): 490-4 (1991).

65. G. G. Gnesin, I. I. Osipova, and L. L. Sartinskaya, "Ultadispersed Nitride Powder-Base Ceramic Tool Materials," *Sov. Powder Metall. Met. Ceram.* 29 (12): 1011-14 (1991).

66. Ekkehard H. Lutz and Michael V. Swain, "Fracture Toughness and Thermal Shock Behavior of Silicon Nitride-Boron Nitride Ceramics," *J. Am. Ceram. Soc.* 75 (1): 67-70 (1992).

67. Ralf Riedel, Martin Seher, and Gerd Becker, "Sintering of Amorphous Polymer-Derived Si, N and C Containing Composite Powders," *J. Eur. Ceram. Soc.* 5 (2): 113-22 (1989).

68. Young-Wook Kim and June-Gunn Lee, "Pressureless Sintering of Alumina-Titanium Carbide Composites," *J. Am. Ceram. Soc.* 72 (8): 1333-7 (1989).

69. Toshihiko Tani and Shigetaka Wada, "Pressureless-Sintered and HIPed SiC-TiB_2 Composites from SiC-TiO_2-B_4C-C Powder Compacts," *J. Mater. Sci.* 26 (13): 3491-6 (1991).

70. H. Endo, M. Ueki, and H. Kubo, "Microstructure and Mechanical Properties of Hot-Pressed SiC-TiC Composites," *J. Mater. Sci.* 26 (14): 3769-74 (1991).

71. Zhijian Shen, Tingkai Li, and Zishang Ding, "Synergistic Toughening Mechanism of Whisker Reinforced Zirconia Toughened Ceramics," *Advanced Structural Materials; Proceedings of the International Conference, C-MRS Int. Symp. Proc,* ed. Yafang Han; (Amsterdam, Elsevier Science Publishers, 1991), 319-24.

72. Jeffrey J. Swab and Steven C. Stowell, "Properties of a TZP/Al_2O_3 Composite After Long-Term Exposure at 1000 °C," U.S. Army Materials Technology Laboratory, Final Report, AD-A246 407, MTL-TR-91-54, 14 pp., 1991.

73. Young-Jei Oh, Tae-Sung Oh, and Hyung-Jim Jung, "Microstructure and Mechanical Properties of Cordierite Ceramics Toughened by Monoclinic ZrO_2," *J. Mater. Sci.* 26 (23): 6491-5 (1991).

74. S. V. Nair, Peter Z. Q. Cai, J. E. Ritter, A. Lightfoot, and J. S. Haggerty, "Mechanical Behavior of Silicon Carbide Particulate Reinforced Reaction Bonded Silicon Nitride Matrix Composites," *Ceram. Eng. Sci. Proc.* 13 (7-8): 81-9 (1992).

75. Timothy V. Lin and P. Darrell Ownby, "Densification and Fracture Toughness Enhancement of Pressureless Sintered Aluminum Oxide-Titanium Diboride Composites," *Ceram. Eng. Sci. Proc.* 13 (7-8): 132-9 (1992).

76. J. Liu and P. D. Ownby, "Boron Containing Ceramic Particulate and Whisker Enhancement of the Fracture Toughness of Ceramic Matrix Composites," *Proceedings of the Conference on Boron-Rich Solids; AIP Conf. Proc.,* (New York, American Institute of Physics, 1991), 574-7.

77. K. Niihara and A. Nakahira, "Strengthening and Toughening Mechanisms in Nanocomposite Ceramics," *Ann. Chim. (Paris)* 16 (4-6): 479-86 (1991).

78. C. Yamagishi, K. Tsukamoto, J. Hakoshima, H. Shimojima, and Y. Akiyama, "Mechanical Properties and Microstructure of β-SiAlON-β-SiC Composites by Pressureless Sintering," *J. Mater. Sci.* 27: 1908-12 (1992).

79. Takao Kanai, Kei Tanemoto, and Hiroshi Kubo, "Reaction-Bonded Boron Nitride-Aluminum Nitride Ceramic Composites," *Jpn. J. Appl. Phys., Part 1* 30 (6): 1235-8 (1991).

80. Y. Akimune, Y. Katano, T. Akiba, and T. Ogasawara, "Mechanical Properties of Air-Annealed SiC-Particle/Al_2O_3 Composite," *J. Mater. Sci. Lett.* 11 (10): 695-7 (1992).

81. M. Hännien, R. A. Haber, and D. E. Niesz, "Mechanical Properties of Silicon Carbide Platelet and Particulate Reinforced Silicon Nitride," *Ceramic Transactions 18, Advanced Composite Materials: Proceedings of the Symposium on Composites; Processing, Microstructure, and Properties,* ed. Michael D. Sacks; (Westerville, OH, American Ceramic Society, 1991), 749-55.

82. Junhong Zhao, Laura C. Stearns, Martin P. Harmer, Helen M. Chan, Gary A. Miller, and Robert E. Cook, "Mechanical Behavior of Alumina-Silicon Carbide Nanocomposites," *J. Am. Ceram. Soc.* 76 (2): 503-10 (1993).

83. Oyelayo O. Ajayi, Ali Erdemir, Richard H. Lee, and Fredd A. Nichols, "Sliding Wear of Silicon Carbide-Titanium Boride Ceramic-Matrix Composite," *J. Am. Ceram. Soc.* 76 (2): 511-17 (1993).

84. Alida Bellosi and Gian N. Babina, "Sintering and Characteristics of Electroconductive Al_2O_3-Based Composites," *Proceedings of the International Symposium on Ceramic Materials and Components for Engines,* 4th, eds. Roger Carlsson, Thomas Johansson and Lars Kahlman; (London, Elsevier, 1992), 389-96.

85. Xiaoxian Huang, Jingkun Guo, Linhua Gui, and Baoshun Li, "SiC Particulate Reinforced Y-TZP Composites," *Proceedings of the International Symposium on Ceramic Materials and Components for Engines,* 4th, eds. Roger Carlsson, Thomas Johansson and Lars Kahlman; (London, Elsevier, 1992), 757-69.

86. X. X. Huang, J. S. Hong, and J. K. Guo, "SiC Particle and Y-TZP Reinforced Mullite Matrix Composites," *Proceedings of the International Symposium on Ceramic Materials and Components for Engines,* 4th, eds. Roger Carlsson, Thomas Johansson and Lars Kahlman; (London, Elsevier, 1992), 795-803.

87. P. Arato, L. Bartha, A. Kuhne, and F. Thummler, "Effect of Post-HIP Treatment on the Properties of Y-Sialon Ceramics," *Ber. Dtsch. Keram. Ges.* 69 (10): 383-4 (1992).

88. W. F. Fischer III, R. A. Haber, and R. M. Anderson, "Mechanical Properties of Alumina Matrix Com-

posites Reinforced With Silicon Carbide Platelets and Particulate," *Ceramic Transactions 19, Advanced Composite Materials: Proceedings of the Symposium on Composites; Processing, Microstructure, and Properties*, ed. Michael D. Sacks; (Westerville, OH, American Ceramic Society, 1991), 773-9.

89. Robert Ruh, Kimberly Y. Donaldson, and D. P. H. Hasselman, "Thermal Conductivity of Boron Carbide-Boron Nitride Composites," *J. Am. Ceram. Soc.* 75 (10): 2887-90 (1992).

90. Bor-Wen Lin, Toyohiko Yano, and Takayoshi Iseki, "High-Temperature Toughening Mechanism in SiC/TiC Composites," *Nippon Seramikkusu Kyokai Gakujutsu Ronbunshi (J. Ceram. Soc. Jpn.)* 100 (4): 509-13 (1992).

91. Dongfang Wang, Jian Li, Zhijian Shen, and Zhiyuan Mao, "The Study of Erosion Resistance and Mechanisms of Silicon Carbide Whisker-Reinforced Silicon Nitride Ceramic," *First Pacific Rim International Conference on Advanced Materials and Processing (PRICM-1)*, eds. Changxu Shi, Hengde Li and Alexander Scott; (Warrendale, PA; Minerals, Metals, and Materials Society, 1993), 721-5.

92. Jeffrey D. Bull, Daniel J. Rasky, and Janet C. Karika, "Characterization of Selected Diboride Composites," *Proceedings of the 16th NASA/DoD Conference on Metal Matrix, Carbon, and Ceramic Matrix Composites*, ed. John D. Buckley; NASA-CP-3175-Part 1, 247-65 (1993).

93. Gen Sasaki, Hirofumi Nakase, Katsuaki Suganuma, Teruaki Fujita, and Koichi Nihara, "Mechanical Properties and Microstructure of Si_3N_4 Matrix Composite with Nano-Meter Scale SiC Particles," *Nippon Seramikkusu Kyokai Gakujutsu Ronbunshi (J. Ceram. Soc. Jpn., Int. Edition)* 100 (4): 536-40 (1992).

94. J. Mukerji and S. K. Biswas, "Preparation of TiN- Al_2O_3 Composite Following a Novel Route," *Adv. Compos. Mater., Proceedings of the International Conference and Exhibition on Advances in Composite Materials (ICACM-90)*, ed. P. Ramakrishnan (New Delhi, Oxford Publishing, 1992), 465-72.

95. J. Mukerji and P. K. Das, "Preparation and Properties of Si_3N_4-TiC and Si_3N_4-SiC Composites," *Adv. Compos. Mater. Proceedings of the International Conference and Exhibition on Advances in Composite Materials (ICACM-90)*, ed. P. Ramakrishnan (New Delhi, Oxford Publishing, 1992), 473-82.

96. Y. G. Tkachenko, D. Z. Yurchenko, G. S. Oleinik, O. A. Shevchenko, and S. A. Satanin, "Self-Reinforced Materials Based on Aluminum Nitride," *Sov. Powder Metall. Met. Ceram.* (9): 85-9 (1993).

97. D. P. H. Hasselman, K. Y. Donaldson, E. M. Anderson, and T. A. Johnson, "Effect of Thermal History on the Thermal Diffusivity and Thermal Expansion of an Alumina-Aluminum Titanate Composite," *J. Am. Ceram. Soc.* 76 (9): 2180-4 (1993).

98. S. A. Mariano, D. Friel, and I. Bar-On, "Elevated Temperature Mechanical Properties of SiC-AlN Particulate Composites," *Ceram. Eng. Sci. Proc.* 14 (9-10): 1077-88 (1993).

99. F. Valin, J. D. Lulewicz, M. Boncoeur, and J. G. Van der Laan, "Boron - Carbon Composites by HIP," *Ceram. Eng. Sci. Proc.* 14 (9-10): 1184-9 (1993).

100. A. Tampieri and A. Bellosi, "Oxidation of Monolithic TiB_2 and of Al_2O_3-TiB_2 Composite," *J. Mater. Sci.* 28 (3): 649-53 (1993).

101. Shinsuke Hayashi, Yasuyoshi Kobayashi, and Hajime Saito, "TiB_2-B_4C Composites Pressureless-Sintered Using Ni and C as Densification Aids," *J. Ceram. Soc. Jpn.* 101 (2): 154-8 (1993).

102. D. P. H. Hasselman, K. Y. Donaldson, J. Liu, L. J. Gauckler, and P. D. Ownby, "Thermal Conductivity of a Particulate-Diamond-Reinforced Cordierite Matrix Composite," *J. Am. Ceram. Soc.* 77 (7): 1757-60 (1994).

103. K. Niihara, A. Nakahira, and T. Sekino, "New Nanocomposite Structural Ceramics," *Nanophase and Nanocomposite Materials*, Vol. 286, MRS Symposium Proceedings (1993), 405-12.

104. Robert Ruh, Larry D. Bentsen, and D. P. H. Hasselman, "Thermal Diffusivity Anisotropy of SiC/BN Composites," *J. Am. Ceram. Soc.* 67 (5): C-83-C-84 (1984).

105. Mark A. Janney, "Microstructural Development and Mechanical Properties of SiC and of SiC-TiC Composites," *Am. Ceram. Soc. Bull.* 65 (2): 357-62 (1986).

106. Carl H. McMurtry, Wolfgang D. G. Boecker, Srinivasa G. Seshadri, Joseph S. Zanghi, and John E. Garnier, "Microstructure and Material Properties of SiC-TiB_2 Particulate Composites," *Am. Ceram. Soc. Bull.* 66 (2): 325-9 (1987).

107. Mark A. Janney, "Mechanical Properties and Oxidation Behavior of a Hot-Pressed SiC-15-vol%-TiB_2 Composite," *Am. Ceram. Soc. Bull.* 66 (2): 322-4 (1987).

108. Wenbang Zhang and F. P. Glasser, "Gel Densification and Ceramic Sintering of ZrO_2-Al_2O_3 Composites" *J. Eur. Ceram. Soc.* 13 (1): 25-31 (1994).

109. Kazuo Mannami, Kohzo Sakai, and Massataro Okumiya, "Electrical Discharge Machining of ZrB_2-Based Composite Ceramics," *Yogyo-Kyokai-Shi* 94 (1): 204-23 1986.

110. C. C. Anya and A. Hendry, "Sintering, Hardness and Indentation Fracture Toughness of Mullite and its Composite with 15 wt% X-Phase Sialon," *J. Eur. Ceram. Soc.* 13 (3): 247-56 (1994).

111. J. P. Singh, "Fracture Toughness of $MgCr_2O_4$-ZrO_2 Composites." *Ceram. Sci. Eng. Proc.* 6 (7-8): 729-35 (1985).

112. Du Pont Lanxide Composites Inc., "Ceramic Matrix Composites" (company brochure, 1991).

113. L. D. Bentsen, D. P. H. Hasselman, and R. Rugh, "Effects of Hot-Pressing Temperature on the Thermal Diffusivity/Conductivity of SiC-AlN Composites," *J. Am. Ceram. Soc.* 66 (3): C40-1 (1983).

114. D. P. H. Hasselman, R. Syed, and T.-Y. Tien, "The Thermal Diffusivity and Conductivity of Transformation-Toughened Solid Solutions of Alumina and Chromia," *J. Mater. Sci.* 20: 2549-56 (1985).

115. T. Nagaoka, M. Yasuoka, K. Hirao, and S. Kanzaki, "Effects of TiN Particle Size on Mechanical Properties of Si_3N_4/TiN Particulate Composites," *J. Ceram. Soc. Jpn.* 100 (4): 617-20 (1992).

116. G. Tilloca, "Ultralow Expansion Ceramics in the HfO_2-TiO_2 System Synthesized by an Hydrolysis and Polycondensation Technique," *J. Mater. Sci.* 27 (20): 5671-5 (1992).

Appendix I

References to Fracture Toughness Test Methods

I-1 W. F. Fischer III, R. A. Haber, and R. M. Anderson, "Mechanical Properties of Alumina Matrix Composites Reinforced with Silicon Carbide Platelets and Particulate," *Ceram. Trans. 19,* Advanced Composite Materials: Processing, Microstructures, Bulk, and International Properties, Characterization Methods, and Applications (Westerville, OH, American Ceramic Society, 1991), 773-9.

I-2 Testing Method for Fracture Toughness of High Performance Ceramics, Japanese Industrial Standard JIS R 1607. Investigated by Japanese Industrial Standards Committee. Japanese Standards Association, Tokyo, Japan (1990).

I-3 S. T. Buljan, J. G. Baldoni, M. L. Huckabee, and G. Zilberstein, "Microstructure and Fracture Toughness of Silicon Nitride Composites," *Whisker Fiber-Toughened Ceram., Proc. Int. Conf.* (Metals Park, OH, ASM International, 1988), 113-23.

I-4 T. Nose and T. Fuji, "Evaluation of Fracture Toughness for Ceramic Materials by a Single-Edge-Precracked-Beam Method," *J. Am. Ceram. Soc.* 71 (5): 328-33 (1988).

I-5 D. C. Larson, J. W. Adams, L. R. Johnson, A. P. S. Teotia, and L. G. Hill, "Ceramic Materials for Advanced Heat Engines," Technical and Economic Evaluation (Park Ridge, NJ, Noyes Publications, 1986).

I-6 D. K. Shetty, A. R. Rosenfield, and W. H. Duckworth, "Indenter Flaw Geometry and Fracture Toughness Estimates for a Glass-Ceramic," *J. Am. Ceram. Soc.* 68 (10): C-282-4 (1985).

I-7 J. C. Newman Jr., "A Review of Chevron-Notched Fracture Specimens", *Chevron-Notched Specimens: Testing and Stress Analysis,* ASTM STP No. 855, edited by J. H. Underwood, S. W. Freiman, and F. I. Baratta, (Philadelphia, PA, ASTM, 1984), 5-31.

I-8 K. Niihara, A. Nakahira, and T. Hirai, "The Effect of Stoichiometry on Mechanical Properties of Boron Carbide," *J. Am. Ceram. Soc.* 67 (1): C1-2 (1984).

I-9 S. C. Danforth and M. H. Richman, "Strength and Fracture Toughness of Reaction-Bonded Si_3N_4," *Am. Ceram. Soc. Bull.* 62 (4): 501-4 (1983).

I-10 R. F. Cook and B. R. Lawn, "A Modified Indentation Toughness Technique," *J. Am. Ceram. Soc.* 66 (11): C200-1 (1983).

I-11 ASTM Standard E399-83, "Special Requirements for Testing of Bend Specimens," American Society for Testing and Materials, Philadelphia, PA (1983).

I-12 K. Niihara, R. Morena, and D. P. H. Hasselman, "Evaluation of K_{IC} of Brittle Solids by the Indentation Method with Low Crack-to-Indent Ratios," *J. Mater. Sci. Lett.* 1 (1): 13-6 (1982).

I-13 K. Niihara, R. Morena, and D. P. H. Hasselman, "Further Reply to Comment on Elastic/Plastic Indentation Damage in Ceramics; the Medial/Radial Crack Systems," *J. Am. Ceram. Soc.* 65 (7): C-116 (1982).

I-14 D. B. Marshall and A. G. Evans, "Reply to 'Comment on 'Elastic/Plastic Indentation Damage in Ceramics: The Median/Radial Crack System," *J. Am. Ceram. Soc.* 64 (12): C-182-4 (1981).

I-15 P. Chantikul, G. R. Anstis, B. R. Lawn, and D. B. Marshall, "A Critical Evaluation of Indentation Techniques for Measuring Fracture Toughness: II, Strength Method," *J. Am. Ceram. Soc.* 64 (9): 539-43 (1981).

I-16 G. R. Anstis, P. Chantikul, B. R. Lawn, and D. B. Marshall, "A Critical Evaluation of Indentation Techniques for Measuring Fracture Toughness: I, Direct Crack Measurements," *J. Am. Ceram. Soc.* 64 (9): 533-8 (1981).

I-17 D. Munz, G. Himsolt, and J. Eschweiller, "Comparison of High Temperature Fracture Toughness of Hot-Pressed Si_3N_4 with Straight-Through and Chevron Notches," *J. Am. Ceram. Soc.* 63 (5-6): 341-2 (1980).

I-18 D. G. Munz, J. L. Shannon Jr., and R. T. Bubsey, "Fracture Toughness Calculation from Maximum Load in Four Point Bend Tests of Chevron Notch Specimens," *Int J. Fract.* 16: R137-41 (1980).

I-19 L. M. Barker, "Short Bar Specimens for K_{IC} Measurements," *Fracture Mechanics Applied to Brittle Materials,* ASTM STP No. 678, edited by S. W. Freiman, (Philadelphia, PA, ASTM, 1979), 73-82.
I-20 G. K. Bansal and W. H. Duckworth, *Fracture Mechanics Applied to Brittle Materials,* ASTM STP 678, edited by S. W. Freeman, (Philadelphia, PA, ASTM, 1979), 38-46.
I-21 A. G. Evans, "Fracture Toughness: The Role of Indentation Techniques," *Fracture Mechanics Applied to Brittle Materials,* ASTM STP 678, ed. S. W. Freiman, (Philadelphia, PA, ASTM, 1979), 112-35.
I-22 B. R. Lawn and A.G. Evans, "A Model for Crack Initiation in Elastic/Plastic Indentation Fields," *J. Mater. Sci.* 12 (11): 2195-9 (1977).
I-23 A. G. Evans and E. A. Charles, "Fracture Toughness Determinations by Indentation," *J. Am. Ceram. Soc.* 59 (7-8): 371-2 (1976).
I-24 B. R. Lawn and E. R. Fuller, "Equilibrium Penny-Like Cracks in Indentation Fracture," *J. Mater. Sci.* 10: 2016-24 (1975).
I-25 J. J. Petrovic, L. A. Jacobson, P. K. Talty, and A. K. Vasudevan, "Controlled Surface Flaws in Hot Pressed Si_3N_4," *J. Am. Ceram. Soc.* 58 (3-4): 113-6 (1975).
I-26 L. P. Pook, "An approach to a Quality Control K_{IC} Testpiece," *Int. J. Fract.* 8 (1): 103-4 (1972).
I-27 W. F. Brown and J. E. Srawly, "Plane Strain Crack Toughness in High-Strength Metallic Materials," *Durability of Adhesive Joints,* ASTM STP 410 (Philadelphia, PA, ASTM, 1966), 13-15.
I-28 T. Miyoshi, N. Sagawa, and T. Sassa, *Trans. Jap. Soc. Mech. Eng.* 51A: 2489-97 (1985).
I-29 D. B. Marshall, B. R. Lawn, and A. G. Evans, "Elastic/Plastic Indentation Damage in Ceramics: The Lateral Crack System," *J. Amer. Ceram. Soc.* 65 (11): 561-6 (1982).
I-30 D. K. Shetty, I. G. Wright, P. N. Mincer, and A. H. Clauer, "Indentation Fracture of W-Co Cermets," *J. Mater. Sci.* 20: 1873-82 (1985).
I-31 K. T. Faber and A. G. Evans, "Crack Deflection Processes-I Theory," *Acta Metall.* 31 (4): 565-76 (1983).